To: Stephen Lee
Enjoy!
Richard A. Postman

ANVILS IN AMERICA

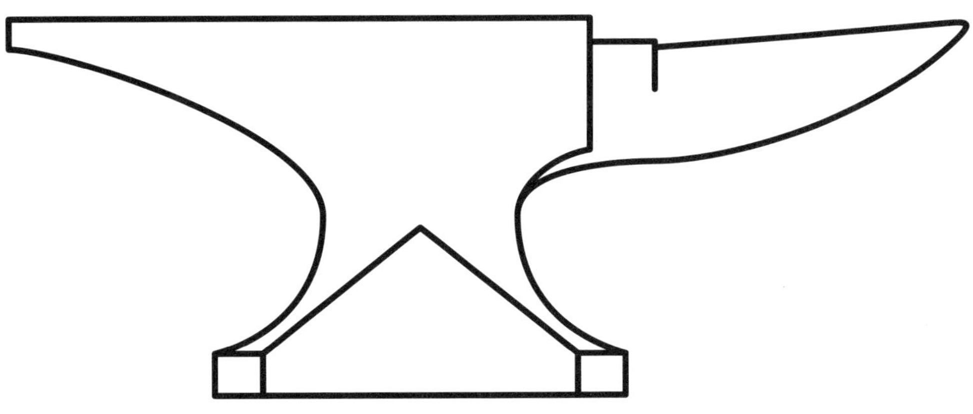

by
Richard A. Postman

The Last Anvil Makers

This print of an original oil painting is essentially a composite of two photographs. One photo taken in color in 1948 portrayed the anvil makers, Mr. Carl M. Wright and his father Mr. Arthur J. (Sy) Wright, anvil makers for the Columbus Forge and Iron Company, forging the horn of a *Trenton* anvil. Carl is holding the anvil and Sy is directing the operation. The steam hammer and the unknown man at the hammer's controls are from another photograph taken at the Columbus Anvil and Forging Company, makers of the *Arm and Hammer* anvil. The date of this photo is unknown, but the steam hammer is an old type developed before 1900.

Sy Wright was the chief anvil maker for the Columbus Forge and Iron Company from 1900 until 1950 when he retired, almost the entire span of wrought anvil manufacturing in the United States. Carl was employed by the company as an anvil maker with his father in the 1930s and retired in 1980, shortly before the company went out of business. Carl and his father were the foremost anvil makers of the company until the last wrought anvil was manufactured in the early 1950s. Indeed, "*The Last Anvil Makers.*"

The painting is the work of the nationally known artist, Nathan Greene, a direct descendent of General Nathanael Greene of Revolutionary War fame. The painting was completed in January, 1996, and is copyrighted by the artist.

Original photographs are by courtesy of Mr. David Tilton of the Columbus Forge and Iron Company and Mr. J. Edward Finneran Jr. of the Columbus Anvil and Forging Company.

ANVILS IN AMERICA

by

RICHARD A. POSTMAN

POSTMA PUBLISHING
Berrien Springs, Michigan

ANVILS IN AMERICA

by

Richard A. Postman

Library of Congress Catalog Card Number: 98-91240

ISBN: 0-9663256-0-5

First Edition - First Printing

Published by
Postma Publishing
10 Fisher Ct.
Berrien Springs, Michigan 49103

This book is dedicated to my wife, Katherine.

Book Number 835 of 1,000 Copies. Date 10/14/98

Printed by I. P. graphics, Plymouth, In. 46563

Acknowledgments

Many people are needed to put together a scattered puzzle such as the one attempted here. I have many people to thank for providing pieces of the puzzle and also for adding a few mysteries. My contribution has been the collecting of data on anvils and putting the puzzle together.

First I must thank my next door neighbor, Mr. Carl Erwin, who found the first anvil I ever bought, an old *Peter Wright,* and who also found the *Fisher* anvil that I had read about in *Practical Blacksmithing.*

Thanks to Mr. Ken Mankel of Mankel's Blacksmith Shop, who actually gave me the push that started my anvil research and who has been a source of encouragement along the way.

Thanks to Mr. Bill Gichner of Iron Age Antiques, who has continued to be a source of information, as well as a provider of some very early anvils to my collection. He is also a great host.

Thanks to Mr. Ken Wilson of Montaque, Michigan, one of my first sources of data and who has kept me informed on his findings over the years. He has also supplied me with a number of anvils.

The above also applies to Mr. Gill Fahrenwald of Olympia, Washington, who continues to provide data, as well as exchanging more than 100 letters over the years. He has also provided much of the puzzle, as well as photos of his miniature anvils.

A special thanks to the following people who have, and in many cases continue to provide help in any way they can:

Mr. John P. Catchings, who provided the first information from England on English anvils, and, with his wife, a home away from home.

Mr. and Mrs. Floyd Daniel, for their hospitality while I recorded their anvil collection, and also for firing the anvil for me.

Mr. and Mrs. Hunter Pilkinton for their wonderful hospitality and for allowing me the use of their library and photocopier, as well as providing considerable data.

Mr. David Shayt of the Smithsonian Institution for all the material and assistance in tracking down pieces of the puzzle.

The following people were involved in American anvil manufacturing or are descendants of those who were. They have provided valuable information, be it materially or verbally, through interviews, letters and phone calls:

Mr. John Boker of H.Boker and Company, New York, New York.

Mr. J. Edward Finneran, Jr. of the Columbus Anvil and Forging Company (the last owner).

Mr. Joseph King of the Columbus Forge and Iron Company.

Mr. John King of Kipco Industries, Inc.

Mrs. Warren B. Sage, granddaughter of Frederick C. Budden of Hay-Budden Manufacturing Company.

Mr. David Tilton of the Columbus Forge and Iron Company (the last owner).

Mr. Carl M. Wright (the last anvil maker) of the Columbus Forge and Iron Company.

The following people have assisted in 1995 and 1996 with invaluable material:
Thanks to Judith Anstee of Leeds, England, for all the material gathered and sent to me on the Mousehole Forge and for locating Mr. and Mrs. Hatfield.
Thanks to Mr. John Fairbanks of Staffordshire, England.
Thanks to Mr. and Mrs. Hatfield of Sheffield, England, owners of the Mousehole Forge site, for more information on the Forge and for allowing me to use Mrs. Hatfield's dissertation on the history of the Mousehole Forge.

Thanks to Mr. Nathan Greene for breaking into your busy schedule to paint *The Last Anvil Makers.*

The following people have provided material assistance and data on anvils in one form or another: Mr. David Baker, Mr. Vance Burns, Mr. Howard Busse, Mr. Jim Campbell, Mr. Fred Caylor, Mr. Frank Christison, Ms. Ruth Cook, Crossley Machine Shop (Fisher & Norris), Mr. Donavin Dunbar, Mr. Kurt Fehrenbach, Mr. Douglas Freund, Mr. Robert Hanson, Mr. Richard Heinecke, Mr. Richard E. Helland, Mr. John Lemoine, Ms. Mayon Lodge of Rochester, New York, Historical Society, Mr. Michael Machnik of the Henry Ford Museum, Mr. Mike McGee, Mr. Ben Miller (Amish blacksmith), Mr. Joe Miller, Mr. Bill Pieh of Centaur Forge, Mr. David Poppke, Mr. Clifton Ralph, Mr. Ken Scharabak, Mr. Kurt Shull, Mr. Keith Sommer, Mr. Jim Wallace, Mr. Larry Wood, Mr. Bob Woznsiak, the smiths and craftsmen at Colonial Williamsburg, Virginia, and a big thanks to all the folks who allowed me to record their anvils.

A very special thanks to Jan McGowan and Robert Meyers for editing the manuscript and to Mr. Greg Hill, a neighbor, who helped me handle the heavy anvils in preparation for photographing.

Last but certainly not least are the members of my family who have put this manuscript on the computer.
Mrs. Betty Johnson (daughter), Mr. Bill Keith (son), and Katherine Postman (my wife).

Credit will be given when possible for any material or photos used in the book.

Introduction

> *It is strange that such little heed has been paid to the anvil, considering that for centuries it was a basic tool throughout the civilized world. However, those things that are common to our everyday living are often taken for granted until it is too late. Why put in literature something that everyone knows? Even the Bible mentions the anvil but once... .* [1] I could not have said it better.

In the fall of 1982, while preparing to teach a newly created class in hot metals which included hand forging, I thought it best that I learn a little of the history or background of the anvil to pass on to the students. Our metal shop had two anvils of one hundred pounds each, bearing the name *Vulcan* and an Arm and Hammer trademark like that on a baking soda box. This trademark was cast in relief on one side of both anvils. I had often wondered how old they were, who made them, and where they were made.

The library at the University where I taught has more than one million volumes. Certainly among all these books there would be one on anvils, or at least a book on blacksmithing containing a section on the history of anvils. I looked through the card file for "Anvil"... nothing! Under "Blacksmithing" I found one book: *American Blacksmithing Toolsmith and Steelworker's Manual.*[2] The index in the front of the book under "Anvil" led me to page 33, on which I found one whole paragraph on anvils. It revealed nothing about the anvil except instructions on how to set up the anvil in a shop. Had I checked another index in the middle of the book (it was two books in one) I would have found pertinent information on the *Hay-Budden* anvil, but I did not discover my oversight until about a year later.

In the years before 1982, I had collected a number of books on metalworking. Though some of them mentioned anvils and their use, they did not include any names, background or history. After my library experience I began looking particularly for books on blacksmithing and wrote to all the listings I found in the *Thomas Register* under anvils. I bought a reprint of the 1888-91 book *Practical Blacksmithing*[3.] and found several pages on how anvils were made, including an American anvil called the *Fisher.* It whet my curiosity for more information and prompted a search for a *Fisher* anvil.

At about the same time, I came across a small advertisement in *Popular Mechanics* magazine for blacksmithing tools. Providence was smiling on my efforts. The address in the advertisement was only about 100 miles from where I lived, so I paid Mr. Ken Mankel's Blacksmith Shop a visit. Mr. Mankel gave me some information on anvils (he manufactures anvils) and the phone number of a man in the state of Delaware who he said would know, if anyone knew, where a book on the history of anvils could be located. That evening I called Mr. Bill Gichner of Iron Age Antiques and asked him where I could find a book on anvils or their history. He gave a little chuckle and said, "There aren't any books on anvils". I said, "I can't believe that," but he assured me that it was true.

With all the books on every conceivable subject in this world, surely you would think that someone would have written a book on an implement as old and important to the Iron Age and so historically relevant to our age. Sure, there are snippets here and there, but nothing of any magnitude has been compiled. I still find it hard to believe. However, the more I talked to smiths and people who own anvils, including anvil collectors, the more I discovered how little is known about them. Thus I began collecting information on anvils to satisfy my own curiosity, later I thought of sharing it with others.

In my search for information, I have recorded over 2,000 anvils (thru 1997) of all makes and patterns, and have looked at hundreds more. Recording includes weight, type of metal, measurements, any markings or trademarks, and any unusual characteristics. This recorded data is the basis for the information given in this publication. I have also consulted manufacturing directories, old trade journals, old hardware catalogs, museums and state archives, and have been fortunate enough to locate and interview a number of the descendants of American anvil manufacturers, as well as those people connected with anvil manufacturing. Additionally, much of the information I have received was passed along to me by other interested people.

In addition to my wish to pass along to interested people the information I have acquired, I discovered another reason for publishing. While studying what little literature there is in this country concerning anvils, it seems that a significant part of it is either false or misleading. I would like to correct as much of this misinformation as possible. For example, in the publication, *To Draw, Upset, & Weld,* the writer states in one sentence, "By the mid-nineteenth century, the *Columbia,* the *Eagle,* the *Hay-Budden,* the *Swedish-American,* and the *Peter Wright* anvils were symbols of dependability."[4] In fact, of the anvils listed only the *Peter Wright* (English) and possibly the *Eagle (*American, made by Fisher and Norris) were even available at that time. *Columbia* (Columbian) was first produced about 1900, the *Hay-Budden* about 1890, and the *Swedish-American* was not an anvil, but the name of an American importing firm, who did not go into business until 1927, and imported the Swedish *Kohlswa* anvil. The book *Practical Blacksmithing*[5] states that the largest anvil ever made was produced by Fisher and Norris for the 1876 Centennial Exposition in Philadelphia, Pennsylvania and weighed 1,600 pounds. Every publication mentioning this anvil since that time states that it weighed 1,600 pounds, when in fact it weighed 1,400 pounds (see Fisher).

In art work representing the Colonial period and the early-nineteenth century in this country, the blacksmith in almost all cases is shown using a late-nineteenth century anvil. We must excuse the artists because they not only have "artist's license," but do not have a correct reference. I have also seen anvils misrepresented in museums as to their age, the material from which they are made, and the process by which they were produced. If a museum is wrong about a subject we are familiar with, we begin to question the information they give concerning subjects we know very little or nothing about.

I hope this work will help correct some of the past errors. Although this work, as any such undertaking, will not be error-free, I have tried to make it as error-free as possible.

I will continue to seek information about anvils and their manufacture, but there should be enough information presented here to satisfy most people interested in the subject. However, the material presented in no measure represents a definitive work on this subject. Not only is it not definitive, my inquiry covers only American-manufactured anvils, imported anvils that were for sale in this country, and a few brought from Europe by smiths for their own use. I have made no attempt to cover anvils made or used worldwide or even in Europe. I leave that for a later time or to someone else.

In the case of many anvils and their manufacture, more questions are raised than answers given. It is hoped that this work will reach people or places that are as yet unknown to me. Perhaps some of these people will contact me with information that will help answer a few of these questions.

Information about anvils and their manufacture is so scant and hard to find that extensive time and a considerable amount of money was and is required to ferret it out. I have worked in machine shops and a forge and know from experience that most management and all labor are primarily concerned with making money. They give little if any thought to the fact that in many cases, they are making a contribution to history as well as establishing roots to look back upon for future generations. I have talked on the phone with executives of one company that had once made anvils, and all that was known was that the company started in the late-nineteenth century, and they were not even sure why the name of the anvil was selected. To my knowledge, no histories have been written about forging companies and particularly about anvil manufacturing in the United States. Getting this information is difficult and also requires some guesswork. It is fortunate that the product of anvil makers is so durable, or we who are curious about such things would have little to research. Thankfully, the anvils themselves can tell a great deal about the people who made them, how they were made, and about those who used them.

Before I conclude I would like to relate the following story as one of the happiest benefits that I have unexpectedly received from gathering and dispensing data on anvils.

To begin, many of the local chapters of the Artists Blacksmiths Association of North America (ABANA), have mentioned my research in their publications and from time to time I receive letters from members seeking information about their anvils. Just before Christmas of 1992 I received a letter from a Mr. Moon (not an uncommon name) from California, requesting information about an anvil he has. I wrote back, answering his questions, and usually that is the end of the correspondence. However, several weeks later I received another letter from him asking for more information about certain types of anvils. At the bottom of the letter he added a "P. S., my grandfather was Ben Postman of Oklahoma via Iowa." I looked at the name, but it did not register at first. The name Postman (originally Postma, of Dutch origin) is very uncommon, and so I said to my wife, "Ben is the nickname for Benjamin, isn't it?" Benjamin Postman happens to be my grandfather's name also. Then the name James Moon meant something to me other than just a stranger. Mr. James Moon is the son of my father's youngest sister, my first cousin! I last saw him in 1941 when we were small boys, while we were visiting our grandparents. He apparently did not remember me, though I remember meeting him. After hearing from me his wife suggested that he take a chance and see if we could possibly be related. I am meeting a side of my family I would never have known, except for our mutual interest in anvils. Don't tell me it's not in the genes!!

It is my wish that this work will further the interest in historical tools, such as anvils and to influence people who own them to appreciate them more, as well as preserve them for future generations.

R. A. Postman
1997

TABLE OF CONTENTS

Foreword

If you are not familiar with metalworking processes and anvil parts, I suggest that you review Chapter I, **Parts of the Smith's Anvil and Their Use,** before reading any further. I also recommend that you turn to the **Glossary** in the back of the book and read through the terms and their meaning.

You can refer to the **Glossary** when necessary, but looking the **Glossary** over first will be of much help to you as you read the various chapters.

It is my desire to make this publication meaningful to those unfamiliar with anvils or metalworking, as well as for those familiar with the subject. This publication was designed to serve as a reference work, as well as for casual reading, in the hope that the reader will derive pleasure as well as information. If one or both of these objectives are realized, I will have accomplished my intent.

R. A. Postman

CHAPTER 1

General Information About Anvils

HAY-BUDDEN AMERICAN

FISHER VULCAN

TRENTON

ACME

ANVILS IN AMERICA

PARTS OF THE SMITH'S ANVIL AND THEIR USE

The anvils discussed in this book have, of course many features in common. So the reader will understand my discussion of the many different types of anvils and their distingishing featurers, I provide a drawing of the standard blacksmith's anvil, as well as a description of its parts and their use. Anvils not following the standard may have fewer or additional features. These will be pointed out as they are discussed.

PARTS

1. **Base -** Includes all parts of the anvil from the waist down.

2. **Body -** The part of the anvil from the waist up can include the *horn, heel,and face.*

3. **Clip Horn -** A small, flat protrusion coming from one or both sides of the *horn* next to the *face* on farrier's pattern anvils.

4. **Edge -** That part of the anvil *face* where it drops off 90^0 into the body.

5. **Drop -** Also called the *blight.* The space from the *face* to the *table.*

6. **Face -** The flat, rectangular, hard surface on the top of the anvil *body.*

7. **Face Plate -** A flat, hardened steel plate that is welded to the top of the *body* and *heel* on most anvils manufactured before 1930 and on all *wrought* anvils before 1908. Usually one quarter to three quarters of an inch thick.

8. **Feet -** The part of the *base* that projects away from the central mass; specifically, the four corners.

9. **Handling Holes -** Also known as *mouseholes.* One to four square or rounded holes of various depths in the *waist,* under the *base* or in the front or rear of the *feet.*

10. **Hardie or Hardy Hole -** Sometimes called the *cutter hole.* The square hole usually found in the *heel,* passing through the *face,* but on occasion found in the *horn* or horn end of the anvil.

11. **Heel -** Also called the *tail* or *quarter.* The protruding top part of the *body* on the opposite end of the anvil from the *horn.* The top of the *heel* is part of the *face.*

12. **Horn -** Also known as the *bick* or *beak*, sometimes referred to as a *pike*. The round, tapered protrusion coming off of the *body*. On *double-horned* anvils, the second *horn* (where the *heel* is normally located) is tapered and flat on top, and is usually part of the *face*. The *table* is a part of the round *horn*.

13. **Pritchel or Pritchel Hole -** Also known as a *spud hole* or *punching hole*. The round hole in the *face*, at the *heel* of the anvil. The farrier's pattern anvil generally has two *pritchels* after 1910.

14. **Shoulder -** On a smith's pattern anvil, it is the side of the *horn* at its junction with the *body*. It is not on a farrier's pattern anvil.

15. **Table -** Also called the *cutting table*. A flat, square or rectangular surface on the top of the *horn*, next to the *body* and below the *face*. It has a soft *face*.

16. **Throat -** The distance from the bottom of the *horn* and *heel* to the top of the *feet*.

17. **Waist -** From a side view, it is the thinnest part of the anvil, where the *body* joins the *base*.

USE

1. **Base -** Supports the *body* of the anvil, serving as a means of anchoring the anvil, giving the anvil mass, and often to impart information, weight, serial number, etc. Late Spanish anvils do not have a *base*.

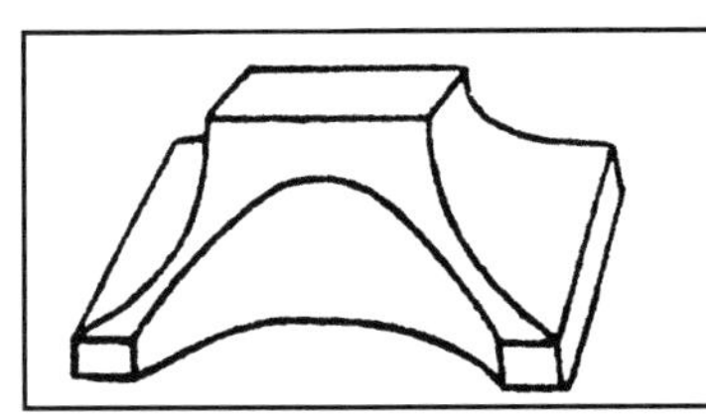

The base for most anvils made between ca. 1850 and 1950

2. **Body -** Contains all the features used on the anvil above the *base*.

3. **Clip Horn -** Used on farrier's anvils to raise metal on the front of the horse-shoe (that protects the front of the horse's hoof, particularly draft horses).

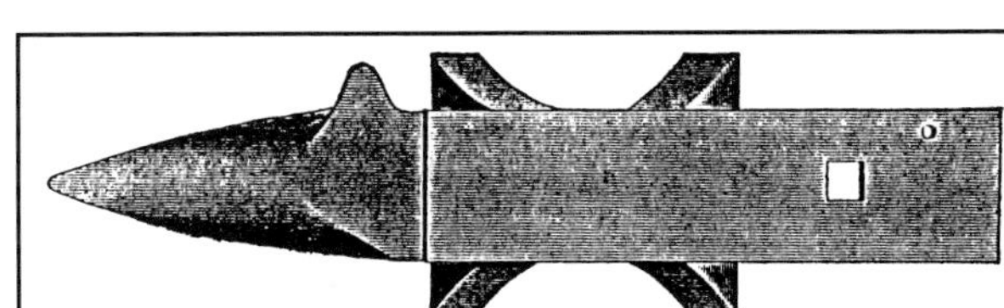

Top view of clip horn off the side of the horn

4. **Edge -** Used for making sharp bends of 90^0 or less. Sometimes it is rounded and used for drawing.

5. **Drop -** It is 90^0 from the *face* and a good place to hold a round rod or for bending rods. Most *continental* pattern anvils do not have this drop.

6. **Face -** The most useful part of the anvil. Used for drawing, upsetting and welding. Also a good place to crack black walnuts, etc.

7. **Face Plate -** Hardened face - see # 6.

8. **Feet -** Gives stability to the anvil *base* and makes it easier to anchor the anvil to a surface.

9. **Handling Hole -** These are used in conjunction with porter bars, to manipulate and position a hot anvil in the process of manufacture. Many of the later American made anvils did not have them, and they are not a part of new anvils today. The hole under the *base* was sometimes used with a spike to anchor the anvil.

10. **Hardie -** Used to hold various kinds of anvil tools, especially a cut-off tool called a *hardie.* Also used for bending rods that fit the hole.

11. **Heel -** On the *London* pattern, it contains the *hardie* and *pritchel*, and is used for bending metal when more than 90^0 is desired.

12. **Horn -** Used for bending radii or curves and for faster drawing of hot metal. Used primarily for making and bending horseshoes on the farrier's pattern.

13. **Pritchel -** Used primarily for punching holes, knocking out nails or pins, or any other use a round hole can be put to.

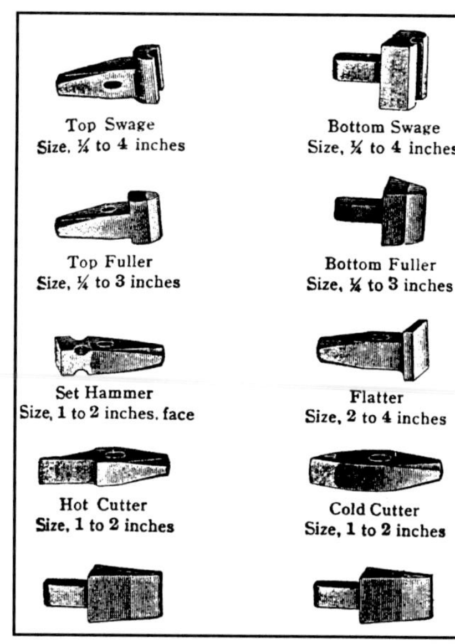

Various standard anvil tools. Square shanks fit in the hardie hole.

14. **Shoulder -** It does not seem to have any use except to produce a smaller *horn.* If it is deep enough it can be used the same way as the *drop.*

15. **Table -** Originally designed for cutting metal with a chisel, rather than using the *face.* It became a traditional feature on the smith's anvil, but is used very little for the original purpose today.

16. **Throat -** Makes it possible to bend metal on the *horn* and *heel* and makes the anvil lighter.

17. **Waist -** See # 16. Makes a good place to wrap a chain around for anchoring the anvil.

THE STANDARD LONDON PATTERN SMITH'S ANVIL FROM THE MID-NINETEENTH THROUGH THE TWENTIETH CENTURY.

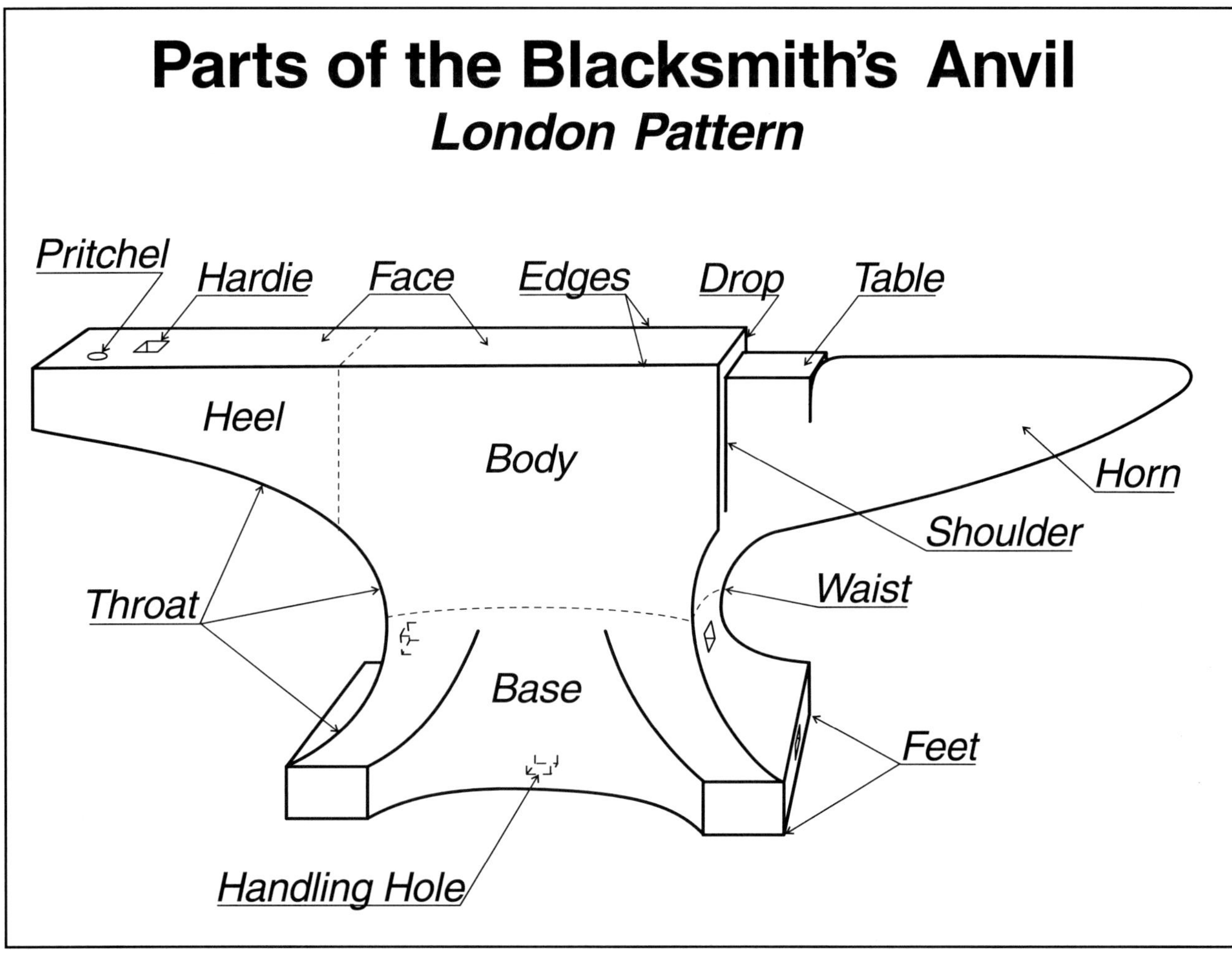

ANVILS IN AMERICA

THE WEIGHT OF ANVILS

The few people who have written about the weight marks on anvils since 1960 suggest that wrought anvils have numbers on them which must be deciphered to determine an anvil's approximate weight. They are referring to the old English "stone" weight system. In fact, all wrought anvils did not use this system. To my knowledge, no American anvil manufacturer used it, and all American-made anvils that have the weight marked are marked in actual pounds. This includes *American, Arm & Hammer, Fisher, Hay-Budden, Trenton,* and *Vulcan,* as well as most of the early Swedish and German anvils. I have found that only the imported English anvils used the "stone" weight system, which they still used in a modified form until about 1990. The "stone" weight system the English anvil makers used did not become a "mystery" until well into the 20th century. One recent writer went so far as to suggest that he had come up with a theory to explain the weight markings on anvils and how to decipher them.[1] His conclusion was correct, but he did not have to come up with a theory for the approximate weight, because any book on blacksmithing that has more than a few paragraphs to say about anvils describes how to convert the markings into pounds. The very fact that the vast majority of anvils coming into this country before the 20th century were English made, using the "stone" weight system, would lead most writers to conclude that this system for marking the weight on anvils was used universally, although it was not.

The weight marked on wrought anvils was not approximate, either, as one writer suggested.[2] Until the 1950s all wrought anvils sold by the pound. When wrought anvils were new, the weight marked on them was within one pound, more or less, of the actual weight. You would not pay for ten pounds of steak and accept nine pounds, would you? Neither did the blacksmith when buying an anvil. However, because of rust, use and misuse, most of the older wrought anvils found today weigh from three to five pounds less than the markings would indicate. Since the 1950s anvils sell by approximate weight, with guaranteed weight of plus or minus ten percent of that advertised.

The markings on English anvils use the "stone," which equals 14 pounds, as the basic weight standard. Eight "stones" make up what is known as "hundredweight," or 112 pounds. "Hundredweight" is usually written "cwt.," and is still used today in England (1990s). Until about 1990 the only English anvil imported to the United States, the *Brooks*, still used the "hundredweight" to indicate their anvil's approximate weight in pounds. *Brooks* marked the weight as cwt. and quarters of cwt., but not as the old anvils were marked. For example, 4 1/2 cwt., cast into the side of this modern anvil, would indicate approximately 500 pounds. An earlier anvil of 500 pounds would have been marked 4 1 24. Remembering the anvils made today do not sell by the exact pound, 4 1/2 cwt. would actually work out to be 504 pounds, well within the advertised plus or minus ten percent. (4 cwt. indicates 4 x 112 = 448 pounds, and the 1/2 cwt. represents 56 pounds; 448+56=504 pounds. This particular anvil could actually weigh between 451 and 549 pounds and still be considered a 500-pound anvil, as marked.) I have a newly made (1983) Swedish anvil, advertised as weighing 350 pounds, that actually weighs 331 pounds. This anvil is marked "161 kilograms" which when converted indicates 354 pounds, so 331 pounds is well within ten percent of the advertised weight. Let it be noted that a casting can be held very close in weight, so one wonders about this anvil maker. I would like it better if anvils sold by the pound as they used to; then we would know exactly what we were getting.

The older English "stone" weight system works as follows, using the 500-pound anvil as an example.

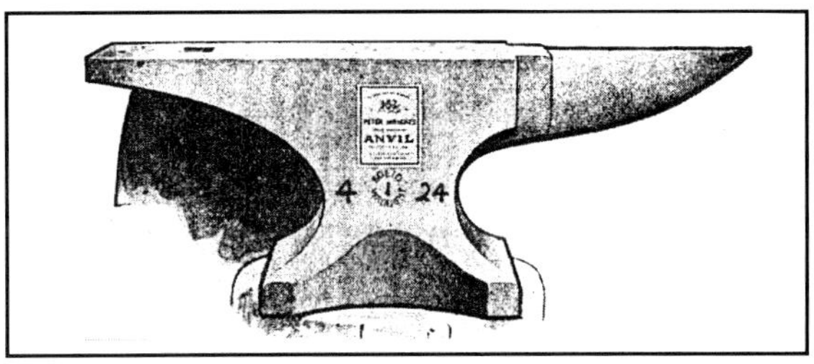

The numbers **4 1 24** on the body are usually found when the horn is to the right as the anvil is observed from the side. The number to the left, in this case **4**, could be any number from **0** on up and indicates a multiple of cwt., or 112 pounds. On an anvil that weighs less than 112 pounds, this number would be **0**. I have yet to record an anvil with **7** or above. In the case of some very old anvils before 1700, if the anvil weighed less than 112 pounds, nothing would be marked to the left. In our example, then **4** cwt. equals 448 pounds (4 x 112). The center number stands for multiple quarters of cwt., or 28 pounds, 2 "stones." The number can be only **0, 1, 2, or 3.** In our example it is a **1** or 28 pounds. The number to the right of center indicates actual pounds, or anything less than 28 pounds. This number can only be **0** to **27**. In our example it is **24**. No fraction of a pound was used. Thus, our 500-pound anvil marked **4 1 24** equals 448 + 28 + 24 = 500. A 112-pound anvil would be marked **1 0 0**, a 140-pounder **1 1 0**, a 141-pounder **1 1 1**, a 312-pound anvil **2 3 4**, and so forth. In each case the first number to the left equals increments of "hundredweight" (112 lbs.), the next or center number is quarters of "hundredweight" (28 lbs.), and the number or numbers to the right are odd pounds (0 to 27 lbs.). These digits or numbers on the English anvils using the "stone" weight system are almost always widely spaced across the upper part of the base, and some of them have a period or dots punched between the digits, thus: **1 . 1 . 16** or, for example **1: 1: 16.** On *Colonial* anvils, those anvils made before the 19th century, the weight marks were cut in with a chisel, and a dot was punched whenever the line of the digit changed direction. On some, a dot was also punched between the digits.

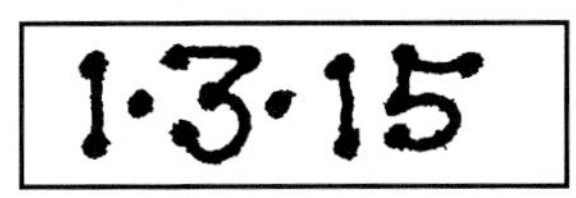

Example

I suppose the punch marks were to help guide the chisel, or maybe they were a decoration. On some of these anvils only the dots remain. In the late 18th and early 19th centuries, number stamps came into common use for putting the weight marks on. These weight marks can be found on either side of the body of the English anvils, but usually they are on the same side as the trademark. If all that can be found on the side of an anvil are numbers that do not match closely the actual weight of the anvil, it is almost certain to be an English make. These weight marks also indicate that the anvil was made for resale on the commercial market by an anvil manufacturer.

The "stone" weight system was not used on all English anvils exported to this country, however. On at least one English make that was imported in the late 19th and early 20th centuries actual pounds were stamped on the side. Of the anvils that I have recorded of this make I have found only one using the "stone" weight system (see *Henry Wright*).

ANVILS IN AMERICA

In a memorandum book of expenses dated February 22, 1774, kept by George Washington, the items listed included an anvil that he bought.[3] The entry reads as follows:

	"C	Q	lbs.	£	S	P
1 anvill	1	2	16	4	6	3"

The anvil would have weighed 184 pounds and cost 4 pounds (£), 6 shillings (S), and 3 pence (P). The C means hundredweight, the Q, quarters of hundredweight, and the lbs., odd pounds. I do not know what this anvil would cost in today's money (1996), but I am certain the anvil was very expensive in 1774.

Some of the cast anvils that were made in this country have the approximate weight cast in relief on them, but the digits used do not indicate the total weight in many cases. For example, on many of the *Fisher* anvils a "20" would indicate a 200-pound anvil, a "35"a 350-pound anvil, a "15" a 150-pound anvil, and so on. This was also true of the *Vulcan* anvil, although they were not advertised by the pound. The *Columbian* cast-steel anvil and most cast-iron anvils did not have the weight marked on them at all and were usually advertised and priced in 25-pound and 50-pound increments.

Until recently the German and Swedish anvils imported for resale in this country used actual pounds to designate the weight, but since World War II they have used the metric system (kilograms). For example, one new German anvil advertised at 275 pounds is marked "125" kilograms. Approximate conversion is 2.2 pounds equals one kilogram. Thus 2.2 x 125 = 275 pounds, and the actual weight of the anvil is 274 pounds.

I do not know of any American manufactured anvil made since the 1960s that has the weight marked on it. The weight is given in the maker's advertisements, but the price, though based somewhat on weight, tends to vary with the anvil's style and by modifications made to the anvil. Almost all of the modern American-made anvils are meant for the farrier's trade and include all types of modifications and features.

The weight on all American -wrought anvils that I have recorded or observed is stamped either on the side below the trademark or on the front of the foot under the horn. The digits are generally close together, but that is not always the case when they are stamped on the side of the body.

Of all the anvils that I have recorded, I have found only one that weighed more than the weight marked on it. It was an American wrought anvil, *Arm & Hammer,* that was marked 175 pounds, but actually weighed 180 pounds. Someone bought a bargain.

General Information About Anvils

The photos shown below, show how the weight would typically appear on various anvils.

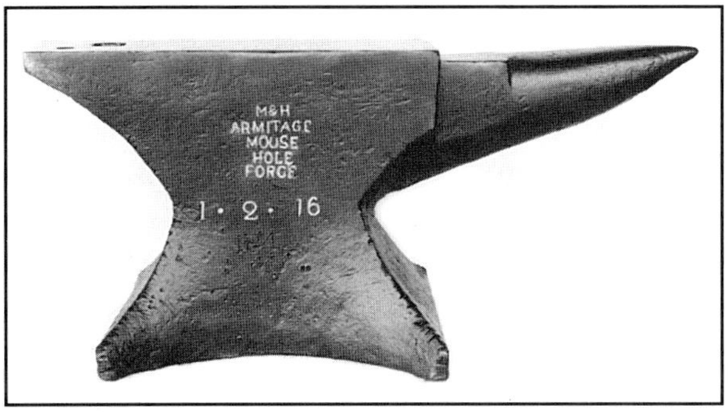

Marks in stone weight on Mousehole anvil. (R.A..P. Photo)

Weight in pounds on Hay-Budden anvil. (R.A.P. Photo)

Marks in stone wight on Peter Wright anvil. (R.A..P. Photo)

Weight in pounds on the front of the foot under the horn on a Trenton anvil. (R.A.P. Photo)

ANVILS IN AMERICA

THE DATING OF ANVILS

Dating anvils can be a very tricky business, especially those made before the 1850s.

In fact only one English anvil manufacturer is known to have dated their anvils before the 1850s (see *William Foster*). Almost all anvils made before the 1850s must be dated by their features only. Some of the pre-1850 anvils other than *William Foster*, have dates on them, but in most cases they were put on when the anvil was bought or to commemorate some event in the persons life. Of course some American wrought anvils can be dated to the year, and some to the day, of manufacture.

Even an anvil's features and shape can be deceptive as to the approximate date of manufacture. Some anvil manufacturers developed advanced shapes and methods of manufacturing while others lagged behind. The lag may have lasted from ten to fifty years, so an anvil that may appear to have been manufactured about 1750 or before may in fact have been manufactured about 1800. Keeping these concepts in mind, I will attempt to give some general guidelines for dating anvils.

I have put all anvils used in this country into three broad categories, based on their shape. These are *Colonial* ca. 1600-1800, *Old English* ca. 1800-1852, and *Modern* 1852-1960s.

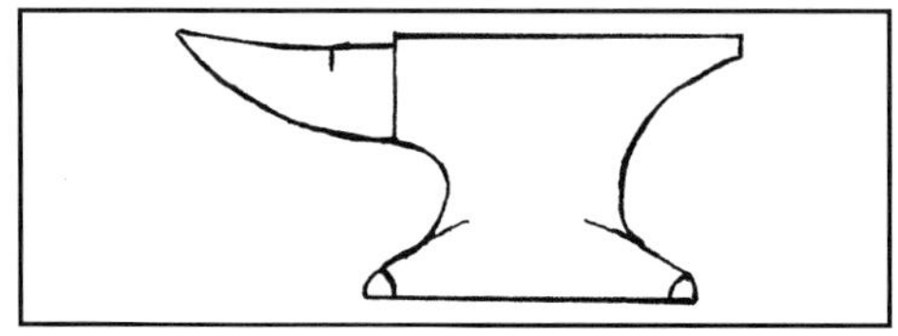

Typical profile of Old English style anvil.

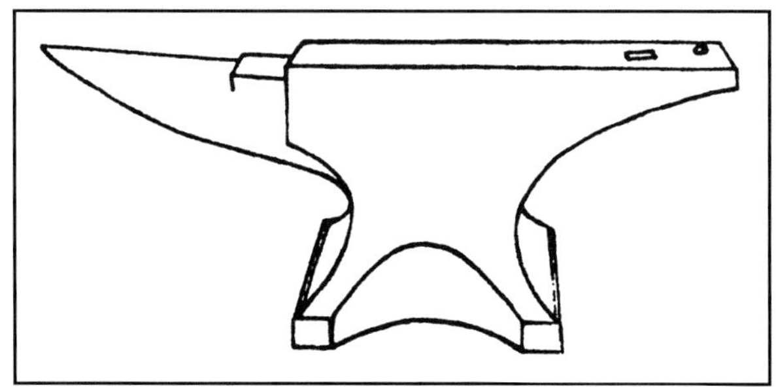

Typical profile of Modern style anvil.

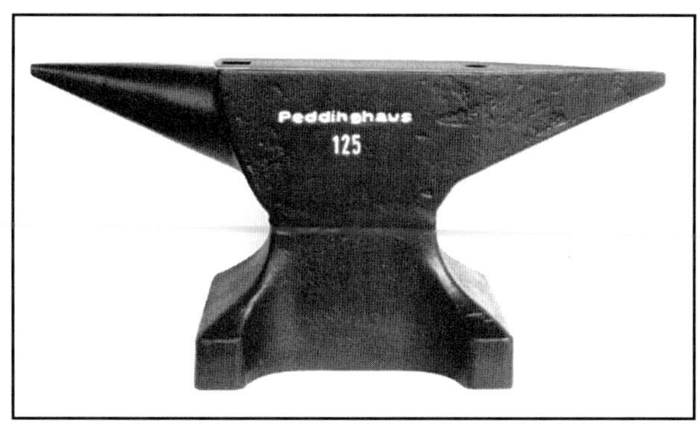

Modern German Continental pattern (R.A.P. Photo)

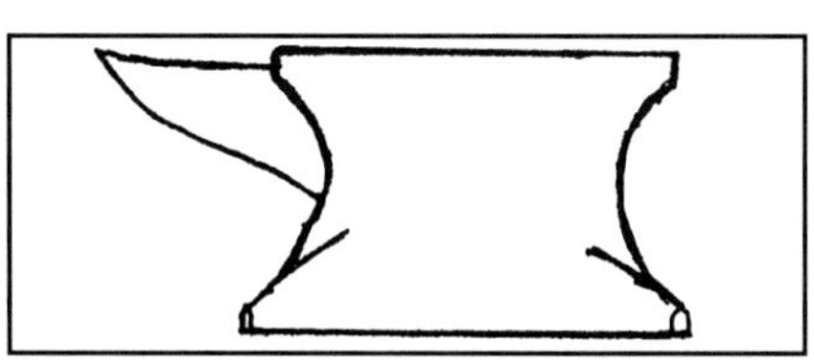

Typical profile of Colonial style anvil.

Modern American pattern. (R,A.P. Photo)

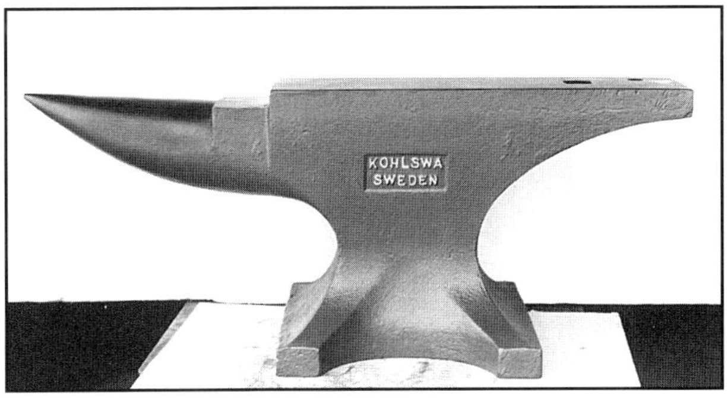

Modern London pattern (R.A.P. Photo)

These categories will be explained in this section. Those manufactured after about 1960 vary considerably according to the manufacturers preference.

By the time the first colonists landed in America the smith's pattern anvil was on its way to reaching the final stage of development, after hundreds of years without change. The shape was being standardized both in England and on the continent of Europe, but differently. The anvil had a hardened steel face plate, coming with or without a horn and with or without a hardie hole. The English-style smith's anvil became known as the *London* pattern and the continental European style as the *Lieges* or *Continental* pattern. The *Continental* pattern differs from the *London* pattern in that it does not have a drop or a table. Its hardie hole is in the face near the round horn; in later years it usually has double horns. Generally the waist of the *Continental* is thicker than that of the *London* pattern.

There were as many as sixty or more variations of the *Lieges* pattern. Toward the end of the 19th and early in the 20th centuries the American anvil manufacturers perfected the *London* pattern, which created what could be called the *American* pattern anvil. The *American* pattern differs from the *London* in having a thinner waist, and the face is longer and narrower for a specific anvil weight.

When the first English landed in America, they brought with them the *London* pattern anvil, which was little more than a block of iron with a steel face and possibly a horn attached. It did curve in at the body slightly,

Top and side view of typical Colonial pattern anvil.

creating the waist. This anvil served our forefathers admirably for at least two hundred years. This style dominated the American market where the English settled. The Spanish and French smiths brought their more ornate *Continental* pattern anvils with them, but their anvils never reached the numbers that the English anvils did. Only a very few can be found from the Colonial period, whereas many of the English *Colonial* style anvils are still to be found.

The oldest reference and drawing we have of a smith's anvil in the United States is in Albert H. Sonn's book, *Early American Wrought Iron*. On page 37, there is an excellent drawing of this anvil with the following caption:

> ***Figure 1.*** *Old Spanish anvil at Fort Marion, St. Augustine, Fla. Believed to have been brought over in the sixteenth century when the first followers of Ponce de Leon settled there. This anvil is reputed to be the oldest in the United States.*[4]

The horn containing the hardie hole is rounded.

ANVILS IN AMERICA

This anvil is the double horned *Continental* pattern with very little base (a Spanish variant) and with a hardie hole. Later many Spanish anvils were made without a base, having a spike under the body to attach them to a support. I planned to go to St. Augustine to see this anvil in 1986 while on a visit to Florida, but before I went I happened to read a note in the book, *Southwestern Colonial Ironwork*,[5] which referred to this anvil. The authors of the book said that they had received a message from the curator of Fort Marion stating that since Sonn had reported on this anvil it had disappeared. I did not go to St. Augustine. In 1989 I heard from a smith in Florida who said that this old Spanish anvil had been found in storage. I have yet to check this out (1995).

The oldest smith's anvil in this country that can be dated and I have seen is known as the *Moses* anvil. It is of the *London* pattern and is in the possession of the Rochester, New York, Historical Society. A date of *"1632"* is chiseled near the bottom of the base, between the feet.

Side-view of the Moses anvil. The date is near the bottom in the center. Part of the weight mark can be seen just above the waist to the right. (R.A.P. Photo)

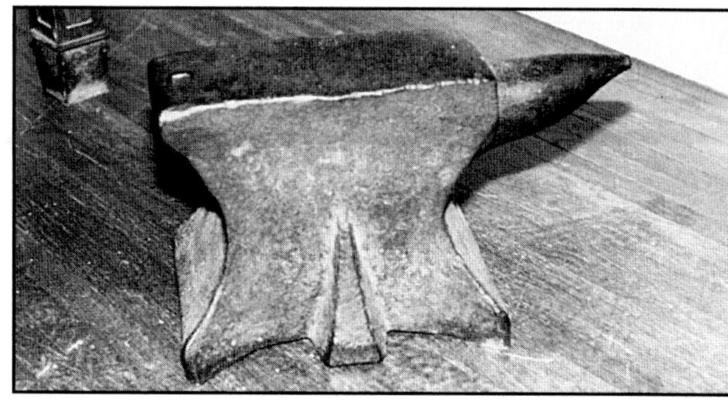

Side-view showing 5th foot or toe. (R.A.P. photo)

Top-view. Photo taken at the Rochester Historical Society (R.A.P. Photo)

According to the record of the Moses family, a Mr. John Moses of Plymouth, Massachusetts, brought the anvil to America in 1632 and the anvil stayed in the Moses family until it was donated to the historical society in 1907.[6] I visited Rochester in 1988 to record this anvil and was hoping to use it as a benchmark for dating other early American anvils. The anvil fits the time period and was probably made earlier than 1632. However I am a little suspicious of the date chiseled into the anvil. In my opinion the date was chiseled into the anvil by a descendant of John Moses, probably sometime in the early 19th century, when a family member realized the anvil's historical significance. However, my opinion is not a reflection on the anvil's age nor whether it was or was not brought over in 1632 by John Moses. For my purpose, the date on the anvil does not help me to deter- mine when the anvil was made.

Here is my rational for concluding that the date was put on much later than 1632. The

date is rather crudely chiseled in. John Moses is reputed to have been a shipwright and undoubtedly of considerable skill.[7] Whoever put the date on was not a skilled craftsman, or if John Moses did put it on, he did a very poor job. Also, the edges of the chiseled marks are rather sharp. If John Moses put the date there in 1632 to commemorate his landing and continued to use the anvil (it has seen considerable use), I doubt that the edges would have remained so sharp. I have observed hundreds of anvils much newer than this one, and the chisel marks on them are not as sharp as these. To illustrate, I located a wrought anvil stamped with a name that was unfamiliar to me. I thought I might have found another anvil manufacturer or a mystery anvil. The anvil also had the weight on it below the name in actual pounds. The anvil's owner allowed me to take it home for study.
I cleaned the stamped name and weight with naval jelly (a mild acid) until I reached bare metal. I then examined the name and the weight marks with a magnifying glass. The indentation for the letters of the name were very smooth with very fine rust pits, whereas the same area of the weight marks was considerably rougher and more deeply pitted. This told me that the name was stamped in at a later date than the weight which was stamped on when the anvil was manufactured. The anvil turned out to be a standard brand manufactured for a hardware company around1900.

Above the "1632" date on the *Moses* anvil I could just make out English weight markings. The date is much clearer than the weight markings. Generally the lower part of an anvil rusts more quickly than the upper part, given the fact that slag, fluxes (acid) and so on, tend to stick to the lower part of the base. This lower part of the base is where the date is located. Enough said.

General Information About Anvils

Probably the first anvil brought into the English colonies was at Jamestown, Virginia, in 1607. Among the tradesmen listed in the settlement at that time was a blacksmith.[8] His anvil could be in someone's collection or in the mud of the James River. I doubt that it went back to England.

Both the Spanish and the *Moses* anvils mentioned were made of wrought iron with hardened steel faces and both have horns and a hardie hole. Some other anvils brought to this country in the earliest days may be much older, but there is no easy way that I know of to determine the exact age of these old anvils of similar style if they are not dated or if no record was kept of them.

For simplicity I have called all anvils in this country made before 1800 *Colonial* anvils, and will refer to them as such rather than by features. From the first English settlement until the 1780s the English smith's anvil did not change at all from my observation. Then in the 1780s or a little earlier, the smith's anvil began to change appreciably in shape and features. About 1780 the table appears on the anvil, the face becomes a little more rectangular, and the horn is longer. In a *Sheffield Directory* of England dated 1787, a Mr. Samuel Alsop is listed as being an anvil, vise and screwmaker.[9] I have recorded three anvils and have heard of others with the name *Alsop* on them. Below the name is the word "Sheffield." These *Alsop* anvils are the same in detail as the other *Colonial* anvils except that the horn is a little longer and there is a cutting table at the base of the horn where it joins the body. On one *Alsop* that I own, the table is 2 1/4 inches long by 2 1/2 inches wide. I will venture to say that the *Alsop* in the directory is the same one who made the *Alsop* anvils that I have recorded.

ANVILS IN AMERICA

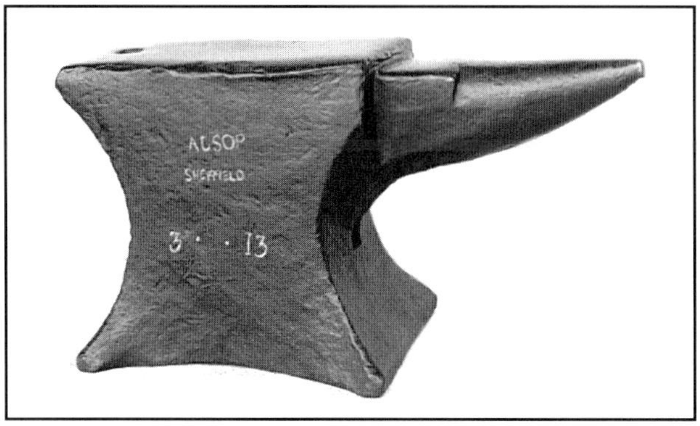

The Alsop anvil. (R.A.P. Photo)

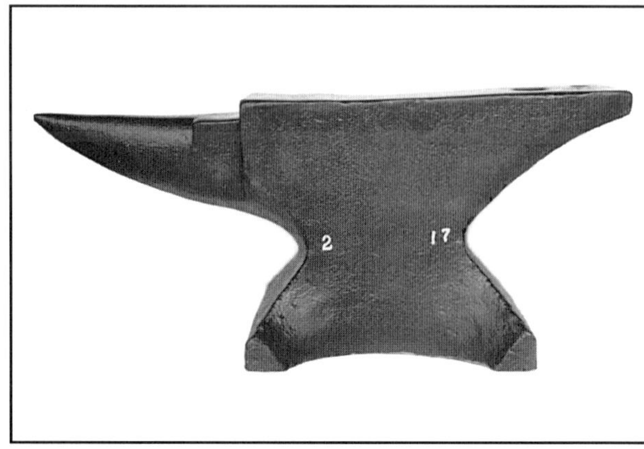

Early-19th century Old English style. Notice the difference in the feet of the Old English and the Colonial Alsop. (R.A.P. Photo)

All English horned anvils after about 1780 have this table, and those without a table of the *Old English* and *Modern* style came to be referred to as farrier's anvils.[10] I might point out that as far as I know, the *Continental* anvils have never had a table.

By at least 1800 the shape of the English anvil changed drastically, becoming what I call the *Old English* style. The horn became longer and larger in diameter, the face longer, and it was not as blocky as the *Colonial*. The heel was longer, the waist thinner and the feet spread out more. The longer and larger horn came down deeper into the body and appears to be part of the body rather than just an add-on, as on the *Colonial* anvils. Of course, as mentioned earlier there was a overlap of anvil style, depending on the manufacturer. Therefore, some of the *Colonial* style anvils could have been produced after 1800 and some of the *Old English* before 1800.

Another thing that changed were the feet. The *Colonial* anvils had feet that came out to a point more or less, but the ones made after 1800 have larger feet that are generally blunt on the end and a sharp edge on the top. One English anvil manufacturer (*Mousehole*) maintained this *Old English* foot style right into the twentieth century.

Hornless Colonial anvil with 5th foot or toe. (R.A.P. Photo)

There is a feature found on some *Colonial* anvils that is not found on the *Old English*. I call it the "5th foot" or "5th toe." The *Moses* anvil has the 5th foot as the photo shows, but this foot does not seem to have any practical purpose. Of the ten or more I have recorded, I have never found any working marks on this foot to show that it was used in smithing. The 5th foot is found on hornless and horned anvils of the period.

In the 1970s, *The Chronicle of the Early American Industries Association* journal carried a series of articles dealing with an anvil known as the *Pomeroy* anvil, which belongs to the

Northampton, Massachusetts, Historical Society. The anvil is of the *Colonial* type, is hornless, and it has the 5th foot. It undoubtedly was made sometime before the middle of the 18th century. The presumed original owner was General Seth Pomeroy (1706-1777), a well known historical figure; thus, the name of the anvil. Articles appeared in five issues of this journal about this anvil and anvils in general.[11] There was no conclusion drawn as to the purpose of the 5th foot. I have not seen the *Pomeroy* anvil, but from the photograph in the journal of June, 1974, it appears to be a typical *Colonial* anvil made of wrought iron with a steel face.

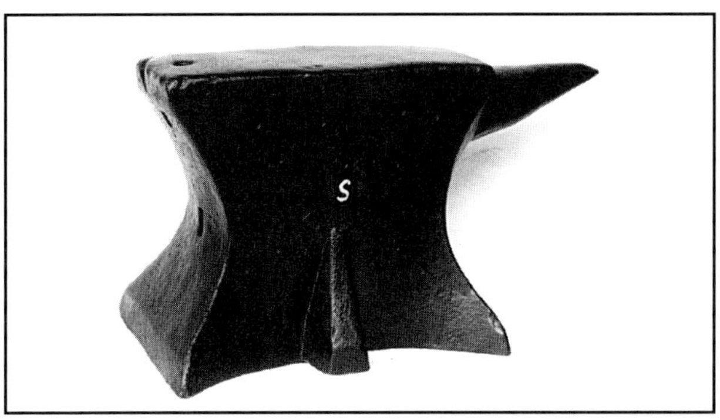

Colonial anvil with 5th foot or toe. (R.A.P. Photo)

This 5th foot was thought by some who wrote the articles to confer a very old age to the anvils that have it.[12] However, I have recorded an anvil with the 5th foot with a date 1810 cut into the body. The date means very little; it may document when the anvil was bought or commemorate some event. What is important is that this anvil has a table on the horn, as well as the larger horn. This anvil cannot be older than about 1780, and I would date it at about 1800.

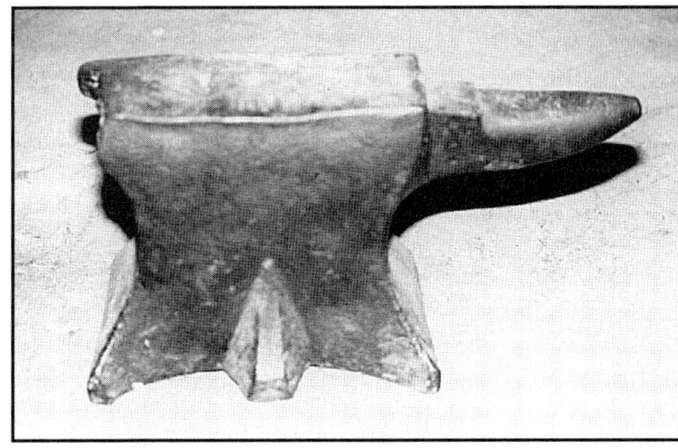

Colonial style with 5th foot, horn and table Museum of Appalachia (R.A.P. Photo)

I suggest that the reason for the 5th foot is this: It is a vestige of what was known on the Continent as the "church window." [13] These "windows" still appear on some German cast steel anvils of recent make. I am sure this concept of the "church window" is taken from an idea of what the stone tablets containing the Ten Commandments of the Bible looked like.

This concept was used as a window motif throughout the Middle Ages, in Europe and is still used today. On an old *Continental* pattern anvils I have studied which had the full "window", I have found that it was a separate piece forged-welded onto one side of the

Single-horned Continental pattern with "church windows" owned by Mr. Bill Gichner. (R.A.P.Photo)

Double-horned Continent pattern with "church windows" (R.A.P. Photo)

anvil's body. This "window" and the 5th foot which is also forge-welded on, had only religious significance and aesthetic value, in my opinion. It may be said that the blacksmith was the master of the four known elements of ancient times, that is, Earth, Fire, Water, and Air. Therefore, the smith was endowed with special spiritual powers, as was his anvil. (see Chapter VII, Anvil Myths). The 5th foot as a vestige of the "window" on some English anvils maintained the idea without the extra work and cost of the full "window." I also suggest that in England by the 18th century, this 5th foot may have become the trademark of an anvil manufacturer, perhaps Mousehole Forge of Sheffield, England. The Mousehole Forge was reported to be the first forge to specialize in anvil manufacturing, and was the manufacturer who standardized the *London* Pattern.[14] It is also known that this forge was in operation before 1700 and possibly before 1600.[15] Everyone of the 5th foot anvils I have recorded is similar, whether they have a horn or not, and of the *London* pattern, suggesting that most of them were manufactured by the same forge. A number of them still have the weight marks, indicating they were sold to the trade and are not home made affairs.

On many of these anvils that I have recorded I have found a large "S" stamped above the 5th foot on the body, which, in my opinion stands for Sheffield. I have seen photos of anvils with the 5th foot that do not have the "S," but they may have been manufactured by another forge or the "S" has disappeared through use and by rust. I cleaned one that I thought did not have the "S", but when the light on that side was just right and I stood at an angle to the side, I could just make it out. I could be wrong about the meaning of the 5th foot and the "S", but until someone proves otherwise or I discover differently, I will maintain my opinion. It is also my opinion that many of the *Colonial* anvils without the 5th foot are as old or older. I suspect that an anvil could be ordered with or without the 5th foot depending on how superstitious the smith was. Of course, the anvil without the 5th foot would cost less too. I date these *Colonial* anvils with or without the 5th foot from the late 1500s to about 1780, and if it has the 5th foot and a table from about 1780 - 1800.

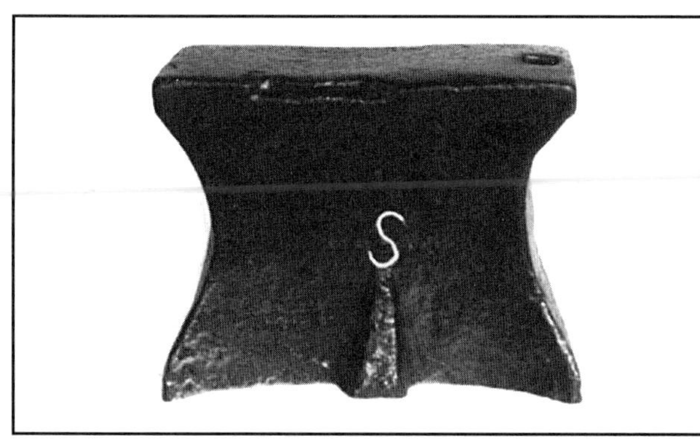

The "S" can be seen above the 5th foot. (R.A.P. Photo)

About 1830 the next feature was added to the English anvil. The pritchel hole, or punching hole became standard from about this time on. It was added to the side of the heel on the anvil's face, generally on the right side, when facing the anvil from the heel end.

Mr. Staley of Dayton, Ohio, has a *Mousehole* anvil of his great great grandfather's. This ancestor punched his name," I.W. Staley", on the side of the anvil, and the date "1830." This anvil does not have a pritchel hole and it was surely made before 1830. This ancestor lived in this country at the time that he bought the *Mousehole* anvil.

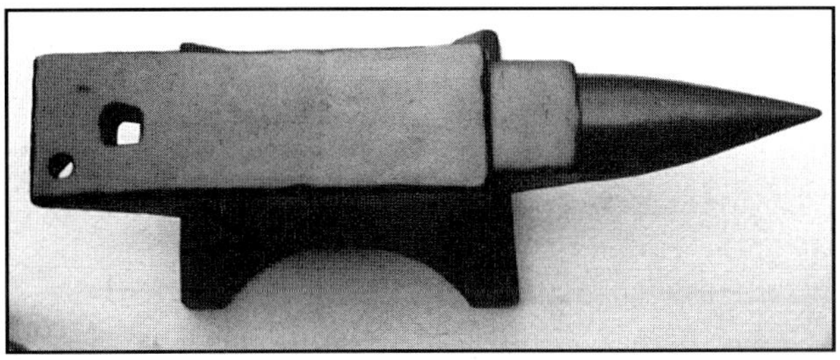

Old English anvil with pritchel. (R.A.P. Photo)

The anvil manufacturer *William Foster* dated his anvils and every one of those that I have recorded, which is dated before 1830, with one exception, is without the pritchel. On close examination it appears that the pritchel was drilled into the "one exception" at a date later than 1830. Every smith's anvil I can date after 1830 has a pritchel hole. Therefore, the conclusion to be drawn is that 1830 is the approximate date for adding this feature.

In 1851, Peter Wright of England received a patent for a two-piece anvil, thereby changing anvil manufacturing forever. In so doing, he put most of his competition out of the anvil business.[16] With the Peter Wright patent, the *London* pattern anvil was perfected as to features, and since this time anvils have changed only in the way they are manufactured and component material. I refer to all anvils manufactured after 1851 that follow the *Peter Wright* pattern as *Modern* anvils.

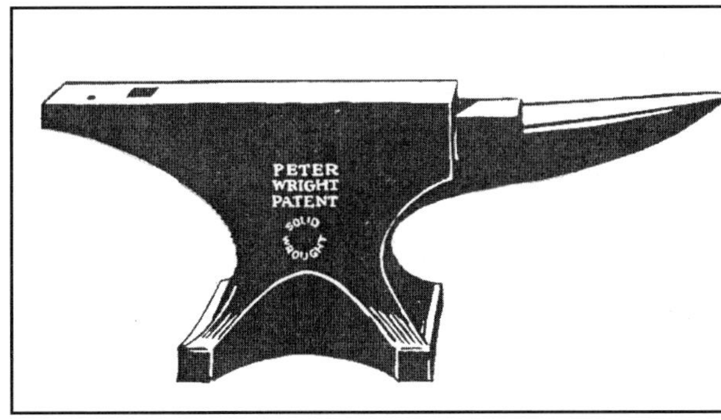

Sketch of modern Peter Wright anvil.

In 1882, the hardware company of S. D. Kimbark of Chicago, Illinois, began developing a distinctive farrier's pattern anvil, manufactured in Germany, and the farrier's pattern has continued to evolve.[17] A second pritchel was added to the Farriers pattern early in the 20th century, about 1910.

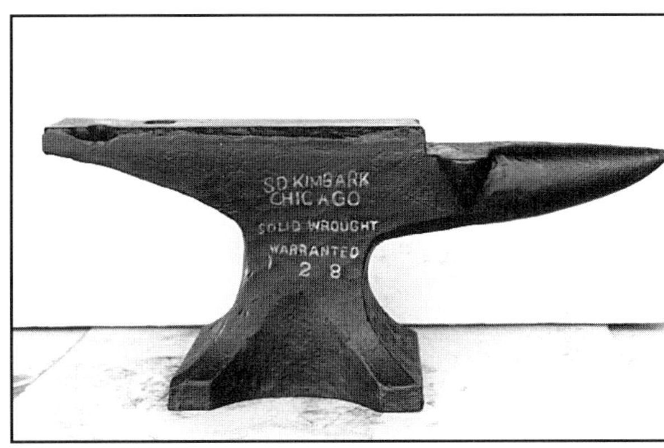

S.D. Kimbark anvil manufactured by Hay-Budden Mfg. Co. (R.A.P. Photo)

The following is a brief summary to help in dating smith's anvils within certain periods.

1. English smithing anvils without a table and of the *Colonial* style can be dated from about the late 1500s to *ca.* 1780, including those with the 5th foot.

2. If an anvil has a 5th foot and a table, it dates between ca. 1780 and *ca.* 1800.

ANVILS IN AMERICA

3. An anvil with or without a hardie hole is no indication of the date it was manufactured. This is also true of the horn or horns. The hardie hole was probably added to the smith' anvils *ca.*the16th century . All smith's anvils made after approximately 1780 have the hardie unless they were a "special."

4. If a *London* pattern anvil has a pritchel hole put in at the factory, it was most likely manufactured after 1830. I have a 5th foot *Colonial* anvil with a pritchel that someone drilled in at a later date, although it has no hardie hole.

5. All *Peter Wright* anvils manufactured before 1851 are very similar to all other *Old English* anvils. All *Peter Wright* anvils manufactured after 1851 have a flat running on top of the foot from side to side under the horn and heel, four handling holes, and are of the *Modern* style.

6. All farrier's anvils having a clip horn were made after 1882.

7. Farrier's anvils with two pritchel holes were manufactured from *ca.*1910 to the present.

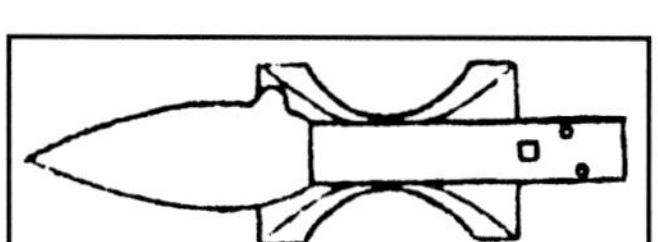

Top view of farrier's anvil with two pritchel holes.

The anvils used in the United States can then be put into six time periods:

1. *Ca.* 1600 - *ca.* 1780 - *Colonial*, steel-faced, with or without a horn or hardie hole.
2. *Ca.* 1780 - *ca.* 1830 - *Colonial* and *Old English*, table added.
3. *Ca.* 1830 - *ca.* 1850 - *Old English*, pritchel hole added.
4. *Ca.* 1850 - *ca.* 1895 - *Peter Wright* and *Modern*, the first modern three-piece anvil, the first American-made anvils.
5. *Ca.* 1895 - 1950 - *Modern*, All sizes and shapes to order; mostly wrought iron.
6. 1950 - *Present,* American made anvils are one piece cast steel, mostly of the farrier's pattern.

For closer dating of particular anvils, refer to the appropriate anvil listed in the index.

The following quotations and comments will give the reader some idea of the early development of the anvil. They also suggest how difficult it is to judge an anvil just by its features, especially before 1800.

In the apocryphal book of the Bible, *Ecclesiasticus*, written by Jesus, son of Sirach of Jerusalem about 280 B.C., there are verses about people who worked with their hands.[18] One of the verses is about the blacksmith.

So the smith, sitting by the anvil
And tiring himself with the rough iron;
The smoke of the fire wasteth his flesh.
And he fighteth with the heat of the furnace;
The noise of the hammer deafens his ear,
And his eyes are upon the pattern of the utensil; He giveth his mind to finish his work
And his sleepless care is to polish it on completion.

These words tell a great deal about these early smiths, but I want to focus on the line, *"The noise of the hammer deafens his ear."* It has to mean that the hammer is striking work on the anvil and the anvil is ringing. Eventually, it either makes the smith deaf or at least he loses some of his hearing. I know from experience how noisy a good anvil can be and have read reports of the bad effect of the anvil's ring on one's hearing. The question arises, what kind of anvil is he working on? As far as I know there was only one kind of metal or possibly two that would have been used for making an anvil at that time, wrought iron and, possibly. bronze. There were types of steel at the time, but it would not have been used in a tool like an anvil. Regardless of what the anvil was made of, it still will not ring unless it has some type of protrusion. I experimented on a couple of hornless cast-steel and wrought-iron anvils I have, and although they do not sound dead like a cast iron anvil, they will not ring. The anvil the smith of old was using must have had a horn or heel of reasonable length.

Most old illustrations of the 13th through the 17th and earlier show anvils that are more or less just blocks without horns.

Various drawings of very early anvils taken from old wood-cuts and engravings.

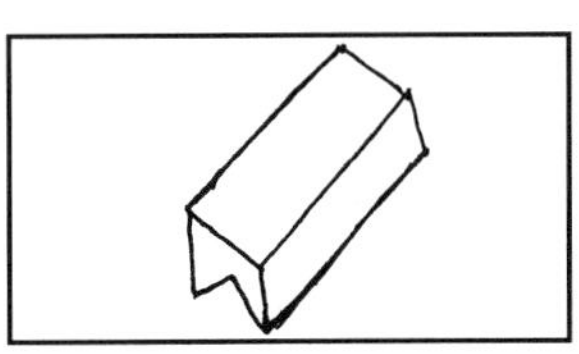

"The Blacksmith and Wheelwright," March, 1918.

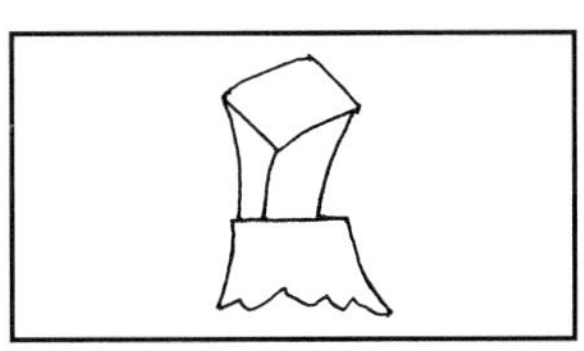

The Blacksmith and Wheelwright," March, 1918.

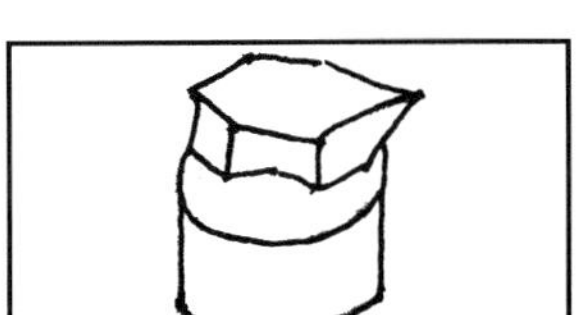

"The Blacksmith and Wheelwright," March, 1921.

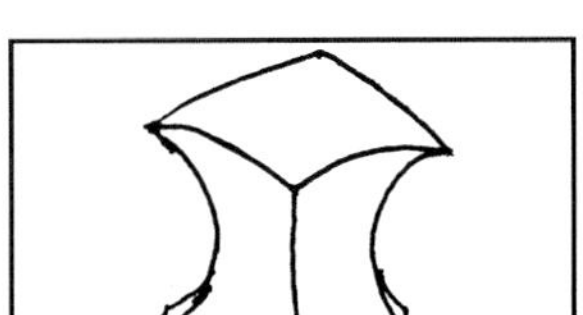

"The Blacksmith and Wheelwright," March, 1921.

However, we can conclude from these verses that by at least 320 B.C. the smith's anvils had horns and were probably made of wrought iron whether they had hard steel faces or not. I have never seen a large bronze anvil, and have no idea whether it would stand up under smithing conditions, although I am sure it would ring if it had a horn.

In *The Antiquaries Journal*, in her article "The Anvils of Bronze Age Europe," Margaret Ehrenbery describes 36 small bronze anvils that are located in museums in Western Europe and were found all over Western Europe and the British Isles (also see Appendix I).[19] Most have been dated from about 1200 to 700 B.C., because of the material found with them.[20] They are not just little blocks with a flat side, but some of them have horns, punching holes, swages of various shapes in them, and a stake for mounting them. Although they are relatively small, they contain about every complexity or feature that has been used on smith's anvils to date.

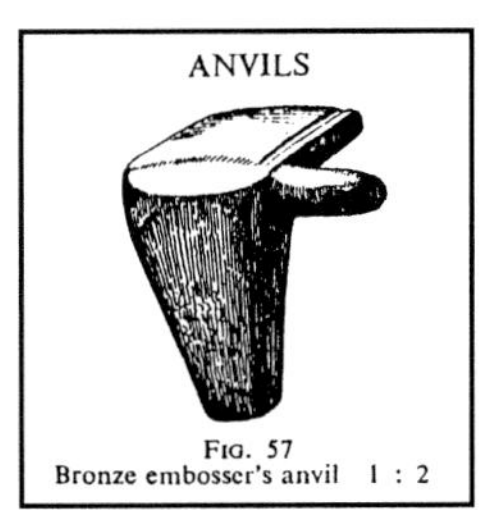

FIG. 57
Bronze embosser's anvil 1 : 2

"The English Tool Book"

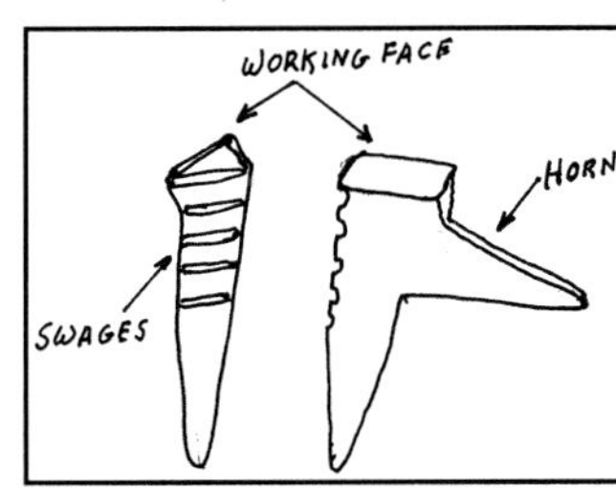

"The Antiquaries Journal"

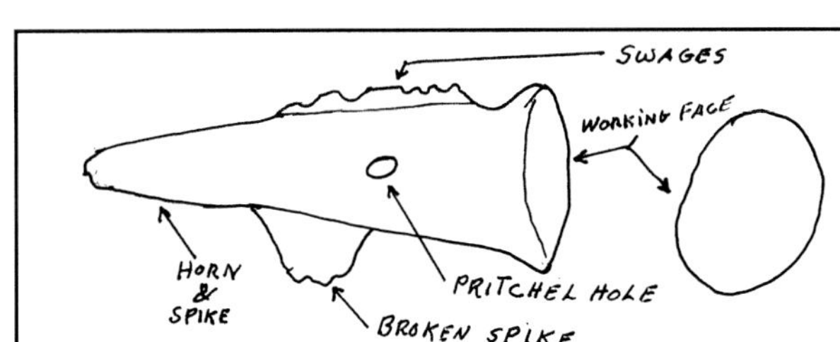

"The Antiquaries Journal" (R.A.P. Drawing)

ANVILS IN AMERICA

Most likely some of these features were incorporated into the larger smithing anvils from time to time as needed. Stakes or bick irons generally served for bending and doing the intricate work. The large anvils were used mainly for heavy drawing and welding, needing nothing more then the blocky shape. It must also be remembered that the finest art smithing we know of was performed during the Middle Ages or shortly thereafter, so their tools were probably not as crude as some artist of old depict them.[21] Therefore, when someone points out a particular feature on an old anvil and says it places the anvil within such and such years, he had better know whereof he speaks.

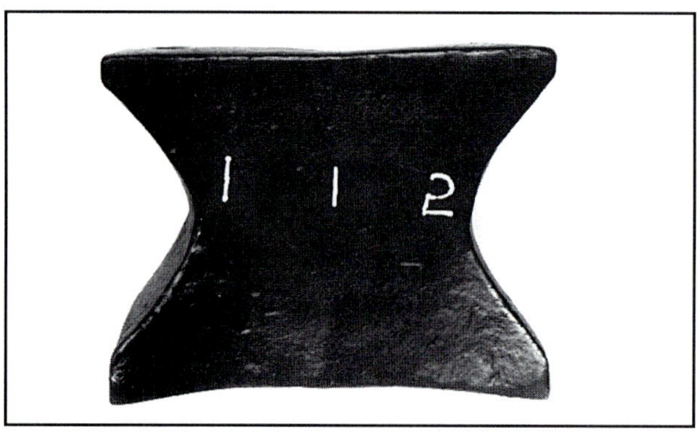

140-pound hornless Colonial anvil with hardie: ca. 1600-1700 (R.A.P. Photo)

An Italian Smithy of the 14th Century

"The Blacksmith and Wheelwright," May 1921.

99-pound hornless Colonial anvil. Notice the difference in shape; ca.1700 (R.A.P. Photo)

16th century armorer's anvil, in the British Museum.

To establish early English specialization in anvil manufacturing, I refer to an article published in the April 1891 issue of *The Blacksmith and Wheelwright* magazine.[22] The journal reprinted an excerpt from a work published in London by T. Waller in 1747, concerned with trades one's children might take up as a life's work. The journal reprinted the part that refers to smithing as a trade for young men. Among the smiths was the anvil maker. The reprint retained the old style "S" which I have changed for easier reading.

General Information About Anvils

Anvil-maker, or Smith, is he who forges, steels, and planishes the faces of all sorts of Anvils, Hammers, & C. for their Brother-Smiths, and other Handicrafts who use them, and these are all such almost who work in any sort of Metal. It is not quite so heavy a Labor as that just mentioned [anchor maker], though some Anvils are very weighty. . .

They [smiths] were incorporated into a Company by the title of Black-smiths, in the year 1571,in the Reign of Queen Elizabeth, and confirmed by King James I. Livery - fine 82 Their Hall is on Lambeth-hill in Thames-street, and their Court-day on the first Thursday of the Month. They have also a Stand in Cheapside, in which they sit to attend the Lord-Mayor on the Day of his installation. Arms. Sable a chevron between the Hammers Argent, handled and crowned or Motto, By Hammer and Hand ---All art doth stand.

This demonstrates that the anvil makers in England were well established by the late 16th century, not long before the founding fathers arrived in America.

ANVILS IN AMERICA

ANVIL PATTERNS AND TYPES

The following quote is a fitting introduction to this section.

> *...There is no question about the art of smithing and the anvil having developed simultaneously.*
> *...From the 12th to the 17th century, smithing all over Europe reached a stage bordering upon perfection, but no attempt was made to standardize the shape of the anvil. The reason for this would seem to be that as a rule, every smith was his own anvil maker, and an anvil being a piece of equipment that lasts a life time or over, it stands to reason that there were few if any expert anvil makers in the Middle Ages.*[23]

In the drawings, woodcuts and art work of this period and earlier, the smith's anvil is depicted as more or less a block of iron in a shape from square to almost round, some with stake-like feet at each corner, some just flat on all sides, with or without horns, and all of no distinct style or pattern. Usually they are depicted as resting on a wood block. Each trade designed their anvils a little differently, but for their heavy work, the anvils were very similar in that they had a large face. Most of the old drawings, woodcuts and art work come from the European continent, but we can assume that the English pattern anvils were the same, at least up to the 16th century. Some of these very old anvil styles can still be seen in *Continental* anvil patterns manufactured in the recent past. The lighter anvil stakes used in crafts in the distant past have not changed at all. These are the ones called *bick irons* or *beck irons* or just *stakes.* They came in all shapes and sizes to suit the work, and were always mounted or driven into some kind of support, whereas the smith's anvil was free-standing in most cases.

Smith's anvils on the Continent changed very slowly if at all. In England however, with its rapid development as a seafaring and trading nation in the 16th century, anvils began to change more rapidly.

An Armorer of the 13th Century

From "Blacksmith and Wheelwright" May, 1921.

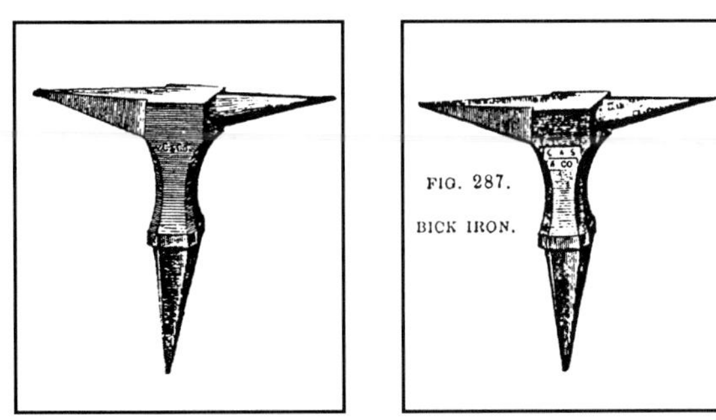

Typical bick irons. Some weigh less than a pound and some up to 300 pounds.

Just why the *London* pattern developed the way it did we do not know, but I am sure it had something to do with England's development as a naval power, where a lighter more mobile anvil was needed. *Continental* pattern anvils even today have the majority of their mass in the body and base, whereas in the *London* pattern the mass began to move toward the anvil's face, giving more face area relative to the weight of the anvil.

The old bick iron from the 16th or 17th century. (R.A.P. Photo)

By at least the middle of the 16th century the *London* pattern began to take shape in England. It was no longer a nondescript block, but was made with a rectangular face. No longer did the sides come straight down; now they flared at the bottom, and the center of the body went in on the long side to produce a waist, thereby reducing its weight while retaining a large working surface on the face. The Mousehole Forge of Sheffield, England, is given credit for the standardizing of the *London* pattern, and in all likelihood provided better than its share of the *Colonial* anvils.[24] In a like manner, after 1850 all anvil makers who stayed in business copied the *Peter Wright* anvil style, which I call *Modern.* After 1780, along with adding of anvil features, the face of the *London* pattern became longer and narrower until about 1900, when the pattern reached its final and most graceful form in the wrought anvils made in the United States.

Roberts Anvil manufactured by the Columbus Forge & Iron Co. (R.A.P. Photo)

Distinctive anvil patterns have developed for particular uses. Most of these have come about only since the late 19th century as new technologies and needs have developed. In the previous section I was speaking primarily of smith's or general purpose anvils. On these anvils any job, except possibly fine watch or jewelry work, could be performed. With the attachments to fit the hardie hole, there was no limit to their usefulness. However, as in ancient times, specialty jobs do not require an all-purpose tool. More often than not a specialty pattern anvil is more efficient and may be less expensive.

The following is a selected list of specialty pattern anvils.

Sawmaker's

The *sawmaker's* pattern needs only a large, flat, hard face, like some of the Medieval anvils. It is used to put a slight surface tension in a circular saw blade. It probably should be called a *saw tensioner* pattern because all it is used for is to support the saw while tension is put in the saw blade with a hammer while cold. These anvils can weigh from 60 pounds for light work to huge blocks weighing up to 1600 pounds.[25] The only really successful cast-iron anvils ever made were of this type, and that is what these *sawmaker's* anvils are made of today. The face of these modern anvils is chilled for hardness and then ground flat or slightly crowned. All the American wrought anvil makers produced these anvils, if only in limited quantity, and made them from the same metals that they used in their smith's anvils. Most of these wrought anvils weighed around 300 to 400 pounds, and they had feet like their standard anvils. Until they went out of business, Fisher and Norris made more of the *sawmaker's* anvils than any other company. Theirs had a cast-iron base with a hardened steel face, the same as their smith's anvils. Incidentally, the Wright brothers (of flying machine fame) used one of these 300 pound Fisher *saw maker's* anvils in their bicycle shop around the turn of the century.[26]

Peter Wright, Mousehole, and *Söderfors* exported *sawmaker's* anvils to this country. I have recorded a *Mousehole*, a *Söderfors* (Swedish cast-steel), and a *Peter Wright* (see Appendix J).

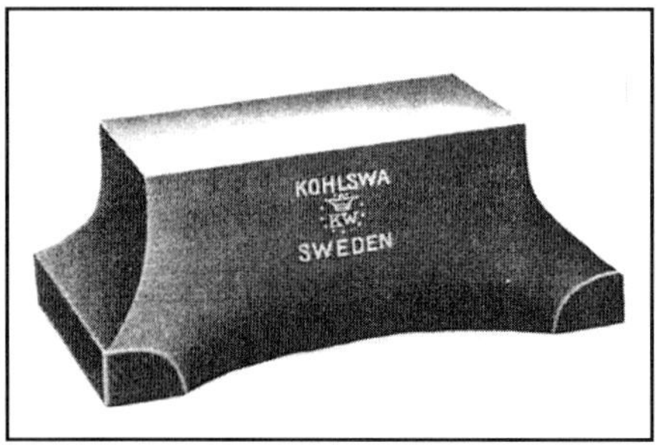

Several Sawmaker's anvils

Coachmaker's

The *coachmaker's* anvil looks just like a smith's anvil except that it has a small flat side horn, usually about five inches long, coming off the body, below the face. There has been some confusion about the purpose of this side horn. In a 1914 article by James Cran,[27] the drawing indicates that this side horn makes the anvil a farrier's pattern. However, recently (1996) I have seen the original poster from which the drawing in the Cran article was taken.[28] The 1914 article had two letters mixed up on the drawings, indicating that the farrier's anvil was the *coachmaker's* and the *coachmaker's* was the farrier's. I was also told by a reliable source that this side horn anvil was a *cutler's* pattern.[29] Personally I could not understand how the side horn could be used for either farrier's or cutler's work, and I went to some length expressing my doubt in the manuscript for this book, but now the error has been corrected. The side horn is ideal for making clips in coach work and clips for art-metal work.

Old English Coachmaker's anvil. (R.A.P. Photo)

Plowmaker's

The *plowermaker's* anvil looks like a smith's anvil except that where the table should join the body, the face is cut back at about a 45° degree angle, making a large angular table. I have recorded three of these and have seen many of them in the trade journals advertisements, but I doubt that very many were made.

The *Trenton plowmaker's* anvils I recorded appears to be a modified farrier's anvil. They have no shoulders and have two pritchel holes, but the horn is the same as a smith's anvil. They are advertised as weighing about 200 pounds (for photo see Chapter VI).

There is one so called *plowmaker's* anvil that is hand held with a round convex working face and weighs less than five pounds. It was probably intended to be used by the plowman to dress a slightly worn plow point while in the field. I would call it a *plow dolly* rather than an anvil (see Chapter VI).

Chainmaker's

The *chainmaker's* pattern anvils that I have seen differ in style with each maker. The one feature that they have in common is a square hole or holes going through the body of the anvil, which was used to hold a square bar. This bar is hinged to another bar or hand-operated swing die on one side of the anvil and a fixed die on the bar on the other side of the anvil. The dies are used to dress up the weld of the chain link. I saw a magazine photograph of one which had a large round horn protruding at right angles from the body of the otherwise hornless anvil[30] (for photo see Chapter 6).

Farrier's pattern

I have mentioned the farrier's pattern before as still evolving, and the majority of the American made anvils today are of this type. Before the 1880s, horseshoes were made and shaped on the smith's pattern anvil. The horn was generally too small for forming large shoes, and the face too wide for working small shoes. Often and especially for draft horses, a piece of metal was forged up in the middle of the shoe to protect the front of the horse's hoof. This clip, or reverse calk (calks protrude from the bottom of the shoe to give the horse traction), was formed at one of the corners of the anvil, usually at the heel, which was never very satisfactory for this purpose. To remedy these deficiencies, several of the English makers removed the table. This produced a larger horn area near the body, but that was about all. In 1882, a large Chicago hardware company, S. D. Kimbark, designed an anvil with a proportionately larger horn in relation to the size of the face. The anvil had a small flat rounded (sometimes rectangular) horn (clip horn), about one inch long, coming off the side of the horn where the table had been, but about an inch from the face. They had this pattern anvil made in Germany until an American anvil maker, Hay-Budden went into business and took the contract.

Here is an article about the *Kimbark* farrier's anvil from 1894.[31]

> ***Farrier's Perfect Anvil***
>
> *The anvil illustrated herewith can be obtained from S. D. Kimbark, Chicago, Ill. It is especially for horseshoer's use, having been originated in 1882 by Mr. Kimbark. Horseshoeing is a distinct branch of blacksmithing, requiring special tools - tools nearly useless for any other purpose.*
>
> *The smaller tools have always been obtainable, but the anvil, the most important implement of all, was such as all blacksmiths used, and rarely could one be found, in the most extensive stock, having any of the features required by the horseshoer. To meet the wants of the craft and facilitate their work, Mr. Kimbark introduced his S. D. K. anvil, and for twelve years has maintained its original high standard, presenting all the essential features recognized by the most practical and enlightened workmen. If the horn on a farrier's anvil is too small, the table too high, the heel too thick, the face too wide or the holes wrongly placed, the anvil is*

not perfect any more than it would be without a clip horn. he S. D. K. anvil preserves all the requisite points, hence its popularity and large sales. For additional particulars, prices, etc. write to the maker, S. D. Kimbark, Michigan Avenue and Lake Street, Chicago, Ill.

From "The Blacksmith and Wheelwright." April, 1898: Courtesy of Mr. Hunter Plikington.

On the farrier's pattern, the horn always blends into the body on the side, leaving no shoulder, and the horn can be so large (swelled horn) that you are sure it would tip down at the horn if the anvil was not securely anchored. I was told that the second pritchel hole, added about 1910,[32] was for the purpose of knocking out a particular nail from the horseshoe that was hard to get to when the shoe had calks on it.[33]

Some of the newer farrier's anvils are unique, having a large hole in the face for bending horseshoes or turning cams on the side of the body for shoe bending. One design made shortly after 1900 had the horn at a right angle to the face.[34] There are a number of modifications that can be added if desired. Farrier's anvils almost never have exceeded 200 pounds and the heaviest one to my knowledge, made in 1995, weighs about 165 pounds. However, Hay-Budden advertised farrier's anvils up to 400 pounds in an 1899 article in the *Iron Age* magazine.[35] I have now (1997) recorded a 250-pound *Hay-Budden* farrier's anvil owned by Mr. David Baker.

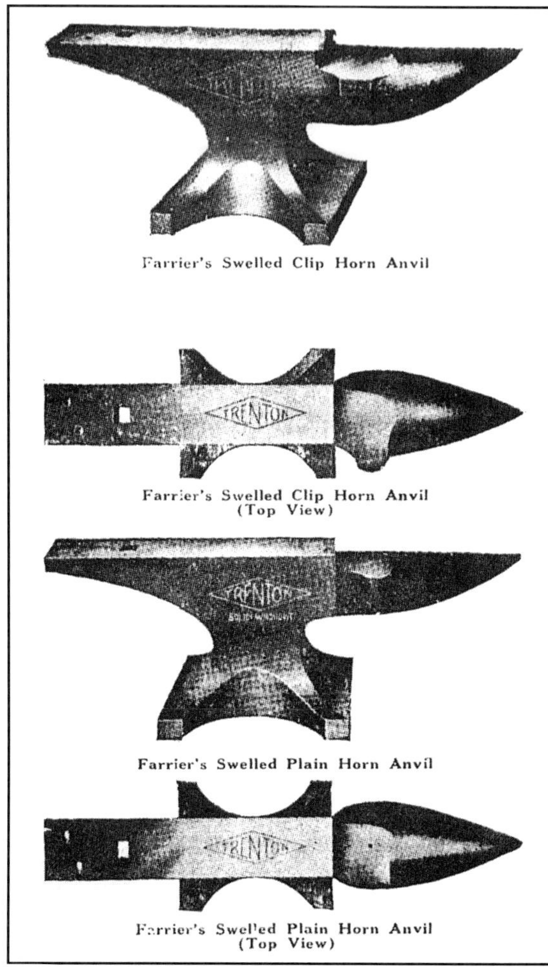

From the Columbus Forge & Iron Co. brochure. Courtesy of Mr. David Tilton.

From the Hay-Budden Co. brochure. Courtesy of John Catchings

The *Continental* pattern anvils were never popular in the United States until recently: the ones found here were brought in by individuals for their own use. However, today many smiths who work ornamental iron are discovering that many of the features of the *Continental* pattern are to be preferred to those of the *London* pattern.

A Continental pattern with extra features: side-face and upsetting block.

The following is a list of tools that are considered anvils by various trades. Photographs and drawings can be found in Chapter VI. Some of those listed are unknown to me.

1. Anvil swage block
2. Cone anvil or cone mandrel
3. Nailer's anvil
4. Cooper's anvil
5. Jeweler's anvil
6. Scythe anvil
7. Railroad or Bridge anvil
8. Axe maker's anvil ?
9. Instrument maker's anvil ?
10. Rope halter anvil
11. Hornless anvil
12. Stake anvils: tinner's, silversmith, etc.
13. Die bold anvil ?
14. Gun maker's anvil
15. Cutler's anvil
16. Craft anvil

General Information About Anvils

The following are some anvil patterns used in Europe and the United States in the 19th and 20th centuries.

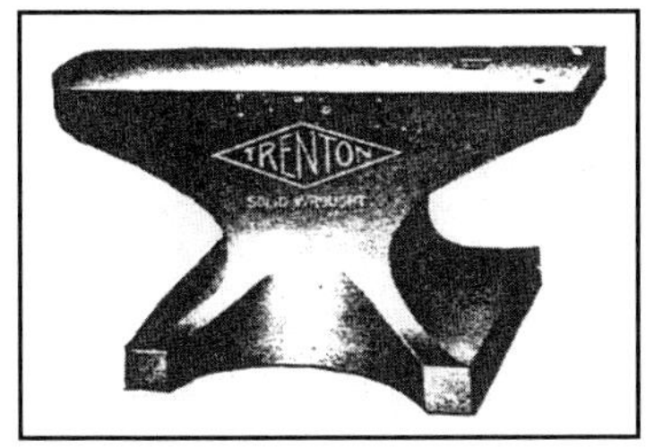

Hornless anvil. From the Columbus Forge & Iron Co. brochure.

Hornless Continental pattern with "church windows."

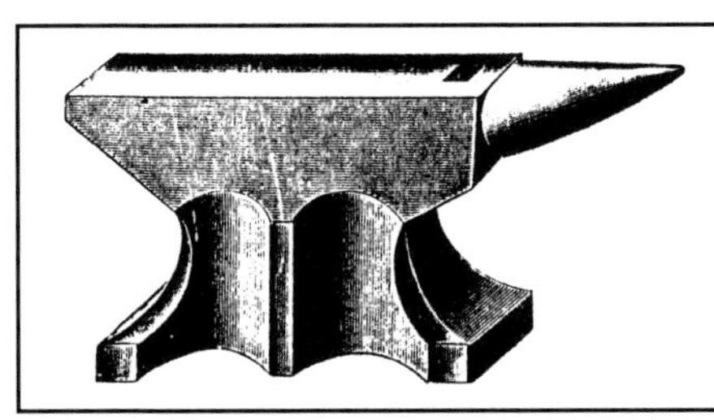

Continental pattern with "church windows" and horn

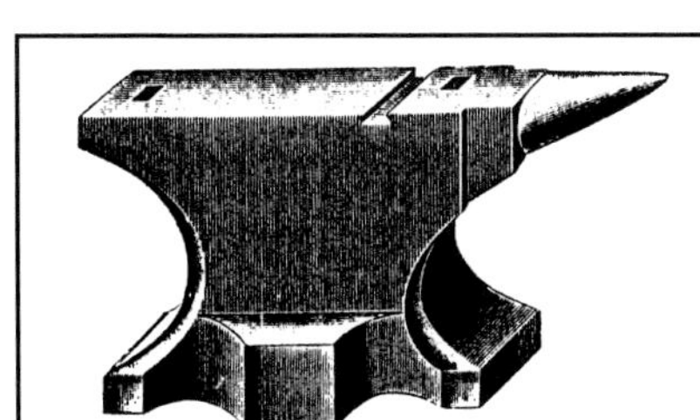

Continental cuttler's pattern with upsetting block.

ANVILS IN AMERICA

Continental pattern with side-face, two horns and upsetting block.

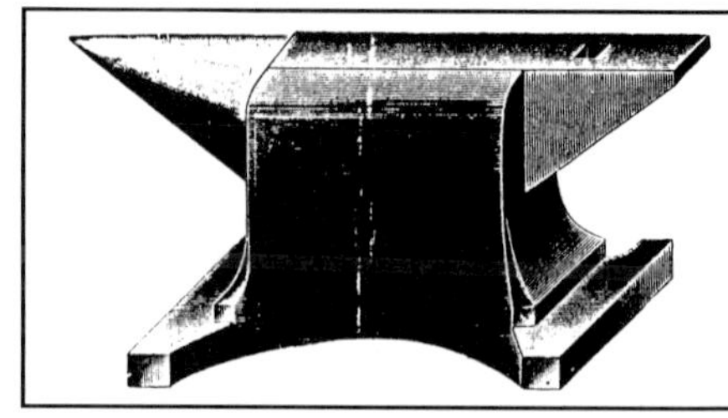

Continental pattern with rounded side

Continental pattern armorer's anvil.

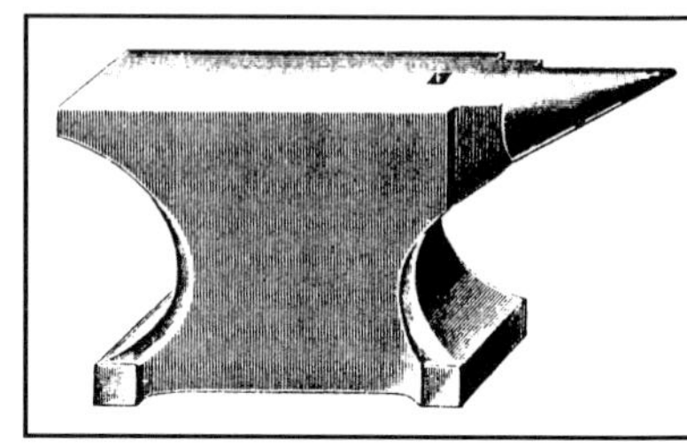

Continental "SOHO" pattern, with only one hardie hole.

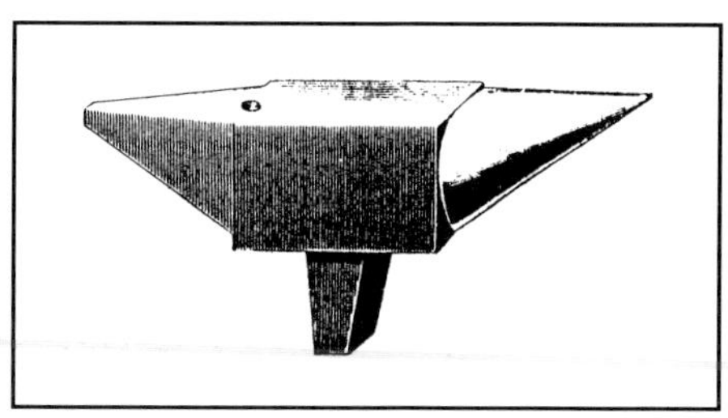

Late Spanish pattern anvil without base

MATERIALS AND METHODS OF ANVIL MANUFACTURING

Materials

Down through the centuries craftsmen have jealously guarded the methods that they have developed to produce a product. Even patent rights were not much protection, so if a craftsman discovered a faster, better, or easier way to reach a desired result, he guarded it to the best of his ability. His secrets were passed on only to his trusted workers. The metalworking field was no exception; in fact, its secrets were more closely guarded than most.[36]

The same is true to some extent for anvil making. Although a few sources such as *Diderot's Encyclopedia* give us some idea of how an anvil was made in the 18th century, and while others have mentioned the metals used, the information is always very general.[37] A forge shop would be hard-pressed to produce a wrought iron anvil made the old way from any instructions available today, and it would require considerable experience and skill for execution. These are probably some of the factors that held back anvil manufacturing as a successful enterprise in this country until almost the 20th century.[38] Therefore, the information given here can be in only general terms as to the materials and methods of anvil fabrication that have been used over the centuries.

In all probability the first anvils were rocks. Even today rocks are still being used for anvils in some places. When iron is nearly white-hot (forging temperature) it is quite soft or pliable, in which case a hard smooth rock would make a fair anvil for working metal. During the summer of 1996 my granddaughter went on a medical mission project to Napal to help build a clinic. While there she had an opportunity to do some sight seeing. While walking with some friends along a street she noticed a local craftsman working on metal pieces. Knowing my interest in blacksmithing she took a closer look. The craftsman was hammering out a rice sickle on a specially shaped rock. She took several photos and bought one of the finished sickles which she gave to me.

I do not know of anyone in this country having used rocks as anvils for metalworking, but we do have records that *meteorites* have been used.[39] For your information there are two basic types of *meteorites*: those made of stony material and those consisting mostly of metals, generally iron.[40] Those used as anvils were made of iron. The most famous of these anvils are the two now in the gem section of the Smithsonian Institution, Museum of Natural History, in Washington, D.C., known as the *Tucson Meteorites* (see Appendix K).

Smith using specially shaped rock to forge a number of different items. A rice sickle in this case. (Katherine Keith Photo)

ANVILS IN AMERICA

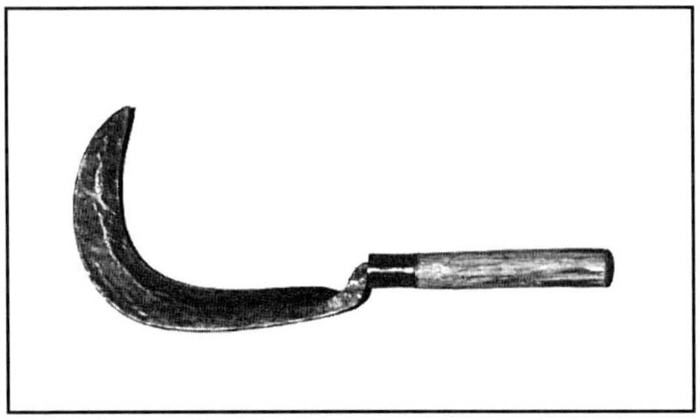

A rice sickle forged by the smith. (R.A.P Photo)

Other than meteorites, anvils have been made from the following materials: cast bronze, wrought iron, wrought iron with a steel face plate, cast iron with a chilled face (hardens the face), cast iron with a steel face plate, cast steel with a hardened face, forged low-carbon steel with a steel face plate, and forged steel with a hardened face. All these metals are basically iron except the bronze. In the following paragraphs they will be discussed as listed.

Bronze

Cast bronze anvils go back to the dawn of metalworking, but the only ones to have survived are small Jeweler's anvils.[41] The large bronze anvils were most likely recast into another product when wrought iron anvils came into common use (see Appendix I).

Wrought Iron

Solid wrought-iron anvils would work well, but they dent easily, and even with careful use they begin to mushroom when used continuously, as wrought iron cannot be hardened to any extent. However, they are easily repaired by simply heating and hammering the misshapened part back into its original shape . The face of a wrought-iron anvil also tends to become more compact and work-hardens somewhat with use, which is a help. These are the type of blacksmith-manufactured anvils used in the southwest United States during the Colonial Spanish period.[42] Apparently, steel faced anvils were almost impossible to obtain in this area until well into the 19th century.[43]

Steel-faced Wrought-Iron Anvils

Wrought-iron and bronze anvils were used until hardened steel-plated wrought iron anvils came into use. How far back these steel-faced anvils go we have no idea, but I would guess they date to at least the Roman period (*ca.*100 B.C.).[44]

Regardless, wrought-iron anvils with hardened steel face plates were the anvils brought over to this country from the beginning of colonization. In some of the material I have read, it is suggested that these early anvils were cast in a mold, instead of being forged.[45] Some cast-iron anvils were in use at that time, but as I mentioned in a previous section they are successful only as heavy blocks. Cast iron (high-carbon iron) melts at about 1800° - 2300° F, whereas wrought iron (very low or no-carbon iron) melts somewhere between 2800° and 3000° F. The cold-blast furnaces and forges used before the late-19th century could not reach or sustain temperatures in the 3000° F range. The reason the early iron workers knew they had a wrought-iron end product (other than cast iron) was that it could not be melted. Therefore, if these early anvils were cast, they would, of necessity, have been made of cast iron. I have yet to discover a cast-iron, English-made anvil in this country. I am sure that after the blast furnace that produced cast iron was developed in Western Europe in the 14th century, iron workers tried to use this new material to make anvils.[46] It is much easier to cast an object than it is to forge it, and less heat is needed for making a cast-iron product. These early anvils were undoubtedly forged, not cast.

You might wonder where the hardenable steel (medium-carbon iron) came from to face these early wrought iron anvils. Before the *Bessemer* process of making steel came into common use in the 1860s, wrought iron had to be produced first, in order to make steel. Incidently *Bessemer* was not trying to make a new product, but was trying to produce a better grade of iron for cannons when he started experimenting with his converter.[47] Hardenable steel contains between 1/2 to 1 1/2 % carbon by weight, and melts between 2400^0 and 2500^0F. In the process of making wrought iron by the indirect method, most of the carbon was burned off and passed from the cast iron stage through the steel stage, but the iron workers of that time had no way of knowing when these points were reached. Therefore, in order to make steel, they worked backwards by adding carbon. They packed wrought- iron bars with a carbon source, usually powdered charcoal, into a ceramic box or crucible which was then sealed. The crucible was placed in a furnace and held at a high heat for a certain period of time until the iron bars absorbed the right amount of carbon to produce a type of steel. The time necessary to produce this steel was learned by trial and error.[48]

The steel produced in this manner was called *blister steel* because of its appearance.[49] It was really only a shell of steel with a soft core of iron. To make a better steel, these *blister steel* bars were cut up, bound together, and then forge welded into a single piece. This distributed the carbon more evenly throughout the iron, producing what was known as *shear steel.*[50] The best grade of this was called *double-shear steel,* where the shear-steel process was repeated. Both *shear* and *blister* steel were used to face wrought-iron anvils.[51] The *blister steel* was used on lower grade or cheaper anvils.[52]

An excellent steel, called *cast* or *crucible* steel, came into use shortly after 1750, but was not used on anvils until the mid-nineteenth century.[53] These steel-making processes consumed a great deal of time, fuel and labor. That is why steel was so expensive, and most tools, such as hammers, chisels, axes, etc., received only a cap of steel which was welded to a wrought-iron body, a process called *steeling.* *Steeling* was used to a great extent for woodworking tools right up until World War I.[54]

A tilt hammer located in the lower left corner. From 1771 Diderot Encyclopedia

ANVILS IN AMERICA

Cast-iron Anvils

As mentioned previously, the blast furnace came into common use in the 14th and early-15th centuries in the Western world.[55] It produced a new product known as *cast iron,* and although the metal was brittle, it had a low melting point and could be cast into any desirable shape at comparatively little cost. Because of this lower cost, compared to wrought iron or steel, many large, block-like anvils were made in this manner. The hammer head and anvil of the old tilt, beam, or trip hammer, was made of cast iron. The special anvil used to finish wrought anvils was made of cast iron also.[56]

An anvil working and finishing block, made of cast iron. From the "Anvil Ring," Vol. 16, #4, 1989.

By quickly cooling the surface of the cast iron while it is solidifying from the liquid state in the mold, a very hard surface will be produced.[57] A fast heat-dissipating metal, such as copper, placed in the mold can produce this quick cooling, or *chilling.* Successful cast-iron anvils, large and small, have been made in this manner up to the present.[58] Cast-iron anvils of the smith's pattern with or without a *chilled* face, were never intended for heavy-duty work, such as blacksmithing. For occasional and light use, *London* pattern anvils, made of cast iron, have served well. When you consider that a cast-iron anvil cost about 1/4th as much as a steel-faced wrought-iron anvil, the cast-iron anvil could be attractive if the work is not of a heavy nature. Many stakes for whitesmithing were and are made of cast iron.

Cast-iron Steel-faced Anvils

The cast-iron steel-faced anvil is an American invention (see *Fisher, Vulcan, Badger, Southern Crescent, Samson* or *American Star* anvils). The manufacture of this type of anvil is described under *Fisher,* Mark Fisher having discovered and perfected the process of welding cast steel to cast iron.[59] As far as I know, the six anvils mentioned are the only ones ever produced by this method. Generally they are not considered as good for general blacksmith work as are the steel-faced wrought-iron anvils and, of course, they do not *ring* like the wrought-iron anvils do. However, I am sure Mark Fisher would have disagreed. It is interesting to note that the *Fisher* steel-faced cast-iron anvils were the first successful anvils made in this country and the last to be produced by a major anvil manufacturer (see *Fisher*).[60]

An early Fisher & Norris cast-iron steel-faced anvil. Old English Style. (R.A.P. Photo)

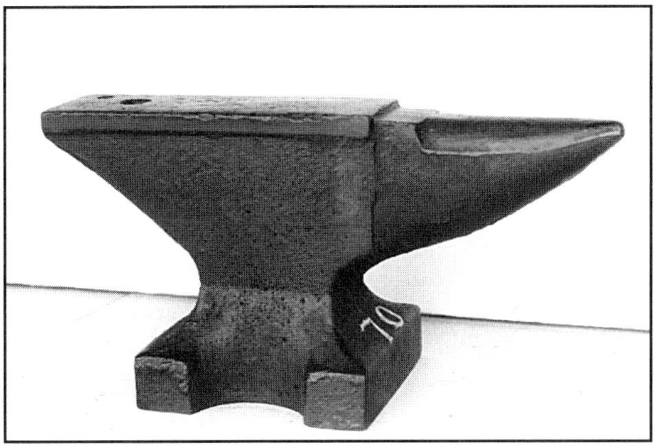

A typical cast-iron anvil. (R.A.P. Photo)

Cast-steel Anvils

Solid cast-steel anvils began to be produced in Sweden around the turn of the 20th century.[61] These were made of high-carbon open-hearth steel and have a hardened face.[62] By this time labor was becoming more expensive than the material involved. Therefore, a less expensive, high-quality anvil could be produced, even though *tool steel* cost considerably more than wrought iron. One American anvil manufacturer (see *Columbian*) picked up this process shortly after the 20th century and produced a very successful anvil.[63] It must be remembered though that a forging is always stronger than a casting when made from the same metal.

Today all anvils, with the exception of a German make, are cast, and most are being made of hardenable steel. The size of cast anvils is limited only by the size of the molding pattern. The largest cast-steel anvil that I know of is the 750-pound *West (*see *West).*

Forged-steel Anvils

As wrought iron became more expensive than the low-carbon steel in the 1930s, one American anvil maker (see *Arm & Hammer*) used a low-carbon cast-steel base, a forged low-carbon steel body, and a high- carbon tool steel face plate.[64] By 1930 the *Trenton* anvils had a solid tool steel body arc welded to a low-carbon cast-steel base with a hardened face.[65] Hay-Budden pioneered this concept in 1908, with the exception of using a forged wrought-iron base (see *Hay-Budden*). This was the first true two-piece anvil. The only forged anvils made and imported today (1996) are from Germany (see *Peddinghous*). This German anvil appears to have a forged base that is electric-wire-welded to a press forged tool steel body with a hardened face. I have heard that some anvils were first cast of tool steel in one piece and then finished to shape by forging.[66] This would lower the cost and still be about the same strength as the forged anvil. However, I have never seen a cast-steel anvil that appeared to be forged afterward. One Swedish anvil manufacturer's advertisement states that their anvils were forged from one solid piece, but I have my doubts (see *Söderfors*).

A number of patents were taken out for one-piece, solid-forged anvils, but to my knowledge, were never applied to any extent.[67] One was taken out as early as 1844 (see Appendix F). This method might work well for a small anvil of 100 pounds or under, but when an anvil weighs over 100 pounds the dies necessary to do the job and the power it would require to forge this size anvil would make it uneconomical. There has never been a large profit in anvil manufacturing and they were just one of many products produced by anvil manufacturers.[68] The steel-faced wrought-iron anvil was manufactured for the longest period of time, in the largest quantities.[69] At least nine out of ten anvils found today are of this type.[70]

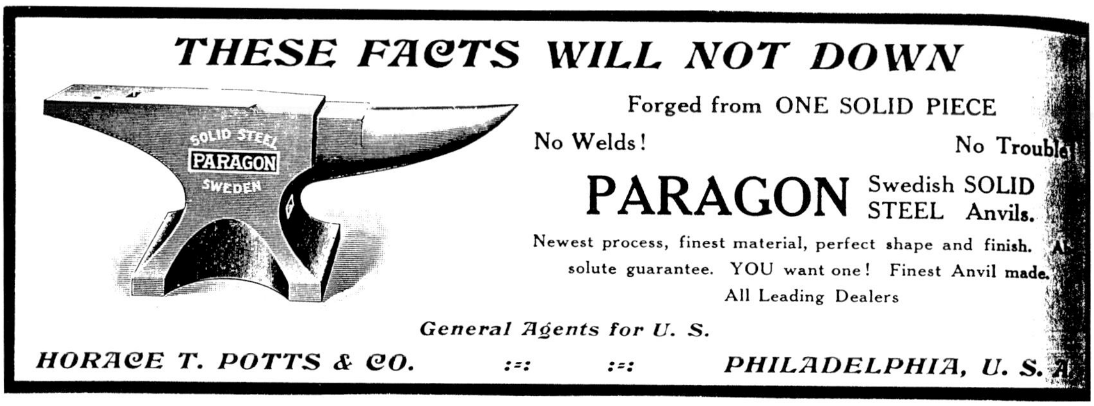

From "The American Blacksmith," March, 1910. An ad for a forged one-piece steel anvil.

Methods of Manufacture

Many of the small anvils made before 1780, weighing under 100 pounds and without a horn, were forged as one piece with a steel face applied. These are the ones that are very blocky, with very little overhang on either end of the face.[71] If a horn was added, it was jump-welded to one end. Most of the larger early anvils were made with from seven to fourteen pieces jump-welded on to a central block, depending on their size.[72]

If you look closely you can see the weld lines. This anvil was most likely manufactured in the first quarter of the 19th century. The horn broke off and it probably had a table. Part of the weight marks are still visible. I would consider this a low-quality anvil for the period.. (R.A.P. Photo)

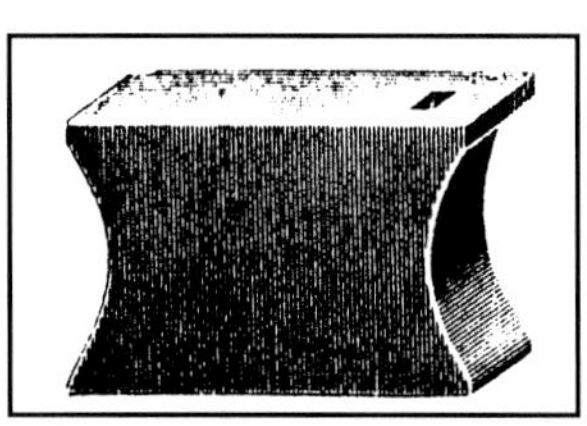

This anvil is not much more than a square block of metal.

The following engraving from the *Encyclopedia of Dennis Diderot* shows an anvil being manufactured from a number of pieces, known as the built-up process.[73]

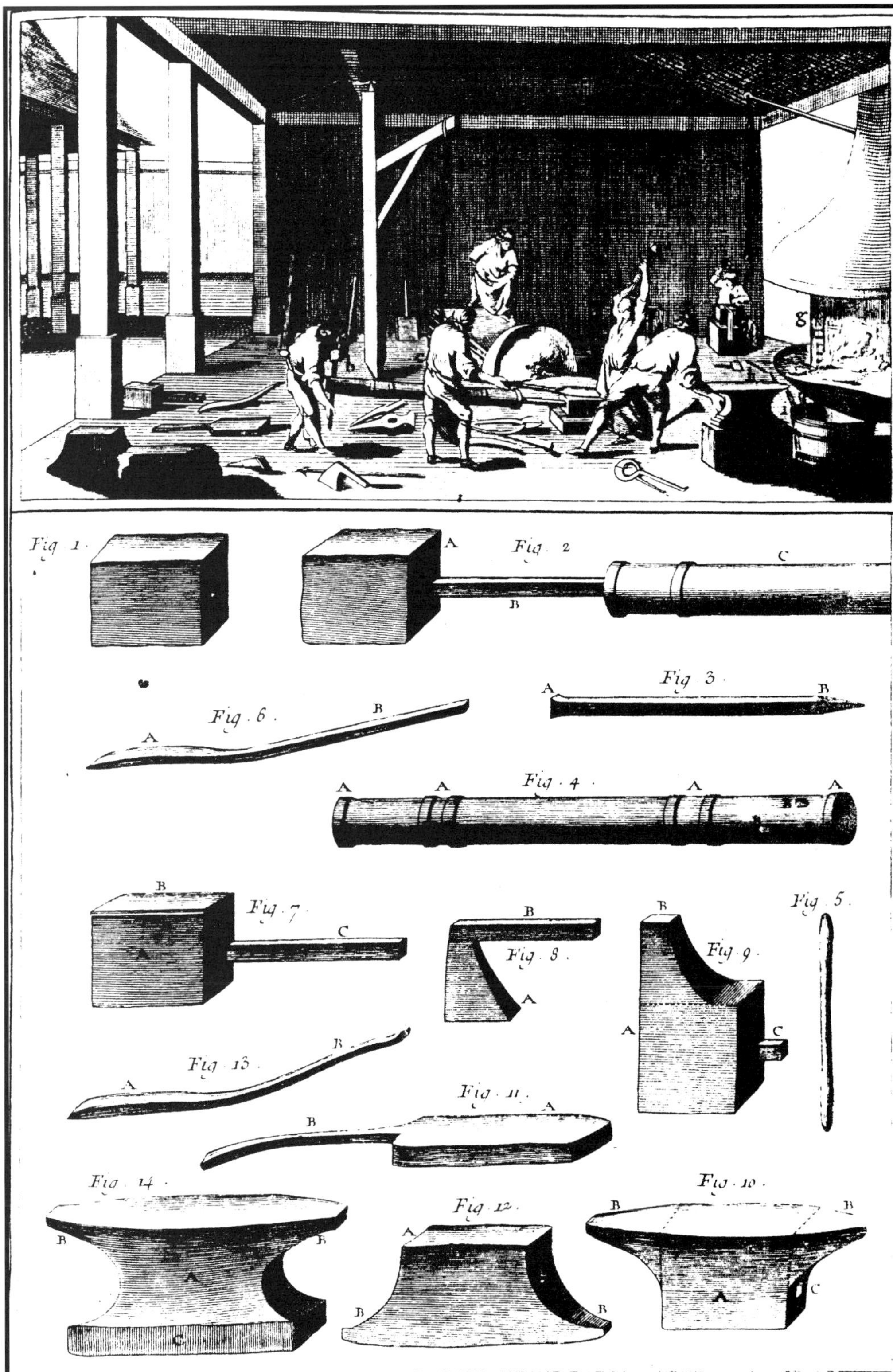
Fig. 1
Fig. 2
Fig. 3
Fig. 6
Fig. 4
Fig. 7
Fig. 8
Fig. 9
Fig. 5
Fig. 13
Fig. 11
Fig. 14
Fig. 12
Fig. 10

ANVILS IN AMERICA

For the most part scrap iron was used to make the majority of the better class of anvils, both then and in later models, even into the 20th century for some English anvils.[74] New wrought-iron was not used due to its inferior quality and cost. Scrap iron, when reworked and forge-welded together, became a better product than the new iron. Generally speaking, the more wrought iron is worked, to a point, the better it becomes. At least one English anvil manufacturer made *Best Scrap* part of their anvil's trademark.[75]

A pile or faggot of wrought-iron scrap ready for the furnace. It will need wire to hold it together until forge welding.

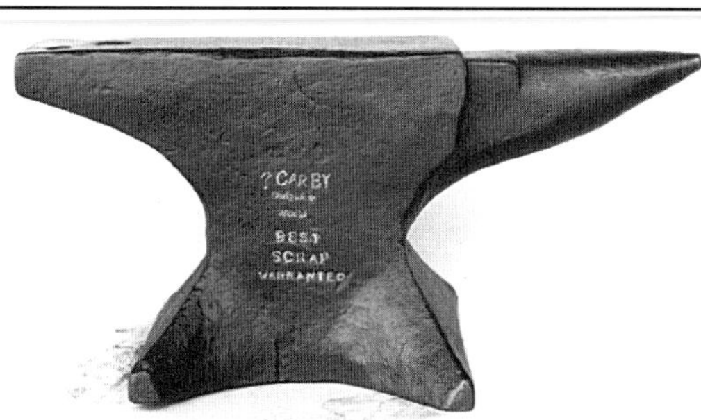

"Best Scrap" as part of the anvil's trademark. (R.A.P. Photo)

Built-up Process

The scrap was piled together and tied with wire, then put into a furnace and brought to a welding heat. It was then placed under a tilt, or trip hammer, and forge-welded into a square mass called a *butt.* This block of scrap became, essentially, one piece after reworking, as well as the basic part of the anvil. Holes were then punched part way into both ends and under one side of this square or slightly rectangular block of iron, as shown. The holes, sometimes referred to as *mouse holes*, allowed iron bars, called *porter bars,* to be inserted into the holes, so that the anvil could be manipulated by three men, while pieces were jump-welded onto the block by other workmen.[76] You can imagine just

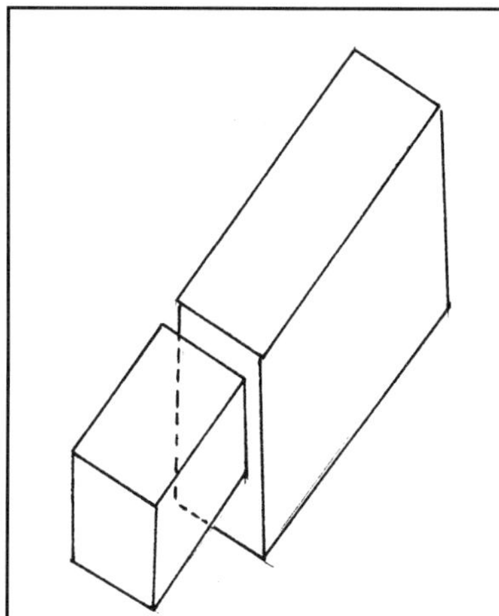

Pieces to be jump-welded together

how much heat a block of 50 to 100 pounds or greater gives off at welding heat (approximately 2400^0 - 2600^0 F). If the anvil was very heavy, a chain from a crane would be attached to the *porter bar* that was inserted in the bottom of the anvil for manipulation. After all the pieces were welded to the block, the whole mass was reheated, placed on a special anvil (cast iron) and the pieces shaped to their respective dimensions.[77] Five to seven men could be involved in making a heavy anvil of this type. The anvil was now ready for the face plate. If the anvil was small, say 80 pounds, the face plate might be put on in one piece. However, because the face plate had to be sledge hammered on manually in the early years, only about 6 inches could be welded on at one time. On a large 400-pound anvil, for example, three or four pieces might be required to face it. The seam, where these plates came together, can often be seen on older anvils.[78]

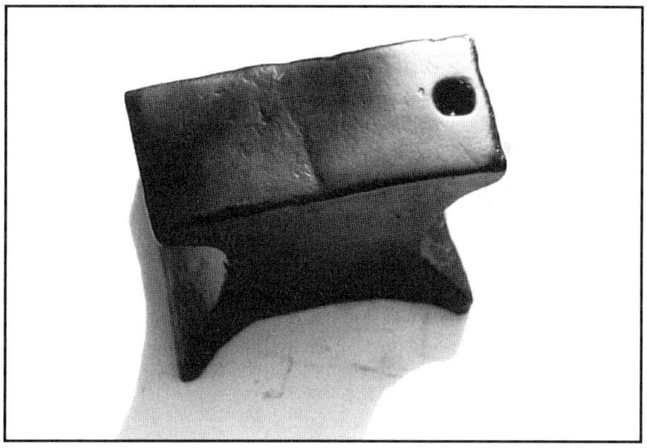

The plate seam is very plain in this photo of a hornless Colonial anvil. (R.A.P. Photo)

Depending on the quality of the anvil to be made, it was reheated and the finishing touches applied. The shape of the body parts was finished with hammers, flatters, and fullers. The face was smoothed and leveled with a flatter and sledge.[79] The face was then heated to the hardening temperature (about 1600^0 F) and quickly cooled by water running over the face. If a hardie or pritchel hole were put in, these would be punched out while the anvil was at forging temperature, but before the face plate was hardened.[80] The face was then ground smooth on a large grindstone. [81]

I found it hard to believe, but one German anvil manufacturer was still producing anvils by this method in the 1920s or later, with the exception that the face plate was put on in one piece[82] (see Appendix P). All this work required very skilled craftsmen, and the proper facilities, as the following article illustrates.[83] Incidentally, by the time this article was written, only the three-piece anvil was being manufactured, with the possible exception of a few anvils made in Germany. This article will illustrate some of the difficulties in producing an anvil. The following quoted article is a response to a previous article in *The Blacksmith and Wheelwright* magazine which I have not found.

"REPAIRING ANVILS"

Not Every Smith Can Do It Successfully

Editor of *The Blacksmith and Wheelwright.*

> *I don't know who B.F. Spalding is that wrote the article in this month's number on 'Repairing Anvils,' but I don't believe he ever did the things himself that he takes a whole page to tell how he thinks it might be done. Perhaps there may be here and there a 'natural genius for this kind of work, and a truant disposition,' in other words, an ingenious tramp that with such directions could patch up after a fashion worn-out 'wrought' iron anvils (none others ever needs it), but it would be a poor job and altogether cost more than a new anvil. Where are the three men in a country shop with the skill of long training to handle it with three bars in the three holes while the smith does his nice work of welding on a piece of 'blister' steel--a bigger job of welding than he ever in his life before attempted? And where is he to get that special kind of steel that he can't find kept in any of the stores in the neighborhood? There is not a hand bellows or hand blower in any shop that will give a 'sufficiently powerful blast' to perfectly weld the steel face, four inches wide by sixteen inches long, of a 150-pound anvil, even if he did 'borrow a bellows for extra fire' from some other blacksmith in the next town. He will find*

it a sweet job, too, to cut off what is left of the old face and shape the body ready and smooth for the steel. The whole paragraph just above the picture of this wonderful genius is true as Gospel, but it is like that other old saying, 'Don't wipe your nose with your elbow.'

What would happen to the thin steel plate during all those repeated heatings to get a weld--of burned steel and worthless edges? Then he gayly directs how to make a file that will give a fine, true surface all over with two men sawing away on it and the 'protuberances,' and, as he says, 'somewhat faster than a planer.' How many blacksmiths have ever tried their hand at making a file that would cut? He is right in saying it will take 'muscle' to work that machine.

His idea of hardening with tub and buckets might work once in a dozen trials, but would give poor and uneven temper--even new anvils with blister steel hardly ever have the same hard temper all over the face. Every Blacksmith knows that a 'butt' weld is about the poorest job there is, even for a two-*inch bar, but to succeed in jumping on as big a chuck as the horn to the body anywhere except in a regular anvil factory is next to impossible, and even then they often drop off after a years work. The best wrought anvils, the 'Peter Wright,' never are made that way, but the whole upper part, horn and all, is forged out in one piece. None but the poorest anvil is nowadays made with the horn butted on. Following such 'directions,' I should say an ordinary blacksmith, not a 'genius' with his three men and their three bars, would take about a week to do the job and have a miserable botch for his pains. Their is a regular concern in Chicago who have all the tools, machinery and the skills, and make a business of repairing wrought iron anvils probably as cheap and as well as it can be done, and this is the price the circular gives for work B.F. Spalding describes as so easy for working blacksmiths of the country, viz.: Steeling, 6 1/2 cents per pound; new horn or heel, 7 1/2 cents per pound; dressing, tempering and grinding, 4 cents per pound. It is a pretty good price to put on an old anvil, but is not one-half what it would cost me or any other blacksmith to attempt such a fool performance. Better send the old anvil to Arthur J. O'Leary in Chicago for fixing up, or buy one of the 'Eagle' anvils that are warranted never to have the face come off or 'settle' or the horn to break, and would cost less money. I have worked on one of them, mostly in forging steel, for eighteen years, and it is as good and true and hard as in the first year.*

Christian Schmidt Trenton, N. J., December 10, 1897.

Modern Process

After 1852 the English anvil manufacture Peter Wright started making anvils using only two pieces (plus the face plate), regardless of size.[84] However, two-piece anvils were being made by others at least 30 years earlier (before 1850), but it appears these companies were not successful.[85] Peter Wright was the company that filed for the patent (see *Peter Wright*). In brief, here is a description of this process.[86]

The base and body were made, as they said, of "selected scrap iron," then forge-welded into the appropriate-sized blocks. The base block was heated and brought to the approximate shape under a trip hammer (by 1860, a steam hammer). Then it was blocked to the finished shape with the use of dies. The body was drawn to shape with the use of loose dies under the trip hammer or steam hammer.

At this point, I am a little baffled. Were the handling holes put in before the body and base were welded together or afterwards? They don't say. All *Peter Wright* (after 1851) and early American wrought anvils have four handling holes, the fourth having been punched into the middle of the foot under the horn. This fourth hole was never put into the multi-piece anvils made by the *built-up* process.[87]

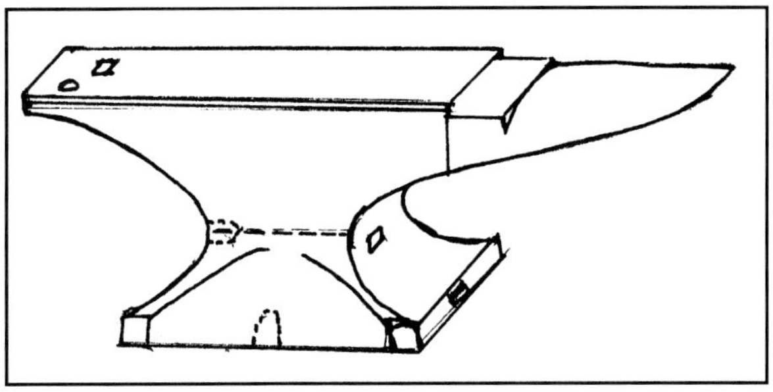

A drawing of an anvil with 4 handling holes. The dotted line indicates where the base and the body are forge welded together.

I was told by one of the last anvil makers that the fourth hole was put there for handling purposes during heat treatment, but the American anvil manufacturers abandoned this hole shortly after the 20th century.[88] The two holes in each end of the waist appear to have been punched in after the base and body were welded together. If this is true, how would the white-hot (2,500° F) pieces of iron have been manipulated without the handling holes to bring the base and body into position under the hammer for welding? Large tongs might have been used. Different anvil makers had their own ways of accomplishing the same tasks, so I will suggest a solution.

After the base block was finished, while the base was still hot, a small rectangular hole was punched into the front of the foot and a larger hole under the base. Small holes were punched into both ends of the body, close to the edge, where the base and body were to meet. In this manner both pieces could be handled with porter bars. As the base piece was placed in position for welding, the bar in the bottom was removed, leaving the bar in front for horizontally positioning of the base, as needed. The body or upper part could be handled by two bars in either end of the waist, or if they did not use holes, it could be handled with two pairs of tongs and be placed in position on the base. After the waist was welded, the holes in the waist could be enlarged, making them extended partially into the base. This procedure may not be totally correct, but at this time, this is my best guess.

With the body and base welded, the next step was the securing of the face plate. The face plate was attached in pieces, as before. But by at least the 1870s, the face plate on the *Peter Wright* anvil was put on in one piece with possibly the use of a steam hammer.[89] The finishing and heat treating were done in the same manner as on the multi-piece anvils, except that the face plate was ground cross-wise on some anvil makes, giving a slightly convex surface. I have recorded a *Peter Wright* anvil, in like-new condition, that still has this slight convex face. After years of use, the convex surface disappears, as the face settles or wears. The convex face would help in "drawing" and "welding." In later years the face plate was welded on before the body and base were welded together.[90]

Almost every wrought-iron anvil manufacturer of the period and later adopted the Peter Wright patented method with the exception of Mousehole Forge. Mousehole Forge continued to export anvils to the United States using the older *built-up* method, until late into the 19th century when they also adopted Wright's method.

Cast Anvils

All cast anvils are made the same way, regardless of whether they are made of cast iron, cast iron with a steel face, or cast steel. First, a foundry pattern of a particular size and style must be made. In earlier times patterns were generally of wood, but today patterns are mostly made with a type of plastic material. Casting limits the number of sizes and styles to the patterns available. A special feature or style anvil could not be made up on short notice, as was the case with forged anvils. Remember, each forged anvil was essentially one of a kind, until the time they were made in closed dies. However, *Fisher* anvils were advertised as being made in over 200 sizes and styles.[91]

After the two piece pattern is chosen, it is placed into what is known as a *flask*. The *flask* is also made up of two parts, the *cope* and the *drag*. Half of the pattern is placed split side down in the *drag* and a damp sand mixture is rammed in around it. The *drag* is then turned over and the other half of the pattern is placed against its mating part. Then the *cope* is joined to the *drag*. Again the damp sand mixture is packed around this part of the pattern. The *cope* and the *drag* are then separated and the pattern removed, then reattached, leaving a hollow impression of the pattern within.

While the pattern is still in the *flask*, several holes are made through the sand to the pattern. The hole or holes into which the molten metal is to be poured is called the *sprue* and the outlet the *riser*. The *riser* allows the air to escape, so that the cavity fills completely. Fisher and Norris patented an arrangement for the iron to enter and exit the cavity through their method of *gating*.[92]

When the setup of the *flask* is completed, the hot liquid metal is poured into one or more *sprues*. The pour is continued until the molten metal appears at the top of the risers. The metal in the cavity is allowed to cool, after which the flask parted and the rough cast anvil is removed. This is the briefest of explanations of the basic metal casting process.

The cast-iron anvils might have had a chill plate placed in the cavity before the pour, where the face would be found, leaving that area of the anvil very hard. This face could be only ground afterward if a chill plate was used, as it was too hard to machine. A rough steel casting would be normalized before the face was machined. Excess metal is trimmed from the rest of the casting, to be used again. The faces of the steel anvils are machined or ground flat, then heat-treated, similar to the wrought-iron anvils, and the face takes a final grind, if necessary. Heat treatment would not be necessary for the cast-iron anvil. If there are hardie or pritchel holes in the heels of cast-iron anvils they are usually part of the foundry pattern. On steel anvils they are usually drilled and broached into the heel before the face is hardened. As stated earlier, a casting is not as strong as a forging made of the same metal. Therefore, the heel and the horn are thicker or reinforced under the horn and heel, especially on the cast-iron anvils. Any writing or trademark found on the cast iron anvils is usually part of the pattern. In the case of some steel anvils the trademark is stamped on after the anvil is complete, the same as on forged anvils.

There is a slight difference in the casting process when a steel-faced cast-iron anvil is produced. After the pattern is removed from the *flask*, a heated tool-steel plate and any other steel parts to be incorporated are placed in the cavity in their proper position and the *flask* quickly closed. The cast iron is then immediately poured into the *sprue*. If the setup and execution are performed properly the molten cast iron will weld itself to the heated steel plate and the other steel parts in the *flask* (see Fisher). After removal from the *flask* the rest of the operations are basically the same as for the cast-steel anvils.

General Information About Anvils

This has been a very brief outline of how anvils were and are made, and the materials from which they are produced. Only the individual manufacturers knew and know the exact details surrounding the construction of their particular anvils. There were many trade secrets employed that are lost to us and may never be known. Additional information on the anvil manufacturing process can be found in the discussions of several of the manufacturers listed in this publication.

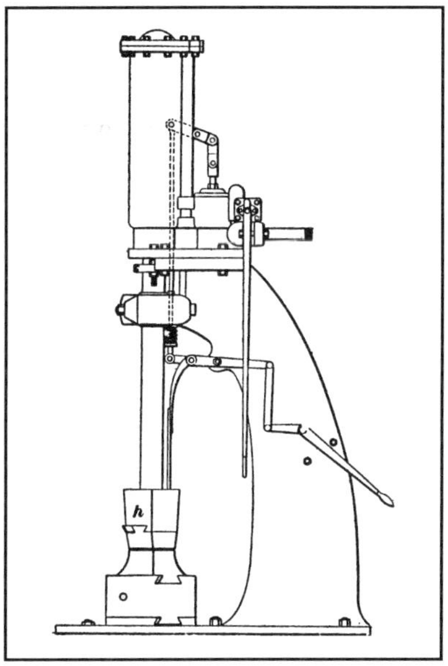

An old style single-frame steam hammer. It was used for welding the faggots together and welding and blocking various anvil parts.

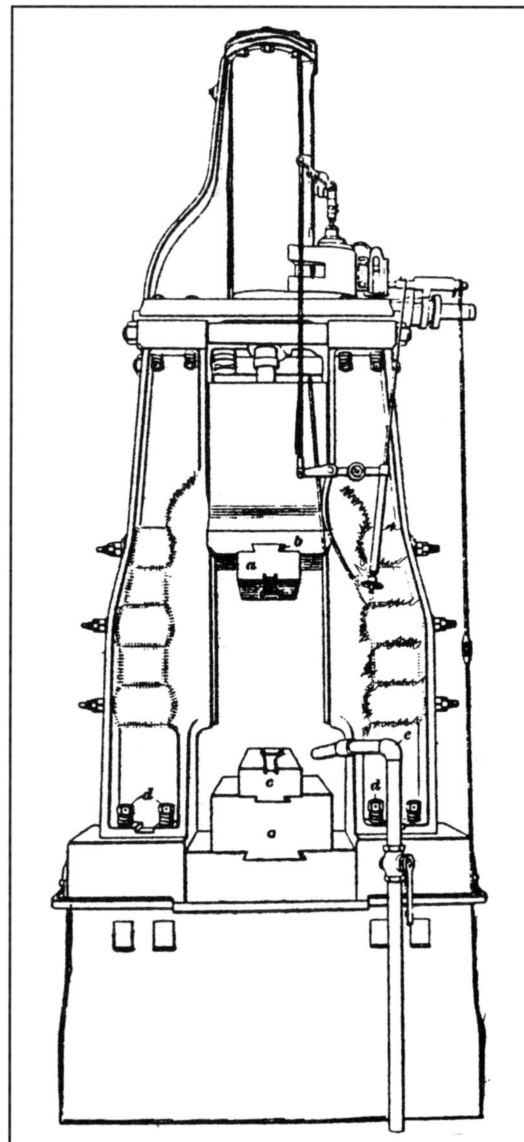

Double-frame steam hammer.

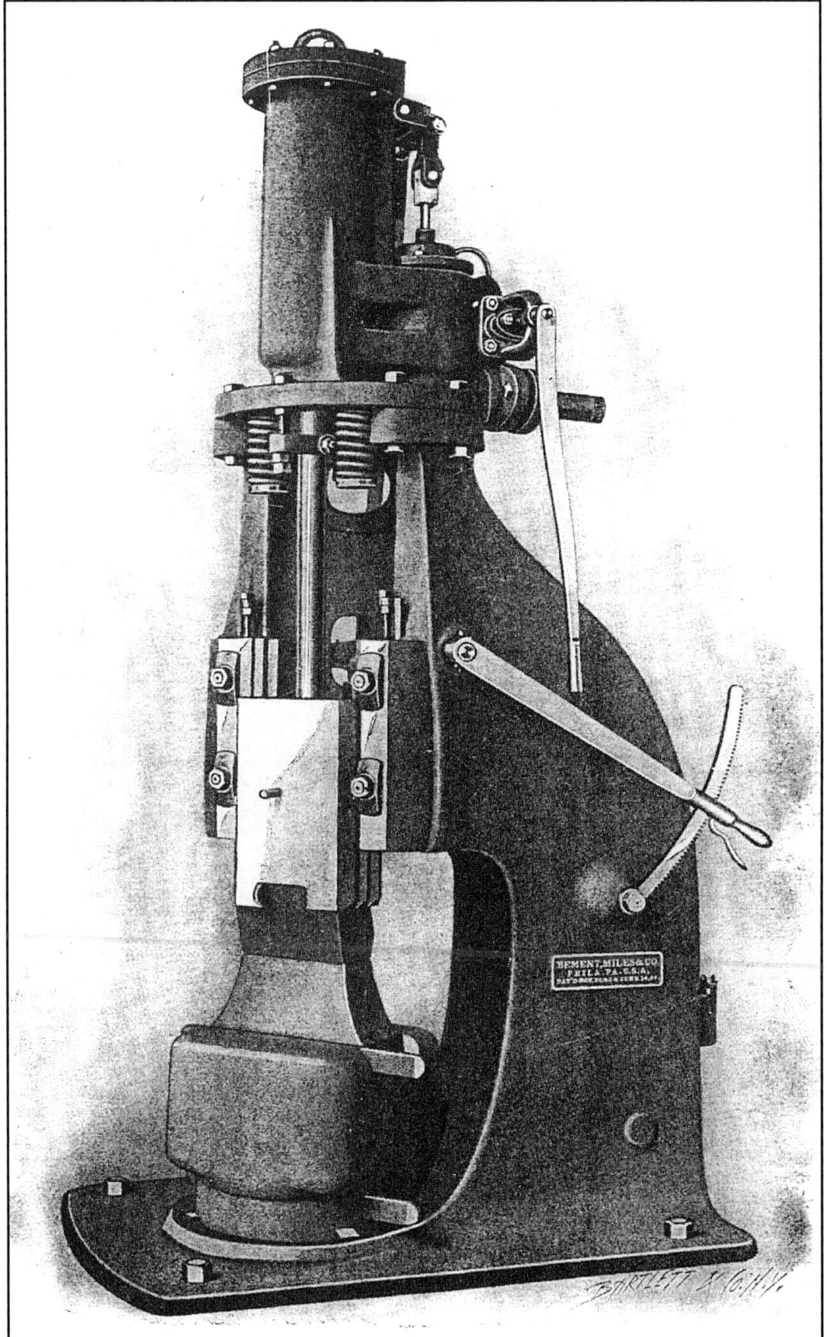

800-POUND BEMENT SINGLE-FRAME STEAM HAMMER.

CHAPTER 2

Early American and Various European Anvils

In all probability, up until the 1850s at least 99% of all anvils used in what is now the United States were imported from Europe. They were either brought by the smith himself or imported for resale. I doubt that even 1% were ever made in this country before the 1850s; however, there are a few references to anvils being manufactured here, though on a very small scale and never on an ongoing basis. Except for one instance, which I will mention later, I am unaware of any American-manufactured anvils before the 1850s still in existence.

These European anvils have come to these shores from at least five countries. First the anvils came from Spain, then from England, France and Germany, and finally Sweden. England was by far the largest contributor of the foreign anvils, starting with the first permanent English settlement at Jamestown, Virginia, and continuing to the present day. Early English anvils are not uncommon, but only a few early Spanish and German anvils are known to me. I have yet to discover an early French anvil, although they were certainly used for many years in the upper Midwest and Eastern Canada.

Until after World War II all German anvils imported for resale here were made-to-order for American hardware companies and carried the hardware company's trademark. I know of only one pre-war German anvil to carry a German trademark (see *H. Boker*), but since World War II they now carry their manufacturer's name or trademark. All early German anvils before the late-19th century were brought here by the smiths. To the best of my knowledge Swedish anvils were not imported into this country until the late-19th century. Today (1997) England and Germany supply most of the anvils coming to the United States, with the exception of a few poor-quality anvils coming from Japan or the Far East.

AMERICAN-MANUFACTURED ANVILS BEFORE *CA.* 1850

> *In looking over the records of early settlers of Virginia we find the following tradesmen listed in the Jamestown Settlement of 1607: tailor, carpenter, goldsmith, refiner, tanner, blacksmith, gunsmith, cooper, bricklayer, mason,... In 1612 came also the shoemaker, shipwright and brickmaker.*[1]

Undoubtedly, the first anvil of any size came with this blacksmith whose name was George Read.[2]

In order to manufacture an anvil it is necessary to have an iron works. The first iron works capable of manufacturing a cast or forged anvil was developed on the Saugus river in Massachusetts between 1644 and 1647, and was known as Hammersmith.[3] The restored site is known today as the Saugus Iron Works, and contains a blast furnace and a refinery, which produced both cast iron and wrought iron at the site.[4] If the Saugus Iron Works ever manufactured anvils, there is no record of the fact, and today their only products still in existence, to my knowledge, are a cast-iron fireplace back and an iron kettle.[5] The literature on the Saugus Iron Works states that it was as technically up to date as any iron works found in Europe at the time.[6] Therefore, it is possible that some anvils may have been manufactured there.

Before Hammersmith, there was an attempt to establish an iron works at Falling Creek, Virginia, on the James River. The iron works was designed to produce wrought iron as well as cast iron, but before the works could go into production, the Indian uprising of 1622 put an end to it.[7] The workers were all killed and the buildings and equipment were destroyed.[8]

I suggest that the following is an example

of an early anvil that was manufactured in America. A number of years ago a farmer in the state of Delaware was removing stones from an old barn foundation dating back to the Colonial period when he came upon a large metal object. The object turned out to be an anvil. In 1993 I had a chance to see and inspect this anvil. In my estimation it is an early cast American anvil.[9] It weighs 90 pounds.

Top View. (R.A.P. photo)

Side view. (R.A.P. photo)

As you can see, the pattern from which this anvil was cast was very crude in design. It certainly would not have come from England. I expected the anvil to be cast iron, but found that when struck with a hammer it had a slight ring to it, indicating a wrought iron or steel content. When struck, cast-iron anvils make only a dull thudding sound. Therefore I suggest that this anvil was produced by an early American iron works, possibly dating from the late-1600s or early-1700s.

The earliest mention of anvil manufacturing in this country comes from a *Boston Newsletter* dated March 4-11, 1716-17.

> ***Anvils*** *- Samuel Bissel, anvil smith, lately come from England, living at New-Port on Rhode Island, makes all sorts of Blacksmith's and Goldsmith's anvils, Bick-irons and stakes and new faces for old ones, at reasonable rates, and may be spoke with or wrote to, at his house or shop near the Topsaile Street in said town.*[10]

I find this ad hard to believe, considering the number of skilled persons required and the capital necessary to build a shop that could produce and face an anvil of any size beyond a few pounds. A forge capable of doing such work as Mr. Bissel claims should show up as an iron works in some historical record, not just this advertisement. However, what a delight it would be to find an anvil with *S. Bissel* stamped on the side.

The following ads are easier to explain.

> ***Iron Foundry*** *- Any person that has occasion for Forge Hammers, Anvils, or Plates, Smith's Anvils, Clothier's Plates, Chimney Backs, Potts, Kettles, Skillets, Cast Boxes, Chaise Boxes, Dog Irons, or any other Cast Iron Ware, may be provided with them by Richard Clarke, at his furnace in the Gore, giving speedy notice (of size and quantity they want) to him there, or to Oliver, Clarke, and Lee, at their Warehouse in King Street, Boston; where they may be supplied with Swivel Guns.*
> Boston Gazette, July 13-20, 1741.[11]

All of the items mentioned, with the possible exception of the "Forge Hammers" and "Smith's Anvils," were generally made of cast iron.

I doubt very much that the "Smith's Anvils" were being manufactured from cast iron. It appears that "Oliver, Clarke, and Lee" might well have been importers and the "Smith Anvils" could have come from England and listed as part of their product line. This was common practice for a number of manufacturers in the past as well as today, to round out a product line.[12]

In the journal *Louisa County* (Virginia), June 26, 1959, the writer mentions an article written in 1759, recommending the purchase of "one anvil of about 2-2-0 [280 pounds] of Byrd make." William Bird was the person who built Pine Forge on Hay Creek, Virginia, in 1744. It is just possible that the "Byrd make" or (Bird) anvil came from this Hay Creek Forge.[13]

The following advertisement came from the *Pennsylvania Chronicle.*

> *Good Forge Hammers and Anvils, Made and Sold by Thomas Mayberry at Hereford Furnace, in Bucks County (1769), and Best Colebrook-Dale-Forge Hammers and Anvils, to be sold by the subscriber, on the Furnace Banks; or by David Potts Merchant in Philadelphia. Thomas Rutter (1768).*[14]

This was not an advertisement for blacksmith's anvils, but for hammers and anvils made of cast iron. They were used as the head and anvil for the old water-powered *helve, tilt,* or *trip* hammers, which were used for working wrought iron. As far as I can tell, these *heads* and *anvils* have almost always been made from cast iron. It might be interesting to note that by 1775 American iron works were producing one- seventh more iron than England and Wales combined.[15]

I have found no mention of American anvil-making after the Colonial period until the mid-19th century. However, there was one person who did take out a patent for a one-piece anvil in 1844, but I am sure it was never put into production (see Appendix F).

SPANISH ANVILS

If any anvils were made in the Spanish-dominated area of this country, they were made by blacksmiths, were small, in all likelihood without steel faces. Such were the ones found in the Southwest.[16] I know of only two Spanish-style anvils brought to this country, not counting the meteorites found in Tucson, that were used as anvils (see Appendix K). One of these two anvils was located in St. Augustine, Florida. A drawing of this anvil is shown in Sonn's Book.[17] The second is located at the Mission San Francisco, an old Spanish mission in Sonoma, California, with the date *1766* stamped on it. To my knowledge, Spanish anvils have not been imported for resale since the Colonial period. It is most unlikely that any of these Colonial Spanish anvils were ever imported for resale, but were brought by the smiths. In the summer of 1996 while visiting in California I had a chance to see and record the anvil at Sonoma. I am now of the opinion that it may be of German origin rather than Spanish.[18] I estimate that it weighs four to five hundred pounds. It is 12 3/4 inches high, the face is 18 1/4 x 7 1/2 inches, and the base is 16 3/4 x 9 1/2 inches. It has the full *church windows.*

Sonoma anvil. (Mr. Gill Fahrenwald photo)

GERMAN ANVILS

Before the mid-19th century the only German anvils coming into the United States were brought by the German blacksmiths who immigrated to this country. I know of only one anvil brought here from Germany before the 1850s. It is presently used as a monument commemorating the memory of a frontier blacksmith's family in southern Illinois.

I had seen photos of this anvil, but did not have the opportunity to personally see or learn of its history until 1995. The anvil is located in the small town of Venedy, Illinois, which owed its existence to a coal mine which opened in 1837 and closed in 1969. The anvil was brought to the area about 1830 and first used at a flour mill, then in a cooper's shop and finally at the mine, when it opened, to dress tools. It was located down in the mine and was used in later years by the miners as something to knock their lamps against to remove the carbon while on their way out of the mine. In the late 1950s, the anvil was finally recovered from the mine and placed in its present position. The anvil weighs about 700 pounds and has the following measurements: the face is 19 1/4 x 8+ inches, by15 3/8 inches high, and the base is 11 x 18 5/8 inches. As you can see it has the full *church windows*. It also has a 1 1/8 x 1 1/4 inch hardie hole. The anvil is covered with gold paint.[19]

After about 1850 a number of American hardware companies and blacksmith tool manufacturers began importing anvils made in Germany, under various company trademarks. These anvils were manufactured to specifications stated by the ordering companies and they were, as far as I can determine, always of the *London* pattern. In the last quarter of the 19th century M.T. Richardson, in his book *Practical Blacksmithing,* says that he did not think highly of the German imports, saying that they were of poor quality.[20]

Venedy anvil. (R.A.P. photo)

I know of one German *Trenton* that was manufactured by the old built-up method where the front feet had broken off from the body.[21] I suspect that most of the early German-made anvils have been scrapped, although I have recorded several mostly under the *Trenton* trademark (see *Trenton).* It seems that English anvil manufacturers , with one exception that I know of (see *Henry Wright),* would export anvils to the American market only under their own trademark or under the name of an English exporting company. Therefore, American companies wanting to sell imported anvils under their own trademark had to go to either Germany or Sweden.[22]

S.D. Kimbark, a large hardware company in Chicago, Illinois, sold German anvils under their own trademark from about 1860 until the 1890s, after which an American manufacturer produced anvils for them (see *Hay-Budden).*[23] I have not been able to find any of these German-made *S.D. Kimbark* anvils, but have discovered several manufactured for them by Hay-Budden.

H. Boker & Company may have imported anvils into this country under their own trademark, *H. Boker,* and after 1860 under the *Trenton* trademark (see *Trenton).*

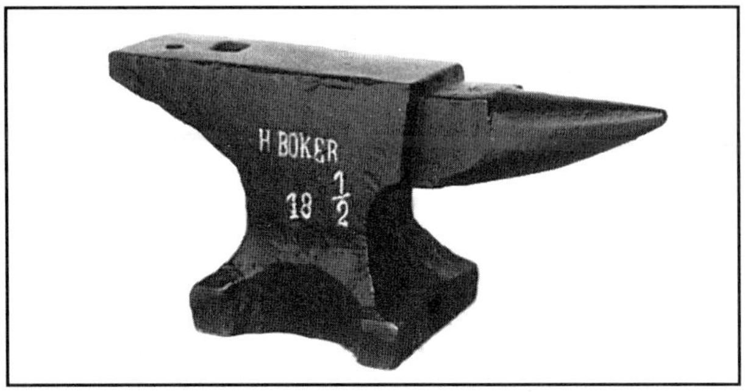

41-pound Boker anvil. (R.A.P. photo)

I have recorded a number of anvils I am sure were manufactured in Germany for various American hardware distributors, such as S.D. Kimbark of Chicago, and a company in Boston, although I am unable to read this company's name. Once a number of American forge companies began manufacturing wrought-iron anvils in the 1890s, the importation of German anvils slowed or stopped completely. Hardware companies began using American-manufactured and English anvils as their anvils of choice.

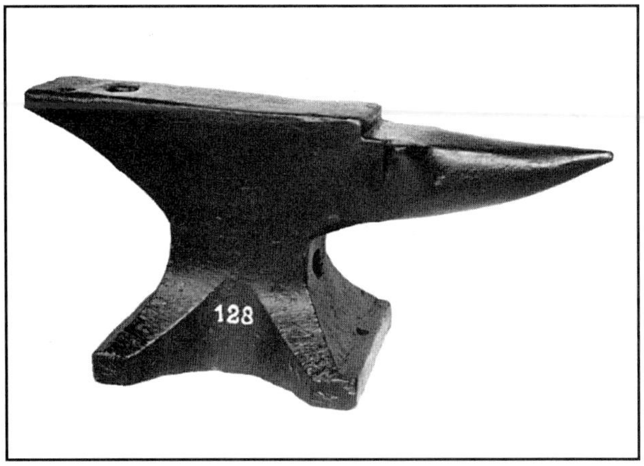

128-pound German made anvil. (R.A.P. photo)

It was not until after World War II that German anvils again came into this country, but this time under the manufacturer's trademark. These were manufactured by the Peddinghaus Corporation. Peddinghaus has its United States headquarters in Bradley, Illinois, and offers anvils in six different weights ranging from 45 to 275 pounds, and a number of different patterns of the *Continental* style. They are a two-piece press-forged anvil, electric-arc-welded at the waist.

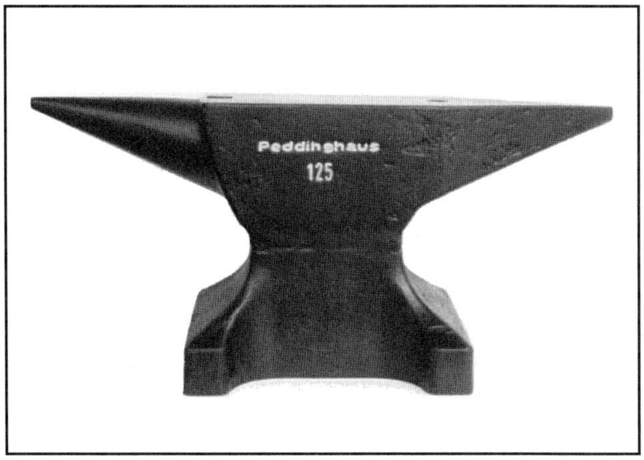

274-pound Peddinghaus anvil. (R.A.P. photo)

Another very old German anvil manufacturer, Refflinghaus, is now selling anvils in this country as well. These anvils are one-piece cast-tool steel of the *Continental* pattern.

These two German anvil manufacturers stamp their names on their anvils, so they are easy to identify. The German anvils manufactured before the 1900s under American trademarks are much harder to identify; unlike the Swedish imports they generally do not bear the name of their country of origin. However, I have heard of one *Trenton* with "Made in Germany" stamped on the side.[24] To identify these early German imports, see if the anvil has a hardware company trademark on it. If it is not one of the American-made brands then it is probably a German make. In the case of the *H. Boker* anvils the name is stamped on the side and the weight is given in kilograms.

ARSM
seit 1866

AUGUST REFFLINGHAUS SÖHNE
Amboßwerk und Gesenkschmiede
5828 Ennepetal-Milspe, Heilenbecker Straße 111 b, Telefon (02333) 71224

Ia-Qualitäts-Schmiedeambosse

Nr. 2 Norddeutscher
mit 1 eingesetzten Horn

Nr. 4 Norddeutscher mit 1 eingesetzten Horn und Stauch
Nr. 5 wie Nr. 4, jedoch mit auslaufendem Horn
Nr. 4 A oder Nr. 5 A wie vor, jedoch mit Voramboß

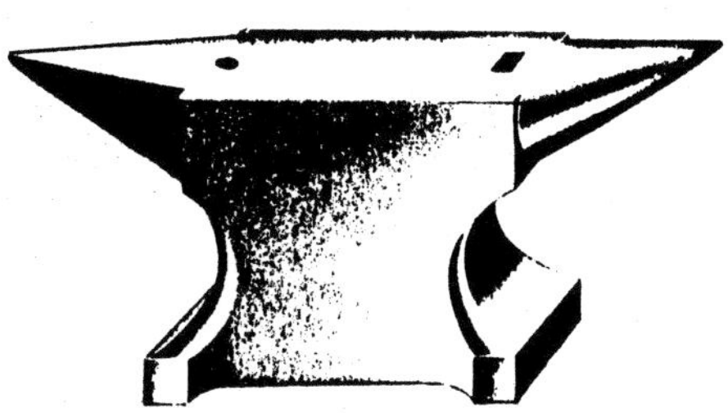

Nr. 8 Norddeutscher
mit 2 eingesetzten Hörnern

Nr. 9 Norddeutscher mit 2 eingesetzten Hörnern u. Stauch
Nr. 13 wie Nr. 9, jedoch mit auslaufenden Hörnern
Nr. 9 A oder Nr. 13 A wie vor, jedoch mit Voramboß

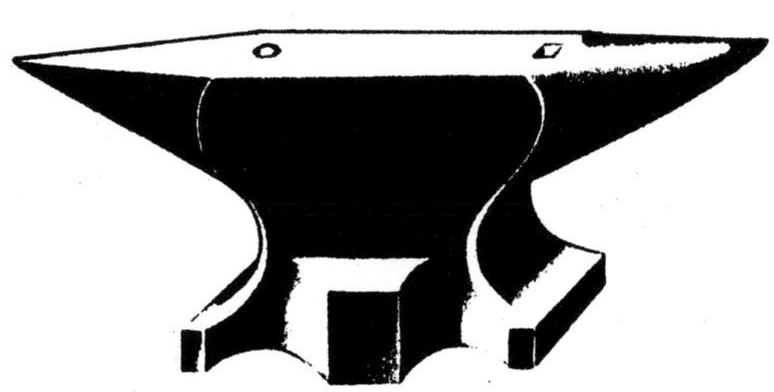

Nr. 57 Süddeutscher
mit 2 auslaufenden Hörnern und Stauch

Nr. 58 Süddeutscher
mit 2 auslaufenden Hörnern, Stauch und Voramboß

Die Formen Nr. 5 und Nr. 13 fertigen wir wegen geringer Nachfrage nur auf besonderen Wunsch. Es muß daher bei diesen Modellen mit einer längeren Lieferzeit gerechnet werden.

– Andere Formen als abgedruckt auf Anfrage –

Various anvil styles from the Refflinghaus brochure.

FRENCH ANVILS

If there are any French anvils left in this country from the Colonial period, they are unknown to me at this writing. We know there were French blacksmiths at the various French trading posts throughout the Midwest for many years. It is fair to assume that these smiths used French-made anvils of the *Continental* style.

The Saint Maurice Iron Works, founded in 1738, in the province of Quebec, could have manufactured anvils, but if they did, how far were they distributed?[25]

The only French anvils I have discovered in this country are from the 20th century and probably came by way of Canada. The one shown here was definitely manufactured in France and was found in a junk yard in Indiana. It weighs 300 pounds and has *P & G, Adennes Qualitece* stamped on the side, as well as the large numbers, *N - 12442*, which I assume to be the serial number. I can find no weight marks.

300-pound French anvil. (R.A.P. photo)

A similar anvil is in a museum in Oregon. It has *Richard Doreval* stamped on the side, with large numbers *N - 12678* stamped near the base.

Mr. Kenneth Lynch, an iron worker and also an early restorer of the Statue of Liberty in New York Harbor, brought a number of French anvils and many French tools from France to

French anvil made by Richard Doreval. (Mr. Gill Fahrenwald photo)

the United States after World War I, which are now fairly well scattered about the country. Mr. Bruce Dembling sent me a sketch of one of these anvils.[26]

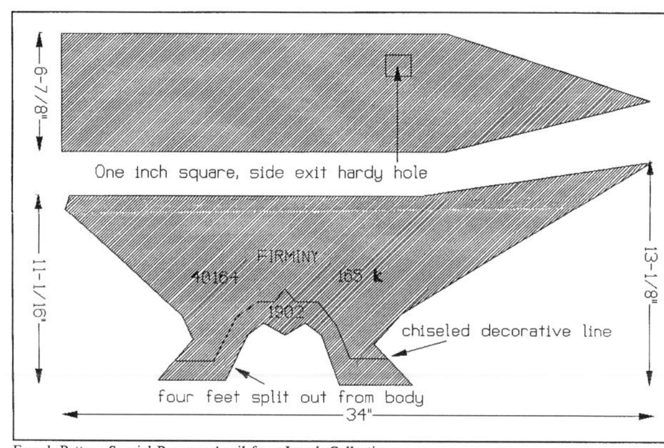

French Pattern Special Purpose Anvil from Lynch Collection

Sketch by Mr. Bruce Dembling.

If you are interested in French anvils I suggest that you visit France (I would love to) and possibly eastern Canada. There is a good collection in the Maison de L'Outil museum in Troyes, France.

SWEDISH ANVILS

Late in the 19th century, Sweden started exporting anvils to this country, both under American hardware company trademarks and under their own names.[27] Sweden was not a newcomer in the production of anvils. One company, probably Söderfors, claimed to have been manufacturing anvils for over 700 years.[28] If that is true they started manufacturing anvils around 1200 A.D., predating any serious anvil manufacturing activity in England by several hundred years. However, every Swedish anvil that I have been able to identify in this country was manufactured in the 20th century or just before. It must be remembered that Swedish iron and steel were of the best quality, as Sweden had excellent iron ore and their iron was sought after by other European countries, especially England. For this reason it is not hard to believe the 700-year statement.

There is one characteristic common to all 20th century Swedish anvils that I have recorded. When sighting these anvils from either the front of the horn or rear of the heel, the body goes in towards the waist from the face, before spreading out to form the feet. A second characteristic that seems common, regardless of the trademark, is the "Made in Sweden" stamped somewhere on the side. The weight is also stamped or cast into the side, and I have found it to be in actual pounds, with the exception of the late *Kohlswa* anvils.

Front view of a Swedish cast-steel anvil. (R.A.P. photo)

In the late 1920s until World War II both Sears Roebuck and Montgomery Ward sold Swedish anvils at various times, but I have been unable to find any examples of anvils with Sears' or Ward's trademarks stamped or cast on them. I suspect that they used the manufacturer's trademark.

The Sears Roebuck catalog of 1932 shows Swedish anvils in four sizes: 70, 100, 125, and 150 pounds. The 70 pound anvil sold for $8.95.

Montgomery Ward catalog, Spring-Summer, 1930. (Mr. Hunter Piklinton)

Up until 1995 I had found only one Swedish-made anvil without the "Made in Sweden" markings. The trademark appears as follows: **"BA????**
?.H &M. Co.
Louisville
170 lbs."

I thought that Belknap might be the company, but H & M is not part of their trademark.

Swedish -American Anvils

The Swedish-American Steel Corporation, an import company located in Brooklyn, New York, imported anvils, as well as other steel products, from Sweden. They were incorporated, August 17, 1927, and dissolved voluntarily December 4, 1961.[29] There were three brands of Swedish anvils coming into the

United States in the middle 1920s and until recently (1995) I was unable to determine which of the three brands were imported by this company. I am now (1997) almost positive that it was the *Kohlswa* anvil.

Kohlswa

According to a Kohlswa Jernverks Aktiebolag's brochure, the company has been manufacturing the *Kohlswa* cast-steel anvil since about 1925. They were imported into this country until about 1960 by the Swedish-American Steel Corporation. That year the Centaur Forge of Burlington, Wisconsin, picked up the line and added a line of farrier's anvils manufactured by Kohlswa Jernverks Aktiebolag to Centaur's specifications, with the Centaur trademark.

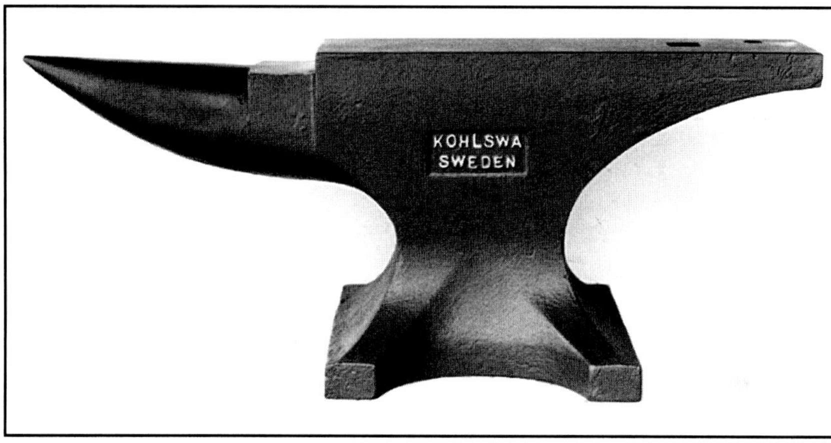

331-pound Kahlswa anvil. (R.A.P. photo)

Various Kahlswa anvils from the company's brochure:

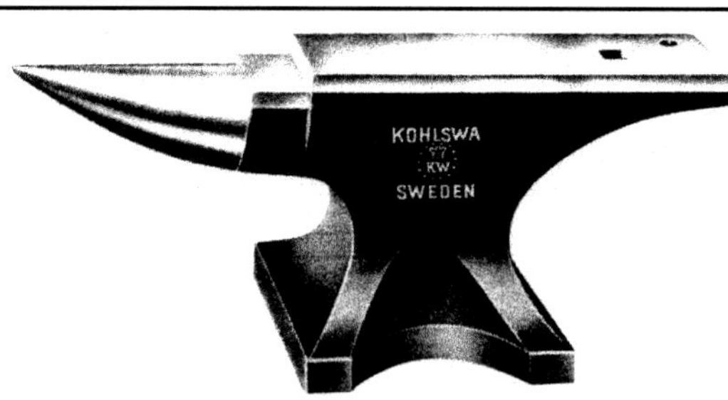

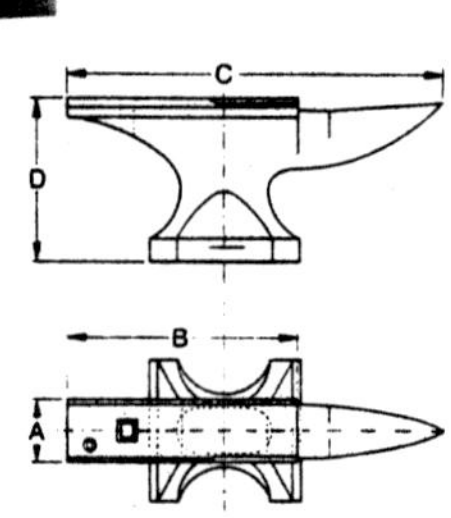

Type A 1

Weight	Dimensions, mm					
Kgs	A	B	C	D	□	Ø
22	78	241	406	194	22	11
32	88	280	470	218	22	11
41	93	324	546	240	22	11
52	97	368	609	250	22	11
75	106	420	687	288	25	14
91	119	457	749	310	32	18
129	128	521	845	340	35	18
161	145	559	902	368	35	18

B 21

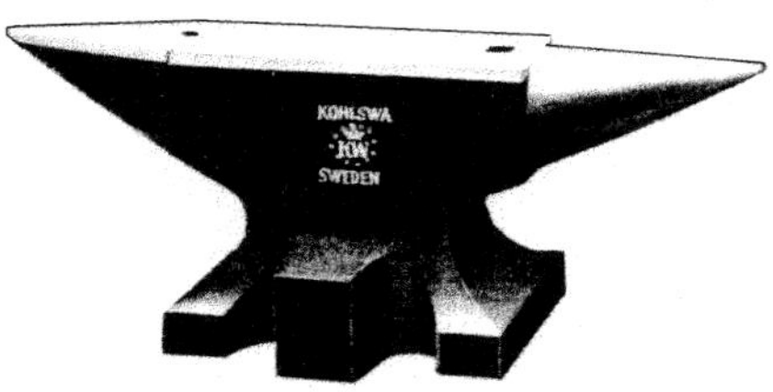

B 22

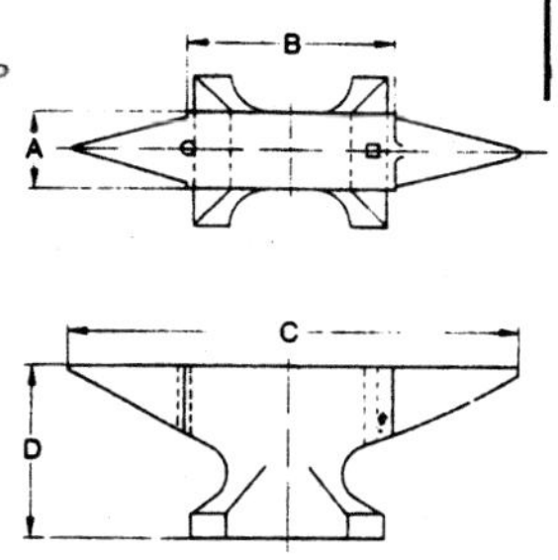

Types B 21 & B 22

Weight, Kgs		Dimensions, mm					
B 21	B 22	A	B	C	D	□	Ø
150	160	144	393	863	326	29	24
185	197	154	422	925	349	31	25.5
225	241	164	450	987	373	33	27

Type A 4

Weight Kgs	Dimensions, mm					
	A	B	C	D	□	Ø
24	78	246	421	191	20	10
32	87	275	471	214	22	12
41	94	299	512	233	25	13
56	105	334	571	259	28	14
70	112	354	606	275	32	15
92	123	390	668	304	32	17
126	137	433	742	337	35	18

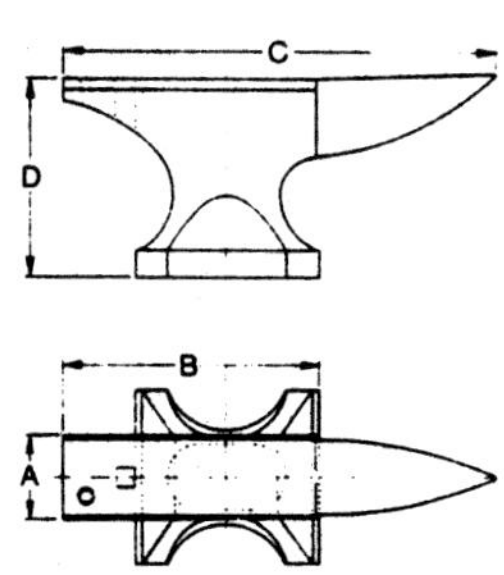

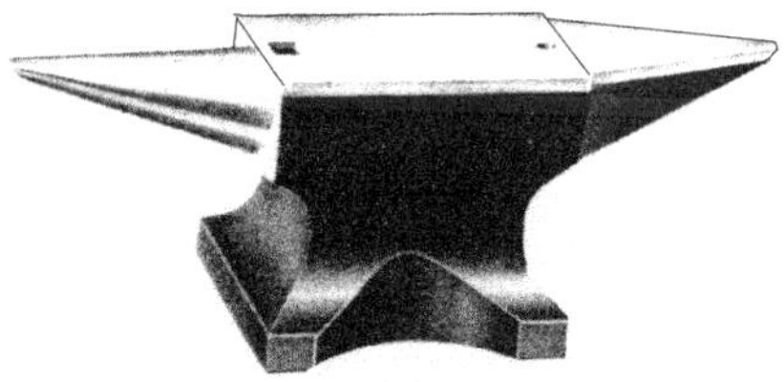

B 25

B 26

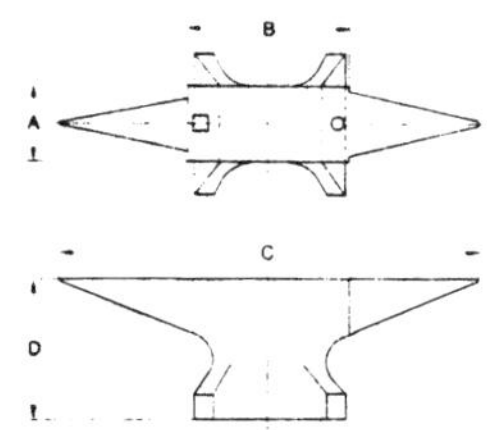

Types B 25 & B 26

Weight, Kgs		Dimensions, mm					
B 25	B 26	A	B	C	D	□	Ø
24	26	91	193	500	176	20	16
38	41	107	226	584	205	23	19
48	—	115	243	630	221	25	20
60	65	122	258	670	235	27	21
69	—	129	272	705	247	28	22
78	84	135	284	737	258	29	23
102	110	145	306	795	279	32	25
133	143	158	335	867	304	34	27
159	170	166	350	910	319	36	29

Type B 25 has also previously been known as E 71.

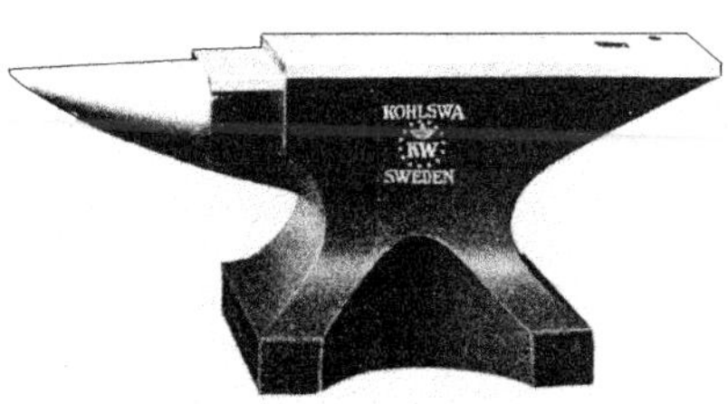

B 27

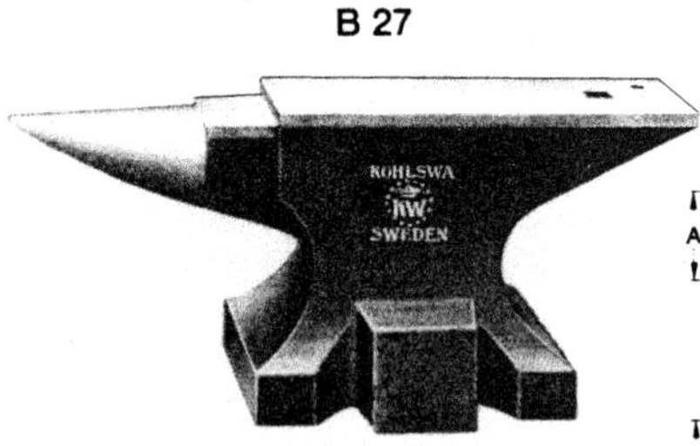

B 28

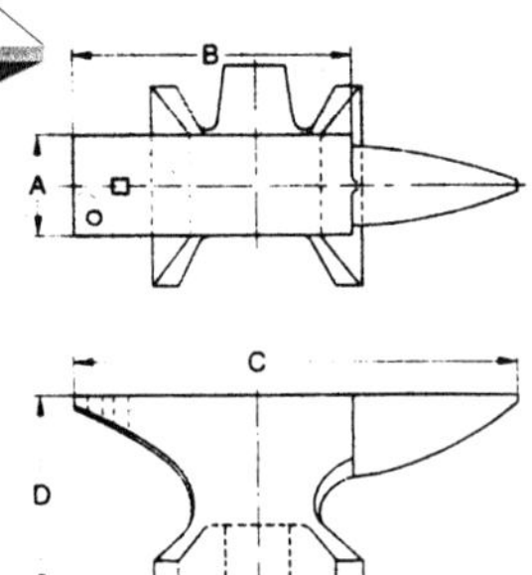

Types B 27 & B 28

Weight, Kgs		Dimensions, mm					
B 27	B 28	A	B	C	D	□	Ø
28	30	91	254	411	181	20	11
40	43	107	297	481	212	23	12
52	—	115	320	518	228	25	13
60	64	122	340	551	243	27	14
71	—	129	358	580	255	28	15
83	88	135	374	606	267	29	15
102	109	145	403	653	288	32	17
153	163	166	461	747	329	36	19
225	240	190	528	855	377	41	22

Type B 27 has also previously been known as E 73.

B 31

B 32

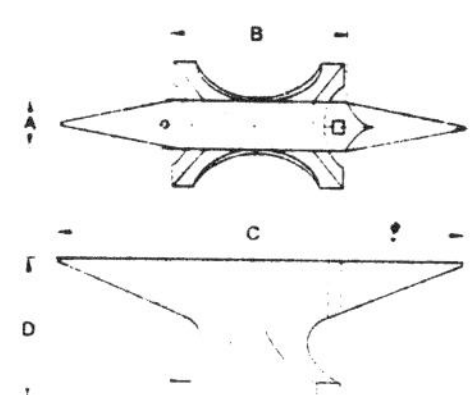

Types B 31 & B 32

Weight, Kgs		Dimensions, mm					
B 31	B 32	A	B	C	D	□	Ø
43	47	79	287	652	232	20	14
61	66.5	87	321	730	259	22	15
90	97	99	362	822	292	25	17
105	115	105	384	874	310	26	19
118	129	110	404	920	327	28	20
141	154	116	423	962	341	29	20
171	187	125	456	1036	368	31	22
214	233	134	491	1116	396	34	24

Type B 38

Weight Kgs	Dimensions, mm					
	A	B	C	D	□	Ø
62	127	240	587	225	23	21
85	140	264	646	248	25	23
109	151	285	697	267	27	24
135	162	307	750	287	30	27
161	173	326	797	306	31	29
187	182	343	839	322	33	30

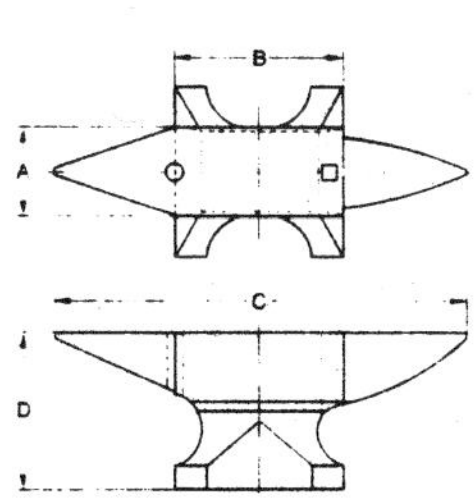

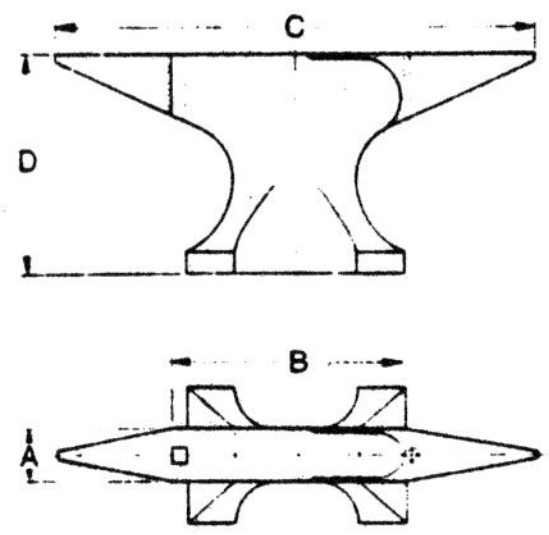

Type C 51

Weight Kgs	Dimensions, mm A	B	C	D	□	⌀
25	58	244	515	231	17	11
48	73	308	648	291	22	14
63	79	336	707	318	25	15
76	83	352	741	333	25	16
87	87	368	774	348	25	17
106	93	394	830	373	28	18
126	99	418	880	395	32	19
145	104	439	924	415	32	20

F 84

F 85

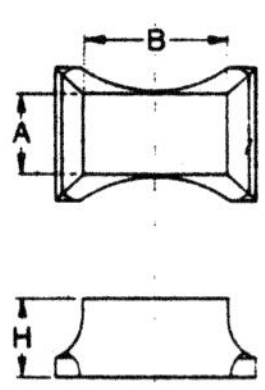

Types F 84 & F 85

Weight, Kgs F 84	F 85	Measurement, mm A	B	H
24		111	202	106
39		132	240	125
49		142	258	135
70		159	290	151
	103	153	278	280
	150	174	314	316

> **KOHLSWA**
> **ONE PIECE CAST STEEL ANVILS**
> have been one of our specialities for more than 60 years. The anvil is cast in one piece of a specially developed steel which is flame hardened on the face to 55–57 HRC.
> The anvil should be handled with proper care and judgment and never be struck on the edges of the face or the points of the bicks.
> Kohlswa undertake to replace free of charge any anvil which can be proved defective in material within a period of 6 months from the time of delivery, providing that such defects have not been caused by unskilled treatment.
> The dimensions and weights given are approximate and can vary slightly from anvil to anvil.

This label was attached to the face of a Kohlswa anvil purchaced new in 1983.

In the late 1980's, many smiths who had brought the new *Kohlswa* anvils began experiencing a problem with soft faces on these anvils. By 1990 Centaur Forge dropped this line of anvils. Centaur still handles anvils made in Germany, England and the United States. The *Kohlswa* anvil is now imported by a company in Canada (1996).

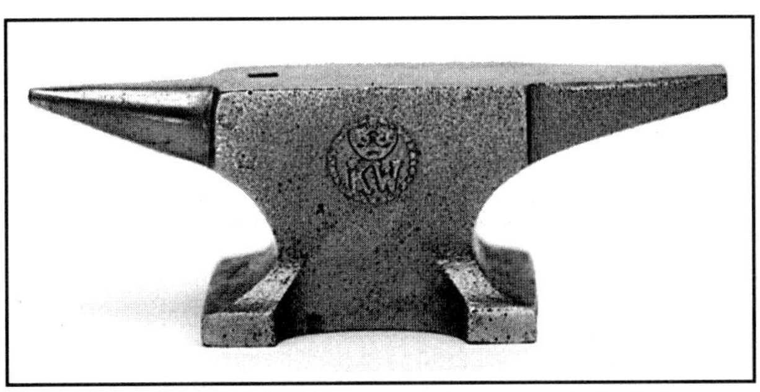

Kohlswa miniature from both sides. (Mr. Gill Fahrenwald photo)

In Centaur's 1985 catalog, there appears an article pointing out the virtues of the *Kohlswa* anvil, which I found rather amusing.

> *All-Steel anvils are cast in one single piece and are consequently perfectly homogeneous steel all through. They are very much superior to anvils with the steel face welded to a soft body. In the welded anvil there is a very marked difference between the steel face and the soft*

iron body, and as the tempering process affects the steel only, leaving the iron unaffected, with the strains and the stress between face and body are liable to some chipping of edges and fracturing of the face and weld, in a very short time.

All-Steel anvils are tempered and drawn very carefully, and owing to the fact that the anvil consists of the same steel material all through, the hardness can be made to decrease gradually downwards from the very hard face to a comparatively soft body, leaving no strains in the anvil.

All-Steel anvils have great elasticity and make the hammer rebound very effectively. A good rebound makes the work easy. No chipping of the edges. No fracturing of the face.[30]

This statement might be true if the *Kohlswa* was competing with new steel-faced anvils, but that is not the case. The last steel-faced anvils were made in the United States and other countries about 1930, with one exception, *Arm & Hammer,* which were last produced about 1950. To my knowledge, by the late 1930s all imported anvils were of one or two pieces and without face plates. *Kohlswa* was competing with and berating a non-existent product in the new anvil market.

All of the *Kohlswa's* that I have recorded have the trademark, *Kohlswa* prominently cast on the side and indicate the weight in kilograms. Most of them also have *Sweden* on them. I have seen several of these anvils with actual pounds on them, but these were probably manufactured before World War II.

If you find a used one to buy, I would recommend that you test it for hardness, if you intend to put it to much use (see Buying a New or Previously Owned Anvil, Appendix B).

Söderfors

As mentioned before, this is the Swedish anvil manufacturer who claims to have made anvils since about 1200 A.D. I have a record of Söderfors selling iron to manufacturers in Sheffield, England, before 1800.[31] There is no doubt that this company has had a long history. I have recorded considerably more anvils of this Swedish company than of the other two companies. They were probably the first Swedish manufacturer to export anvils to this country and possibly continued to do so until World War II.

These anvils were generally sold under the brand name *Paragon,* but they also manufactured anvils for at least one U.S. hardware company (see *Columbian).* They also sold under their own trademark, *Söderfors.*

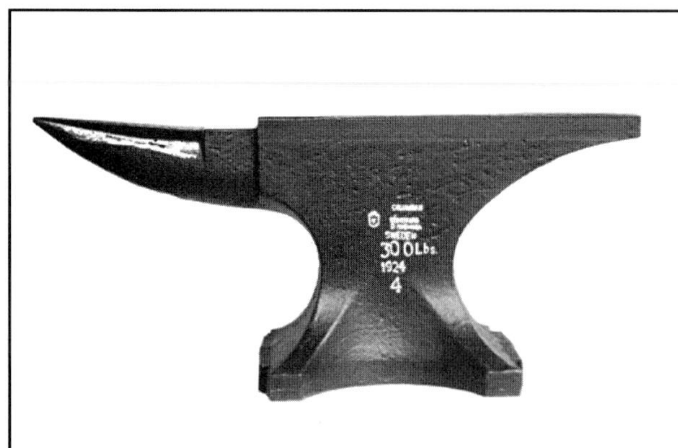

300-pound Söderfors anvil made for the Columbian Hardware Company. Columbian is stampted at the top of the Swedish trademark. Date 1924. (R.A.P. photo)

101-pound Söderfors anvil. Date 1930. Owned by Philip Hecht, D.V.M. (R.A.P. photo)

The earliest advertisement I have comes from the May, 1901, issue of *TheBlacksmith and Wheelwright* magazine, and the last from a 1928 Duluth, Minnesota, hardware company's catalog. In every advertisement, except in the trade catalogs, the sales agent or importer was Horace T. Potts and Company of Philadelphia, Pennsylvania, and the trademark used every time was *Paragon*.[32]

A number of advertisements state that these anvils were forged from one solid piece and not cast. Every one that I have recorded seems to be cast. However, this may be a case where the anvil is first cast and then placed in a closed die and forged to the final shape, leaving a parting line where the two dies met, which gives the appearance of being cast. It has been my experience, however, that a casting leaves a thin parting line where the mold parts, whereas a forging in closed dies leaves a much wider line when the flash is trimmed off. I have never seen an item that was first cast and later forged hot in a closed die, so it is hard to know. Every *Söderfors* that I have seen has a thin parting line under the horn and heel, similar to a casting. If indeed these anvils were forged from one piece of steel, it is extremely hard to prove it by me.

This company was not consistent with the appearance of their trademarks. At times it was cast in relief, and at other times stamped on the side. Some of their anvils appear with dates; the oldest I have found is dated 1902, the newest is 1934. All of the anvils have the weight in actual pounds stamped on the body. Usually, the anvils with the *Paragon* trademark have the outline of a straight razor stamped on their side. Because the anvils are often used on the sides as well as the face, trademarks were sometimes removed, eliminating any chance to say with certainty how the trademark may appear. However, in most cases, *Sweden* can be found somewhere on the side. A shield with a crown, with letters on the shield, is also common. Two examples that I have seen are dated 1902 and 1930. I also have one that has *Best Swedish Solid Cast Steel* stamped in a semicircle around a shield, and *Söderfors* in small letters on the other side.

It is of little consequence whether these anvils are cast or forged, as it has been my observation that they are of excellent quality and hold up well under heavy use. In my recordings I have found all but two in excellent condition. These two were in fair condition, but had chipped edges. I have recorded over 30 *Söderfors* anvils and examined many more. The largest anvil I have recorded weighed 500 pounds and the smallest, 70 pounds; most of them fall into the 100-to-200 pound range.

Söderfors Bruks Akkticbolag, the company's full name, was located in Falun, Sweden, and to the best of my knowledge, exported blacksmith's, farrier's and sawmaker's anvils to this country. I have recorded examples of each. In identifying these anvils, it is best to look for the name *Paragon* or *Söderfors* on the side.

From The Blacksmith and Wheelwright magazine, May, 1901. (From Mr. James Moon)

From The American Blacksmith magazine, May, 1910.

From The American Blacksmith magazine, January, 1912.

From The American Blacksmith magazine, March 1914. After this date Söderfors stopped using "Forged from one solid piece."

From The Blacksmith and Wheelwright magazine, June, 1926.
(From Mr. Keith Sommer)

This is a selection of Söderfors ads that differ slightly from one another.

From the J. C. Christopher Company Catalog, 1926-1930. (From Mr. Hunter Pilkinton)

Paragon Solid Steel Anvils

Blacksmiths' Pattern

Face, table and horn are finely finished and heel and horn are shaped to give best results. When struck, these Anvils have a beautifully clear, bell-like ring.

Weight Pounds	Face Inches Width	Face Inches Length	Length of Horn Inches	Height Inches	Square Hole Inches	Round Hole Inches
50	3	10 1/2	7	7 1/2	5/8	3/8
55	3	10 3/4	7 1/16	7 1/2	5/8	3/8
65	3 1/16	11 7/16	7 5/16	8 1/8	3/4	1/2
75	3 1/4	12 1/2	8	9	3/4	1/2
100	3 1/2	14 7/16	8 11/16	10 1/16	7/8	1/2
125	3 3/4	15 9/16	9 5/16	10 7/8	1	1/2
150	4	16 9/16	9 15/16	11 9/16	1	1/2
175	4 3/16	17 1/2	10 1/4	12 1/8	1 1/8	9/16
200	4 7/16	18 3/16	11	12 11/16	1 1/4	5/8
250	4 3/4	19 1/2	11 5/8	13 3/4	1 3/8	5/8
300	5 1/8	20 3/4	12 1/2	14 1/2	1 3/8	11/16
350	5 1/4	21 7/8	13	15 1/8	1 3/8	11/16
400	5 9/16	22 13/16	13 3/4	15 15/16	1 1/2	3/4
450	5 13/16	23 13/16	14 1/4	16 9/16	1 1/2	3/4

Blacksmith Pattern are made 50 to 450 pounds, as follows:
From 50 to 100 lbs., varying by 5 lbs.
" 100 to 200 " " " 10 "
" 200 to 450 " " " 25 "

Price.................................per Pound ....

Horseshoers' Pattern

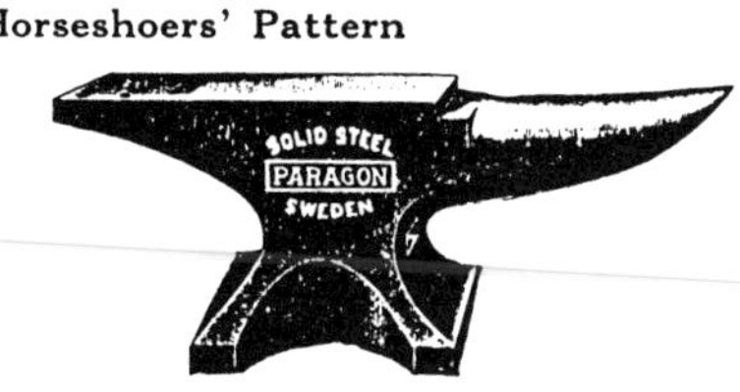

Similar to regular blacksmith anvil, but face is more narrow and longer, horn more pointed and heel not so thick. Exactly suited for easily and accurately shaping a horseshoe or general work.

Made from 125-lbs. to 175-lbs., varying by 5 lbs., both with and without clip-horn.

Weight Pounds	Face, Inches Width	Face, Inches Length	Length of Horn Inches	Height Inches	Square Hole Inches	Round Hole Inches
125	3 11/16	15 9/16	11 3/8	10 1/2	7/8	1/2
135	3 3/4	15 15/16	11 5/8	10 11/16	7/8	1/2
150	3 7/8	16 9/16	12 1/16	11 1/8	7/8	1/2
160	4	16 7/8	12 5/16	11 5/16	7/8	1/2
175	4 1/16	17 1/2	12 11/16	11 5/8	1	9/16

One-inch drop from face to table. Two round holes in heel. One inch thick at end of heel.

Price.................................per Pound ....

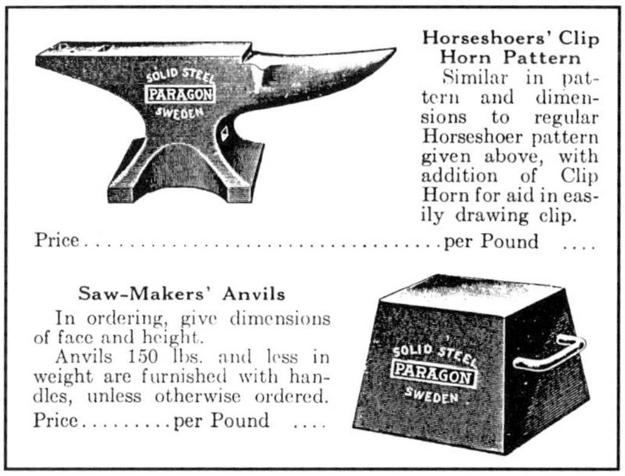

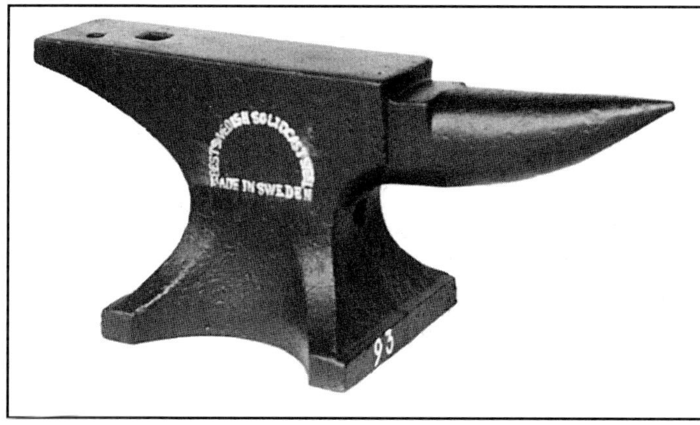

93-pound Söderfors anvil. (R.A.P. photo)

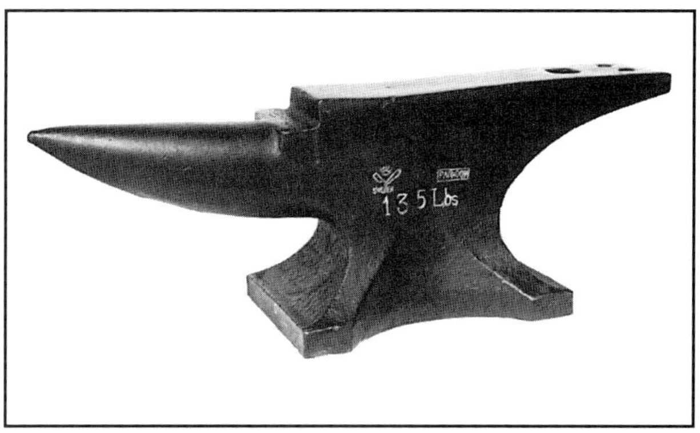

135-pound Söderfors farrier's anvil with clip horn. Notice the straight razor above the "1". (R.A.P. photo)

When the name is missing it is very easy to mistake one Swedish anvil for another. Many *Söderfors* anvils appear to have a separate face plate, but actually do not. These anvils may also have flats similar to the *Peter Wright* anvil, on their feet under the horn and heel. The area under the base may be flat or have a depressed oval, much like the *Trenton* or *Arm & Hammer.* If you can make out the image of a straight razor you have a *Söderfors.*

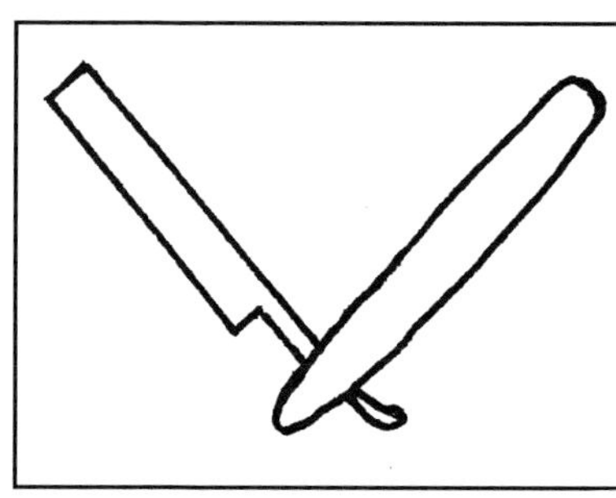

The image of a razor denotes a type of tool steel used in their anvil.

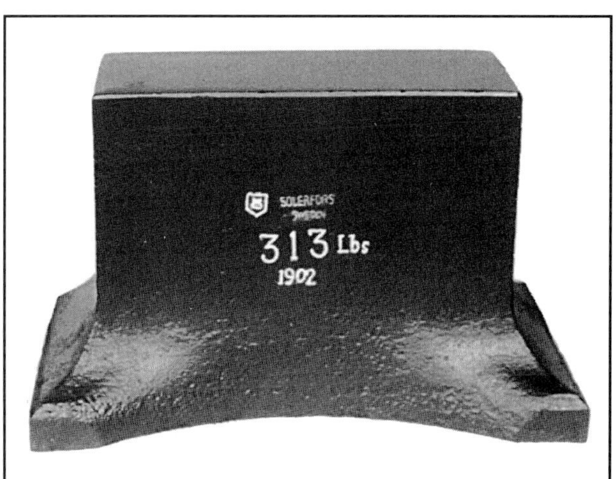

313-pound Söderfors saw maker's anvil, date 1902. (R.A.P. photo)

Since all *Söderfors* appear very much the same, at least since the 1900s, the only way to be sure of their age is to find a date stamped on the side.

In 1996 I obtained a Stock Book of Horace T. Potts and Company, of Philadelphia, Pennsylvania, dated 1915.[33] It illustrates the *Paragon* smith's, farrier's, hornless and various sawmaker's pattern anvils. I quote: "Special prices quoted on application for Sawmaker's, Plowmaker's, Hornless or Special Pattern Anvils of any shape, size or description." They list anvils from 50 to 600 pounds and also anvils from 50 to 450 pounds as being in stock. It is interesting to note that the Stock Book says that the *Paragon* anvils are one piece of steel, but does not say whether they are cast or forged, in which case I would think that they were cast. A photo in the book shows a large group of these anvils stacked in the stockroom and I can see thin parting lines on the front of the anvils' bases. I can also see handling holes.

SISCO

SISCO is the acronym for Swedish Iron & Steel Corporation, who had warehouses located in New York and New Orleans. As of this writing I have no knowledge of when they started exporting to the United States or when they stopped. The only advertisements I have are dated 1913 and 1914, and neither give much information, although they do show the company's trademark, which corresponds to those found on their anvils.

From the Virginia-Carolina Hardware Company, Richmond, Virginia, 1913.
(From Mr. Hunter Pilkinton)

From The American Blacksmith, March, 1914.

I have recorded only five of these anvils that I am sure of. The trademark is as follows:

SISCO
SUPERIOR

The weight is in actual pounds and *Made in Sweden* is marked on every anvil I have recorded. As far as I can tell, they are manufactured of cast steel and have one handling hole in the waist, under the heel. These anvils may have a depressed oval or are flat under the base. Should the trademark be missing, the single handling hole may be its most distinguishing feature. The *SISCO* anvils I have recorded range in weight from 125 to 308 pounds.

Records from the Columbus Forge & Iron Company *(Trenton)* state that between the years 1925 and 1927, the SISCO ordered plowmaker's anvils from them. This suggests that American manufacturers may have been the only producers of this style anvil at the time.

SISCO miniature from both sides.
(Mr. Gill Fahrenwald photos)

North Star

I have found only one advertisement, dated 1930, for this anvil. I have recorded two Swedish anvils with what appears to be a star

308-pound SISCO owned by Gene Harris.
(R.A.P. photo)

shape stamped in the side. Over the weight, which is in actual pounds, is stamped *Made In Sweden.* These anvils were most likely manufactured by one of the major Swedish manufacturers, but there is no positive proof as of this writing. I have recorded a number of Swedish-made anvils, some having *Made In Sweden* marked on them, but it is difficult to say who manufactured them, due to the lack of other markings.

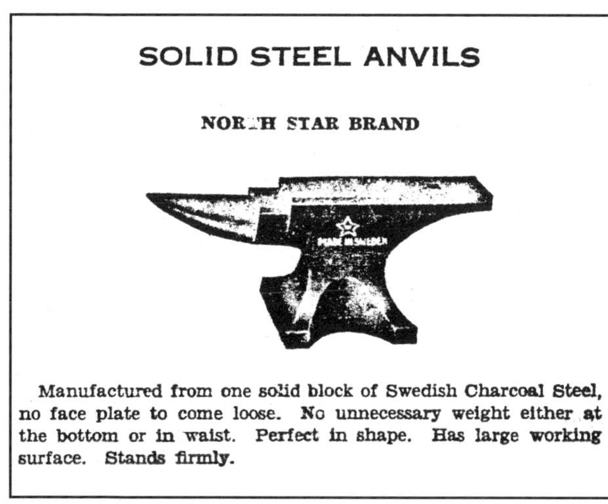

SOLID STEEL ANVILS

NOR.H STAR BRAND

Manufactured from one solid block of Swedish Charcoal Steel, no face plate to come loose. No unnecessary weight either at the bottom or in waist. Perfect in shape. Has large working surface. Stands firmly.

From the Janney, Sensple, Hill and Company catalog, Minneapolis, Minnesota, 1930.
(From Mr. John Cacthings)

OTHER FOREIGN ANVILS

Since World War II, various other anvils have found their way into the American market in small numbers. They are all cast steel or cast iron, and of doubtful quality. Listed are the ones I have chanced upon.

Poland - *Zaklady Stalmech,* anvil of cast steel.

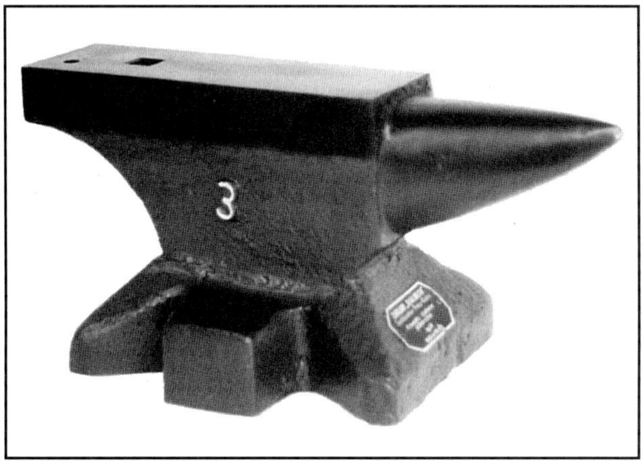

106-pound Polish anvil with upsetting block. (R.A.P. hoto)

Japan - *Jet,* of cast iron.

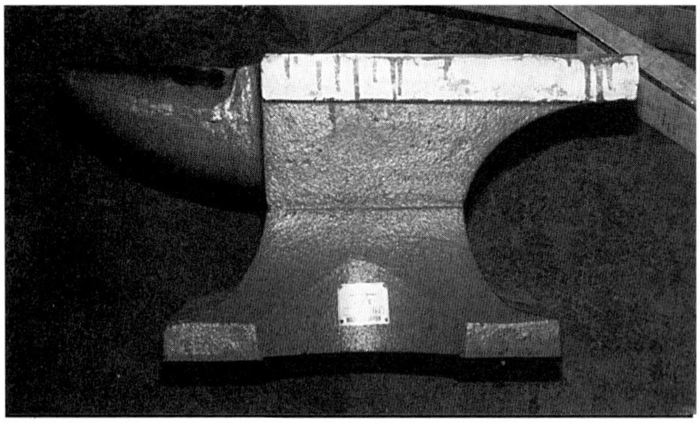

220-pound Japanese anvil. (R.A.P. photo)

Finland - *Milme,* of cast steel.

100-pound Finish anvil. (R.A.P. photo)

China - *Foken,* probably cast iron. I know some anvils were manufactured in Australia, but they were never sold here. It has been reported that anvils were manufactured in Western Canada. These can be found in the northwest United States.

Anvil from Australia. (Mr. Gill Fahrenwald photo)

CHAPTER 3

English Anvils

PETER
WRIGHT

WILLIAM
FOSTER

HADFIELD
AND
SANDERSON

HENRY
WRIGHT

MOUSE
HOLE

WmPARKER
POND FORGE

WILKINSON

ANVILS IN AMERICA

From the very first to the present, English-manufactured anvils have been imported into the United States. They are also the anvils most familiar to us. Two out of three anvils found in a barn, basement or garage, turn out to be a *Peter Wright, Mousehole* or some other English-manufactured anvil. When you consider that between 1600 and 1900 English anvils were about the only ones available in this country, is it any wonder there are still so many to be found? American anvil manufacturers did not really become competitive with their counterparts until the late-19th and early - 20th centuries. Of course starting in 1843 there were *Fisher* anvils, but these were cast-iron steel-faced anvils, considered inferior to wrought-iron anvils of the day. Smiths for the most part liked what they were used to, and when a new product came along that proved no better they stayed with the older. Besides, the *Fisher* did not *ring* and the hammer did not rebound as effectively as with the forged anvils. Even today English anvils are imported into this country, as there are no major American anvil manufacturers still in business.

Earlier in this book I put anvils into three categories by shape: *Colonial* (*ca.*1600 - 1800), *Old English* (*ca.* 1800 - 1850), and *Modern* (*ca.* 1850 - present). I have recorded 200 or more English anvils made before 1850, including 150 manufactured by the "big three," *Peter Wright, Mousehole,* and *Wilkinson.* I have recorded at least 400 of the *Modern* style English anvils manufactured after 1850. In addition, there are hundreds of English anvils that I have looked at but did not record for one reason or another. The following is a list of all three categories of the English anvil makes and trademarks known to me at the time of this writing (1995). The parenthesis indicate where the anvil was made, but was not part of the trademark. They are as follows:

1. Alsop - Sheffield
2. Attwood - Sheffield
3. Booth & Co.-Boyd Mill-Sheffield
4. Bower - Bristoe
5. Brooks - England
6. C?roy - Best Scrap - Dudley
7. Coe?m - Denton - Sheffield - CD ? & CO
8. E. Griffiths
9. Green - (Greenwick)
10. Hadfield and Sanderson - Sheffield
11. Henry Wright (Dudley)
12. Hill - Burmingham
13. J. Wilkinson (Dudley)
14. John ? ? ? so ? -1938 - England
15. Joseph Wilinson (Dudley)
16. Kirkstall Forge - B & B Co - B A
17. L O M A - Froming & Co - Sheffield
18. M. Reading
19. Mousehole Forge - Sheffield
20. Peter Wright (Dudley)
21. Record (Sheffield)
22. Samuel Newbould-Sheffield - Dudley
23. Sanderson - Sheffield
24. Sherman - ?-- & Co - -?-- - field works - Pittsburg
25. Smiths
26. Smith & O'Leary - Chicago
27. T. Tilotson & Co - Sheffield
28. Vaughn England
29. Vulcan Works
30. W. Cole - Dudley Wood
31. William Foster
32. William Parker - Pond Forge - Sheffield
33. Winston or Wilson
34. ? ? ? ?eson & sons - Stourbridge

A number of these listed where the name is obscured, such as # 7 and # 34 are most likely the same as others listed. I believe a number of the others listed were anvils manufactured by major manufacturers under another trademark. Smith & O'Leary, # 26, I know to be an anvil rebuilder in Chicago, Illinois, in the 19th century and Sherman--- & Co.---field, Pittsburg, # 24, is probably one as well. This list has been growing ever since I began writing this manuscript, and I am sure there are other English manufacturers, especially before 1850, and brand name anvils exported to the U.S. that have yet to be discovered.

The following list has come to me from a number of sources as possible additional English anvil manufacturers or trademarks. I cannot confirm these at this time. They are as follows:

1. Bryd
2. John Bullock
3. Joseph Maston
4. Ovendale
5. Rexton
6. Richard King
7. Stanford-Sheffield
8. T.M. Blood
9. Thomas Darwin (or Darwen)
10. Thomas Hill
11. Ward

I determined proof of an anvil's existence by either seeing the anvil myself or obtaining a good photo of it with a readable trademark. I do not take the word of anyone in this matter, because it is just too easy to make mistakes when looking at obscure trademarks or by mistaking one letter for another. I have made this mistake a number of times myself. For instance # 34 may well be *Wilkinson & Sons,* although as far as I know Wilkinson made anvils in Dudley not in Stourbridge. I own this anvil myself and what appears to me to be an "E" may well be an "N". I suspect that # 7 in the "not sure" list, *Stanford-Sheffield* may be *Hadfield and Sanderson.* Some of the names given to me on the "not sure" list were by word of mouth with no confirmed proof.

COLONIAL ENGLISH ANVILS

I have recorded 60 or more anvils manufactured before the 1800s loosely referred to as *Colonial.* Only three of these *M. Reading, Bower-Bristoe* and *Alsop* have trademarks. Most of the others have discernable weight markings, but little else. The anvils with names are well marked and show as much wear as most of the unmarked ones. This leads me to suspect that most of the anvil manufacturers before 1800 did not bother to put their names or trademarks on their anvils, and that the weight markings were used for resale purposes.

The *Bower-Bristoe* is a hornless anvil, with a 1/2-inch hardie hole. It weighs 92 pounds, is 8 3/4 inches high, the face is 9 1/2 x 5 inches and the base is 9 x 8 inches. You will note that the face is only 1/2-inch longer than the base. It has three deep handling holes, as do all wrought anvils manufactured before 1850 that I have recorded, with one exception. However, most handling holes on the other *Colonial* anvils are not as deep as those of the *Bower-Bristoe.* The name *Bower-Bristoe* is marked on both sides, and there appears to be some markings above the *Bower* name on one side. There are no weight markings, which is unusual. It should be remembered that between the late 1500s until about1790 there was practically no change in the English anvils. Therefore, only an approximate date, give or take 100 years, can be given in most cases. Because this anvil has a name on it, I would date it to the early 1700s or a little earlier. In appearance it looks like most other hornless *Colonial* anvils so I did not bother to photograph it when I had the chance.

The *M. Reading* anvil has a horn and a 5/8-inch square hardie hole. It weighs 80 actual pounds, is 9 inches high, the face is 9 1/4 x 4 1/8 inches, and the base is 8 3/4 x 7 1/2 inches. Again, the face is only slightly longer than the base. The horn is 5 1/2 inches long, and on the opposite side from the name the weight is marked as follows: **0-3-1**. This indicates that the weight was originally 85 pounds, 5 pounds heavier than the actual weight. Three to five pounds loss of weight is normal for anvils of this age, if they are not broken in some way or other. You will notice that the weight lines are center-punched and the lines are chiseled in. I would date this anvil from the early 1700s to slightly later.

85-pound M. Reading anvil. (R.A.P. photo)

Opposite side of M. Reading anvil. Notice how the weight marks are punched. (R.A.P. photo)

The *Alsop* has a horn with a table and a 5/8-inch square hardie hole. However, I have recorded three of these anvils (see Chapter 1 Dating Anvils), and one of these does not have the hardie hole. They are all marked the same, where markings still exist. I will give the specifications of the one I own. These anvils are unusual in that they have a table; everything else is typical of the *Colonial* anvil.

It weighs 94 actual pounds, though the weight marks indicate 97 pounds. It is 8 3/4 inches high, the face is 9 3/8 x 4 1/2 inches, and the base is 9 inches long x 7 inches wide in the front and 7 1/2 inches wide in the rear. Again, the face is only slightly longer than the base. The horn is 7 1/4 inches long, including a 2 1/2-inch square table. The base is as long as the face on one of the other anvils and the base is 5/8 inches longer than the face on the anvil without the hardie hole. I would date these three *Alsop* anvils from about 1780 to 1800.

The Sheffield directory for 1787 lists a Samuel Alsop as an anvil maker. The 1797 directory lists Samuel, William and John Alsop all at the same address as anvil makers.[1] William and John were most likely sons of Samuel. Could these anvil makers have produced the *Alsop* anvil? I think so. The directories indicate the Alsop business was growing and they must have been a major anvil manufacturer and feature developer during the late 1700s. Every horned smith's anvil that I have dated after approximately 1790 has a table (see chapter 1 Dating Anvils for the photo of the *Alsop*).

The Sheffield directories for 1787 and 1797 also list a John Bullock, Thomas Darwin or Darwen and a Thomas Hill as anvil manufacturers.[2] I do not understand why Mousehole Forge is not listed, as records prove that they were manufacturing anvils in Sheffield before and during this time period.[3] It may be that, as in the *Thomas Register* of today, you needed to purchase a listing to be included. Mousehole may have felt that they were well enough known, and that they did not need to advertise. Another possibility was the fact that Mousehole Forge was located just outside the limits of Sheffield at that time and therefore were not included. However, there were other well-known tool manufacturers in Sheffield at that time who were not listed in the directories, either.[4]

Early in 1995 there was an auction near Reading, Pennsylvania, that advertised an anvil with a date marked on it of 1614. The entry in the brochure is as follows:[5]

> *"Anvil. Early! Rare! Dated 1614. 18" long, pyramid bottom. Mounted on iron banded 19" high stump. Wonderful look."*

Unfortunately, I received this brochure too late to get further data before the anvil was sold. I mention this anvil only to illustrate how important it is to either see or record an anvil or have good photos and accurate information. The only information I know for sure is the total length and a date that is meaningless without a good photo to go by. What does "having a pyramid bottom" mean? At this time I have no way of knowing.

I have broken down the "nameless" or unmarked anvils of the *Colonial* period into categories based on their outstanding features. They are as follows:

1. Hornless anvils, with and without hardies.
2. Anvils with and without horn and hardies, but with a fifth foot or toe.
3. Anvils with horns and with or without hardies.

A selection of photos representative of these anvils is as follows:

99-pound hornless Colonial anvil with a hardie hole. (R.A.P. photo)

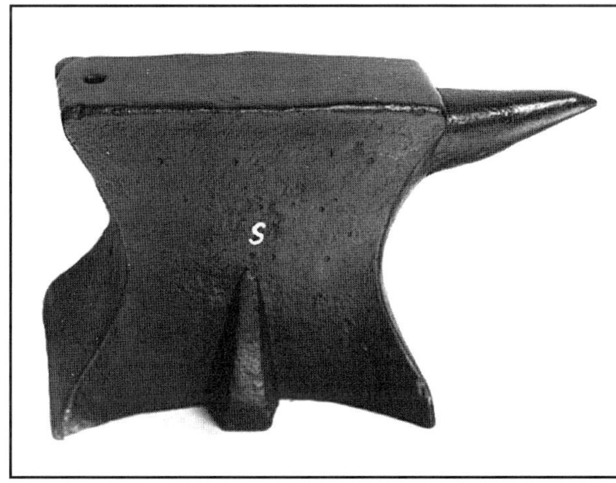

190-pound Colonial anvil with horn and 5th toe. The pritchel hole was added later. (R.A.P. photo)

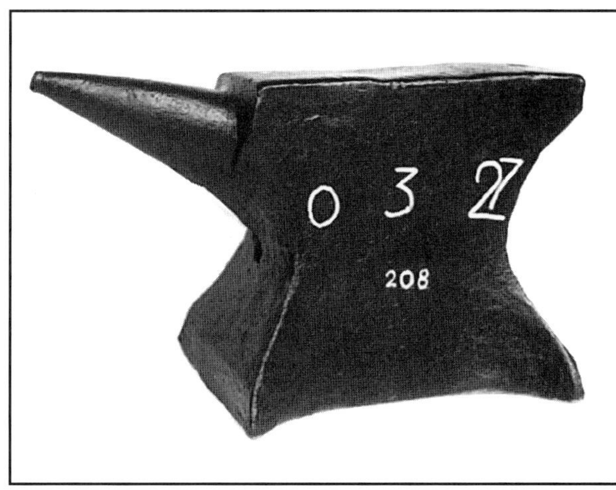

111-pound Colonial anvil with a horn and hardie. (R.A.P. photo)

ENGLISH ANVILS AFTER 1800- "OLD ENGLISH"

The anvils I call *Old English* were those exported to the United States between 1800 and about 1850. They have the most varied trademarks. Why export so many different brand-name anvils? I do not know. You would think that the major manufacturers of the time, Mousehole Forge, William Foster, Wilkinson and Peter Wright, would have been able to easily handle the needed orders. These many brands may have been separate anvil manufacturers, but I find that hard to believe. For example, T. Tillotson and Company was a hardware company that not only sold blacksmith tools, but carpenter tools as well as other items.[6] It is most unlikely that this company would have been the manufacturer of anvils as well. If all the different trademarks that existed during this period of time were indeed produced by separate anvil manufacturers, the competition must have been fierce in England.

About 100 of the 200 or more *Old English* anvils I have recorded have no discernable trademarks, except the weight, a letter or two, or maybe part of the word Sheffield. Most English anvils of the time were manufactured in Sheffield, so that does not help much in determining the actual manufacturer. I therefore rely on certain discernable features, used by the manufacturers, to identify anvils found with no trademarks. For example, Mousehole usually left the underside of the horn sharp, rather than rounded off, and put deep center punch marks between the weight marks. I will describe these features in detail, as I discuss the various manufacturers. One feature that is common for all English anvils from the oldest to the newest; they are all flat under the base.

The following is a list of *Old English* anvils, other than the major manufacturers, that were produced between about 1830 and 1850 or a little later. Most of these are so similar in appearance that without a name they are undistinguishable from one another. They all have these features in common: three handling holes, a hardie, a pritchel, a table and are rounded under the horn.

1. **Attwood** - I have recorded only two of these. One weighs 115 actual pounds, is 10 inches high, the face is 11 5/8 x 3 1/8 inches, and the base is 10 x 9 1/4 inches with a 7 5/8 inch horn, including the table. It is marked as follows:

ATTWOOD
WARRANTED
1 0 5

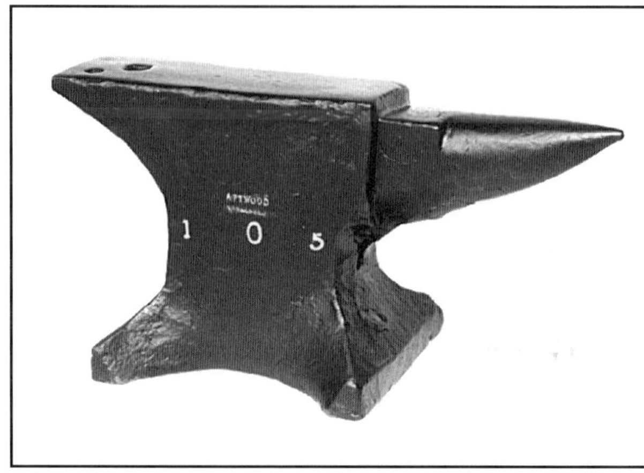

117-pound Attwood anvil . (R.A.P. photo)

2. **Carey or Coley** - I have recorded two of these, 89 and 98 actual pounds. The 89-pound anvil is 9 inches high, but the heel is broken off, so the other dimensions are of little value. The 98-pound anvil is 9 inches high, the face is 11 5/16 x 4 inches and the base is 9 x 9 1/2 inches wide (unusual), with a 7-inch horn, including the table. They are marked as follows:

89lbs: Co?ey 0-3-16 (other side) [7]
DUDLEY WOOD BEST SCRAP
WARRANTED

98lbs: CAREY 0-3-19 (other side)
DUDLEY WOOD
BEST
SCRAP
WARRANTED

This is most likely a manufacturer.

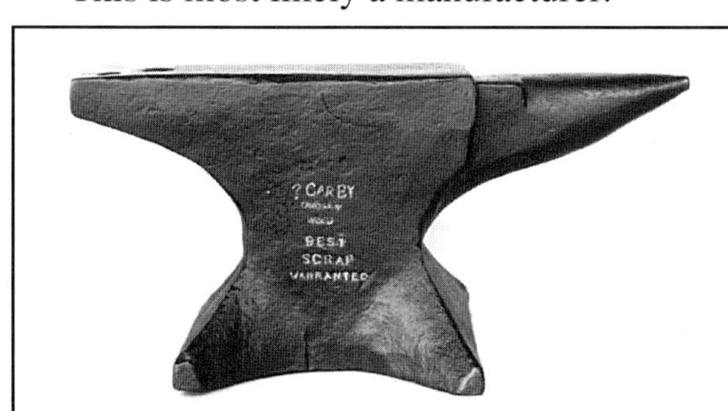

98-pound Carey anvil? (R.A.P. photo)

3. **E. Griffiths** - All I have is a photograph of the trademark for this anvil. The weight was located on the other side of the anvil. It is marked as follows:

E. GRIFFITHS [8]
WARRANTED

4. **Green -** I recorded two of these anvils. The name was located near the base; any other information that may have been there is gone. There was a forge at Greenwick, England, in the early 1850s. One of these anvils weighs 106 actual pounds, is 10 inches high, the face is 12 x 3 3/4 inches, with a base of 10 x 8 inches. The weight was located on the other side of the anvil. The other one weighs 118 pounds, is 10 inches high, the face 12 1/8 x 3 7/8 inches, with a 10 1/2 x 7 7/8-inch base.

#1 GR?EN	#2 GREEN
0 3 12	1 0 6

5. **Hadfield and Sanderson -** I have recorded five of these anvils and have heard of others. This is definitely a manufacturer. The anvils I have recorded range in weight from 102 to 149 actual pounds. The anvil I own weighs 142 actual pounds, is 10 1/4 inches high, has a 12 x 4 5/16-inch face, and the base is 11 x 10 1/4 inches with a 8 1/2-inch horn, including the table. It is marked with large letters as follows:

Hadfield and Sanderson
Sheffield
1 - 1 - 5

The other anvils are marked the same, except one, which was marked **warranted** instead of **Sheffeild**.

145-pound Hadfield and Sanderson anvil. (R.A.P. photo)

6. **Hill -** I have recorded three of these anvils and have reason to believe that Hill was a manufacturer. One of these anvils weighs 224 pounds, but does not have a pritchel. The information given is from the anvil which has the best markings. The face was in very poor condition, and therefore the anvil shows a weight difference. It weighs 177 actual pounds, is 10 1/2 inches high, the face is 14 5/16 x 4 3/4 inches and the base 11 3/4 x 16 1/4 inches. The horn is 8 inches long, including the table. The weight is marked on the other side. The other anvils are marked the same.

I? [9]
HILL
BURMINGHAM
1 - 2 - 9

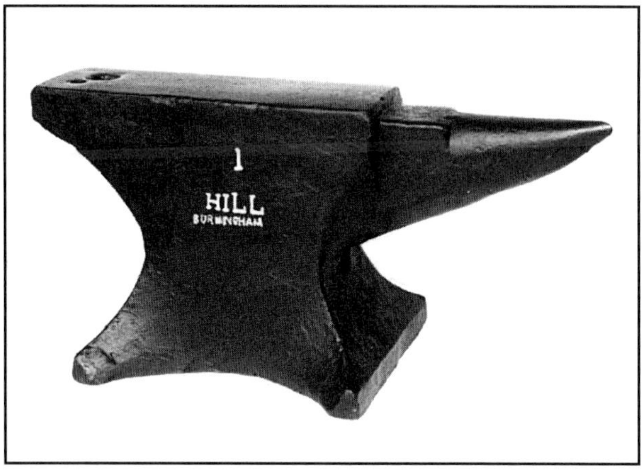

177-pound Hill anvil. (R.A.P. photo)

7. **Kirkstall Forge -** I have recorded two of these anvils; one is double-horned and weighs 250 pounds. I include the dimensions of the anvil with the best markings. It weighs 146 actual pounds, is 10 1/2 inches high, the face is 13 1/4 x 4 1/2 inches and the base, 11 x 10 inches. The horn is 8 1/2 inches, including the table. It is marked as follows:

B S & CO.,[10]
KIRKSTALL FORGE
B.A.
1 - 1 - 16

250-pound Kirkstall double-horned anvil (Robert Armstrong photo)

150-pound Kirkstall anvil. (Robert S. Cerny photo)

8. **Loma, Fromings & Co. -** This is most likely an export concern. I have recorded two of these anvils and know of several others. As with the others, the dimensions are taken from the anvil with the best markings. It weighs 133 actual pounds, is 10 1/4 inches high with a face measuring 13 1/2 x 4 1/2 inches, a 9 1/2 x 9 7/8-inch base (over square), and the horn is 9 inches long, including the table.
The markings are as follows:

LOMA FROMINGS & CO.[11]
SHEFFIELD
1 - 0 - 26

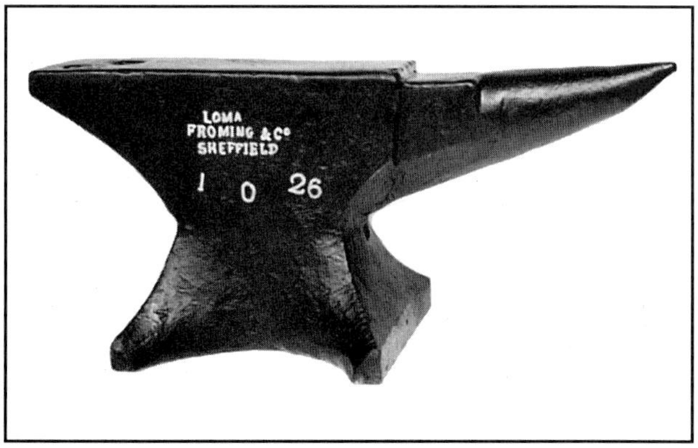

138-pound LOMA anvil. (R.A.P. photo)

9. **Sherman--- & Co.,** - I have recorded only one of these anvils with these markings. It may be an English anvil reworked by an American concern. The anvil weighs 103 actual pounds, is 8 1/2 inches high, has a 12 x 3 7/8-inch face, a 9 x 8 1/2-inch base and a 9-inch horn, including the table. It is marked as follows:

SHE-MAN--- & CO.[12]
----FIELD WORKS
PITTSBURG
1 - 0 - 3

115-pound Sherman anvil. (Steve Roth photo)

10. **Smiths -** I have recorded only one anvil with these markings, but I believe this is a manufacturer.

The feet are different from any other *Old English* anvil, but are not square, as those found on the *Modern* style. They appear to be a transition between the two. It is also unusual in that it has four handling holes and may be a three-piece anvil, having not been manufactured by the built-up process as are most other *Old English* anvils. The anvil's actual weight is 119 pounds, it is 9 1/2 inches high, a 14 1/8 x 4-inch face, has a 9 1/2 x 9-inch base, and a 9 1/4-inch horn, including the table. The first part of the trademark is not clear, but I assume that **SMI** is correct, as it fits. The markings are as follows:

???THS
1 - 0- 11

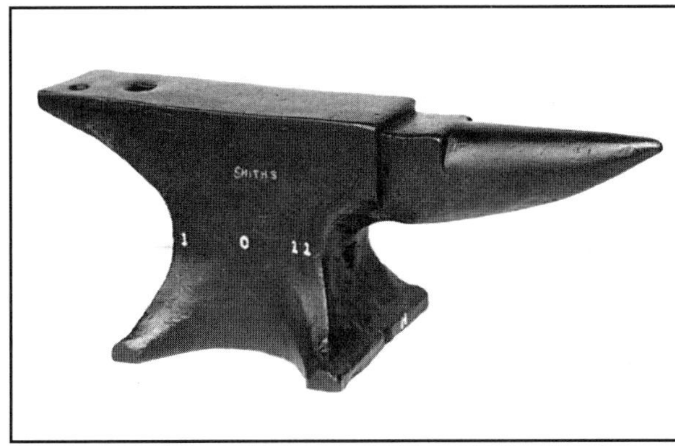

123-pound Smiths anvil. (R.A.P. photo)

11. **Smith & O'Leary** - I found this trademark on an *Old English* anvil, but it was marked *Smith & O'Leary*, *Chicago.* At first I assumed that it was an English anvil made for an American concern.

Later I discovered that Smith & O'Leary was an anvil rebuilder and vise manufacturer, located in 19th century Chicago. The features of this anvil suggest that it was manufactured by the Mousehole Forge with the original trademark removed and *Smith & O'Leary* placed in its stead. The old weight marks were changed to actual pounds. It also appears that the face is not as wide as on the original.

141-pound Smith & O'Leary anvil. (R.A.P. photo)

12. **Tillotson & Co., -** This is most likely an English hardware exporter. In a tool catalog a company named T. Tillotson advertised that they sold various hardware items.[13] I believe that I have recorded two of these anvils. I say this because one anvil has the trademark of a "heart" with no other information. The other anvil has both a name and a heart shape. The last one weighs 150 actual pounds, is 10 1/2 inches high, has a 14 x 4 1/2-inch face, a 11 1/2 x 9 1/2-inch base, and a 7 3/4-inch horn, including the table. The anvil with the most complete information is marked as follows:

TILLOTSON & CO.
❤ SHEFFIELD
1 - 1-14 (On the other side)

The other anvil has only the ❤, and the weight (**1 - 3 - 1**) , which is on the other side.

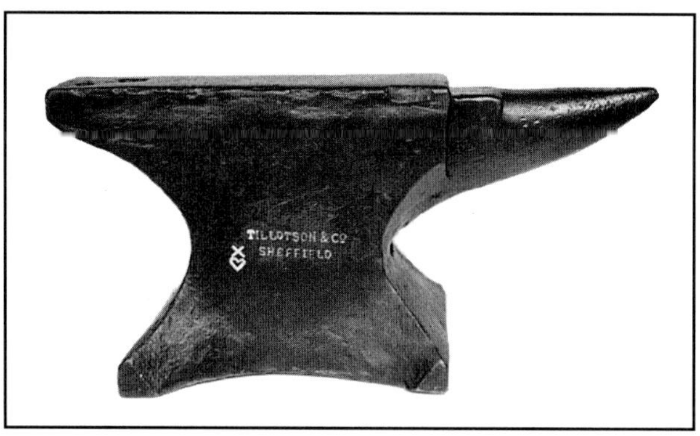

154-pound Tillotson anvil. (R.A.P. photo)

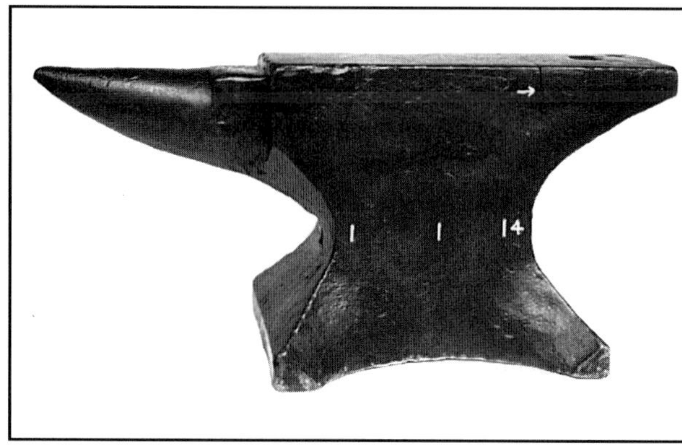

Reverse side of Tillotson anvil. Notice the arrow pointing to the crack where the heel was welded to the body. (R.A.P. photo)

13. **Vulcan Works** - This is probably a hardware exporter. The anvil appears to be a *Mousehole,* as it is sharp under the horn where the horn joins the body. The weight is 115 actual pounds (it origianally weighed 126 pounds) is 10 inches high, has a 12 x 3 7/8-inch face, a 10 9/16 x 9-inch base, and a 7 3/4-inch horn including the table. I have recorded one of these. The markings are poor, but are as follows:

VULC ? ? ? ? R K ?
1 ? ?

115-pound Vulcan Works anvil. (R.A.P. photo)

14. **W. Cole** - This is most likely another hardware exporter. This trademark is unusual in that part of it is shown upside down. I have recorded one anvil with this name. It weighs 98 actual pounds, is 9 inches high, has a 12 x 3 3/4-inch face, a 9 1/2 x 9-inch base, and a 7 1/8-inch horn including the table. The markings are as follows:

W. COLE [14]
DUDLEY WOOD (upside down)
WARRANTED
0 - 3 - 17 (on the other side)

15. **Winston or Wilson** - This *Old English* anvil could be a "re-marked" anvil from one of the major manufacturers. The two names, which are probably the same, are stamped in upside down. It weighs 110 pounds, is 9 5/8 inches high, the face is 11 5/8 x 4 inches, the base is 10 1/2 x 8 1/2 inches, and the horn is 8 inches long, including the table. I thought this anvil might have been manufactured by the Mousehole Forge, but the weight marks are on the wrong side.

The markings are as follows:

?O??E, [15]
WI?SON (upside down)
WINSTON
1 - 0 - 3 (on the other side)

16. **William Parker** - It was not before I had recorded three anvils that I was able to figure out this anvil's full name. Fortunately the letters missing on the first anvil were clearly marked on the second anvil.

One of these anvils has what I call a *side horn.* The information on this anvil is: weight 136 actual pounds, height 11inches, face 12 x 3 3/4 inches, base 11 1/4 x10 inches in the rear and 9 3/4 inches in the front, and the horn is 8 1/4 inches including the table. The *side horn* is 3 5/8 inches long, 1 1/2 inches wide at its base, and 1 1/4 inches at the tip. This would be a carriagemaker's anvil. The marking of these anvils that I have recorded are as follows:

Side horn anvil **Wm??RKER** [16]
POND FO---
SHE------
1 - 0 - 26

The others: **Wm--PKER** [17]
-OND FORGE
SHEFFIELD
0 - 3 - 8

----- [18]
BEST SCRAP
POND FORGE
weight?

92-pound William Parker anvil. Belonging to Frank Prillwitz. (R.A.P. photo)

Without several anvils to compare it would be impossible to determine the identity of these anvils.

17. **Coe??an** - I have recorded two anvils with these markings, unfortunately the markings are not good enough to pin down the manufacturer or exporter. I received the information for one of these anvils by phone. The markings were given, but not the specifications and had it not been for the fact that I already had data on one of these anvils I would not have accepted the phone information as usable. The anvil with specifications weighs 109 pounds (the same weight that was listed on the anvil), is 9 1/2 inches high, the face is 13 1/2 x 3 3/4 inches, the base is 9 1/2 x 8 inches, and the horn, including the table, is 8 inches. The markings are as follows:

COE??AN DENTO B???? & ? [19]
SHEFFIELD CD & CO P or B
0 - 3 - 25

Information on the anvil from the phone call:

CO???N DENTON BUR??KIN & CO
SHEFFIELD CD & CO

18. **????eson &Sons** - I might have assigned these anvils to *Wilkinson,* except for the name below the top letters. I have recorded two of these anvils, but in both cases the manufacturer exporter named is unclear. I will give the information on the largest of these anvils, which also has a side horn. It weighs 297 actual pounds, is 12 1/8 inches high, the face is 19 1/2 x 5 1/4 inches, the base is 13 1/4 x 11 3/4 inches, and the horn is 11 inches long, including the table. The side horn is 5 inches long, 2 1/2 inches wide at the base, and 1 9/16 inches at the tip.

The markings are as follows:

????? [20]
STOURBRIDGE
WARRANT--
EH
2 - 2 - 22 (on the other side)

E SON & SONS
ST - URBRIDGE WT
1 - 1 - 15

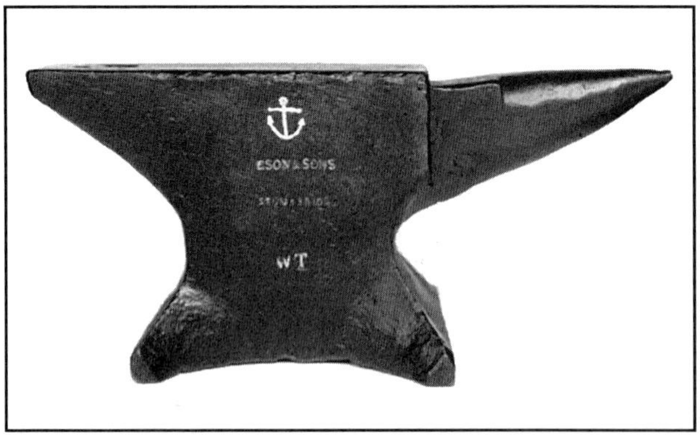

155-pound ?? ESON & SONS anvil. (R.A.P. photo)

Isaac Nash was a large anvil manufacturer in Stourbridge, and in 1927 bought the Mousehole trademark. [21]

Next I will discuss the major English anvil manufacturers, most of whom manufactured anvils from the Colonial period right into the 20th century. I will present them in order of the earliest to the most recent, based on my best guess, from the information that I have at present.

HOLE
FORGE

In introducing the *Mousehole* anvil, I quote from part of an article by James Cran. [1]

Anvil making as an industry was started at the Mousehole Forge, Sheffield,England. How long ago, there is no authentic record, but for well over 200 years anvils have been made for the trade at Mousehole Forge, which for two centuries,at least, was the only works of its kind in the world. The first of whom there is a record of having operated the Mousehole Forge in anvil making is the family of Sir John Burgoune. Then Cocshutt & Armitage, and following them came M. and H. Armitage, who operated the forge for over 100 years. The present owners, Brooks & Cooper, have run the works for upwards of 38 years,and are still making anvils that are hard to beat either in quality, shape or workmanship. Although there are now a number of concerns making anvils, both in England and America, the Mousehole Forge is unique in several respects. They were the first to specialize in Anvil making, they were the first to make any attempt at standardizing the shape of the anvil and they operate their forge to this day with no other power than that developed by an old-fashioned water wheel.

For well over a century there has been practically no change made at Mousehole Forge. The same old-fashioned helve hammer,or 'metal helve' as it is locally termed, is still doing duty and is operated by the already mentioned water wheel. The building itself, with its old-fashioned solid stone walls and low arched windows and doorways, shows but slight signs of the ravages usually worked by time, and are apparently good, barring misfortune, for centuries to come.

Originally Mousehole anvils were made by the building-up process, that is, the corners of the base or feet, the horn and the tail or heel were welded on to a center piece. After this the steel face was welded on in sections. The anvil was then trimmed and finished to the desired shape by the use of hand tools. The face was then ground and hardened and after hardening, the face was again ground and the anvil then received the finishing touches. Some idea of how well this work was carried out, may be had by referring to Fig. 2, which shows a group of standard patterns of Mousehole anvils.

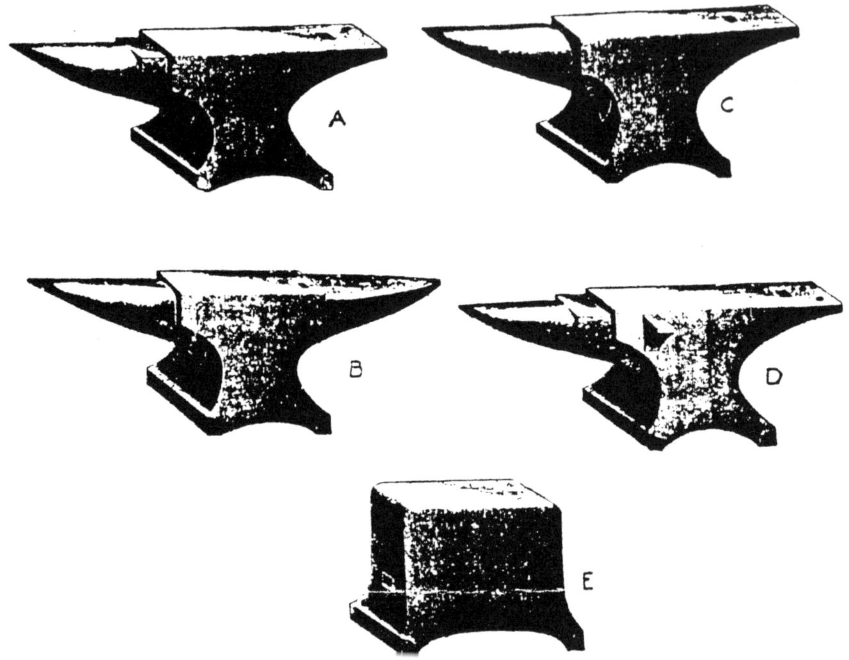

FIG. 2—STANDARD PATTERN MOUSEHOLE ANVILS: A, THE LONDON SHAPE; B, THE DOUBLE PIKED; C, COACHSMITH'S; D, FARRIER'S; AND E, SAWMAKER'S ANVIL

From The American Blacksmith, September, 1914.

Two of the letters are misplaced in "Figure 2." "C" and "D" should be interchanged. The anvil with the "side horn" is the "coachsmith's" and the anvil without a table is the "farrier's."[23]

Undoubtedly, Mr. Cran visited the Mousehole Forge as indicated by the description he gave concerning the architecture of the building itself. This visit probably took place in 1912 or 1913. I do not know where he obtained his information, but I assume that much of what he wrote came from the owners, Brooks & Cooper, and possibly from workmen at the plant as well.

If Mr. Cran is right, and I think he is, the Mousehole Forge was manufacturing anvils by at least 1700 or shortly thereafter, because the *London* pattern was already established by that time and Mr. Cran gives Mousehole Forge the credit for the pattern.

In the summer of 1994 I met an English lady, Judith Anstee, at the historical site where I work part time. She is from the Sheffield area of England (Leeds) and she and her mother were visiting some of the historical sites near where they were staying. I showed them our blacksmith shop and told them of my research on anvils. She took my address and said that she would see if she could find out anything about the Mousehole Forge when she returned home. I forgot about our few minutes of conversation, never expecting to hear from her again. In times past I had written groups concerned with old tool manufacturing in England and had received nothing of real substance on the history of anvil manufacturing. I did not expect someone with no special interest in tools, especially anvils, to come up with any information. How wrong I was!

In January and again in April of 1995, I received material from Ms. Anstee that covers just about everything that is recorded on the Mousehole Forge which is in the Sheffield Central Library, plus other good information. This material seems to contradict some of the information found in Mr. Cran's 1914 article, which I will discuss later.[24] The library material was produced after 1970 from records and from the author's visits to the forge site. Undoubtedly, the writers did not receive any of their information from persons connected with the Forge. The main body of material comes from an article written by Mr. D.T. Smith on the history of the Mousehole Forge,[25] and from an article written by Mr. David Crossley.[26] The following is a time line of operations at the site of the Mousehole Forge, based on these materials and from the dissertation by Julia Hatfield.[27]

1628 - Edward Barber leased "lead mill" property to Thomas Revell of Stamington.

1632 - The site with "two lead mills" was leased by Frederick and Edward Barber to Michael Burton.

1635 - By at least this date the site was being called "Mousehole."

1664 - Edward Barber's will shows that a forge was on the site. It was probably small, the lead mills being the main operation in 1632 and before.

1672 - Barber's trustees sold the forge to George Bamforth II.

1714 - The manor of Wadsley & Owlerton, of which Mousehole was a part, passed to the Burton family, who were connected to the Bamforths by marriage.

1717 - The National List of Ironworkers had the Mousehole Forge producing about 60 tons of wrought iron a year.

1734 - John Cockshutt, "Ironmaster at Worthy Forges," leased Mousehole Forge from the Burtons for 22 pounds per year.

1741 - John Cockshutt was personally managing the Mousehole Forge. Other members of the family were responsible for managing the other forges belonging to them. This was also the first mention that objects other than wrought-iron bars were being manufactured at the forge, possibly even anvils.

1757 - John Cockshutt renewed his lease on the forge, again for 22 pounds per year.

1761 - Iron girders are mentioned, in one account, as being manufactured there.

1762 - William Armitage became Cockshutt's forge manager.

1774 - Anvils being made at Wortly Iron Works, also leased by Cockshutt & Co.

1794 - William Armitage went into partnership with members of the Cockshutt family until about 1816. Anvils are mentioned for the first time as being produced at Mousehole Forge, becoming one of their main products. "C & A, MOUSEHOLE" appears on the side of the anvils.

According to the Sheffield City Museum, by 1500 Sheffield was becoming recognised as a major producer of edge tools and cutlery. To make cutlery anvils are required, good anvils (see Appendix H). Where did the anvils come from? Most likely they were made by the same forges that made the iron and steel for the cutlers. Was the MouseholeForge one of these? They would not have become the major producers of anvils in Sheffield by the last of the 18th century if this were not the case.

1802 - William Armitage purchases land with coal on it.

1814 - Anvils and hammers are listed as the forge's main products.

1816 - William Armitage dies. George and Charles Armitage lease the forge and a third Armitage is made foreman. The Armitage family is sole tenant at this time.

1817 - Charles Armitage dies and in 1819 George tries to sell the lease of the Mousehole Forge, but without success.

1823 - George Armitage dies and the lease is taken over by his sons Morgan and Henry.

1825-1850 - Vises, stakes, tongs and other blacksmith equipment, mason's tools, grinder axles and plate, and other items were added to the forge's product line, in addition to the anvils.

1827 - Morgan and Henry Armitage have taken over the lease. " M & H, ARMITAGE, MOUSEHOLE" appears on the side of the anvils.

1833 - Morgan Armitage dies unmarried and Henry continues to operate the forge.

1842 - Henry Armitage purchases the Mousehole Forge from Lady Burgoyne (a member of the Burton family and widow of Sir John Burgoyne) for 2,100 pounds. In the 1840s Henry purchased other properties, including an additional mill, Grogham Wheels, below the Mousehole Forge on the Rivelin River. A pipeline was run from Grogham Wheels to Mousehole Forge, to help supply air to the forges. Apparently the four water wheels and other machinery required more water than could be supplied by the Rivelin River at that location.

1850 - Henry Armitage dies. Henry Furniss and Thomas Birks manage the forge as trustees of Henry's sons Morgan and George.

1850 - The *Mousehole* anvil is awarded a prize at the Crystal Palace Exhibition, in London.

1864 - A dam of a reservoir breaks and many of the Armitage family are drowned or killed.

1865 - Morgan and George Armitage take over management of the forge.

1867 - Morgan dies and George becomes sole owner.

1875 - George dies childless, leaving the forge to his widow, Annie Armitage, who in turn sells it to William Cooper. " Brooks & Cooper" now appears on the side of the anvil, in place of "M & H Armitage." ApparentlyWilliam Brooks helped finance the purchase of the forge. After 1875, Brooks and Cooper continue to produce anvils and other forged items, for the trades. Brooks and Cooper also retain the rights to use the Armitage trademark as a condition of the sale.

1876 - The partnership between Brooks and Cooper dates from August 1, 1876.

1876 - *Mousehole* anvils are awarded a medal by the United States Commission at the International Exhibition in Philadelphia, Pennsylvannia in 1876 (the United States Centennial).

1880 - *Mousehole* anvils are awarded another medal at the Melbourne International Exhibition in Australia.

1919 - The City Council of Sheffield acquired the Mousehole Forge property and leased it to Brooks and Cooper. The city had tried to acquire the rights for about one hundred years. They wanted to control the water rights to the Rivelin River. However, the Mousehole Forge had held on to those rights from the beginning.

1927 - John Cooper (apparently William Cooper's son) retired and in January, 1928, the firm of Owen - Thomas & Co. took over management of the forge, and according to the authors, continued to manufacture anvils until 1933. Toward the end of the 1920s, records show that the forge was doing a great deal of repair work. Anvil production probably made up only a small portion of their orders by this time. Under Owen - Thomas & Co. the site and products were called "Old Forge," as the name "Mousehole" and trademark "M & H ARMITAGE" were sold to Isaac Nash of Stourbridge.

1933 - Activity at the Mousehole Forge faded away rather mysteriously, so one author stated. Thus ended the production of the last anvils manufactured in Sheffield. It is mentioned that woodworking continued in some of the buildings on the site, into the 1960s. The forge itself stood until 1940, by which time the roof had started to fall in. During World War II, the walls were razed to the foundation and all usable metal removed, most likely for the war effort. (Owen - Thomas & Co. Ltd. actually went bankrupt in 1933 and everyone working at the forge just walked away).

Who are we to believe concerning the history of the beginning of anvil manufacturing at Mousehole Forge? If Mr. Cran is correct, they must have started about 1700 or earlier, as previously stated, although there are no known recorded references to anvil manufacturing at the forge before the 1790s.

The oldest *Mousehole* anvils that I can be sure of have *C & A, Mouse Hole, (Cockshutt & Armitage)* as their trademark. These anvils would date from about 1794 to 1816, and I have recorded two.

As mentioned in a previous chapter, I have located only three *Colonial* anvils with names on them and they are not *Mousehole* anvils.

The Sheffield Directories of 1787 and 1797 do not include Mousehole Forge, although several other anvil manufacturers are listed. However, Mousehole Forge was not in Sheffield proper, but just outside of the city in a rather inaccessible river valley. Other important manufacturers of that time are not listed in the Directory, either. Its very inaccessibility is probably the only reason why any of the ruins remain on the site today. Incidentally, all American wrought anvil manufacturing sites are now parking lots, more or less.

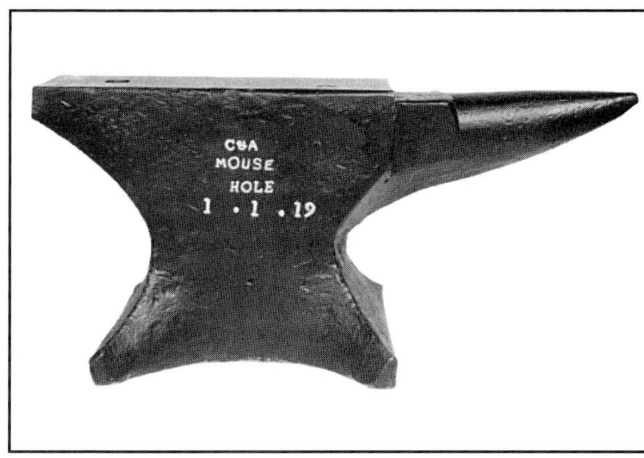

159-pound Mousehole anvil without a pritchel. (R.A.P. photo)

123-pound Mousehole anvil. Notice how the letters and weight numbers are stamped in. It has a pritchel. (R.A.P. photo)

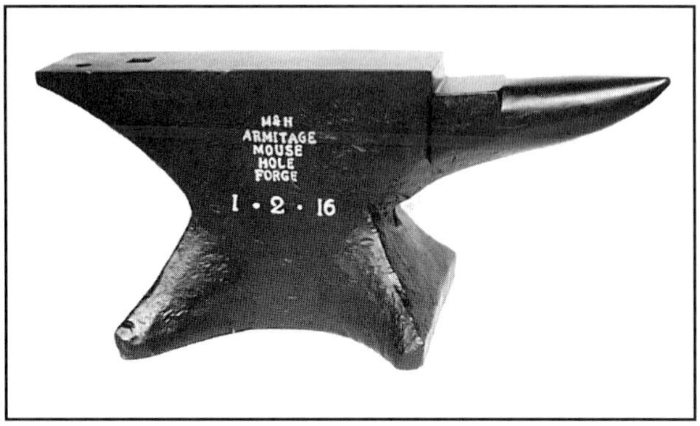

184-pound Mousehole anvil. (R.A.P. photo)

In recording anvils, I have observed that the *London* pattern anvil was standardized at least by 1600, which would support Mr. Cran's contention. Until or unless a *Colonial* pattern *Mousehole* anvil is discovered, everything remains pure conjecture (see Chapter 1, The Dating of Anvils).

As a sideline to this, here is an interesting note. Material that I have received states that the anvil at the blacksmith shop in Gretna Green, Scotland, is a *Mousehole* anvil.[28] I have an old post card that shows the inside of this blacksmith shop and an anvil. The anvil is of the *Colonial* style and on one side is painted "Gretna Green - Smithy - Original Anvil." From the photo, I would date this anvil in the 1600s or at the very latest, early 1700s. If this is indeed a *Mousehole* anvil, it would give credence to Mr. Cran's account of the earlier date for manufacturing anvils at the site (for the story of Gretna Green, see Chapter 7, Anvil Stories).

The name "Mousehole" is also of interest. In *History of the Mousehole Forge,* Mr. D.T. Smith gives his account for the name "Mousehole."[29]

> *The name 'Mousehole' is the old name of the little bit of flat land on the river bend on which, from an uncertain date, some small dwellings and the forge were built. A 'hole' in field names denotes the presence of a place in the river considerably deeper than the average, and no doubt fancy added the 'mouse', descriptive of the old aspect of the river before the making of the Rivelin Valley Road, when Walkley Bank dropped down sheer to the edge of the water just below a very sharp bend; from the path up to Stannington the steep wooded banks seemed to meet overhead and the river to emerge from the hole.*

Postcard from Gretna Green, Scotland. Mousehole anvil? (R.A.P. collection)

One writer suggested that the name Mousehole came from the anvil handling holes, which look much like mouse holes, and were necessary in the manufacture of all wrought anvils before the 20th century. I personally have my doubts about both theories.

As mentioned in the section "Dating Anvils," anvil smiths were well established in

England in the late 1500s. The following is my personal theory on the subject; however, it is probably no better than the others.

In 1596, in preparation for a second attempt to invade England, the Spanish sent a raiding party to the southern coast of England to reconnoiter landing possibilities and to do as much damage as possible. On the Cornish coast in the 1590s was a fishing village named "Mousehole." The following account of the raid is taken from the book, *The Grove of Eagles.*[30]

> *They* [Spanish] *came in by the first light and captured the village of Mousehole. Word came five minutes ago; I turned to claw into my cloths. Every button, every tie took twice the length of an ordinary day. Mousehole! I had never been there but I knew it as a fishing village horseshoed around a tiny harbor.*

This is a historical novel based on fact. The Spanish sacked the town and burned it, then moved inland. In the meantime the English had gathered local forces and were marching toward Mousehole. By the time they arrived, the Spanish had already retreated to their ships. This is most likely the one time in history, after 1066, that England was invaded successfully, even for so short a time.

It is my contention that the anvil and site on the Rivelin River were named to commemorate the town of Mousehole and what happened there. The memory of the event must have been fresh in the minds of the people who started the forge. Normally you do not name a place for an important event that took place ten to twenty years in the past. Incidentally, the town of Mousehole (pronounced"Moussole" or "Mowzel" in the Cornish dialect) was rebuilt and is still in existence today.[31] I am told that it is the residence of a number of artists as well as fishermen, and is considered a tourist attraction. There is also a children's book and video named *The_Mousehole Cat.* These are named for the town. It is possible also that the Mousehole Forge was located somewhere close to the town of Mousehole in the middle of the 17th century and when the owners moved to the site on the Rivelin River, they took the name with them. It is up to you to decide which account you prefer. I am informed by the Hatfields that there was practically no communication or movement from the Mousehole area in Cornwall to the Sheffield area in the early 1700s and before.[32] Therefore, we will probably never know the real origin of the name.

In Mr. Cran's article he states that the only power used at the site was water power, to drive the air to the forges and to operate the helve hammers.[33]

In a brochure put out by the *Ruskin Gallery of Sheffield,* there was a photo of the inside of the Mousehole Forge.[34] Under the photo it states;

> *Anvil manufactured at Mousehole Forge. Taken C 1890, shows the interior of the Mousehole Forge at Malin Bridge. The anvil being faced with steel, the butt (body) of the anvil from the base up to 3/4" from the top is made of wrought iron, a relatively soft material. To insure that the surface is hard and can withstand being worked on, a double shear steel face is welded on to the stock when red hot.*

The photo shows men sledging the face plate onto a large anvil, and other men doing similar work in the background, while others watch. Behind one of the men wielding a sledge hammer I can see what I am sure is a steam hammer. A man is standing before the steam hammer, performing an operation and glancing at the photographer. It does not appear to be a large steam hammer, standing perhaps 10 feet high, but it does show that the forge was using more than water power to produce its products.

Steam requires a boiler. In the year 1907, in papers that concern water rights on the Rivelin River, there appears the statement: "Mousehole forge workings, privately owned, a little steam used."[35] This then indicates that steam probably was in use at the time of Mr. Cran's visit in 1912 or 1913. The author of this statement further writes, "The 'little steam' power noted, may indicate that a small engine was installed, probably at a late period, either to blow the fires or possibly to drive power-tools for finishing processes on the vices, etc., made."[36] In fact, as shown in the photo, the steam was used to operate a power hammer. You would think that the person writing about the use of steam would have found out how it was used. At any rate, some steam power was in all likelihood used continually as long as the forge was in operation after the installation of such expensive equipment.

Next to *Peter Wright, Mousehole* anvils are the most numerous of the English anvils to be found in this country.[37] They were also the only anvil manufacturers who continued to export anvils manufactured by the built-up process after 1852, and were still able to continue to manufacture anvils.[38] What really amazes me is that the Mousehole Forge could produce so many anvils, when all they had were two water-wheel powered helve hammers to do the majority of the work.[39] By late in the 19th century, probably after Brooks & Cooper took over, the anvils were manufactured in three pieces by what became known as the *Peter Wright* method. However, the anvils retained the *Old English* foot pattern. Quality and reputation must have maintained the anvil's popularity, even before conversion to the *Peter Wright* process.

Dating Mousehole Anvils

The majority of the *Mousehole* anvils that I have recorded (over 100) are marked *M & H ARMITAGE.* When you consider that the Armitage family operated, then owned the forge from the 18th through the late-19th century, you would expect that to be the case. The two earlier anvils with *C & A* markings (Cockshutt & Armitage) do not have pritchel holes and are of the *Old English* style rather than the older *Colonial* style. I would date these two anvils in the late 1700s or early 1800s. I also have recorded ten *M & H ARMITAGE* anvils without prichel holes, and these date after or about 1816, but before 1830. The drawings of trademarks at the end of this section will give a more accurate indication of dates according to the trademarks used by the Mousehole Forge.

If my theory in the section on "Dating Anvils" is correct, anvils with the 5th foot and an "S" stamped over it would be the oldest *Mousehole* anvils, and would date from the very early 1600s. However, this is subject to change if future research or evidence prove otherwise.

Identifying Mousehole Anvils

Like all English manufactured anvils, the *Mousehole* anvil is flat under the base. The "stone" weight on every *Mousehole* anvil that I have seen and recorded is just below the trademark, with deep punch marks between the digits, as shown **{1• 1• 1}**. On some anvils only the dots can be found. The underside of the horn comes to a fairly sharp edge, as it extends toward the body. This sharpness under the horn is a characteristic feature of all of the *Old English*- style horned *Mousehole* anvils. Most of the other manufacturers tended to round off the underside of the horn. As for the shape of the *Mousehole* anvil, from the *C & A* anvil on, they remained basically the same until Brooks & Cooper bought the forge. On earlier anvils, under *C & A* and *M & H* trademarks, the anvils have a wider face in proportion to its length than the later ones. After Brooks & Cooper took over, the face became longer and narrower. Most of the earlier *M & H* anvils have some type of markings on the front of the foot, under the horn, such as letters, numbers, asterisks, and so on. These are probably inspector's marks. On one of the *Cockshutt & Armitage* anvils I have recorded, there is a *C & A* on the front of the foot as well as on the side.

Toward the end of the 19th century when the Forge changed over to the three-piece anvil process, a fourth handling hole was added to the front of the foot under the horn. Serial numbers were also added at this time on the front of the foot, but the feet remained the *Old English* style.

I have a theory concerning the manufacture and marking of these later anvils. Generally speaking, as far as the anvils that I have recorded are concerned, the anvils marked *Brooks & Cooper* have lower serial numbers than the later model *M & H, Armitage* anvils. This is my theory: after Brooks & Cooper began placing their names on their anvils sales began to fall off, as commercial outlets tended to be slow in accepting what appeared to be a new anvil brand. As mentioned in the time line, Brooks & Cooper retained the copyright to the *M & H, Amitage* trademark when they purchased the Forge. Therefore, I suggest that they went back to the *M & H, Armitage* trademark after sales had slumped. *M & H* had been a name of quality for almost 50 years, and apparently, the new name *Brooks & Cooper* was not accepted.

The latest advertisement that I have found (1895) and all earlier ads refer to the anvil as *Armitage Mousehole.* However, Owen - Thomas & Co. could not use the *M & H, Armitage* trademark. It is likely that they did not manufacture many anvils, either. Old trademarks of quality products have always been good for sales.

I have recorded one *M & H, Armitage* anvil that looks very much like the *Hay-Budden,* made after 1908. It has the modern style feet and except for the name it would have passed for a *Hay-Budden.* I think this anvil is one of the experimental models that also did not sell well, as I have only found one of its kind.

Brooks & Cooper trademark on a Mousehole anvil. (Jerry Gier photo)

79-pound Mousehole anvil with modern feet. (R.A.P. photo)

The highest serial number that I have recorded is 27274. Could the forge have manufactured over 27,000 anvils after 1875?

In my opinion the *Mousehole* anvils with "England" as part of the trademark, were manufactured after 1910.[40]

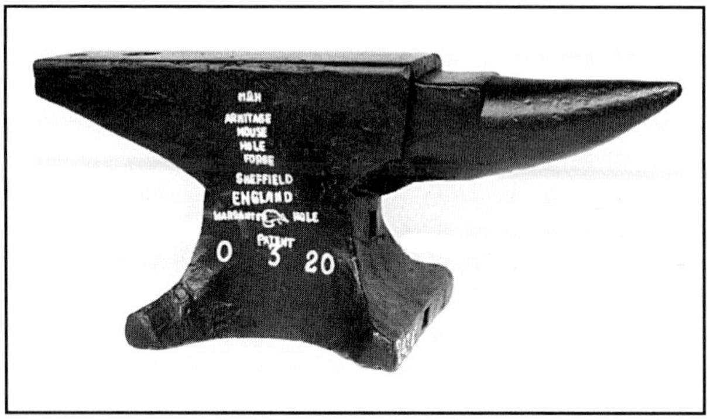

104-pound late model Mousehole anvil. Serial # 24342 with England as part of the trademark. (R.A.P. photo)

I recorded one anvil with the" M & H" upside down, thus:

H ⅋ W

Listed as follows are the trademarks used by the Mousehole Forge, from the oldest to the latest, which I have recorded. Early in 1995 I had a phone call from a friend stating that someone informed him that they had a *Mousehole* anvil with *Henry* instead of "M & H" stamped on its side. I do not count this as a trademark, as the information may not be valid.

The dating of *Mousehole* anvils by their trademarks is subject to change as new information is obtained.

{0·0·0} = weight marks

S

ca. 1600 - 1725

MOUSE
HOLE

0·0 ·0

ca. 1725 - 1780

C & A
MOUSE
Hole

0·0 ·0

ca. 1780 - 1820

M & H
ARMITAGE
MOUSE
HOLE

0·0 ·0

ca. 1820 - 1875

M & H
ARMITAGE
MOUSE
HOLE
FORGE
0·0 ·0

ca. 1830 - 1875

M & H
ARMITAGE
MOUSE
HOLE
FORGE
HOLE
WARRANTED
0·0 ·0

ca. 1880 S/N 49?4

BROOKS & COOPER
MOUSE
HOLE
FORGE
SHEFFIELD
WARRANTED
HOLE

ca.. 1878 S/N 2194

M & H
ARMITAGE
HOLE
FORGE
SHEFFIELD
0·0 ·0

ca. 1895 S/N 8547
Modern Feet

M & H
ARMITAGE
MOUSE
HOLE
FORGE
SHEFFIELD
ENGLAND
WARRANTED
HOLE
PATENT

ca. 1911 S/N 24342

M & H
ARMITAGE
MOUSE
HOLE
FORGE
SHEFFIELD
WARRANTED
HOLE
PATENT
0·0 ·0

ca. 1879 S/N 3561

M &H
ARMITAGE
MOUSE HOLE
FORGE
SHEFFIELD
WARRANTED
PATENT
HOLE
0·0 ·0

ca. 1896 S/N 9106

ANVILS IN AMERICA

Various Mousehole ads and ads where Mousehole is mentioned

ANVILS.

Armitage's Mousehole; the only Warranted Wrought Anvil,........$ per lb.
Peter Wright's Anvil,.. "

Oldest ad 1870, J.M. Warren and Company, Troy, New York, catalog. (From Hunter Pilkinton)

ANVILS—Wrought Iron, Steel Face.

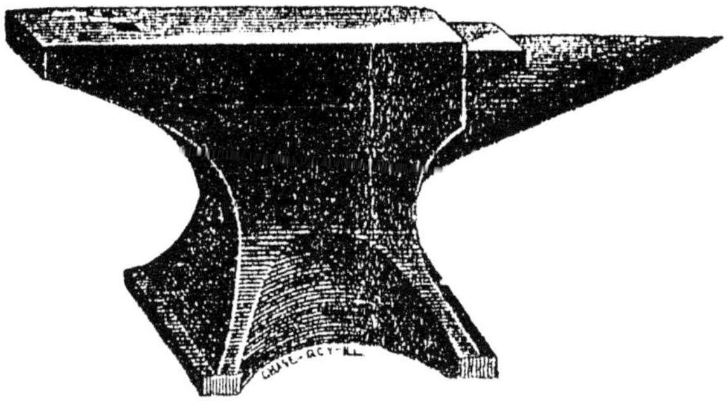

Peter Wright's, per pound.. $0 18
Armitage, or Mouse Hole, per pound.. 18

Anvils over 250 lbs., 1 cent per lb. extra.

EAGLE ANVILS—Cast Iron, Steel Face.

No.	0	1	2	3	4	5	6	7	8	9
Weight, pounds	10	15	20	30	40	50	60	70	80	90
Price, each	$2 25	2 75	3 25	4 00	4 50	5 25	6 00	6 50	7 25	8 25

Anvils weighing 100 to 600 lbs., 10 cents per lb.

Ad from N.B. Nelson Manufacturing Company, St. Louis, Missouri, 1884-1885 catalog. (From Mr. John Catchings)

English Wrought Steel

Wrought First Quality, Crucible Wrought Steel, weights 85 pounds and under
" " " " " " " from 90 to 250 pounds
" " " " " " " over 250 pounds
Wright's SecondQuality, weights from 90 to 150 pounds
Armitage, Wright's pattern, Solid Cast Steel, weights from 90 to 150 pounds

Notice-On foreign anvils the only guarantee is the brand. The word "Warranted" stamped on the anvil does not bind the seller to make good any imperfections. The refusal of our Government to refund the duty is one reason why the English makers warranty virtually amounts to nothing.

Ad from W. B. Belknap and Company, Louisville, Kentucky, catalog, 1892. (From Mr. Hunter Pilkinton)

ANVILS.

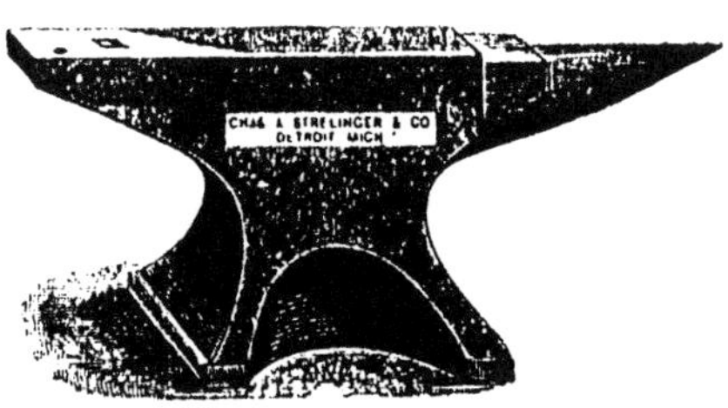

FIG. 284.

In wrought iron anvils, the Wright, Wilkinson and Armitage "Mousehole" are the best known; they are all English made. We believe there is but one American concern, and their output is small. Have been informed (by the house who sells them) that they are equal to the best English.

The price has been in the neighborhood of twelve cents per pound, varying from one-half to one cent more or less for a great many years.

Ad from Charles A. Strelinger and Company, Detroit, Michigan, catalog 1895.

The followiong three pages are from Mrs. Julia Hatfield's unpublished dissertation and were found in the old office building of the Mousehole Forge when the Hatfields bought the property in 1983. The property had been vacant for fifty years.

Page from advertising brochure ca. 1890. Notice the edge where a mouse has chewed.

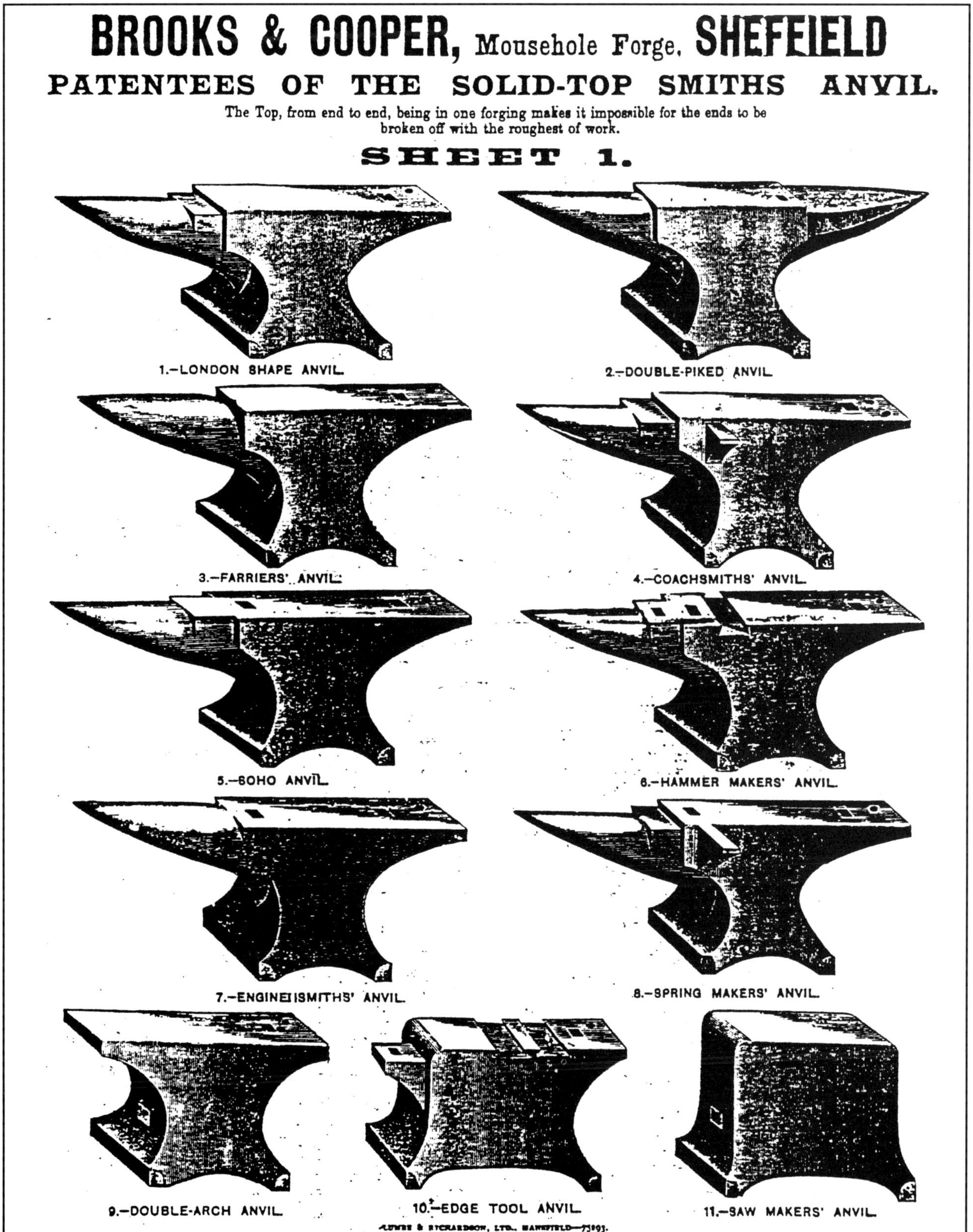

Some of the anvil patterns produced by the Mousehole Forge.

Mousehole Forge
Sheffield 18 Aug 1892

BROOKS & COOPER,
SUCCESSORS TO
M. & H. ARMITAGE & CO.,
MANUFACTURERS OF
GENUINE MOUSEHOLE ANVILS,
VICES, HAND, SLEDGE AND ENGINEERING HAMMERS,
GRINDERS' AXLES & PLATES, MASONS' & EXCAVATORS' TOOLS;
Also of Best Scrap Iron, Shear and Cast Steel, and all kinds of Forge and Millwork.

CORPORATE MARK — HOLE — ESTABLISHED OVER A CENTURY.
TRADE MARK — M. & H. ARMITAGE. MOUSE HOLE FORGE.

NONE ARE GENUINE BUT THOSE MARKED.

No.

ARMITAGE MOUSEHOLE ANVILS.

THE TOP IN ONE FORGING FROM END TO END.

Smiths' ANVILS, above 1 cwt. and under 4 cwt. @	25/-	per cwt.
" " ¾ to 1 cwt.	27/-	"
" " ½ to ¾	31/-	"
" " Small, 30 to 40 lbs., 13/- 40 to 50 lbs., 14/- 50 to 56 lbs.	15/-	each.
" " If Double Piked	1/-	per cwt. extra
Saw ANVILS,	28/-	"
EDGE TOOL and FILE ANVILS	32/-	"
All ANVILS above 4 cwt. and under 6 cwt.	2/-	" extra.
COPPER or SILVERSMITHS' STAKES, POLISHED IRON and STEEL	1/-	per lb.
Do. Do. Solid STEEL	1/6	"
COOPERS' and SMITHS' STAKES, IRON and STEEL	-/4	"
ANVIL STANDS and SWAGE BLOCKS	9/6	per cwt.

QUOTATIONS FOR ALL KINDS OE ANVILS MADE.

STAPLE VICE, Solid Box, made from Scrap Iron, 40 lbs. each and above	32/6	"
Do. Second Quality	31/-	"
Do. With Spherical Washers	2/6 Ea	extra.
Parallel VICES, 3 4 5 6 7 8 inch JAWS. each. 27/6 33/- 44/- 53/- 70/- 80/-		
VICE BOXES and PINS	-/8	per lb.
TUE IRONS, 12 to 16 ins., 1/-; above 16 to 20 ins., 1/1; above 20 to 24 ins.	6/3	per inch.
FORGE and TILT HAMMER HEADS	74/8	per cwt.
HURSTS and TAIL HOOPS	65/4	"
GRINDERS' PLATES (Rough) to 10 in., per pair: 5/- 12 to 23 in., 20/- 24, 22/- 25, 24/- 26 ins.	26/-	"
" AXLES	25/-	"
" NUTS and WASHERS	45/-	"
" AXLES and PLATES (Finished)	50/-	"

Terms: Monthly = less 2½%

ALL KINDS OF REPAIRS DONE ON THE SHORTEST NOTICE.

The cost sheet of Mousehole products as of August 18, 1892.

Sheffield Local Register:
Decease of Mr. H. Armitage.Mousehole Forge.May 19th.1850.Aged 45.
Decease of Mr.Thos.Armitage.Mousehole Forge.May 13th,1852.Aged 65.

The same ad as on page 92, but found about 1940 by C.H. Lea when he was taking photos of the site. Notice that a mouse worked on this one too. (Sent by Ms. Judith Anstee)

PETER WRIGHT

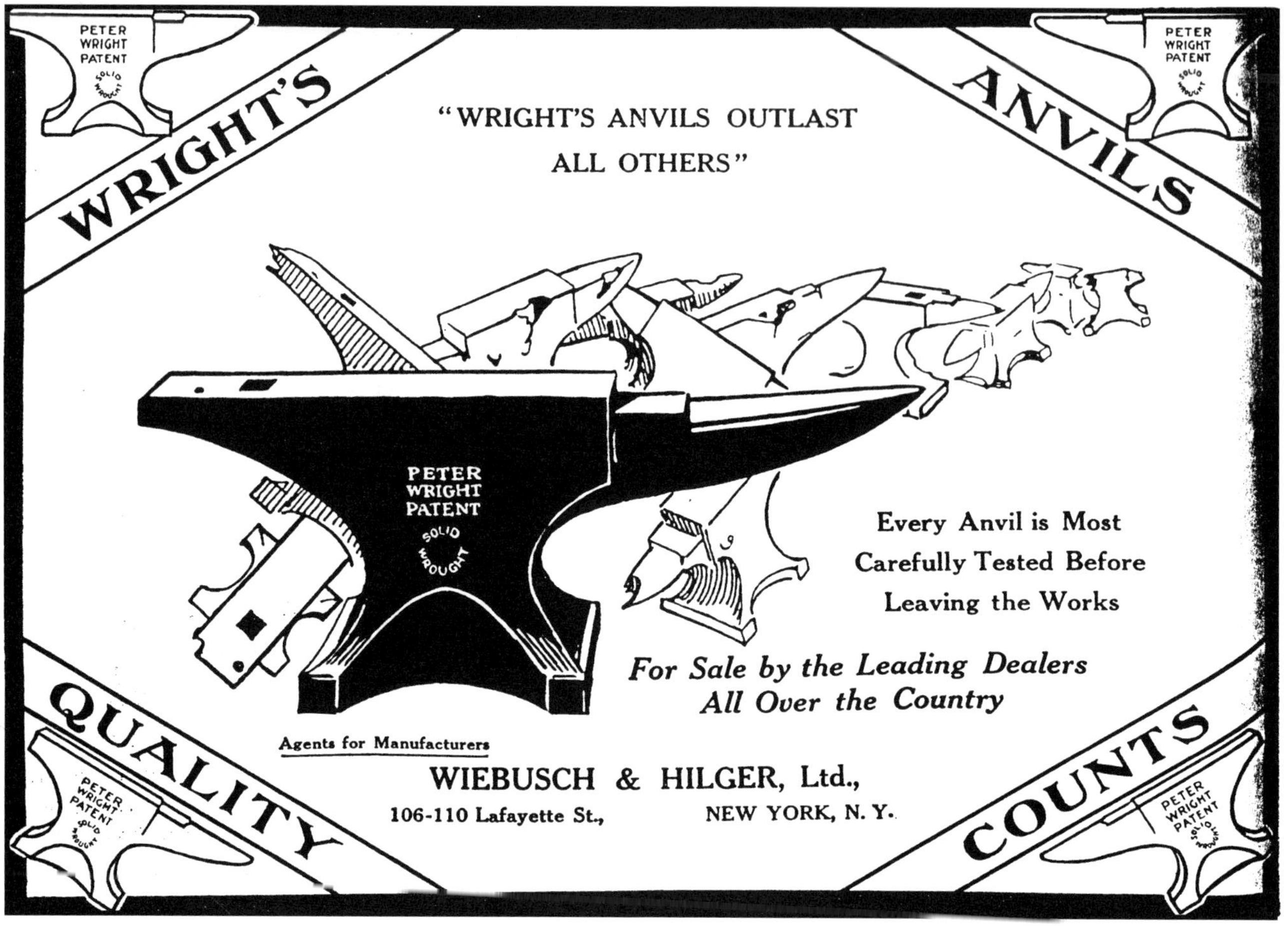

From The American Blacksmith, January, 1910.

What I know of the history of the *Peter Wright* anvil comes from a couple of articles written in American blacksmith journals and from the anvils I have recorded. I have not, to my knowledge, recorded a *Peter Wright* older than about 1830, although any number of the *Colonial* anvils could have been made by Peter Wright & Sons of Dudley, England. Incidentally, Dudley is very near Burmingham, England, a section known as the Black Belt because of all the coal smoke caused by heavy industry.

The following article is taken from the journal, *The Blacksmith and Wheelwright,* July, 1904, entitled, "The Manufacture of Anvils in Great Britain."[41]

> *The Hardware Trade Journal of London, England, recently published an account of a visit by a reporter of that paper to the Constitution Hill Works, Dudley, where Peter Wright & Sons have for over two hundred years been engaged in the manufacture of anvils and vises.*

'We reproduce this article with illustrations as likely to be of interest to American blacksmiths, large numbers of whom use Peter Wright anvils.'

The Hardware Trade Journal says:

'The name of Peter Wright & Sons, of Dudley, is so extensively known in connection with blacksmith's anvils and vises that it will need no introduction from us. To have a reputation as an old-established house is, in these days of new inventions, a doubtful blessing. In many cases the fact that a firm was established twenty, fifty or one hundred years ago only goes to prove to the mind of the outside world that their methods are obsolete their goods old-fashioned, and that for really up to date service a concern of more recent origin must be found. This is one side of the question.

This view, however, is not always correct, and there are many hundreds of instances throughout these islands that point the other way. In the case of Peter Wright & Sons, every year of the two hundred during which they have been in existence has seen modern improvements, superior machinery and scientific methods introduced into the production of their manufactures. It is therefore correct to say that every anvil at present leaving Constitution Hill Works is not only the outcome of the very latest in the way of scientific invention, but that each of those articles has behind it the whole of the experience of a couple of centuries. It will undoubtedly, therefore, be of interest to give brief description of the process of manufacture as it is carried on today.

The firm are [is] turning out anvils of a finer quality than ever before, and worthily maintain the reputation which they have secured in every part of the world.

It is interesting, in passing, to note that this is one of the branches of industry which we appear still to retain, in face of much talk of foreign competition. For example, in the United States of America there is no better known blacksmith's anvil than that manufactured by this house. The same remark would apply to almost any quarter of the globe.

In the old days all anvils were 'built up', that is made by welding together a number of separate pieces onto a central block of common iron. Each of these vertical butt welds was, of course, a grave source of weakness, and such anvils frequently broke up. Under the firm's patent, which was the invention of Peter Wright (the great-grandson of the founder), all their anvils are made from only two forgings, which are called the 'top' and the 'bottom' respectively, and requires only the one weld at the waist, even this being, in addition, carefully 'tied' in.

The finest wrought iron scrap is carefully picked over by hand at the forges, and then passed into large furnaces, where it is either balled or piled, and then thoroughly worked under powerful steam hammers into blooms. After going through reheating furnaces these are forged with great care under heavy steam hammers, so as to form the tops and bottoms from which the anvils will be made.

These forgings are now paired together, and owing to their design, the greatest strength and working surface is secured for each anvil, according to its weight. These are then handed over to the anvil makers who have been trained to their respective duties all their lives, and have become specialists at their work. This trade is very largely hereditary, and many of the men are representatives of families that have worked for Peter Wright & Sons for generations. These forgings now undergo a further process of hammering and trimming, which contributes great toughness and soundness. They are then welded together at the waist under powerful hammers, which insures that absolute homogeneity of the whole anvil, which is of such vital importance.

The next step is to weld on the steel for the face, and this, as may be imagined, is a delicate and most important process. The steel employed is of the finest grade, combining all the different qualities necessary for the heavy work that an anvil face has to stand. The firm is constantly experimenting by means of chemical and physical tests, in order to secure the very best material for their purpose. The face of the wrought iron body is heated up to [the] welding point in a large smith's hearth, and the steel slab in another special hearth, both being in charge of foremen, who always superintend this process. Specially washed fuel is used, and a true weld is made without the aid of any patent compounds or mixtures. To make a sound weld of the face of an anvil of say, 700 pounds, is of course, a really fine piece of work, requiring the most careful methods and highest degree of skill. The anvil now has the bick and table trued up, the holes put in, and the finishing touches made, after which it is heated to the correct temperature and plunged into great hardening tanks, where streams of water give it the final temper. Long experience has determined the degree of hardness which proves the best and safest to insure good, all-around wear.

The final process is to grind the steel face, which is done on huge grindstones, and the anvil, after being polished, is passed on to the warehouse.

Throughout this time the anvil has to undergo various exhaustive tests at the hands of one of the principals, for hardness, squareness and freedom from any flaws. Constant practice enables a trained ear to detect even small defects by the ring given out, and it is all this care which causes a Peter Wright Patent Solid Wrought Anvil to ring like a bell. After approval, the trademarks are stamped on every anvil, these consisting of the words 'Peter Wright Patent' and also 'Solid Wrought' (the latter being in the form of a circle). In addition to these the green label is affixed, this also being duly registered. The manufacturers state that large orders are constantly being shipped to all parts of the world in [the] face of tariffs and high freights.

Our illustrations show in Fig.1 the process of the top and bottom forgings which compose the anvils being welded together. Fig.2 is a photo showing the anvil as it is nearing completion and having the finishing touches put on by the hammerman. Fig.3 illustrates a blacksmith's anvil in the finished condition.

Wiebusch & Hilger, 9 to 15 Murray street, New York, are the agents for the manufacturers in this country.

Unfortunately the copy of this article that I have is not good enough to reproduce the photos (Figures 1,2,3). Wiebusch & Hilger limited, importers, 9 to 15 Murray St. New York, New York, incorporated January 19, 1888, and voluntarily dissolved June 27, 1932.[42]

Another article in the same journal dated July, 1909, essentially covered the same material, but I have not secured a copy of it yet.

There are a number of mistakes and misleading statements in the previous article, but before I critique the article I will present part of another article about *Peter Wright* anvils, taken from a larger article entitled, "History and Development of the Anvil." [43]

> *For many years, piling or building up was the only known process of making anvils, and it was not until Peter Wright, an anvil manufacturer of Dudley, England (at one time a workman at Mousehole Forge), conceived the idea, and patented it in England, of making anvils in two pieces. This happened about the middle of the last century and proved an epoch in anvil making. Under the new process of making anvils, a great deal of the fame and business that had for so long centered around Mousehole Forge was diverted to the works of Peter Wright, and the result was that the Peter Wright anvil became the most popular in the market and is still well known in all quarters of the world.*
>
> *The Peter Wright process, which is almost invariably used in the manufacture of anvils, may be described as follows: The base or bottom part is formed by piling up scrap iron and welding it into a solid mass. While still hot it is placed in a die and blocked to shape; the power used being steam The whole of the upper part, including the horn and tail, are forged in one piece from scrap iron, the greater part of the shaping being done at the same heat at which the scrap is welded. Next the steel face welded on , the hardie and pritchel holes punched, the tail or heel squared and the horn finished to shape. The two pieces, base and upper part, are then of about the shape shown in Fig.3 and are ready to be welded together at the waist. This is done by jumping. The two pieces are heated in an open fire; when the welding temperature has been reached the two pieces are placed as near as possible in alignment under a steam hammer and a few blows join them solidly. The steam hammer is then used as a vise to hold the anvil while the rough burs around the edge of the weld are hammered in. The anvil is then thrown on the floor where it can easily be turned in any position or direction and finished as far as forging goes, with hand tools.*
>
> *For a long time the steel faces of anvils, even under the Peter Wright process, were welded on in sections of about 6 inches in length. Thus it will be seen that usually two and sometimes three pieces of steel were required for an anvil face. Now, the whole face, regardless of size, is welded on in one piece; this is an American innovation and will be referred to later.*
>
> *...There is still another English anvil, namely, Wilkeyson's, that ought to be mentioned, but it is practically the same as the Peter Wright anvil and manufactured by the same process. It is needless to go into details.* [Author's note: "Wilkeyson's" is the wrong spelling.]
>
> *Up to 1886 these three English concerns mentioned* [Author's note:

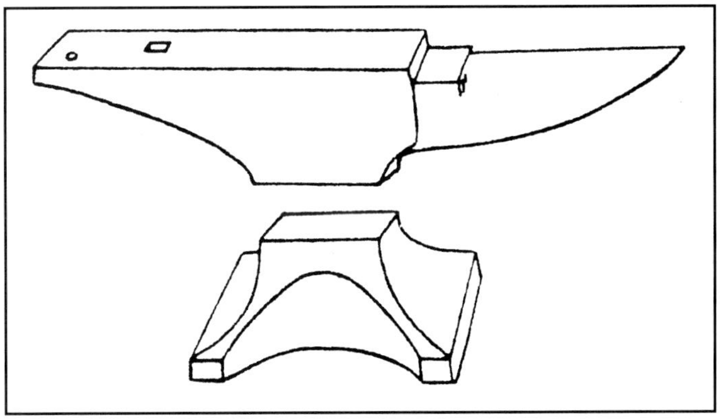

Figure 3.

Mousehole, Peter Wright, and Wilkinson] did practically whole of the anvil *business of the world, and kept large forces of anvil smiths, helpers and finishers employed.*

If you noticed, there is some confusion between the 1904 and 1914 accounts as to the founding of Peter Wright anvil manufacturers. The 1904 article states that Peter Wright & Sons had been in business for 200 years, placing their beginning about 1704. It also states that Peter Wright was the great-grandson of the founder. However, if Peter Wright the "first," assuming he was Peter Wright, started the business about 1704, then we would have to conclude that he was at least in his late twenties or early thirties. It seems incredible that there were only three generations between the founder and the "inventor." Also as mentioned in the Cran article, why would Peter Wright the "inventor" work for the Mousehole Forge when he could have learned the same trade at his father's forge in Dudley? Incidently, Peter Wright was not the inventor of the two-piece anvil, as stated. He patented a process, but in 1820 a Mr. Richard King of England invented the two-piece anvil. The following account was located by Mr. David Shayt in the archives of the Smithsonian Institution where he is a specialist in Crafts & Trades (1989). He forwarded it to me. The front cover of the booklet reads as follows:[44]

Transactions of the Society, Instituted at London for the Encouragement of Arts, Manufactures,and Commerce; with the premiums offered in the year 1820.

Vol. XXXVIII, London. sold by the housekeeper, at the society's house in the Adelphi; and by all booksellers. Price Fifteen Shillings MDCCCXXI, [Author's note:1821].

No. IX

Improved Construction of Anvils

The smaller, or Vulcan Silver Medal and Five Guineas were this Session voted to Mr. Richard King, of Thames Street, Bankside, for his improvement in the Construction of Anvils. The following communication has been received from the Candidate, and Models of the Invention have been placed in the Repository of the Society.

8 Thames Street Bankside *22nd March, 1820*

'Sir;

Having long experienced the difficulty attending the making of sound anvils (used by smiths in all departments of the trade) by the present system and the frequency with which the quarters and beak irons break off, I am induced to request that you will please to lay before the Society for the Encouragement of Arts, Manufactures, and Commerce, the model of an improvement suggested by me, in the manufacture of that article, which will alleviate the frequent occurrence of breaking in those manufactured in the common way; and of which I also send a model, to show more clearly the nature of my improvement.

If the Society should be pleased to consider it worthy their Countenance and support, I shall feel myself honored, and will attend to explain whenever they may be pleased to take it into consideration.

I am ,Sir,
&c.&c.&c.
Richard King

A. Aikin, Esq.,

Secretary, &c.&c.'

The usual way of making large anvils is, first to forge a nearly cubic mass for the body of the anvil, and then to unite the beak and quarter with this block, by means of plain vertical surfaces, secured by welding. The feet are also welded on in the same manner. The face is made of steel, and is laid on the top of the body of the anvil and of the quarter. The several parts above mentioned are shown in Plate VI, fig.5. It is obviously, however, impossible, on this construction, that the welding should be effectual for more than an inch or two from the external line of junction: hence a succession of heavy blows on the face of the anvil will necessarily sooner or later cause these vertical joints (especially of the beak and quarter) to give way.

Mr. King follows a very different process. He forges the body as well as the beak and the quarter of one piece of iron, and the foot or support of another; these are united by means of a horizontal plane joint and welding, as shown fig.6, from which it is manifest that the usual causes of the destruction of anvils are avoided, with the farther advantage of increased simplicity in the manufacture.

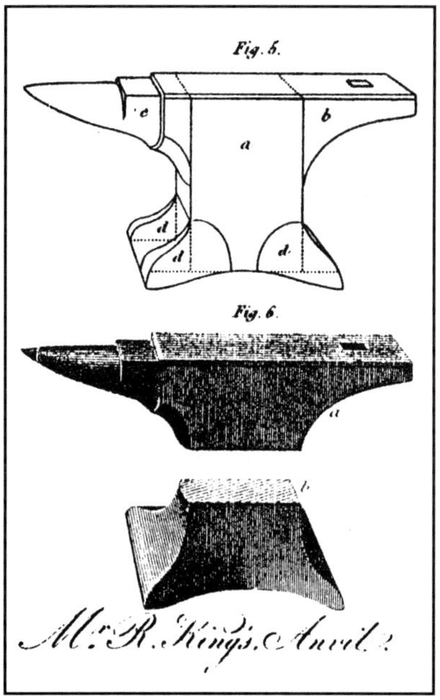

Figure 5 & 6.

The article was not clear enough to reproduce, from the original except the figures 5&6.

It appears from Mr. King's letter to the Society that he was either an anvil maker or worked for one. This was 32 years before Peter Wright came out with the two-piece process.

Apparently Mr. King's process was used because a treatise published in 1842 in England on the"Manufactures In Metal" states in one section;

> *...Some anvil-smiths, instead of thus uniting the beak and quarter as well as the feet, by welding pieces to the block, forge the upper part of the body, including the beak and the quarter, out of one piece of iron, and the lower portion, with its projections, out of another piece, as represented below. These parts, represented in Fig.19, are then welded together by the horizontal plane, in consequence of which mode, by avoiding all vertical junctures, the anvils are said to be less liable to fracture, as well as being by the course much simplified in the manufacture....*(see Appendix H).

ANVILS IN AMERICA

This sounds like it was taken from Mr. King's letter. It would be amazing if an *Old English* anvil was found with "R. King " or "Richard King" stamped on the side. In any event, two piece anvils were being produced ten years before Peter Wright's "invention." It is possible that Peter Wright was also producing two-piece anvils some time before he applied for the patent.

Cran also mentions that welding the face plate on in one piece "is an American invention and will be referred to later."[46] He is referring to Hay-Budden Manufacturing Company, which started making anvils about 1886.[47] However, he is wrong about the face plate being an American innovation. M.T. Richardson in his book *Practical Blacksmithing* did not know Hay-Budden existed when he edited articles from his *Blacksmith & Wheelwright* magazine for this book *Practical Blacksmithing.* [48] He states:

> *The common grade of English anvils and all those of German make weld the steel on in two or three pieces according to the size of the anvil; the best English brand, however, of late years has the face in a single piece of shear steel.*

He must be speaking the of the *Peter Wright* anvil.

Near the end of the article Cran speaks of an anvil name, *Wilkeyson's.*[49] For awhile I was trying to locate a *Wilkeyson* anvil, but I kept finding *Wilkinson* anvils and I thought I was looking for two different makers until I found the following ad in the *Charles A. Strelinger and Company* catalog of 1895.[50]

> *In wrought iron anvils, the Wright, Wilkinson, and Armitage Mousehole are the best known; they are all English made....*

These are the same three anvils mentioned in Cran's article; therefore I was looking for an anvil that did not exist because of the writer's or a printer's error.

Reproduced here is the patent issued to Peter Wright in 1852.

A. D. 1852 ....N⁰ 13,902.

Manufacture of Anvils..

Wright's Specifications.

> ***To All To Whom These Presents Shall Come, I,*** *Peter Wright, of Dudley, in the county of Worcester, Vise and Anvil Manufacturer, send greetings.*

Whereas *Her present most Excellent Majesty Queen Victoria, by Her Royal Letters Patent under the Great Seal Of Great Britain bearing date at Westminster, the Twentieth day of January, in the fifteenth year of Her reign, did, for Herself, Her heirs and successors, give and grant unto me, The said Peter Wright, Her especial license, full power, sole privilege and authority, that I the said Peter Wright, my exors, admors, and assigns, and such others as I, the said Peter Wright, my exors, admors, or assigns, should at any time agree with, and no others, from time to time and at all times during the term of years therein mentioned, should and lawfully might make, use, exercise, and vend, within England, Wales, and the Town of Berwick-upon-Tweed, my Invention of* ***"Improvements in the Manufacture of Anvils;"*** *in which said Letters Patent is contained a proviso obliging me, the said Peter Wright, by an instrument in writing under my hand and seal, particularly to decribe and ascertain the nature of my said Invention, and in what manner the same is to be performed, and to cause the same to be enrolled in Her Majesty's High Court of Chancery within six calendar months next and immediately after the date of the said in part recited Letters Patent, as in and by the same, reference being there unto had, will more fully and at large appear.*

Now Know Ye, that, in compliance with the said proviso, I, the said Peter Wright, do hereby declare that the nature of my said Invention, and the manner in which the same is to be performed, is particularily described and ascertained in and by the following description thereof, reference being had to the Drawings hereunto annexed, and to the letters and figures marked thereon (that is to say):-

According to the plan hitherto adopted of making anvils, these articles are constructed of several pieces, which are formed by forging into the proper shapes, and are afterwards welded together to form a complete anvil. Anvils of the ordinary construction are usually formed of from nine to fourteen pieces, according to the size; and the consequence of constructing them in this manner is, that they very reliable to be damaged in use by some of the numerous joints giving way, thereby allowing portions of the anvil to be knocked off by the heavy blows to which they are frequently subjected. Now the object of my Invention is to manufacture anvils in such a manner as greatly to diminish their liability to be broken in the manner above described; and this object I propose to attain by making anvils in one, or at the most in two, solid pieces, which are formed or moulded in moulds or dies instead of being forged.

In carrying my Invention into effect for making anvils in two pieces, one mode of proceeding is this: - I prepare two hollow dies, one of which is in form the counterpart of the intended form of the upper portion of the anvil (see Fig.1), and the other counterpart of the intended form of the "butt" or lower portion (see Fig.2). These dies or moulds may be made of cast iron. Now having brought a piece of iron suitable for the purpose to a welding heat, and worked it into something approaching the shape of the part of the anvil which it is destined to form, I place it in the corresponding die, and by the application of a series of blows delivered by a forge or other large hammer I cause the mass of heated iron to fill the die, and thus to assume the shape thereof. The same mode of proceeding is adopted both for the upper and lower parts of the anvil, and when the two pieces are completed, I weld them together, so as to form a block having only one joint in it, and then I face and finish the anvil in the ordinary manner.

Another mode of working which I adopt in manufacturing an anvil in two pieces by means of dies is as follows: -Instead of the upper and lower parts of the anvil being made separately, and connected together by welding, as above described, the anvil may be made in two halves, counterparts of each other, as if the anvil were cut through longitudinally, such halves being made in moulds or dies of corresponding shape in the manner already described, and afterwards welded together so as to form a block having one longitudinal vertical junction, as shown at Fig.3. Or otherwise the upper part may be made in a die as first described, and then welded to a block constructed by the old process of forging, the upper part being that which it is most important to form in one solid piece. In making the smaller sizes of anvils I make the whole in one piece. For this purpose I use plates of faggotted or piled iron, and having heated them in ordinary furnace to a welding heat, I work the mass of iron into a shape approaching the shape of an anvil. I then place it in a die of the form shown at Fig.4, and then hammer it into the shape of the entire anvil, and I face and finish it in the usual manner.

It will be obvious that in making anvils in the several ways above described the essential feature of the process is the use of moulds or dies, and the compression of the metal therein. And I must observe, that although by the use of moulds or dies I am enabled to manufacture anvils in one or two solid pieces, with due regard to economy, yet of course I do not mean hereby to limit my Invention to the making of anvils in one or two pieces, the substance of it being, as I have said, the use of dies and compression, and not the particular number of pieces moulded; I accordingly hereby declare that I claim under the above in part recited Letter Patent the mode which I have described of manufacturing anvils, by which I can make them in one or two pieces by forcibly compressing the metal of which they are to be composed in a die or dies, and I do not claim any other of the matters or things above described.

In witness whereof, I, the said Peter Wright, have hereunto set my hand and seal, this Seventeenth day of July, in the year of our Lord One thousand eight hundred and fifty-two.

Peter (L.S.) Wright

Signed, sealed, and delivered by the within-named Peter Wright, in the presence of

John Bolton,
Sol^r, Dudley.
J. H. Smith,
His Clerk.

The patent and patent drawings are courtesy of the Smithsonian Institution.

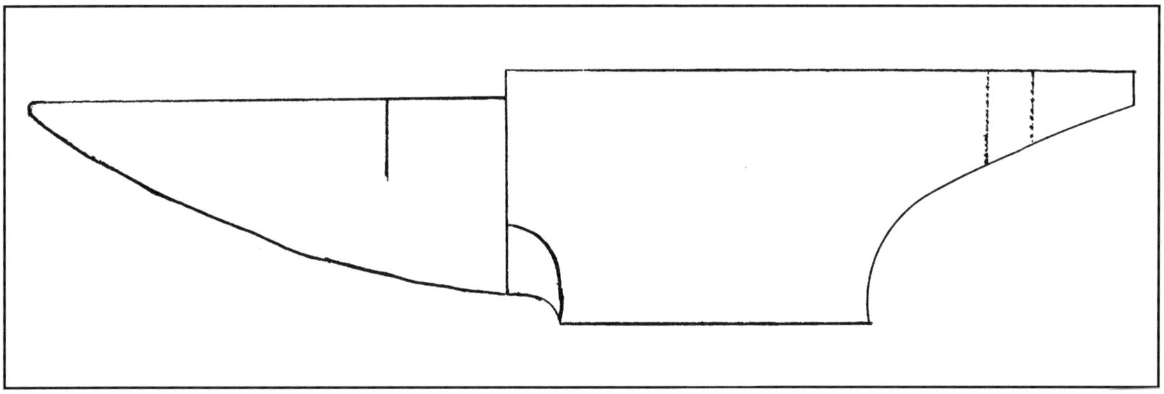

Figure 1.

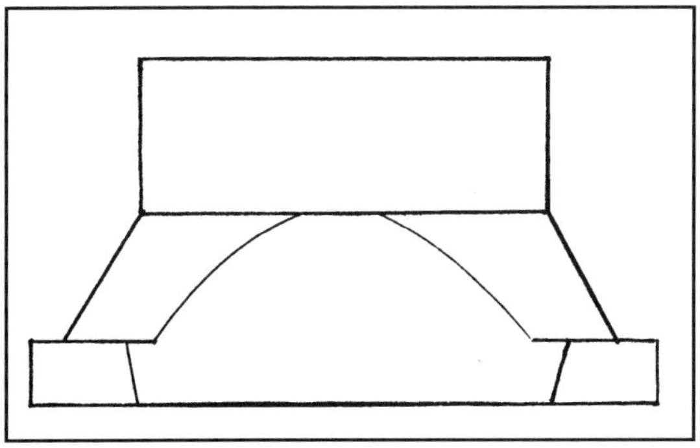

Figure 2.

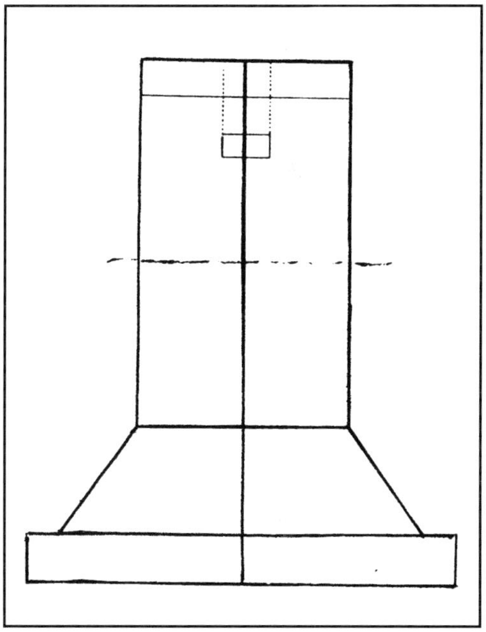

Figure 3.

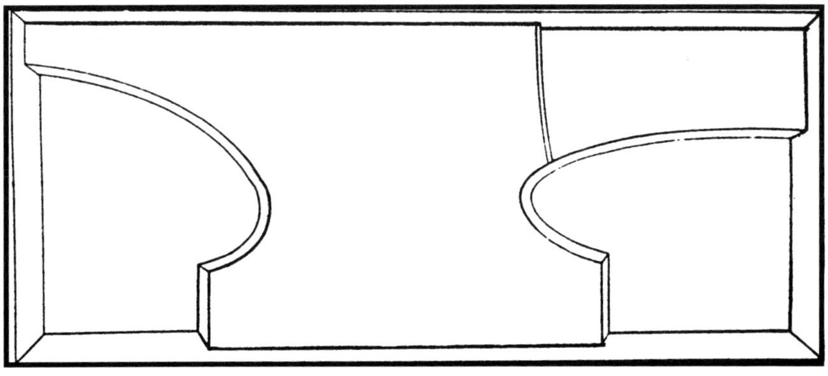

Figure 4.

The patent sounds very much like Richard King's letter except for the use of dies to put the finishing shape on the anvils. From the two descriptions of *Peter Wright* anvil making in 1904[52] and 1914,[53] I do not get the impression that dies were used except very loose ones, that is, a bottom die approximating the correct shape and a flat top hammer die striking the anvil body part being worked in the lower die. I have recorded more *Peter Wright* anvils than any other (probably 200 or so) and have looked at hundreds of others. When their weights are the same they are not consistent by any stretch of the imagination with their measurements, meaning that the body or upper part was worked in open dies by very skilled hammer-men.

Did Peter Wright steal another man's idea? Did he buy out Richard King? Did Richard King get a patent on his idea? Hopefully future research will answer these questions. In any event Peter Wright started making anvils under his own patent and in so doing put most of the other anvil makers out of anvil production in England. Another reason that might have given Peter Wright & Sons an edge on the competition is the use of steam hammers. This hammer was invented about 1830, and it was very likely that Peter Wright put them to use at an early date. It would seem from my recording of English anvils that only the Mousehole Forge and Wilkinson survived as anvil makers after 1852. Let it be remembered that the Mousehole Forge was the oldest continuous anvil maker in England, with a good reputation and loyal customers. Wilkinson not only made anvils, which were but a small part of their business, but made cannons and swords, as well as other products. This is probably what saved these anvil makers when the new Peter Wright patent was issued in 1852. Between the years of about 1852 until the beginning of the 20th century Peter Wright & Sons held in all probability 60 to70% of the wrought anvil market in the United States.

As stated before, I have recorded more *Peter Wright* anvils over the last ten years (1984-1994) than any other make. I may go so far as to say that there are probably more *Peter Wright* anvils left in this country than all other wrought anvils combined. For the past four or

"Peter Wright" Solid Wrought Blacksmiths' Anvils

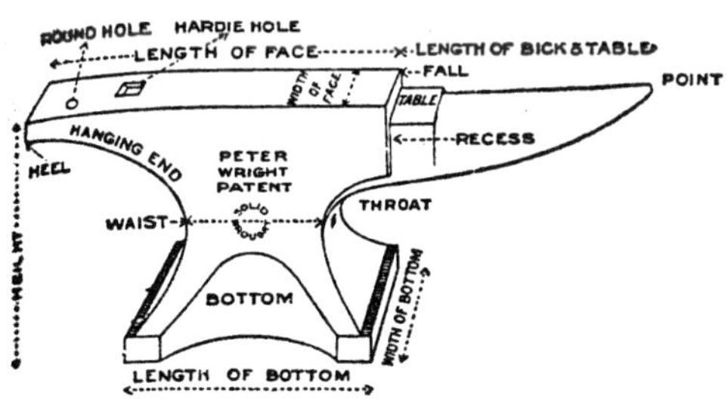

"Known the world over." Made from two solid wrought iron forgings called "top and bottom," which are welded together at the waist of the anvil under powerful steam hammers. The face of the top is heavily steeled with Sheffield steel of a special analysis, carefully tempered to stand the greatest wear. It is generally recognized that an anvil, the body of which is made of wrought iron, gives a more satisfactory resistance than an anvil, the body of which is made of steel, either forged or cast.

Finished anvils are subjected to an additional hardness test by the latest scientific instruments.

Weight Pounds	Face, Inches	Bick and Table, Inches	Square Hole Inches
80	12 x3 1/4	8 1/2	3/4
90	12 1/2x3 1/2	9	3/4
100	13 x3 1/2	9 1/2	3/4
110	13 x3 3/4	9 1/2	7/8
120	14 x3 3/4	10	7/8
130	14 1/2x4	10	1
140	15 1/2x4	10 1/2	1
150	16 x4	10 1/2	1
160	16 x4 1/4	11	1
170	16 x4 1/2	11	1
180	16 1/2x4 1/2	12	1
190	17 x4 1/2	12	1
200	17 x4 3/4	12	1 1/8
225	17 1/2x4 3/4	12	1 1/8
250	19 x4 3/4	12 1/2	1 1/8
280	19 x5	12 1/2	1 1/4
300	21 x5	12 1/2	1 1/4
350	22 x5 1/4	13	1 3/8
400	23 x5 1/2	13	1 7/16
450	24 x5 3/4	14	1 1/2
500	24 x6	14	1 1/2
550	25 x6 1/4	15 1/2	1 1/2
600	26 x6 1/2	16	1 1/2

NET PER POUND **$0.25**

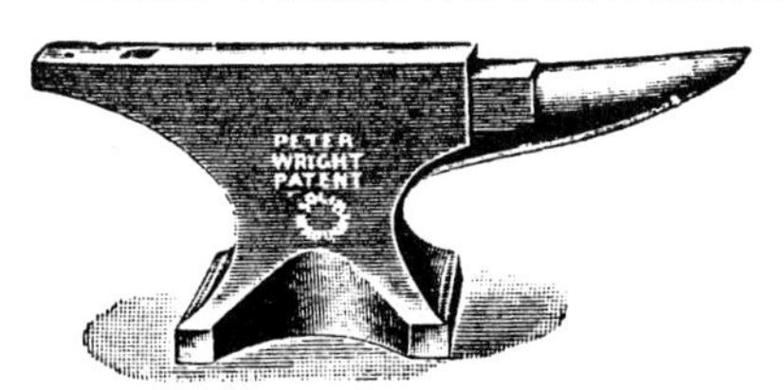

Guarantee

Fully guaranteed as to defects in materials and workmanship. Guarantee does not cover, however, damage arising through ordinary wear and tear, nor for damage caused to anvils through improper testing the face for correct temper with sharp or edged instruments.

Advertisement in the 1937 Master-Carr Supply Company catalog Chicago, Illinois.

five years I have only recorded those *Peter Wright* anvils on which there is something different. It may be well to note that all American wrought anvil makers and those imported copied the Wright process and style, and even the cast anvil makers copied the *Peter Wright* style.

As of this writing it would seem that *Peter Wright* anvils were being imported until the late 1930s. The last advertisement I have found comes from the *McMaster Carr Supply Company* catalog, Chicago, Illinois, dated 1937.

However McMaster Carr, like many other hardware catalogs, often leave an ad in the catalog long after the product is no longer manufactured.

The earliest advertisement I have is in the catalog of *J.M. Warren & Company*, of Troy, New York, dated 1870. It lists the *Peter Wright* and the *Mousehole*, but does not give prices and uses a composite drawing.

In an advertisement in the *W.B. Belknap & Company* catalog of 1892, two qualities of *Wright*

ANVILS.

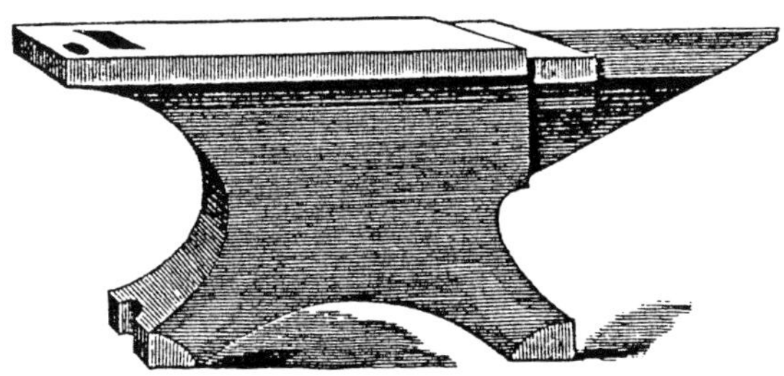

Armitage's Mousehole; the only Warranted Wrought Anvil,........$ per lb.
Peter Wright's Anvil,... "

From the J.M. Warren and Company 1870 catalog, Troy New York. (From Hunter Pilkinton)

anvils are listed. I do not understand this ad, as every other ad lists only one quality, "the best." However, it might be explained by the fact that the ad does not use the name "Peter."

In the February 1891 *The Blacksmith & Wheelwright* magazine a notice was included.[54]

English Wrought Steel

Wrought First Quality, Crucible Wrought Steel, weights 85 pounds and under
" " " " " " " from 90 to 250 pounds
" " " " " " " over 250 pounds
Wright's Second Quality, weights from 90 to 150 pounds
Armitage, Wright's pattern, Solid Cast Steel, weights from 90 to 150 pounds

Notice-On foreign anvils the only guarantee is the brand. The word "Warranted" stamped on the anvil does not bind the seller to make good any imperfections. The refusal of our Government 'to refund the duty is one reason why the English makers warranty virtually amounts to nothing.

W. B. Belknap and Company, catalog, 1892. (From Mr. Hunter Pilkinton)

ANVILS IN AMERICA

The attention of our readers is directed to the announcement on another page of Peter Wright's Patent, Solid Wrought Anvil. The manufactures direct attention in their reputation attained by the anvils has led makers of an inferior anvil to imitate the brand. They caution buyers to see that each anvil is stamped with the full trademark as shown in their advertisement, on one side, and has the green label affixed to the other. These anvils can be obtained from all the principal hardware dealers in the United States.

This following advertisement in the May, 1893, issue of the same magazine is what this statement is referring to. I do not have the 1891 *Peter Wright* advertisement.

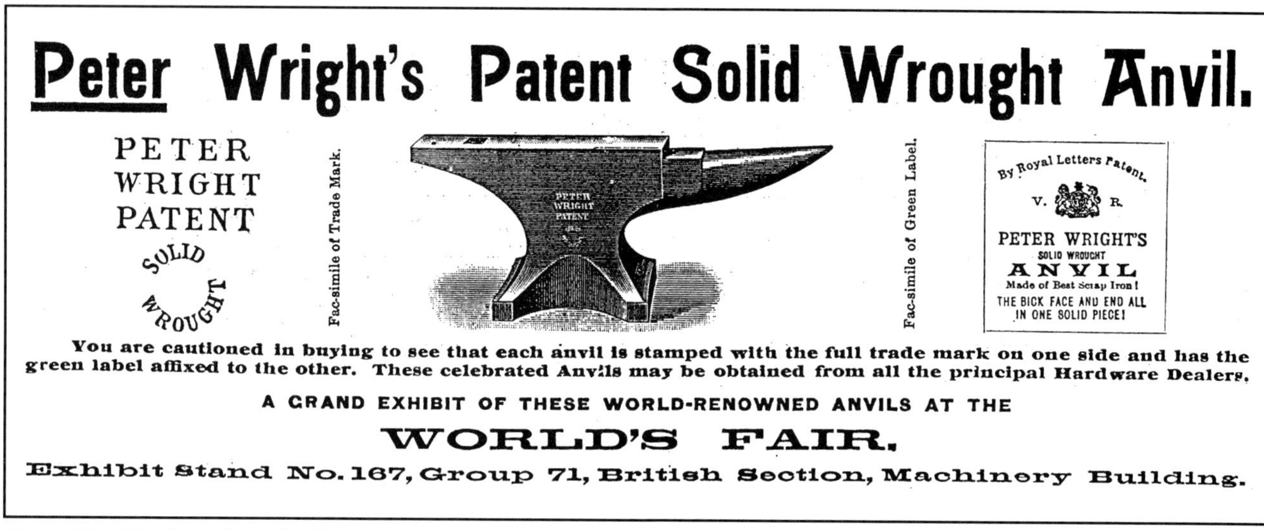

From The Blacksmith and Wheelwright, May 1893.

This imitator had to be the *Henry Wright*. I have recorded a number of these anvils and without the name *Henry Wright* on the side, they look exactly like the *Peter Wright*, including the "Solid Wrought" in a circle (see *Henry Wright*). Could this be the reason that only the word *Wright* was used in the 1892 advertisement, and was *Henry Wright* considered "Wrights Second Quality," anvil?

Identifying *Peter Wright* Anvils

The only way to identify *Peter Wright* anvils made before 1852 is to find the name on them, because they look like all other *Old English* anvils. Between 1830 and 1852 the trademark looked like this*: P. Wright, Warranted.* The weight was on the other side.

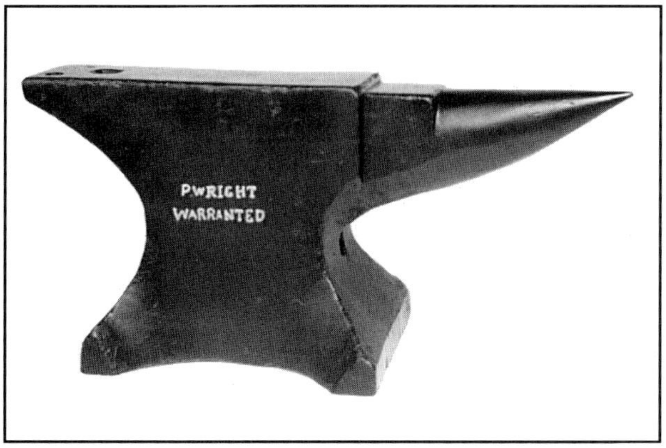

110-pound Peter Wright, pre-1852. (R.A.P. photo)

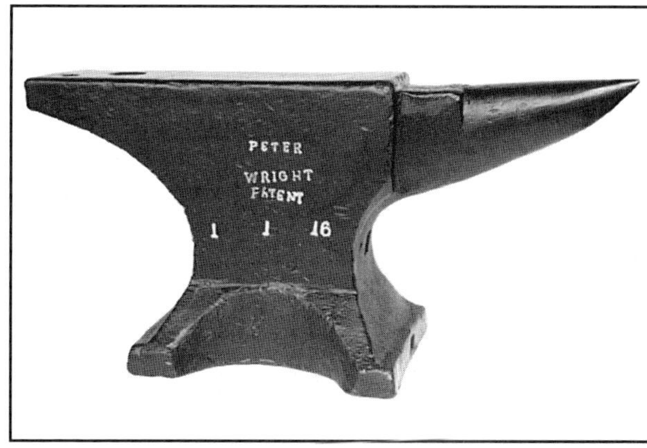

156-pound Peter Wright, ca.1852-1860. (R.A.P. photo)

The *Peter Wright* two-piece anvil made shortly before or after 1852 looks entirely different and becomes what I call the *Modern* anvil. The horn and the heel are a little longer, and the waist is thinner, but the major change in appearance is in the base or, more particularly, the feet. The tops of the feet are no longer sharp or rounded, but flat, and there is a step, or flat, running across the top of the feet all the way across, under the horn and the heel. There is also a sharp demarcation, almost a line, between the base and the lower part of the waist on each side of these anvils. The "line" may be the product of dies. An additional handling hole was added, in the front of the foot under the horn, giving these anvils four such holes. I have never recorded a *Peter Wright* two- piece anvil that did not have these four handling holes. This anvil was made between 1852 and about 1860 and bore the trademark "**PETER, WRIGHT, PATENT**." The weight was now below the word "**PATENT**." There is no "**SOLID WROUGHT**" in a circle on the side.

I was looking through the book, *Gardner's Photographic Sketch Book of The Civil War.*[55] Plate 16 showed a traveling Union forge. On a keg next to the forge stood an anvil, and upon close inspection I identified it as a *Peter Wright* made between 1852 and about 1860. The plate is marked, "Pamunkey, Va., May 1862."[56] In the same book on plate 77, I identified a *Fisher* anvil. Under the plate it says, " Forge Scene, Front of Petersburg, August, 1864"[57] (see Fisher).

The *Peter Wright* anvils made from about 1860 to about 1885 were nearly the same as those after 1852, except that the horn was a little longer and they were more uniform in shape with smooth blending curves under the horn and heel and down both sides. The "line" was gone. The steel face plate was put on in several pieces on the anvils weighing over 150 pounds. The words "Solid Wrought" were added to the trademark, stamped just below the word "Patent." In many cases the "Solid Wrought" is all that can be recognized as it is in the form of a circle. Also, the middle weight number, quarters of hundredweight, is generally in the middle of the circle. If one of these anvils with more than one plate on the face is well used, you can usually see where the plates join. It leaves an impression of a line across the face of the anvil. If a *Peter Wright* has two or more plates on the face you can be pretty sure it was made about 1860 and before 1885.

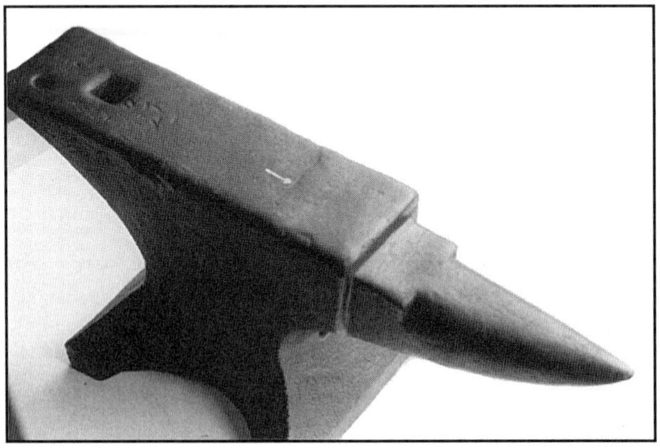

241-pound Peter Wright, pre-1852. Notice the plate separation indicated by the arrow. The plate separation is not generally prominent. (R.A.P. photo)

Between about 1885 and 1910 the only change I have observed is the one-piece face plate. A single face plate on a *Peter Wright* anvil would indicate a date after 1885, though it is usually difficult to detect.

139-pound Peter Wright. Made ca. 1885-1910. (R.A.P. photo)

Over 90% of the *Peter Wright* anvils that I have recorded were made before 1910. In 1910 England passed a law requiring that all china, etc., for export purposes be marked with the word "England." I assume that this law applied to the anvils as well, because all the later *Peter Wright* anvils bear the word "England" just below the word "patent" as do the later *Mousehole* anvils. Therefore, I conclude that the anvils that have the word "England" on them were made after 1910. These *Peter Wright* anvils with "England" on them have a higher, thinner waist and the horn is more rounded on most of them. Otherwise they look about the same as the previous models. These usually have a letter between the feet below the weight marks, also.

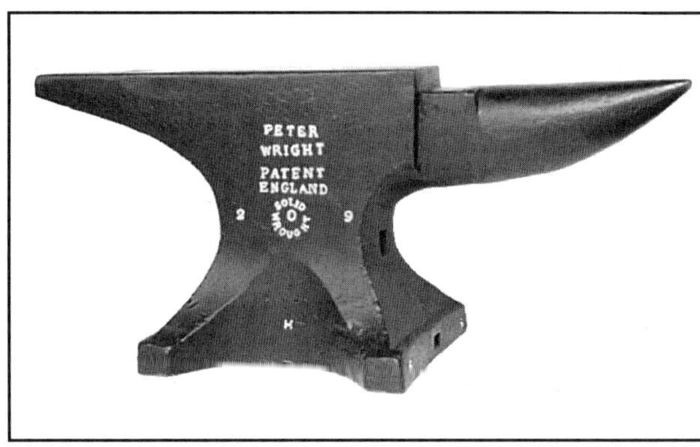

233-pound Peter Wright. Made after 1910. (R.A.P. photo)

This is conjecture on my part, but sometime around 1930 the *Peter Wright* changed to a true two-piece anvil. That is, they are without a face plate. I have recorded only one of these and it was a farrier's pattern. If this anvil had a plate, it was the best cover-up job I have ever seen. This anvil does not have "solid wrought" in a circle on it.

128-pound Peter Wright farrier's anvil without a face plate. This anvil was in use at the time I photographed it so I did not get to record it and have never met the owner again. (R.A.P. photo)

Peter Wright did a wonderful job of blending the face plate weld line with the body, and it is even hard to find the edge of the plate on many of the older ones. I have a large *Peter Wright* smith's anvil (457 lbs.) with "England" on it. I have not been able to determine for sure whether it has a face plate or not, although I am pretty sure that it does, because the entire face is slightly concave from front to rear from heavy work being performed upon the face. An anvil with a solid tool-steel body with a hardened face will not compress like a steel plate over a wrought iron body. But you cannot tell whether it has a plate or not by looking.

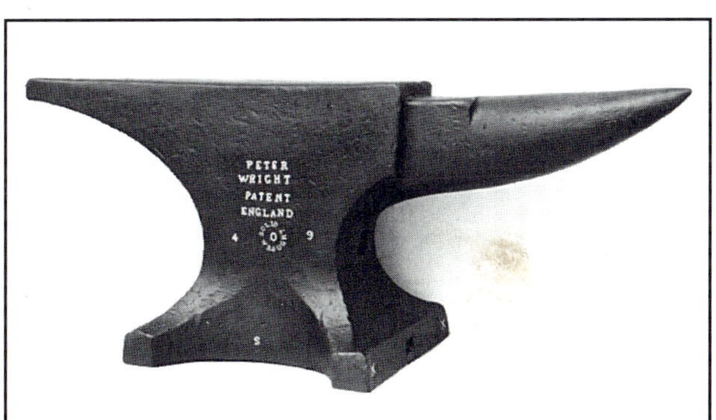

457-pound Peter Wright anvil. (R.A.P. photo)

Peter Wright did the best job of finishing their anvil of any make ever produced. Maybe that is the reason that they continued to sell in this country even after better American wrought anvils were on the market, selling for one to two cents less per pound.

It is rare that you cannot find part of the trademark on any *Peter Wright* anvil. However, even without the trademark, should a wrought anvil have four handling holes and the flats across the front and back of the feet under the horn and heel, you can almost always be sure that it is a *Peter Wright*. The entire trademark and weight are on the same side on those after 1852. This is also true of most wrought anvils. Every *Peter Wright* I have recorded or seen has the weight stamped on using the "stone weight" system.

On the front of the feet under the horn are usually all types of markings, such as numbers, letters, asterisks, anchors, and Roman numerals. I am not sure why they used so many different marks and I have not come up with any pattern for the marks that would give me a clue to their meaning, but I know they were not put there for fun. My best guess is that they are inspector's marks.

Peter Wright made smithing anvils weighing from about 40 to about 600 pounds.[58] It was reported in *Practical Blacksmithing* that they made one weighing 960 pounds for the 1876 Centennial in Philadelphia, Pennsylvania.[59] I assume it was a *Peter Wright* because I doubt that the Mousehole Forge could make an anvil that large. The smallest *Peter Wright* that I have recorded weighed 51 pounds and the largest weighed 458 pounds. One hundred to about 150 pounds seems to have been the most popular blacksmith size in the United States.

Peter Wright and Sons was also a major manufacturer of leg or blacksmith's vises and many of these can still be found in this country.

ANVILS IN AMERICA

Toward the end of the nineteenth century an American hardware company, S.D. Kimbark, of Chicago, Illinois, started developing a special anvil particularly suited for farriers.[60] They had their anvil pattern made in Germany at first, but by the 1890s they were made in the United States.[61] The main features were a narrower face, a large diameter horn, and a very short flat horn, called a clip horn, coming off the side of the table at a right angle to the face.

This anvil became very popular with farriers. Peter Wright and Sons it seems, tried for a piece of this American market. Their first attempt was to add the clip horn to the side of the table without changing anything else. I have recorded only two of these, which indicates that they were not received very well. The afore mentioned *Peter Wright* farrier's anvil without the plate is a beautiful farrier's anvil in every detail, and could compete very favorably with any American farrier's pattern anvil. Peter Wright and Sons learned their lesson, but by that time the anvil market was almost gone.

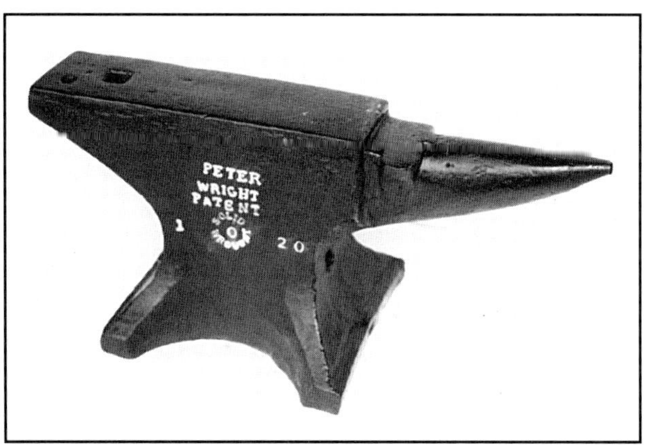

132-pound Peter wright anvil with a clip-horn coming off the side of the table. (R.A.P. photo)

I recorded a *Peter Wright* anvil that had a short clip horn coming off of the face plate just above the table at a right angle to the face plate and was part of the face plate. Was this an attempt at a farrier's pattern anvil also?

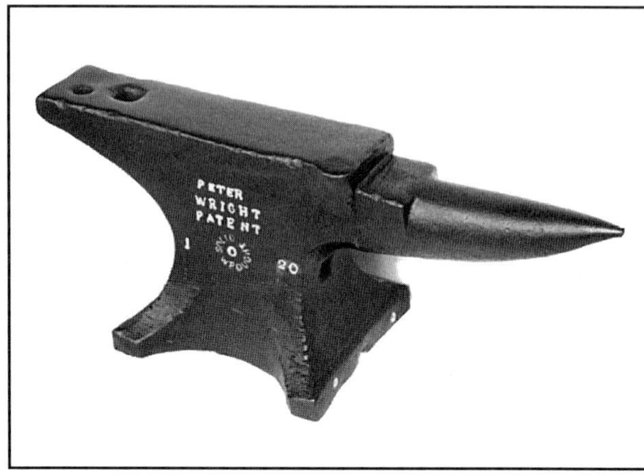

132-pound Peter Wright anvil with a clip-horn coming off the side of the face. (R.A.P.)

A friend of mine has several *Peter Wright* anvils on which the trademark is over twice the size of the normal ones. I wonder why? These things happen from time to time and it must be remembered that wrought-iron anvils made before the use of closed dies are almost as individual as people.

To my knowledge Peter Wright & Sons never used any but their own trademark on the *Peter Wright* anvils.

I am sure Peter Wright & Sons made as many different patterns of anvils as any of the other makers, including sawmaker's, double-horned and so on, but the only one that sold well in the United States was the smith's anvil.

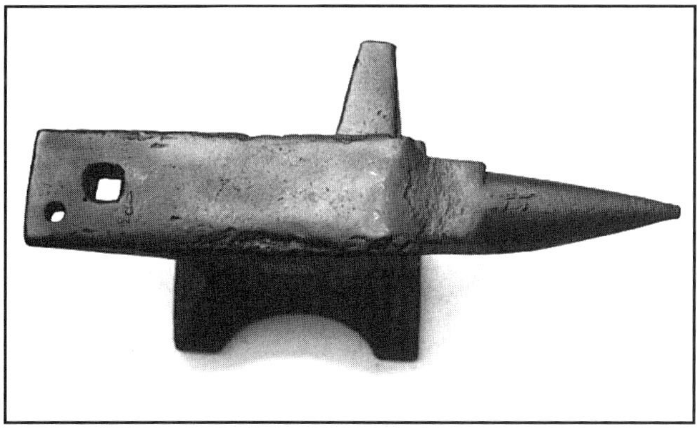

121-pound Peter Wright coachmaker's anvil. (R.A.P. photo)

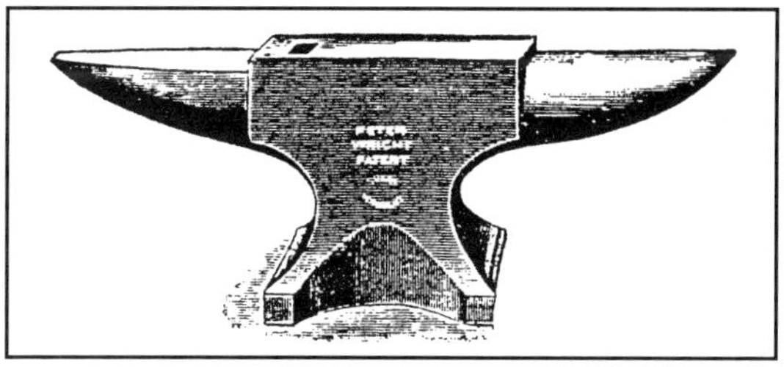

Doubled-horned Peter Wright anvil.

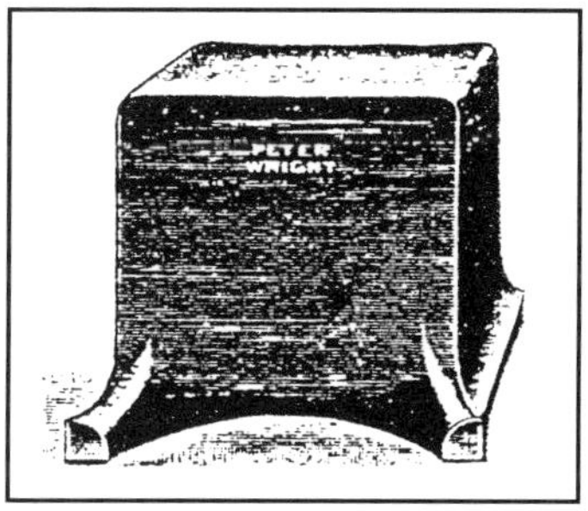

Sawmakers's Peter Wight anvil.

Courtesy of Alvin Sellens

Just recently (1995) a family friend found a leg vise with both the trademark of *Peter Wright* and that of *Wilkinson* on it, indicating that possibly Wilkinson bought out Peter Wright.[62] A fellow blacksmith told me that he had heard that Wilkinson had bought the Peter Wright works about 1914, but continued production as before and dropped the *Wilkinson* anvil.[63]

The following selected Peter Wright advertisements cover about forty year span from the oldest to the latest.

ANVILS—Wrought Iron, Steel Face.

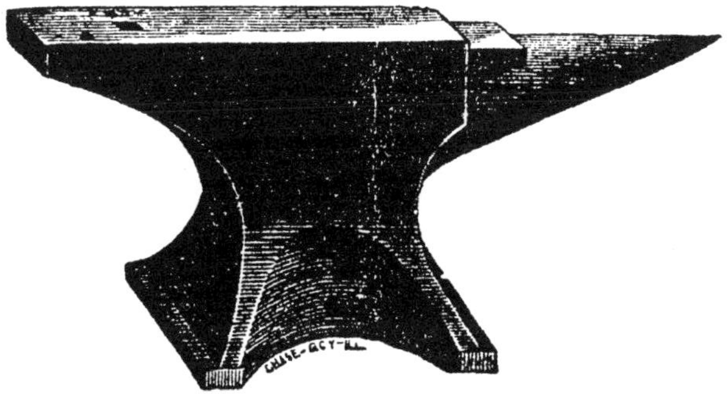

Peter Wright's, per pound $0 15
Armitage, or Mouse Hole, per pound 15
Anvils over 250 lbs., 1 cent per lb. extra.

EAGLE ANVILS—Cast Iron, Steel Face.

No.	0	1	2	3	4	5	6	7	8	9
Weight, pounds	10	15	20	30	40	50	60	70	80	90
Price, each	$2 25	2 75	3 25	4 00	4 50	5 25	6 00	6 50	7 25	8 25

Anvils weighing 100 to 600 lbs., 10 cents per lb.

N. O. Nelson Mfg. Co. St. Louis, Missouri catalog 1884-1885. (From Mr. John Catchings)

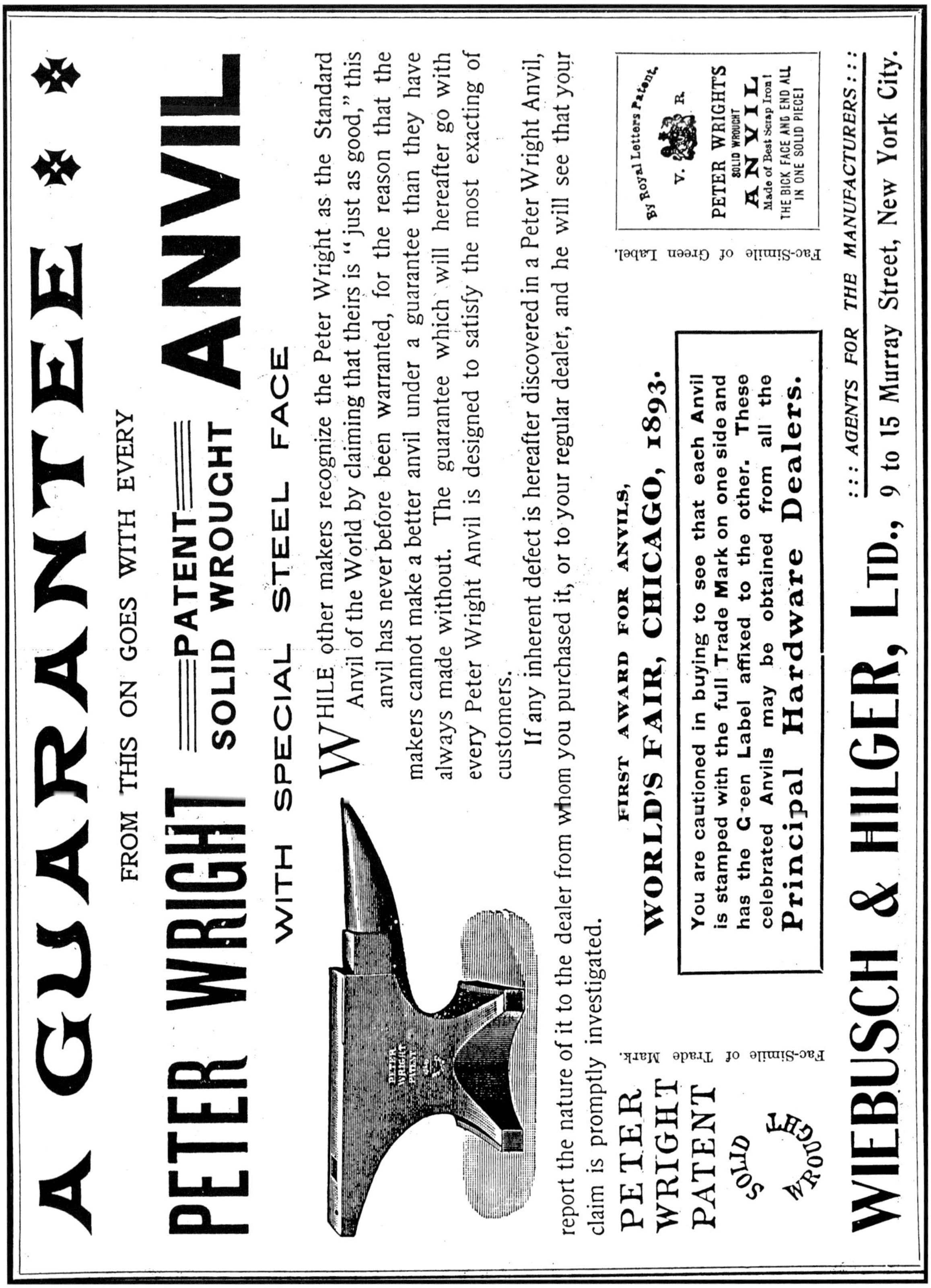

The Blacksmith and Wheelwright August, 1901. From 1893 to 1901 the ads in this journal were the same.

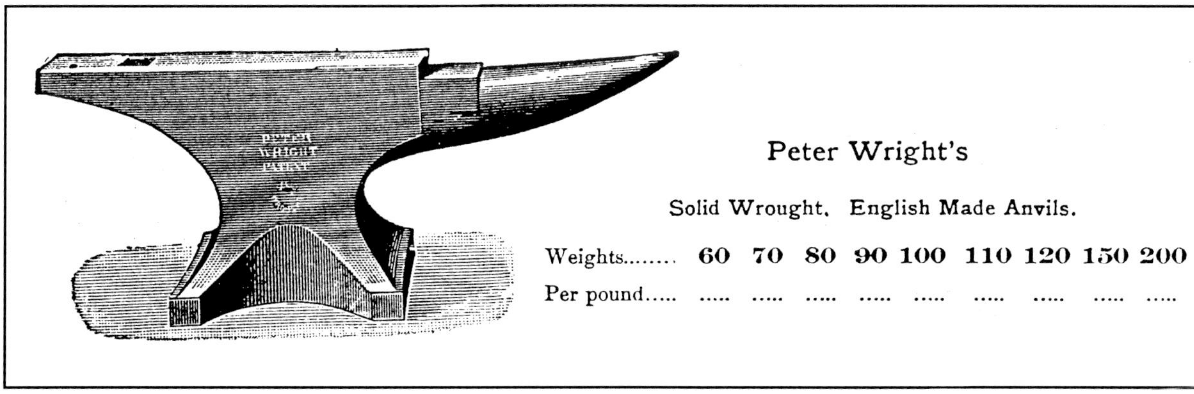

C.M. McClung and Company catalog 1903-1904, Knoxville, Tenn. (From Mr. John Catchings)

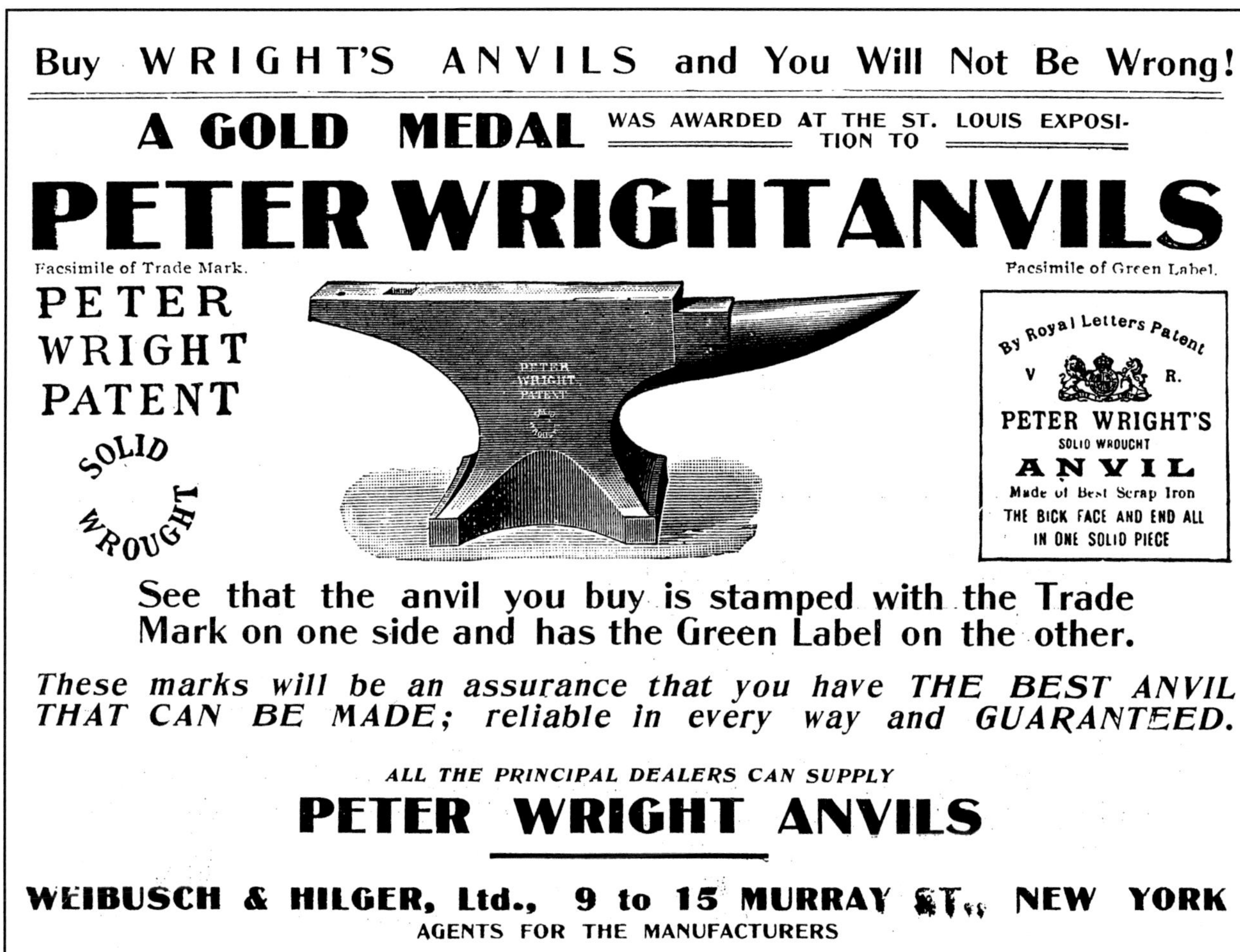

Horseshoer's Journal, September, 1905. (From Mr. Gill Farhrenwald)

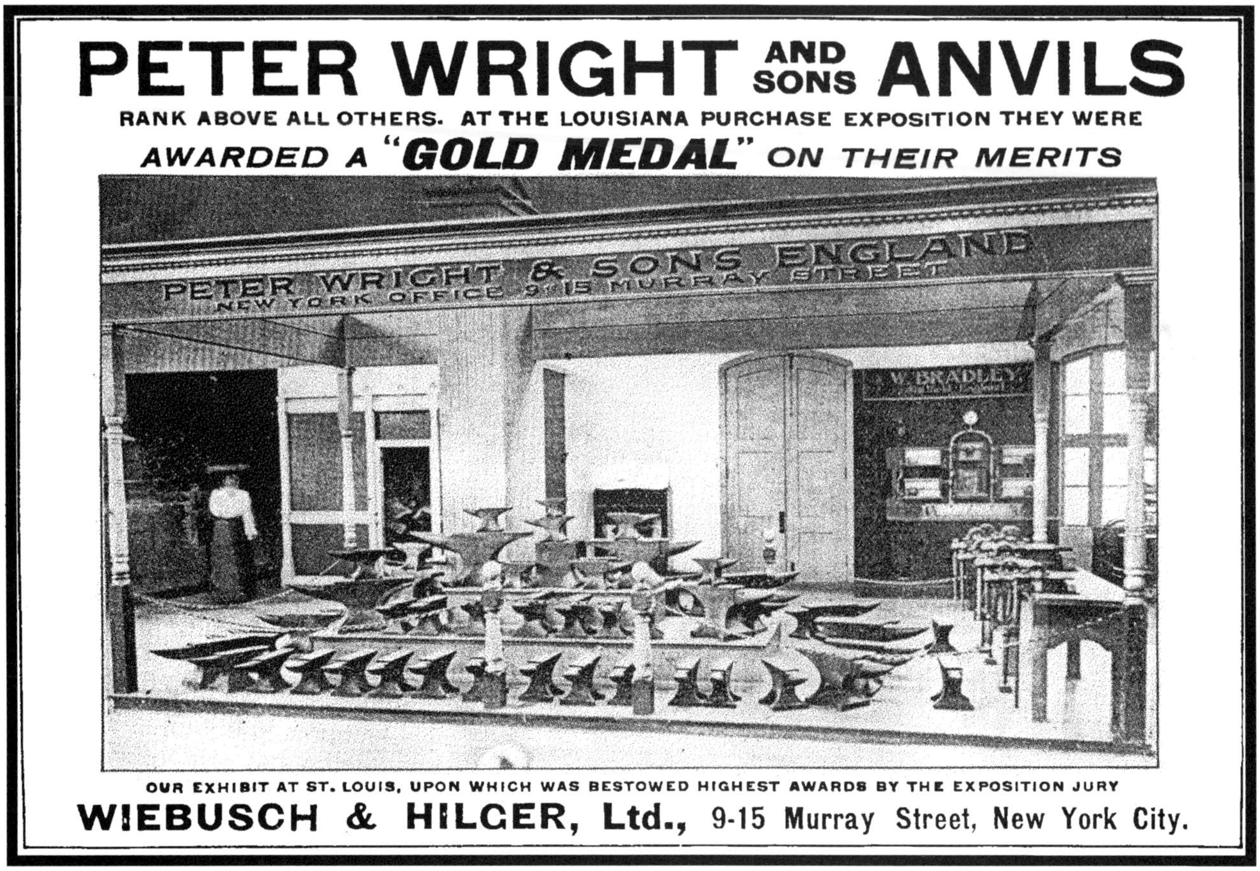

The American Blacksmith, July, 1906.

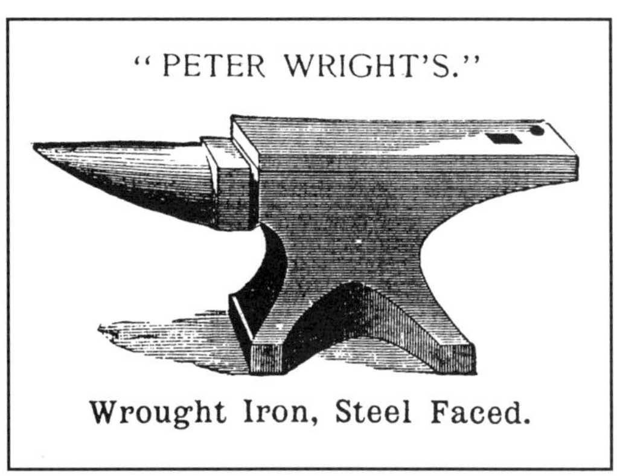

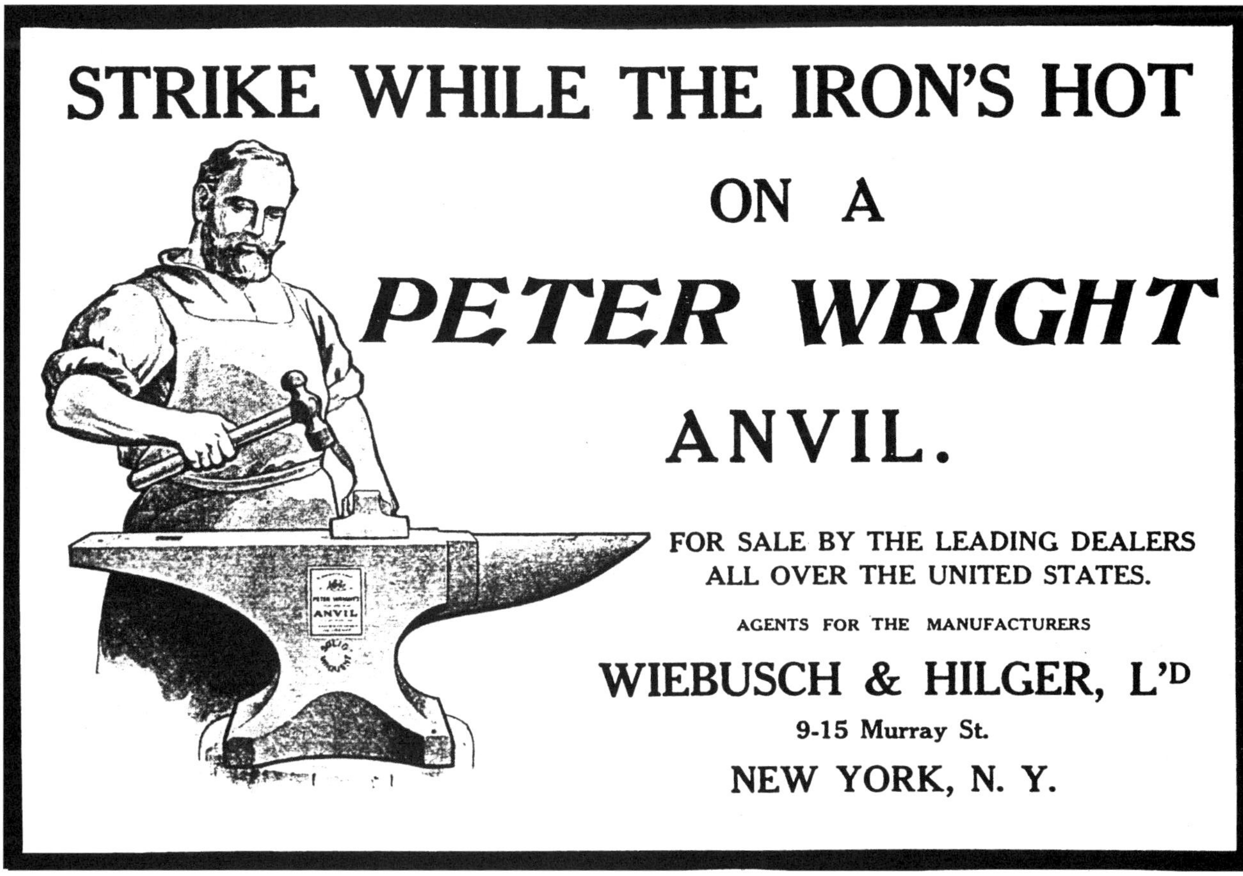

The American Blacksmith, September, 1907.

PETER WRIGHT'S

These Anvils are faced with special Sheffield shear steel, converted from Swedish iron. The brick, face and end are all in one solid forging, so that it is impossible for them to break.

Code Phaltiel

84 to 350 lbs.	Base	**Per lb.,**	**$0.35**
70 to **84** lbs.	Advance	**Per lb.,**	**.01**
60 to **69** lbs.	Advance	**Per lb.,**	**.02**
50 to **59** lbs.	Advance	**Per lb.,**	**.03**
45 to **50** lbs.	Advance	**Per lb.,**	**.04**

The American Blacksmith, September, 1909.

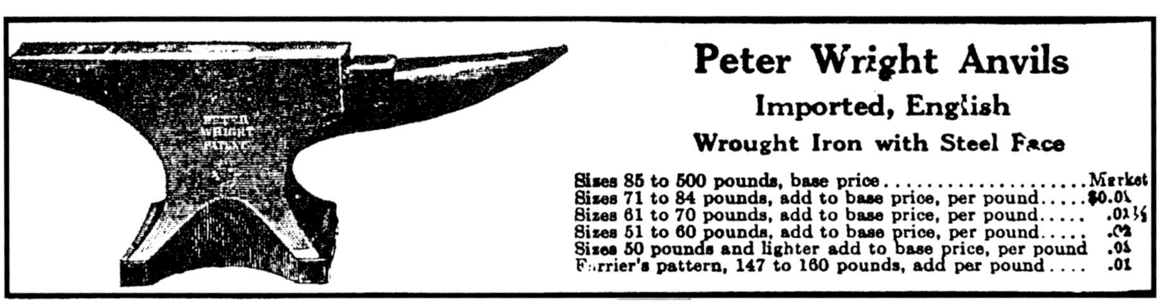

The Secret of Success

OF THE

PETER WRIGHT ANVIL

Is the care bestowed upon its manufacture both in mode of construction and material employed.

ONE OF THE ANVIL GRINDING MILLS AT PETER WRIGHT'S WORKS.

For Sale by the Leading Dealers all over the United States

Agents for the Manufacturers

Wiebusch & Hilger,

Limited

NEW YORK CITY

The Blacksmith and Wheelwright, October, 1909. (From Mr. Roy Price)

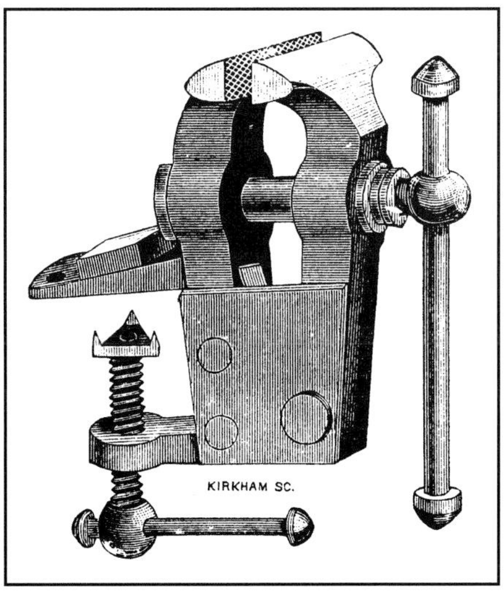

The American Blacksmith, January, 1912.

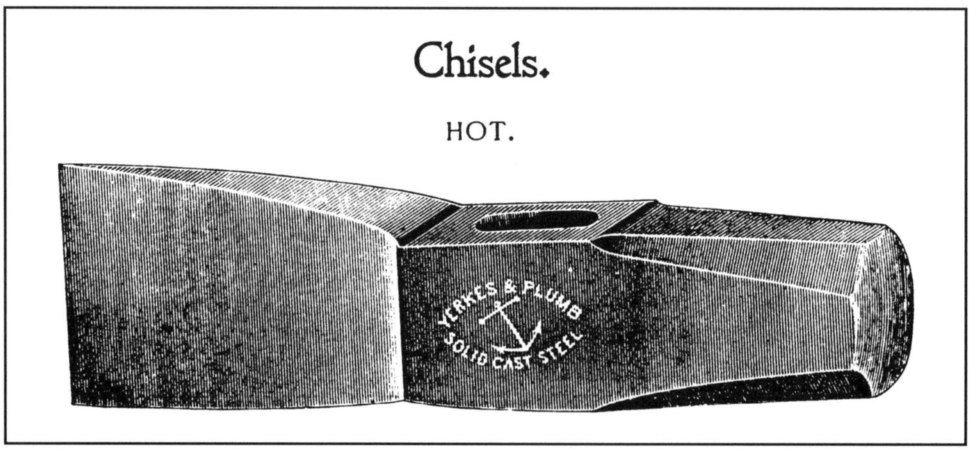

The American Blacksmith, October, 1912.

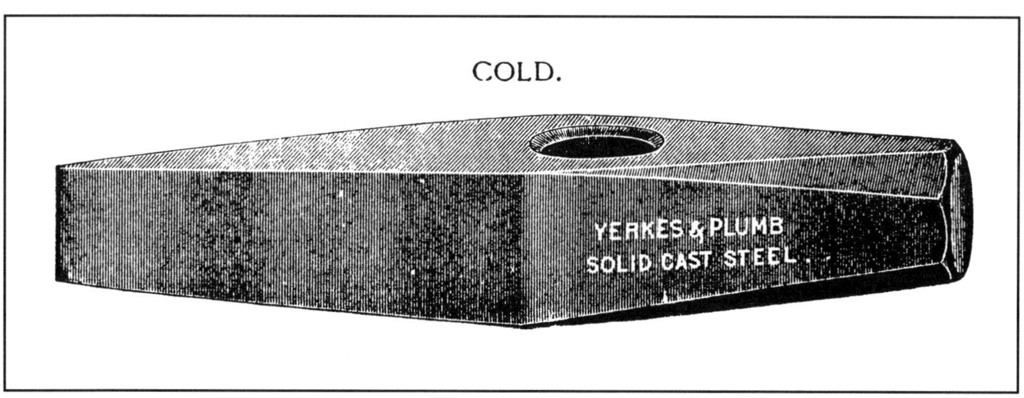

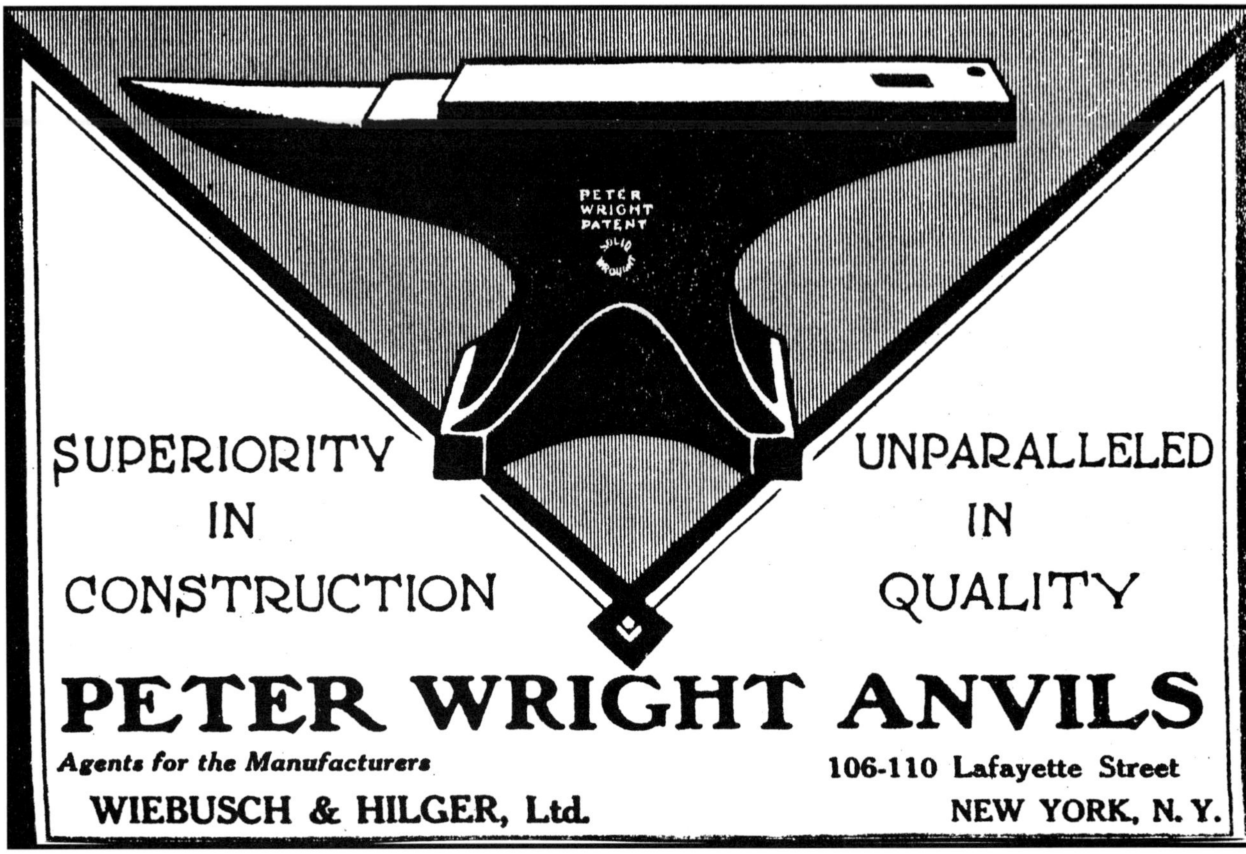

The American Blacksmith, February, 1913.

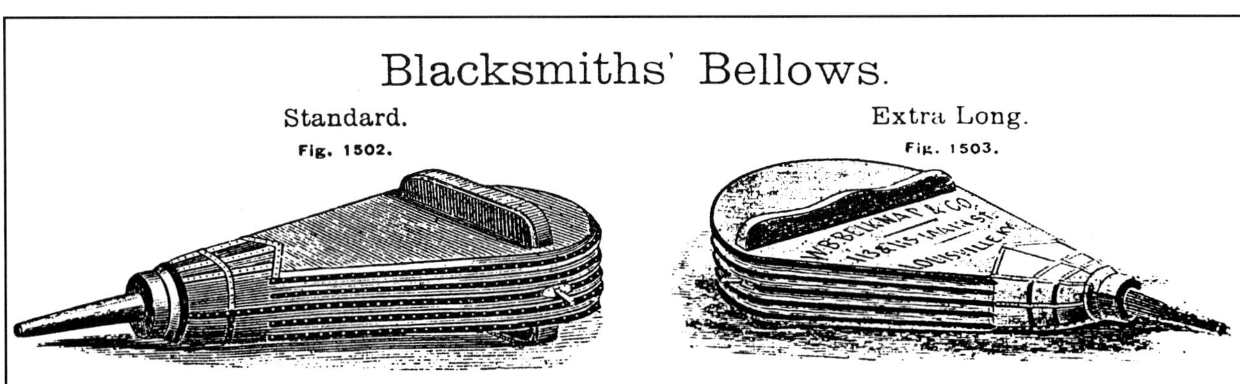

The American Blacksmith, March, 1913.

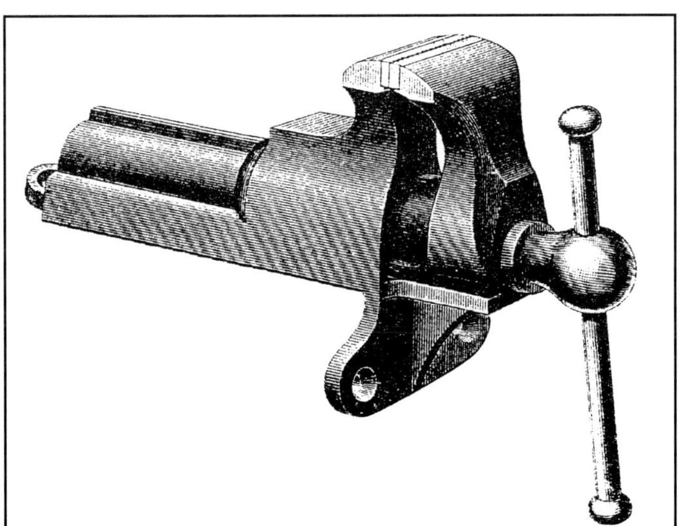

The American Blacksmith, April, 1913.

The American Blacksmith, March, 1914.

With the coming of World War I shipments of English anvils slowed and so did their advertising.

WILLIAM FOSTER

All I know about *William Foster* anvils comes from the anvils that I have recorded, 27 to date (1995) and others that I have heard of. They were probably manufactured in Sheffield, England, although there is no indication on any that I have recorded as to the place of manufacture.

It is too bad that the other anvil makers did not follow William Foster's practice of putting the date of manufacture on their anvils. Every *William Foster* that I am sure of has a date or some part of a date left. I have recorded dates from 1816 to 1864, most falling into the 1830s and 1840s.

All I have recorded are multi-piece anvils, although I am not sure about the one dated 1864. They must have gone out of the anvil business some time in the late 1860s or 1870s. It is possible that they stopped exporting anvils to the United States about that time.

Most *William Foster* anvils are smith's anvils, but I have also recorded a couple of the double-horned ones and a number with the side horn coming off the body at a right angle or coachmaker's anvils.

On all I have seen the trademark is always on the side with the horn to your right, and the "stone weight" is always on the other side.

I have never discovered an advertisement for this anvil.

From 1816 until 1864 these smith anvils all appear very similar and are all of the *Old English* style. Without their trademark or some part of it, the *William Foster* would be indistinguishable from any other make of that period.

Generally there is the name, then the date, and near the bottom on the side, between the feet, a letter or letters which are probably inspector's marks. Some of the anvils from the 1830s and 1840s have one or two crowns above or beside the name. The one dated 1864 has three crowns as part of the trademark.

There are two anvils dated 1816 and 1817, one of which has a pritchel hole and one does not. The anvil dated 1817 does not have a pritchel hole. The next one I have recorded is dated 1835 and has the pritchel, as do the rest after that date. On closer inspection it appears that the pritchel hole in the 1816 anvil was drilled in at a later date. They all have a hardie and table. Up until 1843 the trademark on them is as follows: "William Foster," date, and initials.

On the 1816 and1817, only the date and part of the word "Foster" remain, and there may have been only the word "Foster" and then the date.

The double-horned *William Foster* anvils always have the horns stepped down from the face, the same as on the smith's anvils. This is true of all the double-horned English-made anvils I have observed. Of course one horn is flat and the other round on top.

Of the 30 anvils (1997) I have recorded, five are the side-horn carriagemaker's pattern.

96-pound William Foster carriagemaker's anvil dated 1835. (R.A.P. photo)

William Foster carriagemaker's anvil dated 1835 from the front. (R.A.P. photo)

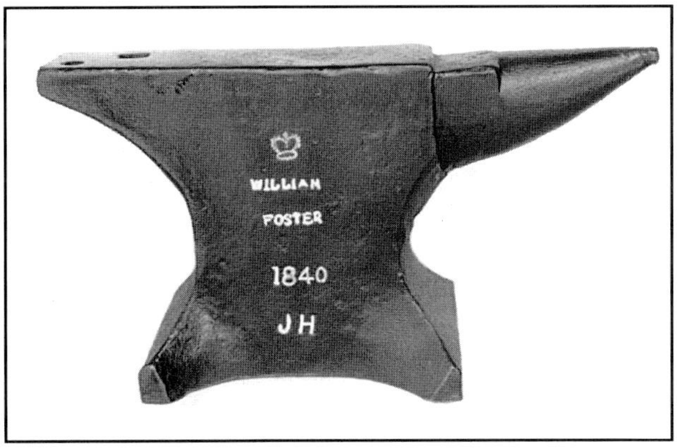

81-Pound William Foster anvil dated 1840. (R.A.P. photo)

108-pound William Foster anvil dated 1851. Photo taken at Mr. Bill Gichner's. (R.A.P. photo)

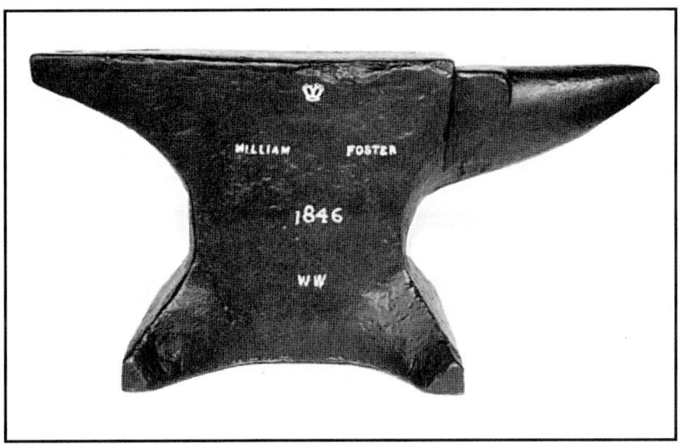

160-pound William Foster anvil dated 1846. (R.A.P. photo)

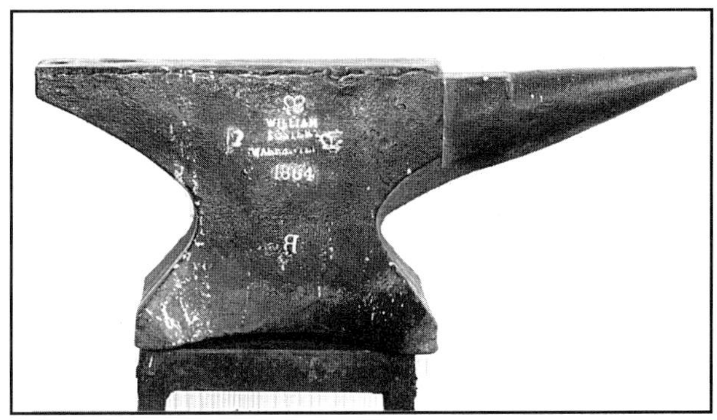

165-pound William Foster anvil dated 1864. This anvil also has the word "Warrented" as part of the trademark. (Mr. Kenneth Martin's photo)

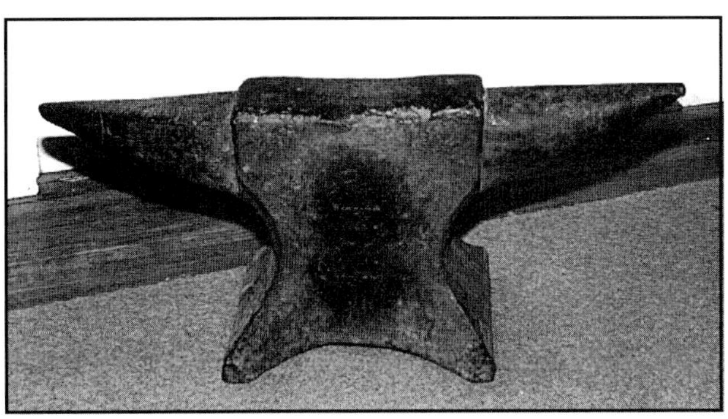

160-pound William Foster double-horned anvil dated 1848. Mr. Hunter Pilkinton collection. (R.A.P. photo)

WILKINSON

This is the anvil maker that Mr. Cran, in the 1914 article, spelled "*Wilkeyson.*" Wilkinson is also the one I was told bought out Peter Wright & Sons, which from a recent discovery on a vise appears to be true.[64]

I have obtained only two mentions of *Wilkinson* in hardware catalog advertisements, and none in blacksmith journals. This leads me to believe that by the late-19th century their market share in the United States was so small that the company did not feel it worthwhile to advertise here.

ANVILS.

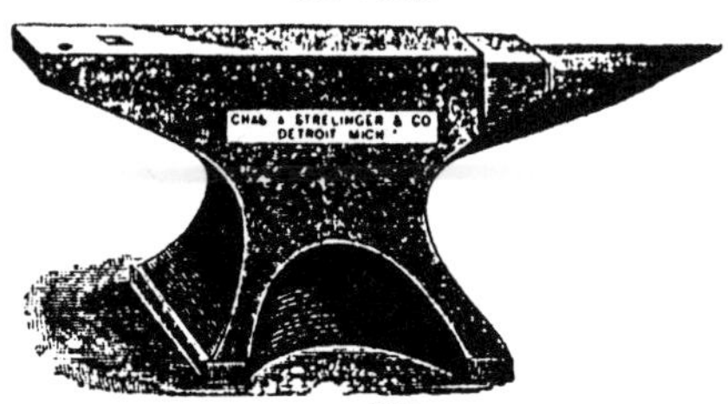

FIG. 284.

In wrought iron anvils, the Wright, Wilkinson and Armitage "Mousehole" are the best known; they are all English made. We believe there is but one American concern, and their output is small. Have been informed (by the house who sells them) that they are equal to the best English.

The price has been in the neighborhood of twelve cents per pound, varying from one-half to one cent more or less for a great many years.

Charles A. Strelinger and Company, catalog, 1895, Detroit Michigan.

Anvils.

Fig. H 428.

"Peter Wright" Anvils, 100 lbs. each and over.......................................per lb..........
"Wilkinson" " " " .... " ....
"Eagle" " " " .... " ....
"Trenton" " " " .... " ....

Prices of Anvils weighing less than 100 lbs. will be quoted on application.

M.N. Buck Mfg. Co. catalog, 1900, St. Louis, Missouri. (From Mr. Hunter Pilkinton)

Five trademarks from the oldest to the newest.

JOSEPH
WILKINSON
WARRANTED

J. WILKINSON & SONS
QUEENS
DUDLEY

WILKINSON'S
PATENT
SOLID ANVIL

WILKINSON'S
PATENT
SOLID WROUGHT

WILKINSON &
CO
WARRANTED

This is the only anvil maker beside Peter Wright who continued to manufacture anvils and to change their anvil style from the *Old English* to the *Modern* style. They also went from a built-up anvil to a two-piece anvil, following Peter Wright's example.

I have recorded 20 of these anvils (1995). The oldest ones are very blocky, with a very wide waist, and are close to the *Colonial* style. I have not recorded any that did not have a pritchel or hardie hole, so I have to assume that the oldest ones recorded date to about 1830.

The trademark appears to have changed five times and remained very uniform with each change. However, the "stone weight" might be found on either side, but usually on the opposite side from the trademark.

The smallest *Wilkinson* anvil I have recorded weighs 85 pounds and the largest 570 pounds. Most fall in the 150-pound range.

As part of the later trademarks there appear what I would call "crossed hot dogs" or "crossed sausages." I have been told that these are crossed cannons as well as crossed swords. I suppose they are one of the latter two, but you could not prove it by me. Without part of the trademark these anvils will appear like any other

Old English anvil, except the latest ones which have four handling holes and look very much like an early American anvil.

I have been told that there are more *Wilkinson* anvils in Australia than any other make.[65]

A letter from the Black Country Museum in Dudley, England, says that the Wilkinson anvil works was in Dudley, England (1996).[66]

160-pound early Wilkinson anvil. (R.A.P. photo)

119-pound J. W. Wilkinson anvil. (R.A.P. photo)

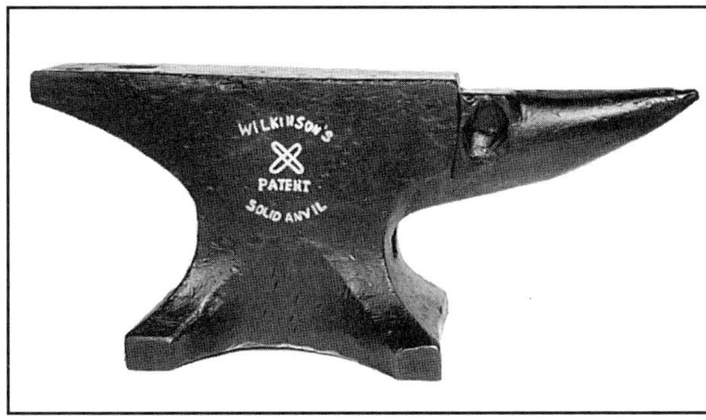

122-pound late Wilkinson anvil. Notice the feet. This anvil had a clip horn on the side of the table. (R.A.P. photo)

HENRY WRIGHT

Who was Henry Wright? One of Peter Wright's sons? Whoever he was, the anvil bearing his name looks just like the *Peter Wright* anvil in every respect and the finish appears to be as good.

I have obtained only two advertisements for this anvil, one dated 1890 and the other 1891.

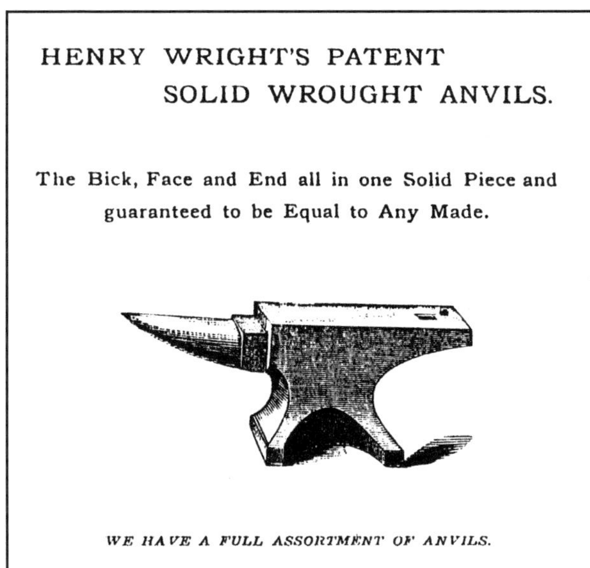

Butts and Ordway, Heavy Hardware, catalog 1890. (From Mr. John Catchings)

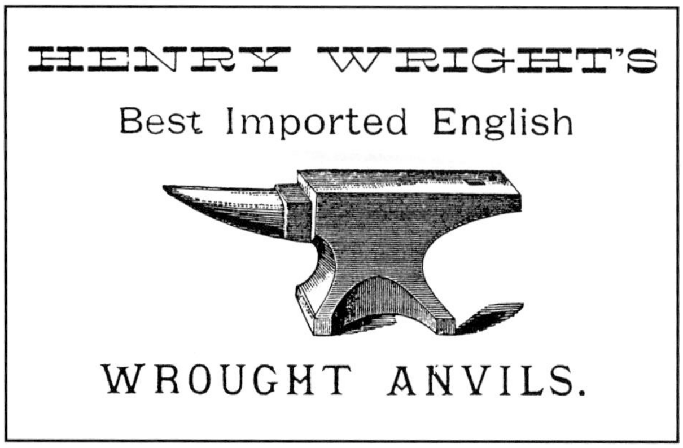

Charles C. Lewis Co. catalog, 1891, Sprinfield, Mass. (From Mr. Hunter Pilkinton)

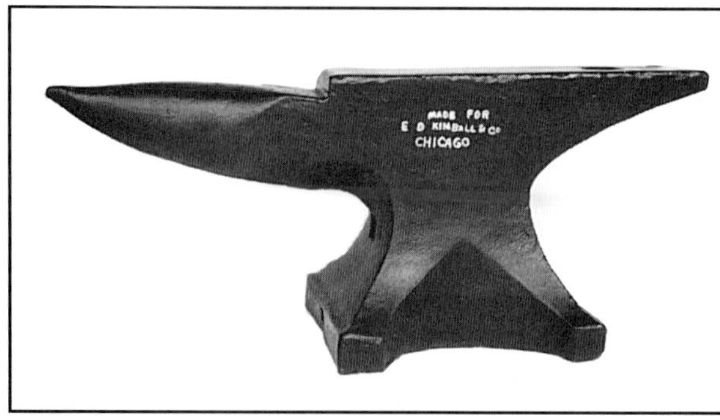

165-pound farrier's anvil made by Henry Wright for E. D. Kimball and Co., Chicago, Illinois. May have been made in the 1890s. (R.A.P. photo)

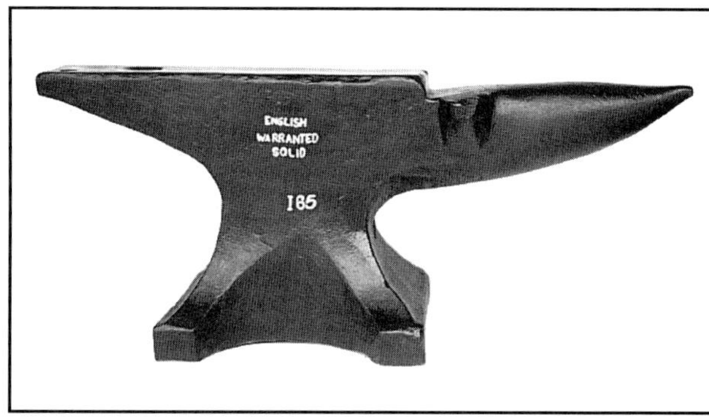

The opposite side of the 165-pound Henry Wright anvil. (R.A.P. photo)

79-pound typical Henry Wright anvil. (R.A.P. photo)

It was in 1891 that Peter Wright warned of an imitator. Undoubtedly this was Henry Wright.[67] I have recorded only eleven of these anvils as of 1995. It is my guess that the company started in the late 1880s and lasted until about 1914. They made anvils not only under their own trademark, but under at least one American company's trademark, *E.D. Kimball & Co.,* of Chicago, Illinois. This is the only English anvil maker that I know of to have done so. I suspect that they were hard-pressed for business. I have recorded two changes in the trademark for these anvils. The earlier trademark follows:

HENRY
WRIGHT
DUDLEY
WARRANTED
SOLID

The other is as pictured in the photo of the 79 pound *Henry Wright.*

The weight in most cases is in actual pounds. On only one have I found the "stone weight," and it was on one of the earlier anvils. The weight was on the opposite side from the trademark on this anvil.

I assumed that Dudley was the town of manufacture from the trademark on several of the earlier anvils, the same as that of Peter Wright & Sons. Dudley as the place of manufacture was confirmed in a letter from the Black Country Museum in Dudley, England, in 1996.[68]

MODERN ENGLISH ANVILS

Shortly before World War II an English anvil maker produced a solid cast-steel anvil. As far as I know this anvil was never imported for resale into this country, but came by way of the Far East in 1994. It weighs 554 pounds. It is marked as follows:

193?
JOHN ???SO ?
ENGLAND

554-pound cast-steel English anvil.
(Mr. Gill Fahrenwald photo)

After World War II, only one anvil was made in England. It is of cast-steel, and is still being made. It was made under the name *Brooks* first and handled by the Glendale Forge, Thaxted, Essex, England. Most of these anvils have only the word "England" cast into the side and some bear the name *Brooks*. I was told it depended on which plant did the casting, as there were two foundries. One foundry put "England" on and the other put "Brooks, England". The weight can be in "stones," "kilograms" or actual pounds.

Later advertisements referred to these anvils as "Vaughan/Brooks"and then "Vaughn."

The anvils made today (1995) are called *Vaughan*, and are handled by Vaughan's (Hope Works), Baker House, King Street, Lye, Nr. Stourbridge, West Medlands DY9 8UT., England.

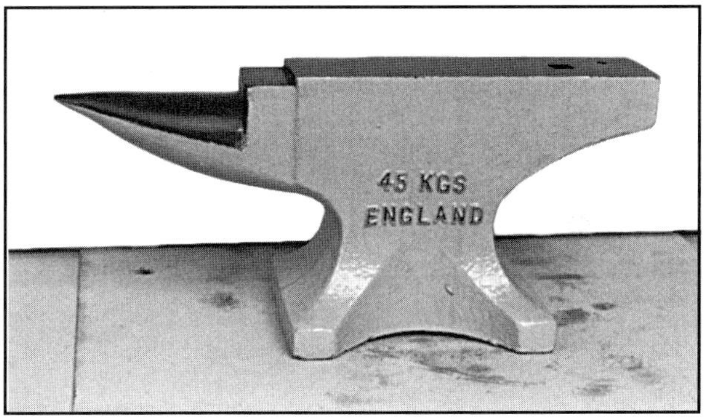

100-pound Brooks anvil. Notice the thickness of the heel.

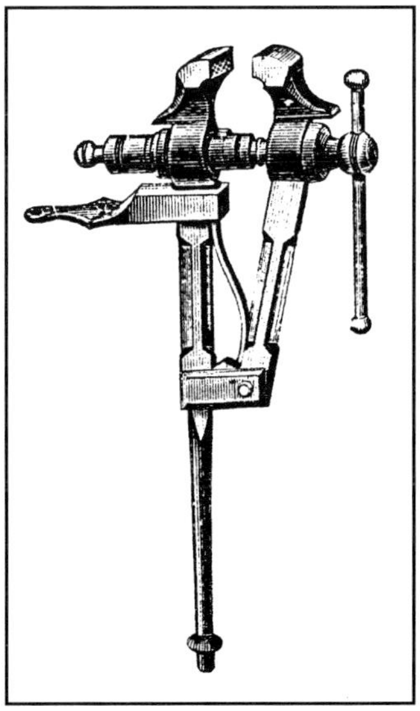

BROOKS ALL-STEEL ANVILS

When you purchase all-steel anvils from Kennedy-Foster Co., Inc., you are guaranteed satisfaction because these British-made anvils are cast all in one piece and consist of perfectly homogeneous steel throughout.

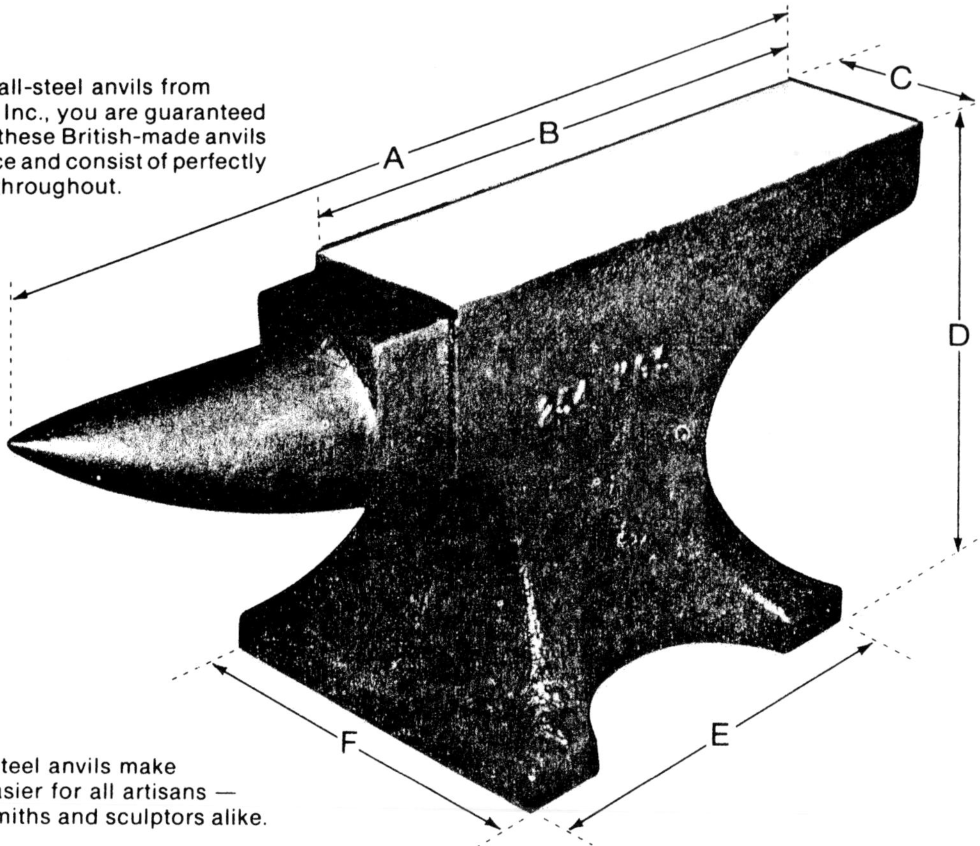

Kennedy-Foster all-steel anvils make creative iron work easier for all artisans — horsepeople, blacksmiths and sculptors alike.

WEIGHT	A	B	C	D	E	F	Square Hole	Round Hole dia.
	Inches	Inches	Inches	Inches	Inches	Inches	Inches	Inches
11	9⅜	6	1⅞	4¼	4⅜	3¾	½	¼
28	12¼	7½	2⅝	6⅜	6	5⅜	⅝	⅜
40	15	10	3	6⅞	7	6	⅝	⅜
56	15⅛	9½	3	7	7¾	7	⅝	⅜
84	19⅛	11¾	3½	8⅝	8	6⅞	⅝	⅜
100	20½	12½	3½	9¼	9	7½	¾	⅜
140	23½	14	4⅜	9½	9¾	9¼	1¼	½
168	25¼	15	4	11	10¼	9½	1¼	½
196	27	16½	4¼	11½	10¾	9½	1¼	½

DISTRIBUTORS

KF KENNEDY-FOSTER COMPANY, INC.
855 Bloomfield Avenue, Clifton, NJ 07012
201/471-2400

An advertisement for the Brooks anvil. Ad is probably not valid any longer.

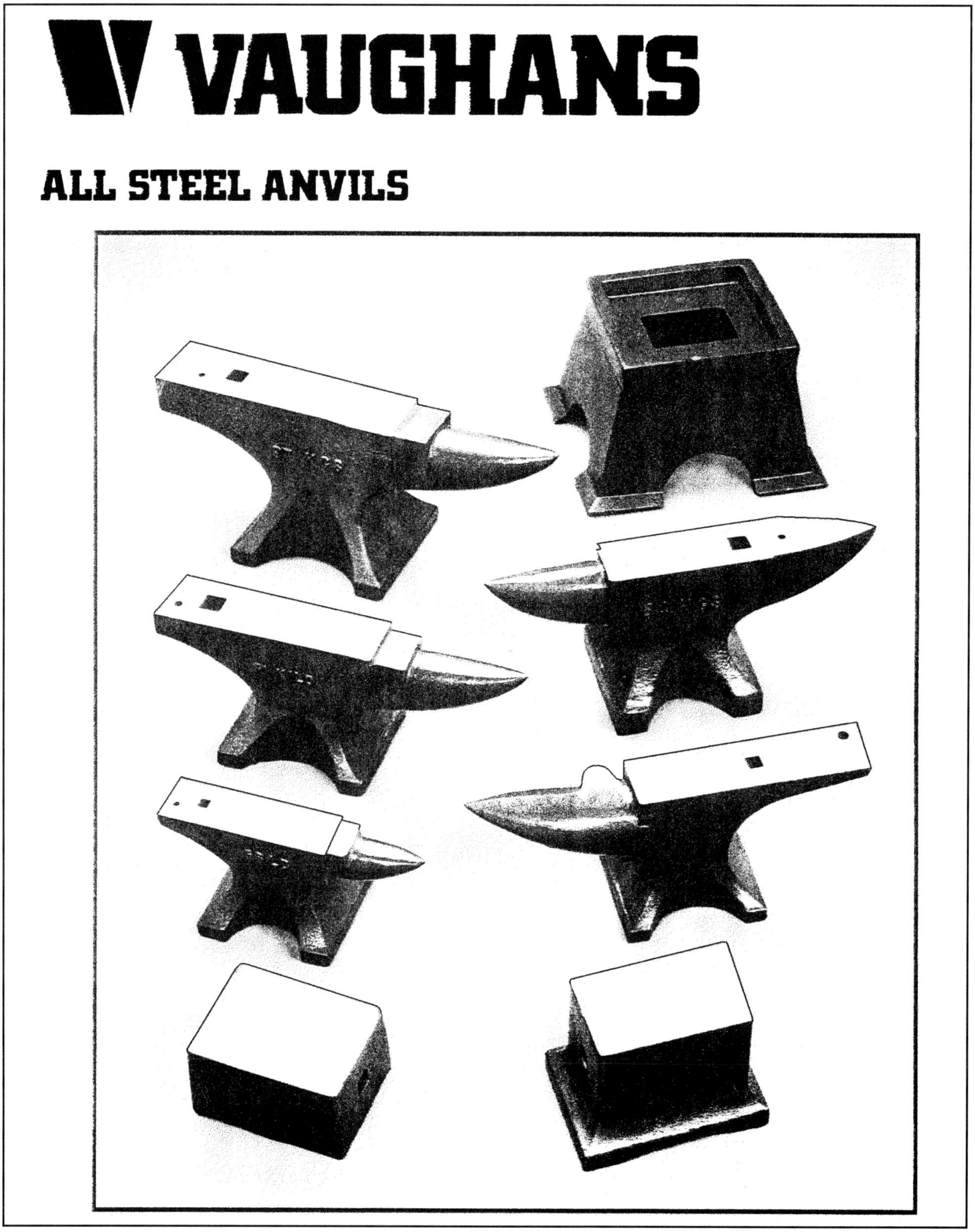

Anvils made by Vaughan from their brochure. Notice that the double-horned anvil looks like the 554-pound cast-steel anvil.

ANVILS IN AMERICA

Record of Sheffield, England, (famous for their wood vises) makes three small anvils of 11, 33, and 55 pounds. They are made of ductile iron. The smaller two are craft or bench anvils, but the 55-pound anvil would serve for light forging. Grizzly Imports, Inc. is one of the distributors in the United States. (1997)

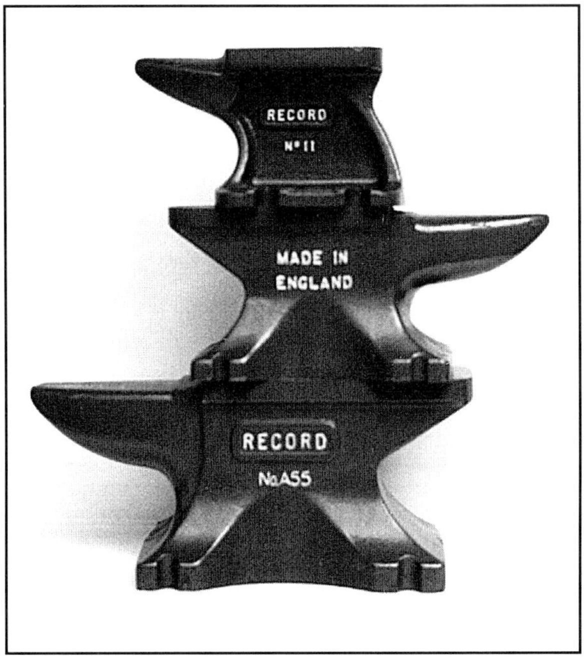

11-, 33-, and 55-pound Record anvils.(R.A.P. photo)

CHAPTER 4

American Cast Anvils

SOUTHERN
CRESENT

CENTAUR

C

N C

W

H S I

FULTON

WEST

MANKEL

CAST-IRON ANVILS

I am sure that cast-iron anvils have been made since cast iron was first produced in Central Europe in the 14th century.[1] For many purposes cast-iron anvils are perfectly satisfactory. They are very easy to make; all that is needed is a pattern in the shape of the anvil wanted. Cast iron has the lowest melting point of the major ferrous metals, about 1880^0 - 2200^0 F, therefore requiring the least amount of fuel to melt.[2] It could be poured into a mold directly from the blast furnace or readily remelted from the "pig*s*" and then poured.

The head and the anvil of the old water-powered tilt, or helve hammers, were made of cast iron. The platform, or base, on which wrought anvils were welded and worked by hand were of cast iron. Any thick square-like piece of cast iron works well for an anvil, so long as the sharp edges are not struck or there are no thin parts. Many smithing-type anvils on which heavy blows are not required are made of cast iron, even in the *London* pattern. Of course, all cast iron, as with steel, is not the same. To say cast iron made in the 14th century was as good as that of the 20th century would probably be a gross overstatement. Most cast-iron anvils made today are made of malleable iron or ductile iron. Ductile iron is almost equal to mild steel in tensile strength.

Most cast-iron smith's anvils, until recent years, were nameless. That is, whoever had them cast did not bother to put a trademark on them. I call these "no namers." It is my opinion that some of the anvil makers who produce cast-iron anvils with steel faces also produced cast-iron anvils which bear no indication as to the maker. They did not want the low-quality cast anvils to become confused with their better quality steel-faced anvils. On some "no namers" the weight can be found exactly in the same spot as on the steel-faced cast anvils. Sears Roebuck and Company and Montgomery Ward and Company, until recent years, always carried these cast iron "no namers" in their catalogs. Any iron foundry can make them and most of them probably did. Some have chill-hardened faces and some do not. The quality varies as much as the shapes.

The following cast-iron anvil manufacturers did put their trademarks on their anvils. I am sure there are others that I have not discovered yet.

680-pound cast-iron "no name" anvil.
(Mr. Thomas Moore photo)

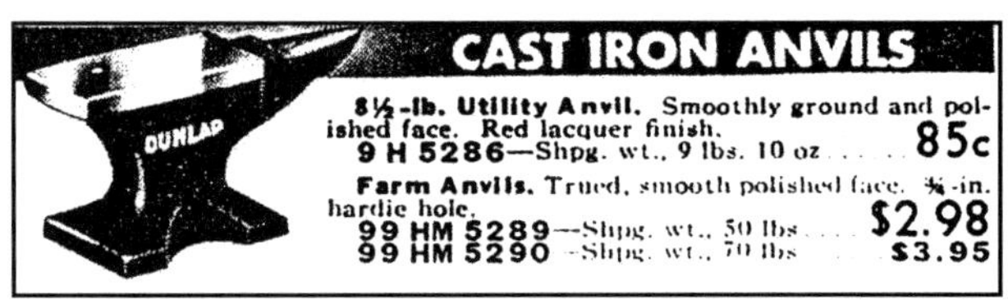

ALAMO

The Alamo Iron Works of San Antonio, Texas, has been making cast-iron anvils since 1878.[3] They are made in five weights, either of cast iron or ductile iron, as the brochure shows. The trademark is the same as on the brochure, with the weight in pounds on the side of the anvil.

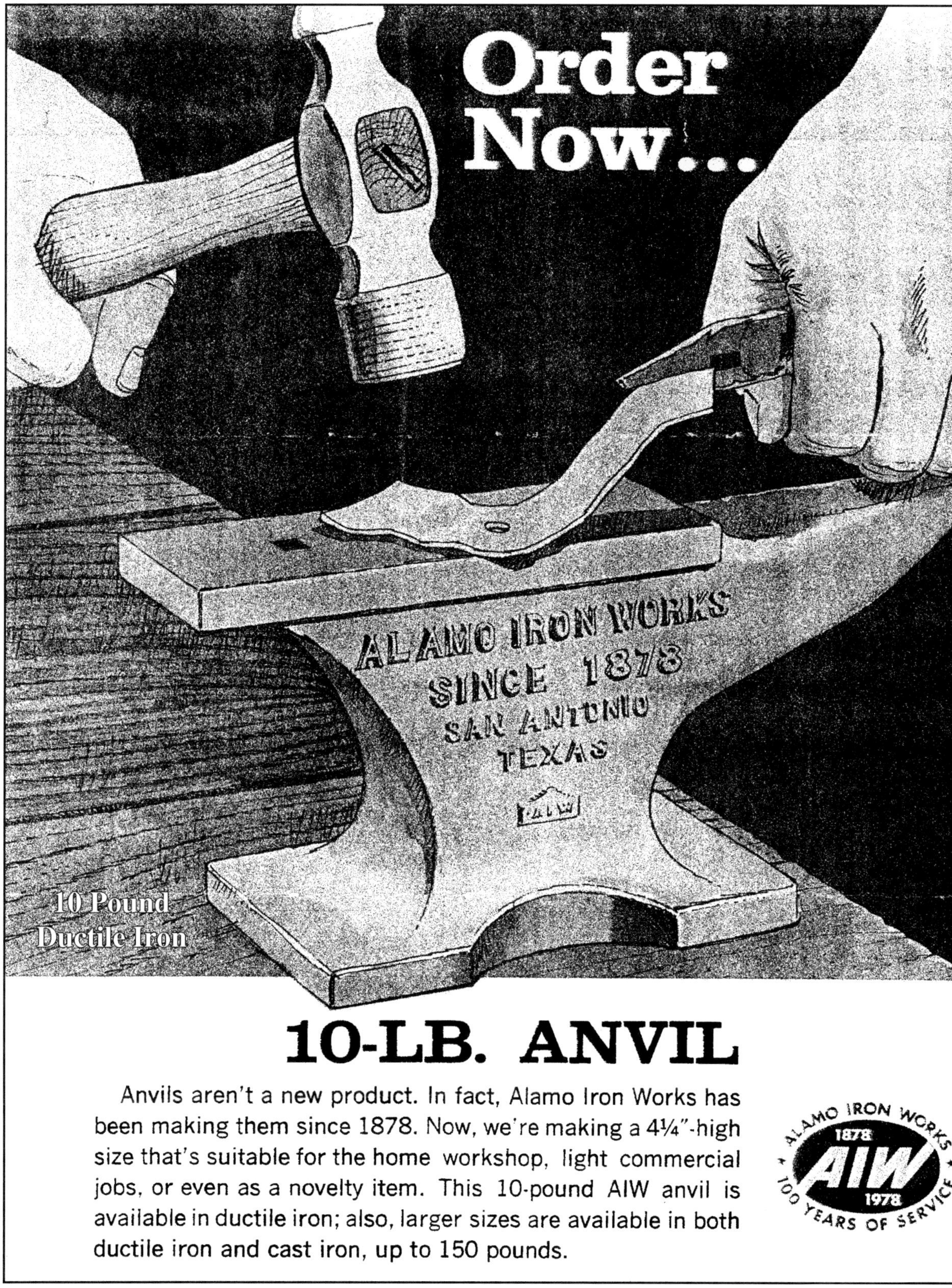

Alamo brochure for 10-pound anvil.

STOCK NO.	CASTING MATERIAL	WEIGHT (LBS.)	WIDTH FACE	DIMENSIONS (INCHES)			
				LENGTH FACE	LENGTH HORN	SIZE ROUND HOLE	SIZE SQUARE HOLE
804-7-10	Ductile Iron	10	2¼	5⅜	3⅞	—	⅜
804-7-1	Cast Iron	50	3	8½	6	½	¾
804-7-2	Cast Iron	75	3¼	10¾	7¼	½	¾
804-7-3	Cast Iron	100	3¾	12¼	7½	½	¾
804-7-8	Cast Iron	150	4	14	8½	½	¾
804-7-4	Ductile Iron	50	3	8½	6	½	¾
804-7-5	Ductile Iron	75	3¼	10¾	7¼	½	¾
804-7-6	Ductile Iron	100	3¾	12¼	7½	½	¾
804-7-7	Ductile Iron	150	4	14	8½	½	¾

ALAMO IRON WORKS

101 Montana • Box 231 • San Antonio, Texas 78291 • (512) 223-6161

ALAMO IRON WORKS 1878 AIW 1978 100 YEARS OF SERVICE

Alamo anvil specifications.

10-pound Alamo anvil. (R.A.P. photo)

75-pound Buffalo anvil. (R.A.P. photo)

BUFFALO FORGE

In 1896 the Buffalo Forge Company, makers of industrial products including blowers and forges for blacksmiths, came out with a cast-iron anvil. To the best of my knowledge they manufactured them in small numbers until about 1916.[4] These were said to be made of semi-steel, with a chilled face and made in weights from 50 to 150 pounds.[5] I have recorded one anvil of 75 pounds and have heard of others. The *Buffalo* anvil differs from most smithing anvils in that the horn blends into the body like a farrier's anvil. It also has hold-down lugs as part of the base for mounting.

The trademark is cast in bold relief on one side and the weight in pounds on the other side as follows:

Buffalo Forge Company
Buffalo, N.Y.

The anvil that I have recorded was badly defaced and the pritchel hole was as large as the hardie. Either the pritchel was made that way or someone drilled it out.

The *Buffalo* anvil was listed in the index of *The Blacksmith and Wheelwright* magazine of April, 1896, but not in their general *Buffalo Forge Company* catalog for 1896.

BUFFALO FORGE COMPANY

Forge-Shop Tools

Anvils and Blocks

In addition to a complete set of blacksmith tools which we are now offering, we manufacture a complete line of semi-steel anvils in sizes varying from 50 to 150 lbs., and also cast-iron anvil blocks suitable for the various size anvils. These semi-steel anvils are considerably cheaper than regular steel anvils and are entirely suitable for such light work as is usually done in a Manual Training Forge Shop, but when preferred we can furnish a steel anvil of any standard manufacture.

Ad from the Buffalo Forge Company, Buffalo, New York, 1916 catalog.

BAY STATE

I have never seen an anvil of this make that was recognized as such. This anvil is shown and listed in the *Simmons Hardware Company* catalog of 1921-22 and others. The copy of the advertisement that I have in my possession is too dark to reproduce; therefore I will quote from it.[6]

Trademark **Bay State**. *Farmers' cast. Contains a large percent of steel, which makes them Hard and Tough; Body and Horn japanned finish; Face is true and well polished; Square Hardy, near the Heel.*

They came in 50-,70-, and 100-pound weights and listed for 22 cents per pound.

CHAMPION

The 1936-37 *Champion Blower & Forge Company* catalog lists a cast-iron anvil. The illustration shows the company trademark on the foot only. Having never personally seen one of these anvils I cannot say how they are marked and do not know how long they were manufactured.

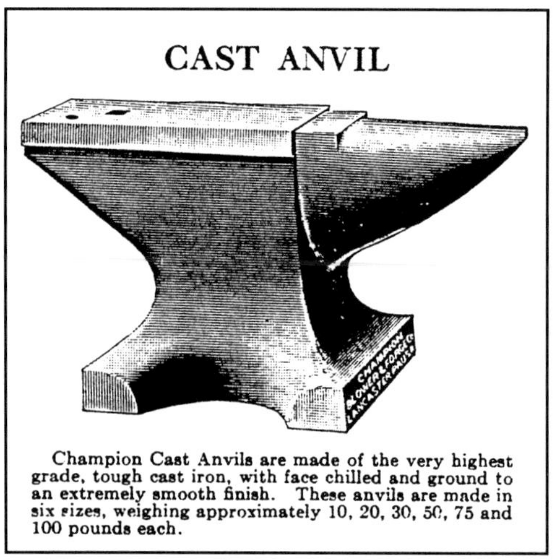

GEM

I have recorded one of these anvils but do not know at this time who manufactured them. The anvil was advertised in the 1903-04 *C. M. McClung & Company* catalog of Knoxville, Tennessee, and in the 1934 and 1940 *Belknap* catalogs. This would seem to indicate that the *Gem* anvil was made over a long period of time. The 1903 advertisement has the anvil listed in 75- and 100-pound increments and in the 1934 and 1940 catalogs at 100 pounds only. The anvil that I have recorded weighs 100 pounds. The large letters "GEM" are cast in bold relief on the side, but no weight is marked. It also has lugs on the base for mounting. The pritchel hole is right behind the hardie hole.

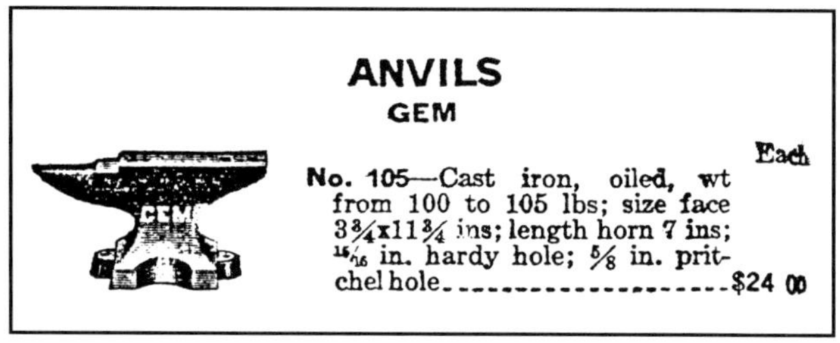

Ad from Belknap Hardware & Mfg. Co., Louisville, Kentucky, 1940 catalog. (From Mr. John Catchings)

100-pound Gem anvil. (R.A.P. photo)

ROCK ISLAND

The Rock Island Manufacturing Company of Rock Island, Illinois, not only produced bench vises, but at one time anvils as well. As far as I can tell from illustrations found in advertisements there is no trademark, only weight which is under the horn. The 1925 *Faeth Company* catalog of Kansas City, Missouri, lists them at 12 cents per pound and the *Rock Island Manufacturing Company* catalog of 1923 also listed them. This catalog also listed an 8 1/2-pound "Household Anvil."

CAST IRON ANVIL
(Stock No. 1036)

Cast Iron Anvils. Made of a special very tough cast iron or semi-steel, painted black. Faces are ground, polished and lacquered, are well proportioned and while not recommended for regular blacksmiths' work, makes a good serviceable anvil for the farm and small mechanic.

Wt., Lbs.	Height, In.	Width, In.	Length, In	Per lb.
50	7¾	3⅛	14½	$0 12
70	8½	3½	18	$0 12
100	9¾	3½	20	$0 12

Ad from The Faeth Co., Kansas City, Mo., 1925 catalog. (From Mr. John Catchings)

PHOENIX

The Phoenix Hardware Manufacturing Company of Buffalo, New York, produced a cast-iron bench or crafts anvil as well as an anvil vise of a fairly large capacity, making the anvil portions rather large. I have seen these anvil vises without the moveable vise portion, which many consider anvils in their own right. In a 1941 Manufacturers Directory the company is listed under "Blacksmith's Anvils."[7] Sears Roebuck and Montgomery Ward carried these anvil vises in their early catalogs.

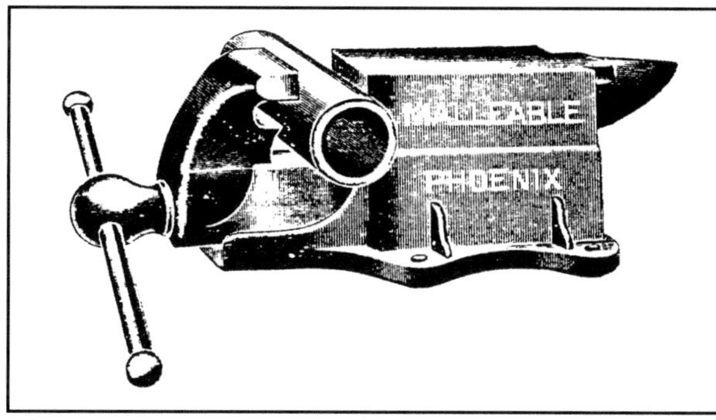

Phoenix anvil vise.

GIPSEY

The Bonney Forge & Tool Works of Allentown, Pennsylvania, made a cast-iron bench anvil and produced an anvil vise combination under the name *Gipsey.* The anvil portion was quite large and could serve as a small anvil for light work. Here again, I have seen an example of this anvil vise without its moveable portion. The company also manufactured a combination anvil, vise and drilling machine weighing 45 pounds.

BRIDGE ANVILS

I mention bridge or railroad anvils here, not in reference to a manufacturer, but due to the fact that most of them were made of cast iron and almost all of the ones that I have seen were also "no namers." These will be discussed later.

CAST-IRON ANVILS WITH STEEL FACES

This type of anvil is peculiar to the United States. I have no knowledge of any other country producing them, although there was an English patent taken out for one (see Appendix F). They are an American invention of Mr. Mark Fisher. In the late 1830s or early 1840s Mr. Fisher discovered a process in which cast or crucible steel (the best grade of steel) could be welded to cast iron.

Six American manufacturers used this process to make anvils. Although this type of anvil is considered a lower-grade product compared to wrought or all-steel anvils, they were the first anvils to be successfully made and marketed in this country. They were also the last anvils to be made by a leading manufacturer. That speaks for itself when you consider that several of the steel-faced cast-iron anvils sold for almost as much as the wrought-iron anvils. I will cover the six companies that I know of in alphabetical order. They are *Badger, Fisher & Norris, Samson, Southern Crescent, Star,* and *Vulcan.*

BADGER

In July of 1900 the Racine Steel and Iron Manufacturing Company of Racine, Wisconsin, incorporated as the American Skein & Foundry Company. [8] About this time they added an anvil to their product line, calling it the *Badger,* after the state animal.

I have yet to locate one of these anvils, although they were manufactured for about 50 years. However, I have located a number of the screw jacks that were manufactured by them during this period of time. These have the "A.S. & F. Co." (American Skein & Foundry Co.) markings, as well as an oval trademark with a relief of the *Badger* animal in its center. I have to believe that this is how the trademark would appear on the side of their anvils.

American Skein & Foundry Co. screwjack. (R.A.P. photo)

The earliest advertisement that I have obtained comes from the *George Worthington Company* catalog of 1909. I doubt that the anvil illustrated actually represents the *Badger* anvil. This ad shows the company offering smithing anvils in 18 different weights from 20 to 250 pounds. The last advertisement that I have seen was in the *Miller Brothers Hardware Company* catalog. There is no exact date on the catalog, but it was issued sometime in the 1930s. The illustration in this catalog is probably close to what the actual anvil and trademark look like.

BADGER

The face of this anvil is of solid tool steel, welded to the cast iron body of the anvil, tempered and polished.

Point is made entirely of tough untempered steel, the horn has a covering of same material.

Nos.	2	3	4	5	6
Wt. each, lbs.	20	30	40	50	60
Each	$8.00	9.00	11.50	12.00	13.00

Nos.	7	8	10	12	15
Wt. each, lbs.	70	80	100	120	150
Each	$14.50	16.00	20.00	24.00	30.00

Miller Brother's Hardware Co., ca. 1930's catalog.

I do not know when the company stopped manufacturing the *Badger* line of anvils, but it was either in the 1950s or early 1960s because the Illinois Iron & Bolt Company bought the American Skein & Foundry Company.[9] The Illinois Iron & Bolt Company began manufacturing the *Vulcan* anvil in Racine, Wisconsin, and apparently dropped the *Badger.* Production of Vulcans continued until about 1969. In 1971 a Japanese firm bought out the foundry and closed its doors forever.[10]

BADGER, BLACKSMITHS'.

Fig. 1521.

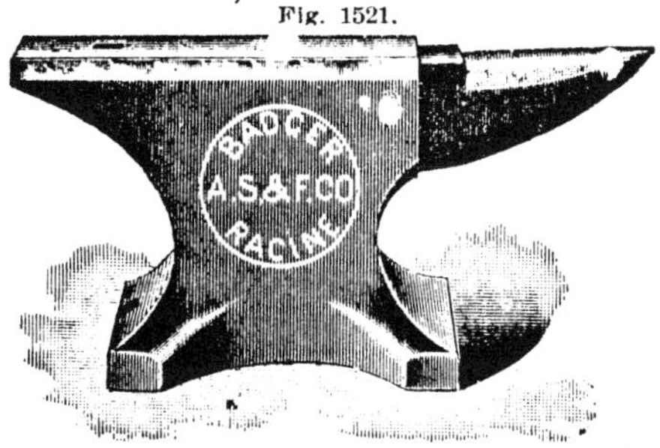

Nos.	2	3	4	5	6	7	8	9	10	11
Weight, lbs.	20	30	40	50	60	70	80	90	100	110
Each	$4 00	4 50	5 25	6 00	6 50	7 25	8 00	9 00	10 00	11 00

Nos.	12	13	14	15	16	17½	20	25
Weight, lbs.	120	130	140	150	160	175	200	250
Each	$12 00	13 00	14 00	15 00	16 00	17 50	20 00	25 00

The Face of the Badger Anvil is Solid Tool Steel, welded to the Cast Iron Body of Anvil. The Horn is covered with, and its point is made entirely of tough untempered Steel.

Above Anvils for shipment from factory only.

The George Worthington Co., 1909 catalog.

If any of my readers know of the whereabouts of a *Badger* anvil please contact me.

ANVILS
BADGER

Steel face, welded to cast iron body, horn is covered and point is made entirely of tough untempered steel.

Nos.	**2**	**3**	**4**	**5**
Approximate wt. each, lbs	20	30	40	50
" size of face, inches	2¾x6	2¾x7½	3x7½	3x8½
" length over all, inches	10¾	12	12½	13
" height, inches	5½	6¼	7¼	7½
Size of Hardie hole, inches	½	½	⅝	⅝
" " Pritchel hole, inches	3/16	3/16	3/16	¼
Each	$8.50	10.00	11.50	13.50

California Hareware Co., Los Angeles, California, 1921 catalog.

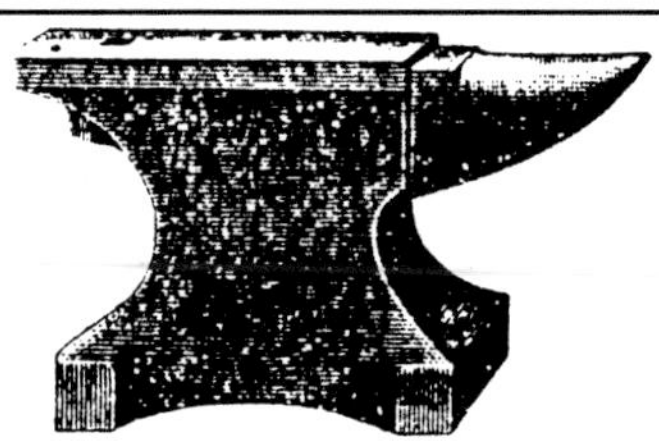

BADGER.

Solid Steel, Polished Face, Welded to a Cast Iron Body, Natural Finish; The Horn is Covered with and the Point is made Entirely of Tough Untempered Steel; Square Hardy Hole near the Heel.

Nos.	**50**	**70**	**100**
Dimensions of Face, inches	3¼x8½	3½x10	3¾x12¼
Height over all, inches	7½	9	10
Length over all, inches	13	16½	19
Size of Hardy Hole, inches	½	¾	¾
Weight each, lbs.	50	70	100
Each	**$13.50**	**16.50**	**22.00**

Packed Loose; Face and Horn Burlapped.

Simmons Hardware Co., St. Louis, Mo. 1930 catalog.

FISHER & NORRIS

More has been written about the *Fisher* or *Eagle* anvils than all other American anvils combined. This is probably justified by the fact that Fisher & Norris of Trenton, New Jersey, was the first successful anvil manufacturer to exist in this country, as well as the last of the major anvil manufacturers to stop producing anvils. The company was founded about the year 1843 and produced anvils for 136 years, until December 14, 1979.[11] A lot of 100 anvils for the United States government were the last anvils made.[12] During that 136 years the company probably produced in excess of 400,000 anvils. By 1909 the company had advertised having 300,000 in use. They produced more anvils than any other American anvil maker.

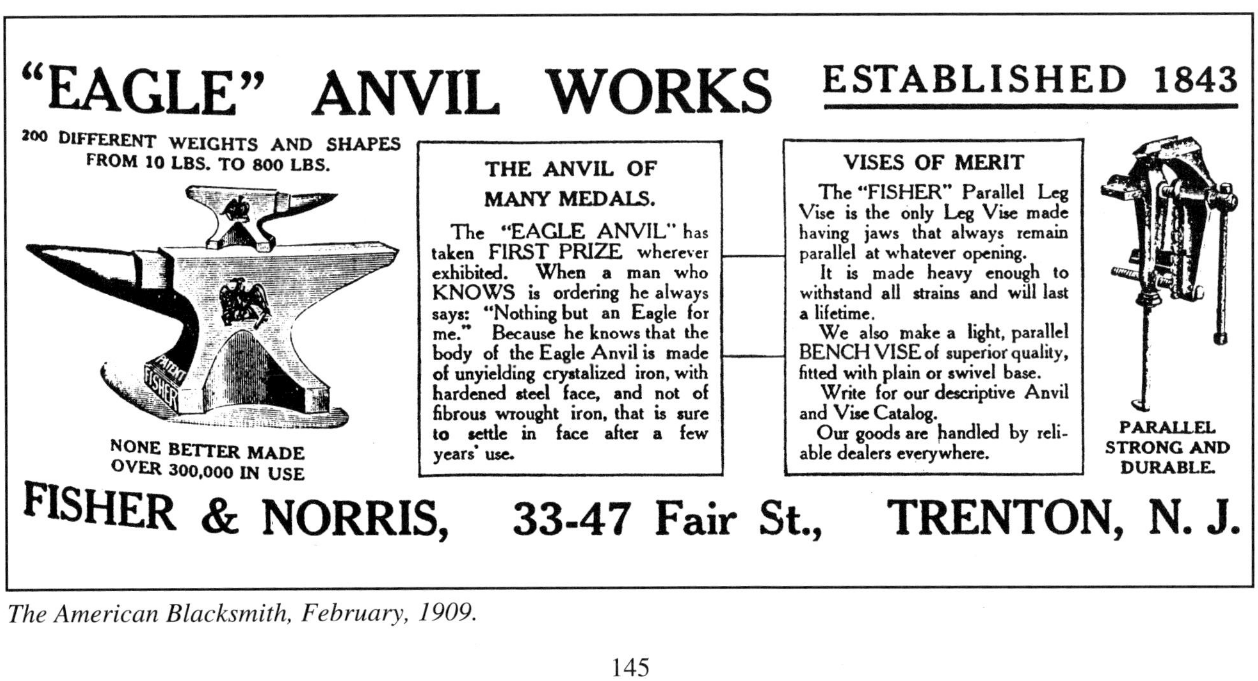

The American Blacksmith, February, 1909.

The fact that these were cast-iron steel-faced anvils and not the more popular wrought-iron anvils, preferred by most blacksmiths, makes this number even more remarkable. Most smiths and wrought-iron anvil manufacturers considered the *Fisher & Norris* a lower grade anvil. One consideration was the fact that these anvils did not "ring" when struck and did not rebound the hammer as well as anvils made of wrought iron. Fisher would have come back with a rebuttal of their own, but more on that later. "Ring" or no "ring," many smiths, industries and the United States government continued to buy and depend on the anvil that came to be known as "Old Reliable."

> *Eagle Anvils are probably the most widely known anvils in the world. For seventy years, Fisher & Norris have been the sole makers of these anvils. Today they are used in thousands of progressive shops in every country, and on account of their durability and the complete satisfaction they always give they have become known as the "Old Reliable."*
>
> *Eagle Anvils are made only in Trenton, N.J., by Fisher & Norris who have been established there since the year 1843.*[13]

Fisher & Norris was the only major American anvil manufacturer in the United States until the 1890s. That and the fact that they were cheaper than the imported English, German and Swedish wrought anvils helped their sales greatly. They did have one other advantage over the pre-20th century wrought anvils. They did not settle in the face or "belly" like all wrought-iron anvils would do in time when put to heavy use. The only disadvantage one leading American wrought-iron anvil manufacturer could say of them was "they were too thick in the heel."[14]

The major disadvantage that I see with this anvil is that should the face crack or a piece of the face breaks off in use, they can then become very difficult to repair. It is cheaper to buy a new one than to try to have one fixed. The company guaranteed them against breakage caused in manufacture or in use, but that is no help for someone today. The following, "Directions for Returning" a cracked or broken *Fisher & Norris* anvil, will explain their guarantee.[15]

Established 1843.

Directions for Returning "EAGLE" Anvils

Our Anvils are the only ones in the market that are **Fully Warranted.**

The "Peter Wright," and all other brands of so-called "Wrought" Anvils, must be purchased at the customer's own risk; consequently, when the Face, which in these foreign-made Anvils is only a low grade of "Shear Steel," turns out to have soft spots, or soft all over, or so brittle that it crumbles and breaks along the edges, or proves to be imperfectly welded, cracking and peeling off from the body, he can get no redress - they will not take it back, nor replace it with another. Sometimes the whole Tail breaks, or the Horn drops off. Then he is forced to pay out more money for another that is just as likely to go in the same way.

Now, while we cheerfully give a broader and more liberal guarantee to the purchaser of our "Eagle" Anvils than is ever given on any other tool in the trade, we require compliance with certain conditions, as follows, viz.:

I - The Anvil claimed to be defective must be returned to us for our inspection before shipping another to replace it. We must insist upon this for a number of reasons. There are certainly irresponsible concerns in this country who haveat different times made and sent out worthless imitations of our Anvils, which the purchaser, supposing to be genuine "Eagle" Anvils has upon their failure reported and expected to be made good by us - we must see them first, for identification. Instances have occurred where our Anvils have been purposely broken by malicious persons, who have then falsely claimed that they have been defective. We can always detect this if all the parts are returned to us. Besides the above reasons our examination enables us to discover any faults in manufacture, to be corrected in the new Anvil which we send you as soon as the old one is received by us. The so-called "Trenton" anvils are not made in Trentonnor by us. See that the "Eagle" and "Fisher" are on the Anvil if ours.

II - State in what particular the Anvil proves to be imperfect.
This is necessary because sometimes Anvils are returned to us which at first sight appear to be in perfect condition.

III - MARK ON THE CUT at head of sheet at what part of Face or Horn the defect has been found, and mail the same to us. This is because it facilitates our inspection as to defects in the Cast Steel face of seams, softness or imperfect welding.

IV - Mark the Anvil in Paint, thus: "F. & N., Trenton, N. Jersey," also your initials and "Old Iron."
This is very important because, if nothing but a "tag" is put on an Anvil, it gets torn off at the first handling, and consequently the Anvil may never reach us, or if it does, there would be nothing on it to show who has sent it. Marking it "Old Iron," and having it so designated instead of "Anvil," gets lower freight rate.

V - Mail to us, at the same time, the Railroad Receipt. Without this the Anvil will be delayed in reaching us, as there are four different Transportation Lines that bring freight to Trenton.

VI - If the Anvil is broken, be careful to send all the pieces together. This is because it enables us to learn the cause of failure, and to correct it in the Anvil which we send you in replacement.

As in all manufactures, it will sometimes occur that among the many thousands of Anvils yearly made by us, notwithstanding our most rigid inspection, there will occasionally escape one with defects of manufacture. If defects in manufacture develop within one year after shipment from factory, it is our wish and intention that every such one shall be returned to us and replaced to the purchaser.

Every genuine "Eagle" Anvil has the word "FISHER" stamped in.

FISHER & NORRIS

Trenton, N. Jersey, June 1st, 1911.

The above material was supplied by the Smithsonian Institution.

I have recorded far fewer of these *Fisher "Eagles"* than I have of the American wrought anvil manufacturer who made the fewest anvils (see Arm & Hammer). Considering the large number of *Fisher* anvils manufactured I suggest that the majority of the *Fisher* anvils may have gone to the scrap yard. I find very few smiths or hobby smiths of today (1995) using these anvils. The phrase, "no ring - no want" may have some truth to it.[16]

I know of three publications that give brief histories of the Fisher & Norris company. The best of these, in my opinion, is in Alexander Farnham's book, *Early Tools of New Jersey and the Men Who Made Them.*[17] H.R. Bradley Smith's book *Blacksmith's and Farrier's Tools at the Shelburne Museum* is also good.[18] M.T. Richardson in his book *Practical Blacksmithing* gives an interesting account of the *Fisher & Norris Eagle* anvil and how it was made.[19] For those of you who do not own these books or can not get them from a library or from someone you know, I will give a brief review of the company's history as gleaned from these authors, as well as from an interview that I obtained in 1984 from the present owners and a little of my own conjectures.[20]

Mr. Mark Fisher was said to have operated a tannery in Levant, Maine, in the 1840s. He must have been interested in metalworking because after a fire in which his tannery burned to the ground, while "digging around in the debris, he found where a piece of steel and iron had welded together."[21] For many years men had been trying to weld cast iron to steel without much success, and here Mr. Fisher finds it welded together in a hot tannery fire. After experimenting for a while he came up with the solution. Mr. Fisher decided to apply this new knowledge to the facing of cast-iron anvils by welding on a crucible steel face plate (the best steel available). Maybe he had a cast-iron anvil at the tannery and the piece of steel he found was welded to a part of this anvil. Whatever the case, he most likely started making and experimenting with anvils in Levant or possibly Newport, Maine, and by 1847 took out a patent with William Morton, Jr. (probably a backer) on the welding of cast iron to steel and cast iron to wrought iron.

UNITED STATES PATENT OFFICE

William Martin, Jr., and Mark Fisher, of Newport, Maine.

Improvement in Welding Cast-Iron to Malleable Iron or Steel

Specification forming part of Letters Patent No. **5,331,** dated October 16, 1847

To all whom it may concern:

Be it known that we, William Martin, Jr. and Mark Fisher, of Newport, in the county of Penobscot and the State of Maine, have made a new and useful Improvement in the Manner of Uniting or Welding Cast-Steel and Steel of other kinds or Wrought-Iron with Cast-Iron; and we do hereby declare that the following is a full and exact description thereof.

Various attempts have been made to unite or weld cast-steel and steel of other kinds or Wrought-iron with cast-iron by the casting of the iron in a molten state upon the steel or wrought-iron. The process by which this has been attempted has been to prepare a plate or other duly formed piece of cast-steel or of other kinds of steel or of wrought-iron by brightening the part with which it was intended the cast-iron should unite and coating it with calcined borax or other analogous flux. The piece thus prepared has then been laid in its proper place in the mold in which the casting was to be made, and the molten iron has then been poured in upon it. The results of this mode of procedure have been altogether uncertain and unsatisfactory. In some cases the union has appeared to be perfect, but in a very large majority of cases the union has been extremely partial, and has most frequently failed entirely. The consequence has been that after much loss the procedure has been abandoned, which, if successful, would have been of great value.

In our improved process we prepare the plate or piece of cast or other steel or of wrought-iron in the manner above described, brightening its surface, coating it with calcined borax, and laying it in the flask or mold. Thus far there is not any novelty in our manner of proceeding. Our invention consists in the manner of forming the gates or jets through which the molten iron passes into the flask, which is such as that it shall not be poured directly onto the piece of cast-steel with which it is to unite, but that it shall flow over its prepared surface, and be made thereby, as it were, to wash it. By this means fresh particles of the molten iron are brought into successive contact with every portion of the face of the steel or wrought-iron, while the flux protects the metal from oxidation all foreign matter is completely removed, and the union of the two metals is perfect.

In the molding of some articles it will be necessary to form a recess in the sand to receive the waste metal that has been allowed to flow over the cast-steel; but where there are a number of similar articles in the flask every competent founder will be able so to arrange the jets as that the molten metal will pass from one prepared piece of steel or iron to the other in the first instance, and as the flasks fill the articles will be protected without the necessity of any receptacle for waste metal, all which will be readily understood by the molder.

Having thus fully described the nature of our improvement in the process of uniting or welding cast-steel or steel of other kinds or wrought-iron with cast-iron, we do hereby declare that we do not claim the uniting of the two portions of metal with each other by merely pouring the cast-iron upon the prepared piece of steel or of wrought-iron, this having been frequently attempted; but we limit our claim to improvement to the so forming the jets through which the molten iron passes into the flasks as that the molten metal shall flow laterally over the surface of the prepared wrought-iron, for the purpose and substantially in the manner herein fully made known.

We. Martin, Jun.
M. Fisher.

Witnesses as to the signature of We. Martin, Jr.:
This. P. Jones,
Lema. Williams.
Witnesses as to the signature of M. Fisher:
H. L. Dankly,
Chase P. Mason.

By at least 1853, Mr. Fisher had moved to Trenton, New Jersey, and acquired a new partner and backer in the person of Mr. John H. Norris. Thereafter the company was known as Fisher & Norris of 37 to 45 Fair Street, Trenton, New Jersey. In 1853 Fisher & Norris received an additional patent for the polishing of anvils. There may have been earlier *Fisher* anvil patents acquired before these, but I have no information on them.

UNITED STATES PATENT OFFICE

Mark Fisher and John Norris, of Trenton, New Jersey.

Apparatus For Polishing Anvils

Specification of Letters Patent No. 10,066, dated October 4, 1853.

To all whom it may concern:

Be it known that we, Mark Fisher and John H. Norris, of Trenton, in the county of Mercer and State of New Jersey, have invented certain new and useful Improvements in Facing and Polishing Anvils and other Irregular Conical Surfaces and the horns thereof made of cast-iron with wrought steel or iron faces or in any other way; and we do hereby declare that the following is a full, clear, and exact description thereof, reference being had to the accompanying drawings, which form a part of this description, in which -

Fig 1, is a perspective view of the whole apparatus. Fig. 2, shows the anvil in a different position from Fig. 1.

Our invention consists in applying an emery wheel or other grinding wheel or stone with a universal action to an anvil so hung and supported as to be revolved on the axis of the conical horn, while it can be turned horizontally by the swing rests by which it is supported, adapting itself to the inequalities of the casting.

The construction is as follows: We form an upright frame, consisting of two upright posts (a) properly braced, in the top of which there is supported the journals of an axis (b) on which are two pulleys, one of which (c) receives a band from the driving power; the other (d,) communicates motion to another pulley (e,) or a shaft (e^1) at the ends of two arms (f,) whose center is the axis (b,); these arms (f,) hang down and swing upon the axis (b,); two other arms (g,) are jointed to the ends of arms (f,) by means of shaft (e,) which passes through both; and these lower arms (g,) bear on their lower ends the axis of the emery wheel (h,), driven by a pulley (i,), by being connected by a band with pulley (e^2,) on shaft (e); the polishing apparatus thus constructed can be brought into a variety of positions, as required, by certain inequalities of the casting, and can also polish the plane surface of the face of the anvil.

In front of the upright posts (a,) and at a proper distance therefrom, there is a sliding standard (k) made to slide in a proper groove (l,) in a right line in front of the posts; this standard (k,) not only slides, but it has also a swiveling motion; and from one side near the bottom an arm (m) projects nearly horizontally, its outer end branching out into a curved crosspiece in form nearly resembling the letter (T); on the two ends of the cross piece are rollers that rest on the platform below, so that as the standard (k,) swivels the end of the cross-arm shall trundle around in a circle, while it is capable of sliding along with standard (k,) the whole length of its groove (l,). In the top of the standard (k,) there is a recess into which the small point of the horn of the anvil fits, laying over far enough to support the weight of the anvil, and at the same time projecting a little above the surface or the top of the standard so as to be polished to its extreme point; a puppet head projects up from the arm (m), near the cross through which a centering mandrel (m,) passes in a line with the axis of the conical horn, bringing the surface of the same to a horizontal line, as clearly illustrated in the drawing, so as to bring all parts of the anvil properly under the horizontal polisher for its action thereon. There are handles projecting from the ends of the arms (g,) by which the polishing wheel is guided, and the anvil is turned by a crank upon the center mandril (n,) as it is worked off. What we claim therein as new and of our invention is-

Suspending the anvil in the sliding and vibrating frame, and polishing part of the apparatus, and operating them as herein described.

M. FISHER.
JOHN H. NORRIS

Witnesses:
Mary M. Evans,
Saml. Evans.

Fisher & Morris,

Polishing Anvils.

No 10,066. Patented Oct. 4, 1853.

Fig; 1.

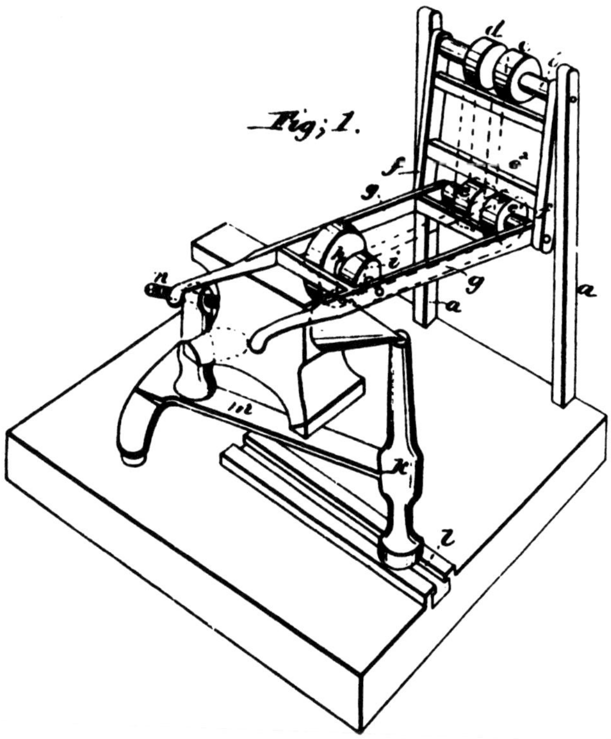

Fig; 2.

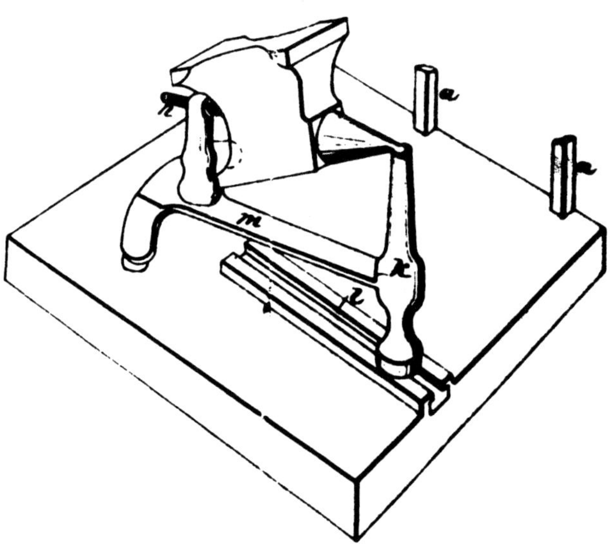

Both patents and drawings courtesy of the Smithsonian Institution

Mark Fisher, the founder, died in 1871 and John Norris in 1872, leaving Clark Fisher, the son, manager of the Eagle Anvil Works. Shortly after 1900 Clark died, and his daughter Harriet Fisher assumed the role of manager and owner. One account states that Harriet Fisher operated the company until 1961, at which time she sold it to Crossley Machine Company.[22] Another account basically states the same information, but identifies the owner as Mrs. Silvano A. Andrew, most likely one and the same person.[23] In the December, 1910, issue of *The American Blacksmith*, Mrs. Clark Fisher is listed as the company director[24] and from another source, Harriet Fisher is said to have been in charge of operations from 1903 to 1939.[25] Whatever the truth, the Eagle Anvil Works remained in the Fisher family until 1961.[26]

In addition to anvils, Fisher & Norris manufactured patented railroad rail joints, a unique leg or blacksmith's vise and other items that required the welding of cast iron to steel or iron.[27] Through the years the *Fisher & Norris* anvils evolved in shape and size. The *Fisher* anvils that I have recorded dating before the 1860s are stubby with short horns and heels compared to the wrought-iron anvils of the same period. Along with an increase in size, the anvil's proportions also improved as time passed. In 1877 major changes took place in the anvil's design. The feet were changed from the *Old English* style to the *Peter Wright* style, and lug mounts were added to the front and rear of the base on the larger anvils. Fisher & Norris evidently received a patent for these lugs. Unfortunately, I do not have a copy of this patent at this time. A number of these anvils manufactured shortly after 1877 had "Patent Apr. 24, 1877" across the top of the front of the foot under the horn or on the back foot under the heel. I suspect that the 1877 patent refers to these lugs. Anvils over 500 pounds are without these lugs, as they are not going to jump around when worked on, anyway. Before 1877 most of the smaller anvils (under 200 pounds) had a cut-out in the middle of the rear foot for mounting.

The year of manufacture was also added in relief on most of the anvils after 1880. The date is generally found on top of the foot under the heel. However, the date is not found on all later anvils, especially ones over 300 pounds. It took

Vise ad from The Blacksmith & Wheelwright, December, 1917.

time to make the date changes to all the patterns, plus there were fewer orders for the larger anvils. It was not cost-efficient to make these changes continuously, when approximately 200 patterns were involved, and some were sold only infrequently.

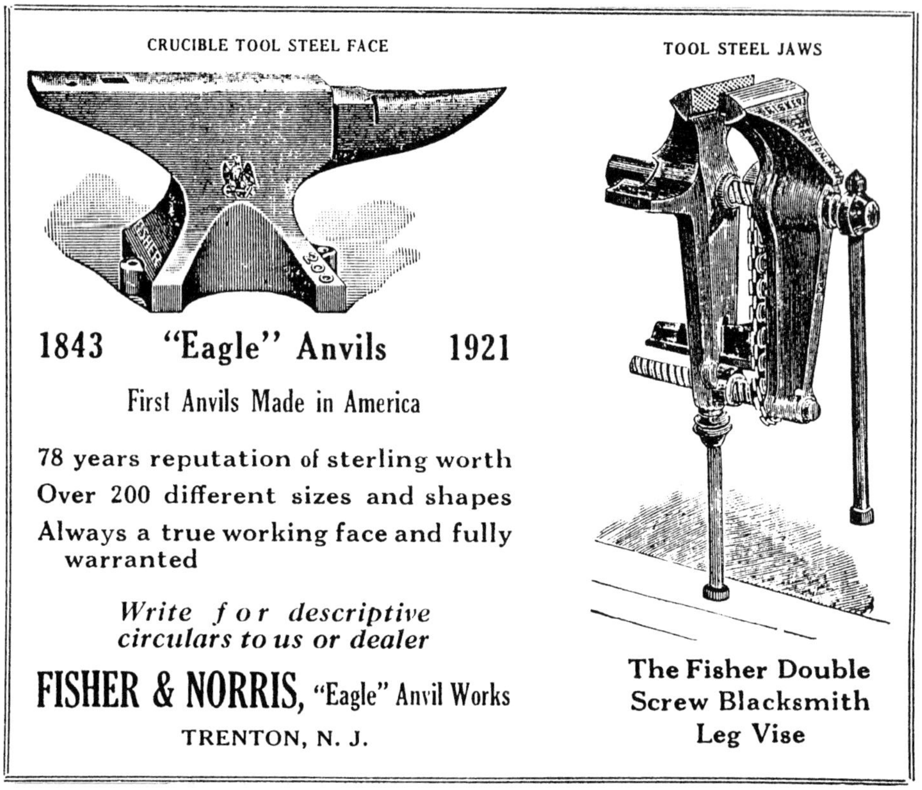

From The Blacksmith & Wheelwright and Tractor Repair Journal, May, 1921.

Here is the correction to an error that has crept into all the written accounts about Fisher & Norris. The company is credited for manufacturing the largest anvil in the world, which was exhibited at the Centennial Exposition in Philadelphia, Pennsylvania, in 1876 as a 1600-pound smith's anvil.[28] Fortunately the Smithsonian Institution has two photographs of the Fisher & Norris exhibit. I obtained copies of these photos, compliments of Mr. Shayt of the Institution. The photos reveal not a 1600-pound anvil as originally thought, but a 1400-pound anvil instead. However, this is still the largest smith's anvil to have been manufactured. The 200-pound error probably lies in the reporting on the display. Whoever the author or newspaper reporter was who originally wrote an article on the display may have mistaken the "4" for a "6" when taking notes. Whatever the truth may be for this error, the point is that Fisher & Norris was still the company that manufactured the single largest smith's anvil ever recorded. A 1906 advertisement in *The American Blacksmith* states that the company manufactured smith's anvils up to 1300 pounds as standard practice, though in their latest ad they changed the maximum weight to 700 pounds.

Fisher & Norris display at the 1876 Centennial Exposition. Courtesy of the Smithsonian Institution.

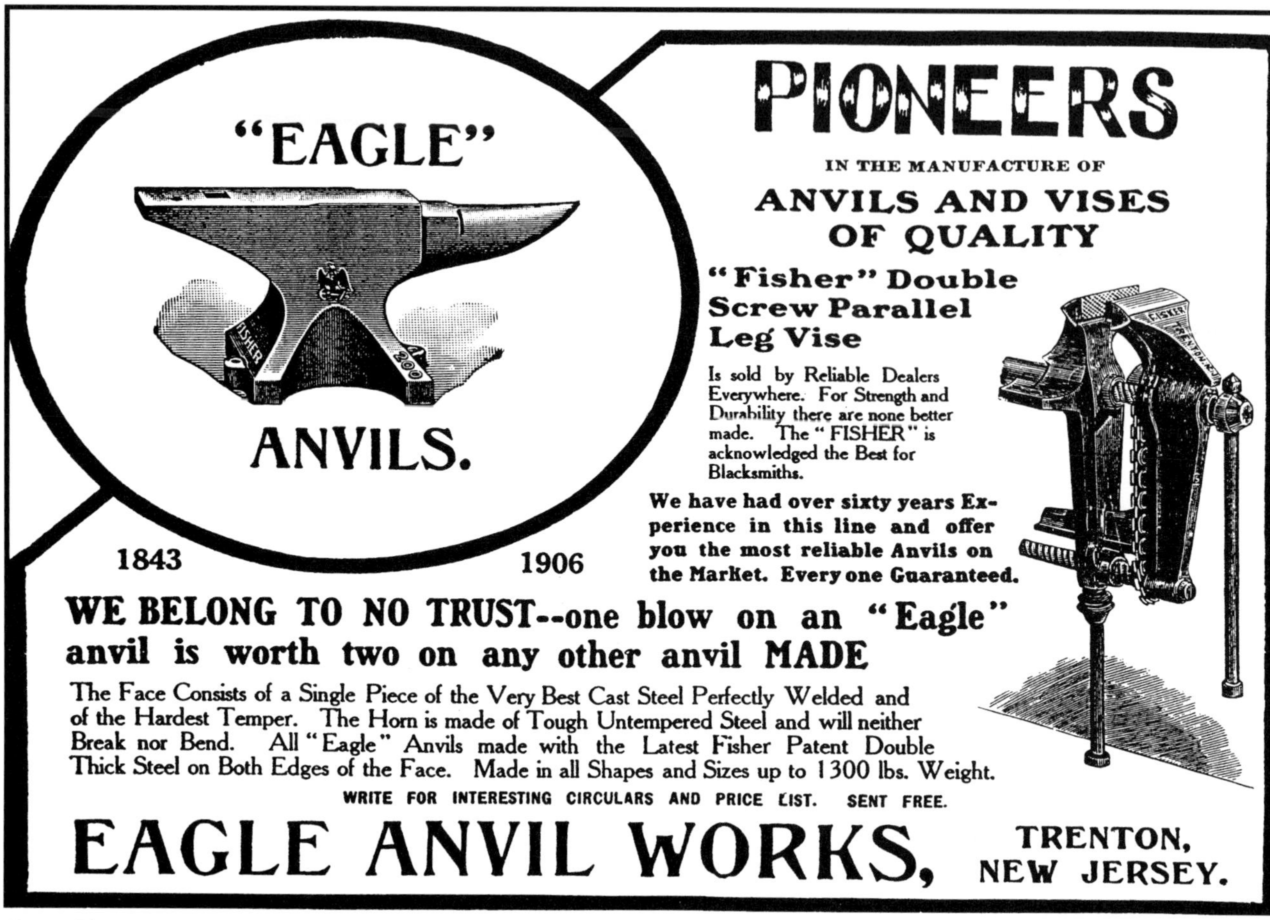

From The American Blacksmith, July, 1906.

The "Eagle" holding an "Anchor," cast in a number of different styles, was the company's trademark, along with the name "Fisher" on the front of the foot under the horn. Fisher & Norris stated in their advertisements that the "Eagle" had to be there if it was a genuine *Fisher* anvil.[29] However, some of the anvils I have recorded are missing the "Eagle" while bearing the rest of the Fisher trademarks, and are definitely *Fisher* anvils. At first I thought that someone left the "Eagle" off of the pattern by mistake, and you do not scrap a perfectly good anvil for such a mistake. As time passed, I kept recording more of these "Eagle-less" *Fisher* anvils. Finally a reliable source came up with an answer that makes sense and is most likely true.[30] After the Civil War Southerners, for the most part, held the United States Federal government in contempt. (I know because my mother's side of the family is very Southern). The "Eagle" represented the Federal Government. I was told that the Federal Cavalry, when raiding towns in the South, made a point of finding the blacksmith shops and destroying the smith's anvils whenever possible. This mischief made it difficult for the smith to shoe horses for the Southern Cavalry or to make other repairs for the Southern forces. As a result many anvils were needed in the South after the War. Fisher apparently discovered the dislike of the "Eagle" by Southern smiths who were reluctant to buy their anvils, and removed this symbol from anvils being shipped to the South. I have no idea how long this practice lasted and you can be sure no one put anything in writing, but I would guess it lasted close to the beginning of the 20th century, if not longer. If you find an anvil that is a *Fisher* in every respect except the missing "Eagle," you probably have one of these Southern *Fisher* anvils.

The weight was generally cast in relief on anvils made after about 1880 on the top of the foot, away from the body, next to the horn. The anvil weights were in 10-pound increments under 200 pounds, 25-pound increments up to 300 pounds, 50-pound increments up to 500 pounds and 100-pound increments for all anvils over 500 pounds. Most of the very heavy anvils over 500 pounds that I have seen do not show the weight. Generally speaking, the last digit is left off of the weight mark on many of these anvils: that is, 10 = 100 pounds (see The Weight of Anvils).

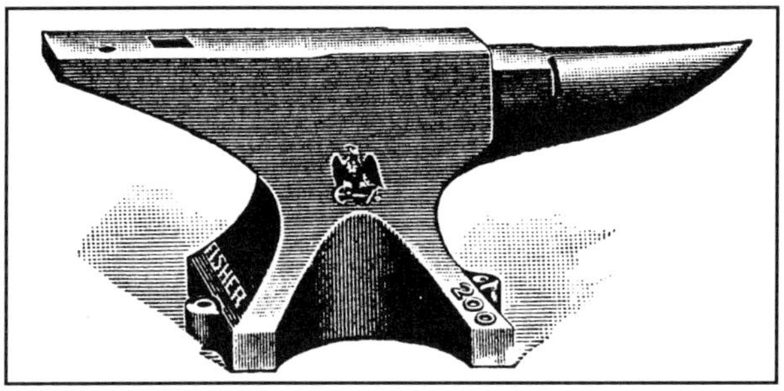

Engraving of a 200-pound Fisher anvil.

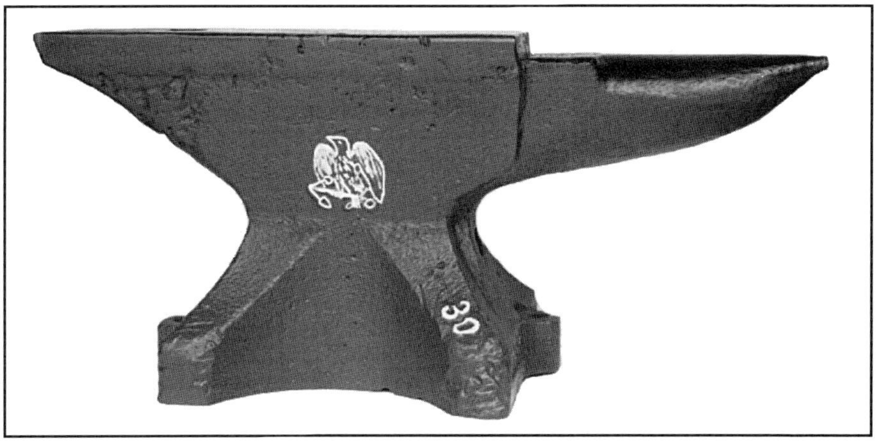

300-pound Fisher, dated 1911. Notice that the last zero is missing. (R.A.P.photo)

The very early *Fishers,* in most cases, show the "Eagle" in a circle in slight relief and later in bold relief without the circle. A *Fisher* that I recognized in Gardiner's *Photographs of the Civil War* was mounted on a stump with the heel toward the camera.[31] I could tell it was a *Fisher* because the side of the "Eagle" stood out in such bold relief. After 1876 most of the "Eagles" appear in flat relief, but the "Eagle Anvil Works" was not always consistent with the "Eagle" styles. All *Fisher* anvils with "Eagles" that I have come in contact with have the "Eagle" shown on the side as you face the anvil with the horn on your right. On one 60-pound anvil dated 1882, I observed that the "Eagle" was in a recessed square. On another *Fisher* with the "Eagle" within the recessed square, dated 1887, were found the following patent dates on the front of the foot under the heel. "Patent Apr. 24, 1877, May 13, 84, June 26, 87." Foundry marks were not consistent either, and can be found on various parts of some *Fisher* anvils.

100 -pound very early Fisher anvil. No markings except the "Eagle" in a circle, pre-1860. (R.A.P. photo)

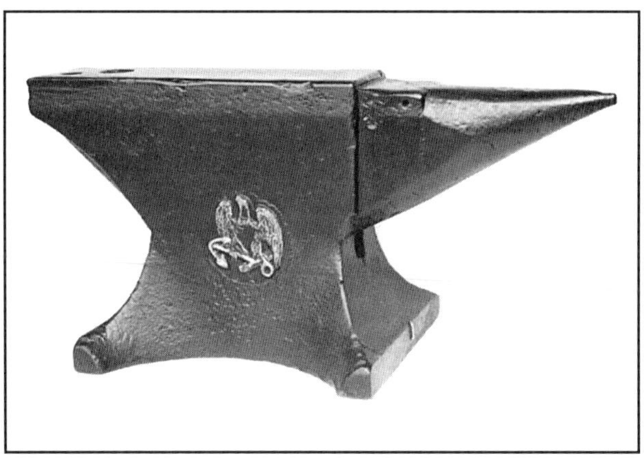

135-pound very early Fisher anvil. No markings except the flat Eagle. I date it ca. 1870. (R.A.P. photo)

95-pound Fisher anvil. The Eagle is recessed. Only the name Fisher appears on the front of the foot. I date this anvil ca. 1877. (R.A.P. photo)

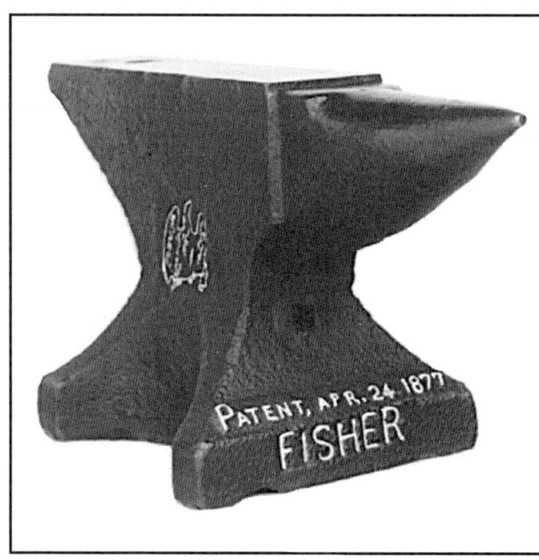

150-pound Fisher anvil. It has the Eagle recessed, Fisher and the patent date. There is no date and no lugs. I date this anvil ca. 1880, but before 1882. (R.A.P. photo)

70-pound Fisher anvil. It has the Eagle recessed, Fisher, the weight, no lugs and is dated 1882. (R.A.P. photo)

130-pound Fisher anvil. Typical of all smith patttern Fishers after 1890. Dated 1908. (R.A.P. photo)

100-pound Fisher anvil. Typical of all late model Fishers. Dated 1913. (R.A.P. photo)

The trademark or markings were, of course, a part of the casting. Depending upon how well the mold filled out, these markings were either clear or broken, sometimes leaving parts of the "Eagle" or date missing.

It can be seen clearly from photos taken at the 1876 Exposition that Fisher & Norris made numerous styles of anvils beside the smith's pattern. Many of those shown are gold and silversmith's stakes, as well as sawmaker's anvils. The only chainmaker's anvils that I have recorded, until recently (1994), were made by Fisher & Norris.

Another view of the 1876 Centennial Exposition display. Courtesy of the Smithsonian Institution.

110-pound chainmaker's anvil. Dated 1907. This is the smallest of this pattern anvil that Fisher & Norris made. (R.A.P. photo)

150-pound Fisher farrier's anvil. Dated 1888. (R.A.P. photo taken at Mr. Robert Hansen's)

About the year 1887, Fisher & Norris started making a farrier's pattern anvil with a very distinctive trademark. To date I have recorded two of these anvils and have heard of others.

The following Fisher & Norris brochure was sent to me by Mr. James Campbell of Pennsylvania, and it lists many of the sizes, weights and patterns of their anvils. In my opinion the brochure dates back to the late 1930s. Under the "Our Products" heading, they list "Export Anvils." They apparently made the European double-horned patterns and an anvil they called *Insonora,* which must have been for the Spanish or South American market. Some of their advertisements state that they shipped anvils world-wide.

270-pound "Insonora" anvil. (Mr. Raymond E. Strom photo)

THE FISHER BLACKSMITH
Tool Steel Face
"EAGLE" ANVIL

U. S. Patents

Manufactured in Weights from
8 to 1,000 pounds

BY

FISHER & NORRIS
"EAGLE" ANVIL WORKS

Trenton, New Jersey
U. S. A.

Cable Address: "Anvil", Trenton

Established 1843

"EAGLE"
BLACKSMITH'S ANVIL

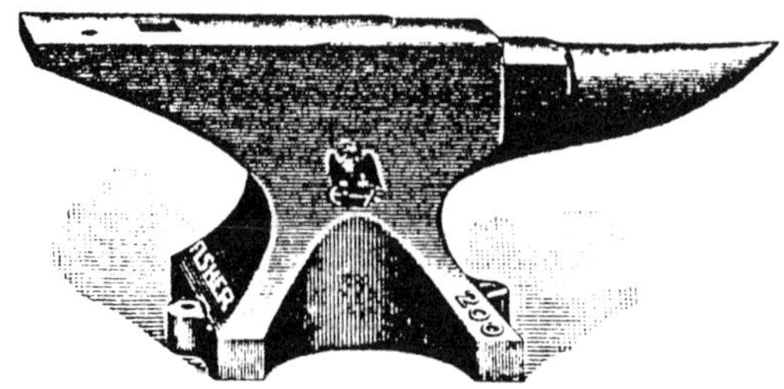

U. S. PATENTS:
200 DIFFERENT WEIGHTS AND SHAPES
FROM 8 LBS. TO 1,000 LBS.

LIGHT WEIGHT ANVILS OR MINIMS

DIMENSIONS:

SIZES UNDER 80 LBS.

Wt. Lbs.	FACE				HORN
	Length Inches	Width Inches	Hardie Hole Sq. In.	Pritchel Hole	Length Inches
8	5¾	1¾	⅜	¼	3½
10	6	2	½	5/16	3¾
20	7¼	2¼	½	5/16	4¼
30	8¾	2½	½	⅜	5
40	8¾	3	⅝	7/16	6¼
50	9¼	3	⅝	7/16	6½
60	10	3	⅝	7/16	6½
70	10½	3¼	¾	½	6½

"EAGLE" ANVILS

EAGLE ANVILS are manufactured at the oldest Anvil works in this country having been in general use for nearly a Century, and have at all times maintained an excellent reputation.

THE FACE is one piece of High Grade Tool Steel, perfectly welded to the body by our own process, accurately ground and tempered to suit the majority of Blacksmiths. After many years of steady use it remains perfectly true, and does not settle, break nor become loose from the BODY, which is a solid piece of unyielding Gun Iron of a special analysis. THE HORN is a solid piece of tough untempered forged steel, of unequalled strength.

ANVIL PRICE LIST

80 lbs. to 1000 lbs. sizes....c. per lb. Base

EXTRAS FOR LIGHT WEIGHT ANVILS

70 lb. size...........	½c.	per lb. extra
60 " "	1c.	" " "
50 " "	2c.	" " "
40 " "	3c.	" " "
30 " "	5c.	" " "
20 " "	8c.	" " "
10 " "	15c.	" " "
8 " "	20c.	" " "

Avoid deafness by working on an "Eagle" Anvil. It does not ring.

"EAGLE" ANVILS

DIMENSIONS:

SIZES 80 LBS. TO 1,000 LBS.

Wt. Lbs.	FACE				HORN
	Length Inches	Width Inches	Hardle Hole Sq. In.	Pritchel Hole	Length Inches
80	11	3¼	¾	½	7½
90	11	3¼	¾	½	7½
100	12	3½	¾	9/16	8½
110	12¾	3½	¾	9/16	8½
120	12¾	3½	¾	9/16	8½
130	14	4	¾	9/16	8½
140	14½	4	⅞	9/16	10
150	15	4	⅞	9/16	10
160	15	4¼	1	9/16	10
170	15	4¼	1	9/16	10
180	15½	4¼	1	9/16	10
200	16½	4¾	1⅛	⅝	11½
225	16½	4¾	1⅛	⅝	11½
250	18½	5¼	1¼	⅝	12½
275	19¾	5¼	1¼	⅝	12½
300	19¾	5¼	1¼	⅝	12½
350	20	6	1⅜	¾	13
400	21	6	1⅜	¾	13
450	22	6	1⅜	¾	13½
500	23	6½	1½	13/16	14½
600	24¼	6¾	1½	13/16	14½
700	25½	7¼	1⅝	13/16	14½
800	26¼	7¼	1⅝.	13/16	16¾
1000	27¼	7¾			16¾

We make special anvils of any weight or shape desired from 8 to 1000 pounds.

IRON ANVIL STANDS

Solidly built and neat in appearance. Made to fit the base of anvil and has lugs for fastening the anvil to stand with bolts.

IDEAL FOR MANUAL TRAINING SCHOOLS.

OUR PRODUCTS

"EAGLE"	BLACKSMITHS' ANVILS
"	HORSESHOERS' ANVILS
"	PLOW MAKERS' ANVILS
EXPORT	DOUBLE HORN ANVILS
"	"INSONORA" ANVILS
"EAGLE"	SAW MAKERS' ANVILS
"	CHAIN MAKERS' ANVILS
"FISHER"	BLACKSMITHS' VISES
"	MACHINISTS' VISES
"EAGLE"	SWAGE BLOCKS & STANDS
"	ANVIL STANDS
"FISHER"	ENGRAVERS' STAKES

OUR GUARANTEE

All our Anvils are given a most severe test, and are inspected before shipment. Should an Anvil, upon being used, prove defective, and is returned to us for examination within one year from date of shipment, we will replace it free of charge, providing same is not found defective through abuse.

FISHER & NORRIS

"EAGLE" ANVIL WORKS

37 to 45 Bloomsbury St. TRENTON, N. J.

U. S. A.

It must be remembered that Fisher & Norris had to have a pattern for every anvil size and style that they manufactured. Forged anvils made in open dies could be changed at will, with the anvil smiths adding or removing any features at their discretion. Fisher & Norris must have kept a large force of pattern makers at their disposal in order to replace worn-out patterns, create many new patterns to add to their growing line, and to meet special-order requests.

The following excerpt is taken from the book *Practical Blacksmithing.*[32] This gives us a good description of how the *Eagle* anvil was manufactured.

American Cast Anvils

In 1847, the late Mr. Mark Fisher, believing in the possibility of welding cast steel to a high grade of cast iron, which had up to that time been unknown, discovered a perfect and successful process by which the two metals could be welded together in any desired dimensions. (The largest anvil in the world was made by the Fisher Eagle Anvil Works for the Centennial Exposition in 1876. Its weight was 1600 lbs). The value of this process for anvils was apparent, as there could thus be obtained a perfect working surface of the best quality of cast steel, capable of hard and uniform temper on a body, which from its crystalline and inflexible structure, would never settle or get out of shape in use --one of the defects, liable to occur by continued hammering in [on] anvils with a fibrous wrought iron body, under the steel.

It also enabled a steel working surface to be applied to the horn, which previously had been only of plain iron.

The first manufacture of these anvils in this country, began under his patent in 1847, and is now recognized as a better article than the old-fashioned imported kind, over one-half of the anvils used in this country, it is said, being made by this process, and so certain and successful is it that they are the only ones in the market, fully warranted against breakage, settling of the face, or failure in any respect.

It is needless to say, that the ordinary cast iron would not answer for a tool, subject to such severe usage as an anvil.

The metal employed, must have a strength equal to that in gun castings, a certain elasticity to stand the strain of high heating and sudden cooling of the tempering process and perfectly sound in all parts. Though many so-called "cast" anvils have, from time to time, come upon the market, only one concern, in the country, and that the original one, operating under the Fisher patents, has continued to produce anvils with all the qualities described as necessary in these tools. The mode of manufacture is naturally quite different from that of wrought iron anvils.

The steel used is one piece, for the face, of the best tool cast steel.

The anvil is cast bottom side up, having this steel horn, placed in the "drag" or lower part of the mold.

Before filling in with the metal, which is not only to form the body of the anvil, but also to effect in its passage, the perfect welding required, the steel face and horn are heated to a bright color, and every part of their exposed surface is covered by the molten metal. After the necessary annealing, this rough anvil is removed, trimmed, planed true, and put into its finished shape, the cutter-holes made exact, and it is then ready for the hardening and tempering process. This last is the critical test, for both iron and steel must be heated to a high point and then suddenly plunged into the cold hardening liquid. Should there be any spot between the two metals not perfectly welded, the steel will separate, or the whole anvil will crack and fly into pieces; so that if it passes this stage successfully it is reasonably sure to be perfect, and therefore the makers can safely give a full warranty to the purchaser.

Recent improvements have added much to the value of this make of anvil. By extending the steel part of the horn down into the body, all danger of breakage of the horn, where it joins the main part is prevented. Also both edges of the steel face are made of double thickness, which prevents crumbling or splitting off of those places most exposed to severe usage, so common with the old fashioned anvils.

ANVILS IN AMERICA

Two peculiarities distinguish the American from the foreign anvil. They are more solid from crystalline structure of the body, and therefore do not bounce back the hammer or sledge, thus retaining all the effect of blows in the piece worked on, and the steel face always retains its original true surface for the same reason. Also there is very little "ring" in them, and this peculiarly is sometimes urged as an objection by those accustomed to the wrought iron anvils.

One of their ads ends like this: "Eagle anvils do not make music, they make money for their users,..."

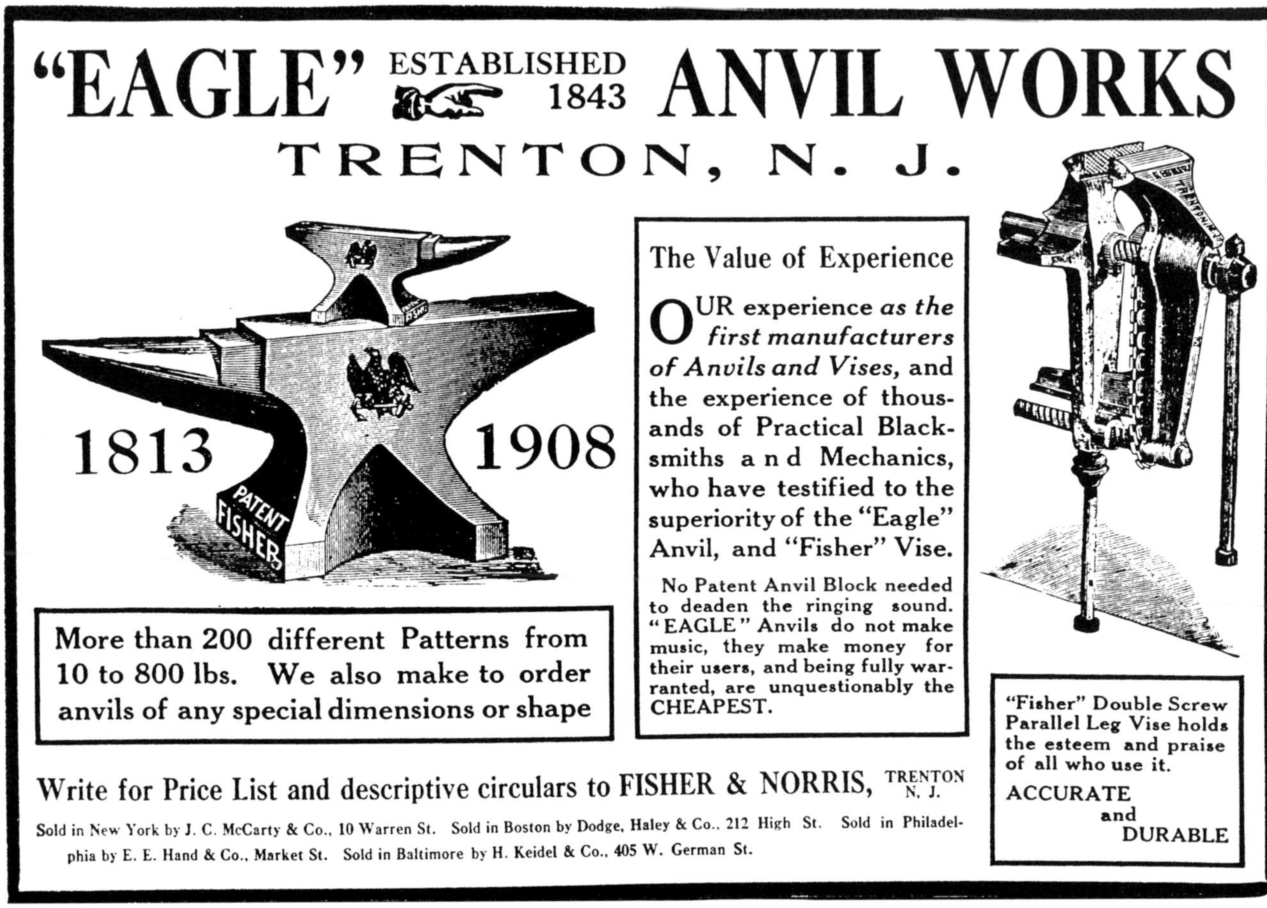

From The American Blacksmith, January, 1909.

The following two articles appeared in *The Blacksmith & Wheelwright* magazine. The first one is taken from *Practical Blacksmithing* by M.T. Richardson, but had appeared in the magazine of which Richardson was the editor. The second article is a letter from Fisher & Norris found in the July, 1888, issue of the same magazine and is a commentary on the first article.

Mending an Anvil

I will try to describe a job that was done lately in the shop I am working in. The base of a wrought-iron anvil had been broken off as shown in Fig. 59. Not wishing to throw the anvil away, the boss told us to try to repair it, and we did it in the following manner.

We first looped a piece of 1 1/8-inch iron around the end of the anvil, with a flat spot just above the loop on which to catch a hook, so as to enable us to handle it better. We next put what I call a binder of 5/8 round iron around the beak iron to prevent the porter bar from slipping off. Next we got a piece of iron, something the shape of the piece "A", in Fig. 59, with a bar welded on the side for handling. This piece was about as wide as the body of the anvil. We then put the anvil in the fire to prepare it for welding, which was done by cutting away the uneven places and scoring it with a chisel. We then put the anvil in the fire for a weld, building the fire up especially for it. The piece to be welded on was brought to a heat in a separate fire. When all was ready the anvil was carried out of the fire by the aid of a bar of iron run through the loop, and turned into position by use of the hook and the flat spot on the bar. The piece was then welded on and put into shape with a big fuller, which left the job as seen in Fig.60. The side was then scored and the anvil put back into the fire for a side heat, while the piece "B" was made. It was brought to a heat by the time the anvil was hot, and then they were brought out and welded and put into shape like the end piece. The other side was then put through the same process, and the whole touched up with fuller and flatter, which left the job in good shape as shown in Fig. 61, and as good as new.

By Apprentice

In bringing the anvil to a welding heat the temper would have been removed from the face. In the previous Fisher article, the writer says, putting the temper in or hardening the face is the most critical operation. The "Apprentice" never mentions this. The "Apprentice's" article was probably in the May, 1888 issue of the magazine.

Why 'Wrought' Iron Anvils Break[34]

To the "Editor of *The Blacksmith & Wheelwright*"

The ingenuity shown by an 'Apprentice' in the patching up of one of the old-fashioned so-called 'Wrought' anvils has interested us. But it must strike the practical iron worker that only the poorest material can be put into such anvils when a solid lump breaks off from the bottom, by the simple jarring of work on the face, though we know such breakages are of common occurrence. After he had finished his job it may be a satisfaction to an 'Apprentice' to know that the only real wrought iron there was,was the piece he so skillfully put on.

As a matter of fact the makers of 'wrought iron anvils' use spongy, slaggy, half worked iron, a grade even below 'muck bar,' with no strength or value for anything else. They do this for the double reason of cheapness and because it welds easier to the also cheap and uneven tempering 'Shear' steel of which they make the face.

Using such stuff in these foreign anvils is also the cause of settling under the face and breaking off of the horns, and, as in this case, even of the base. It is in no sense 'wrought' iron such as the blacksmith uses in his work. Anyone can satisfy himself of this by knocking off the corner of one of them and trying to draw it out as he would other iron.

We have often made this experiment and found that under test it would not carry more than six or seven *thousand pounds to the square inch.* [Author's Note: good wrought iron has about 45,000 pounds per square inch tensile strength.] [35]

The common notion is that the 'ringing' of these anvils is a sign of some excellence, but if they were of real wrought iron, tough and fibrous, they would not ring when struck, with any such sound as the 'wrought' anvil so-called.

If such breakage had occurred with an Eagle Fisher anvil, the 'Apprentice' would not have to exert his ingenuity. His boss would simply have reported it to us and got a new and perfect anvil free of charge, no matter how long it had been in use, but we have never heard of an Eagle anvil, out of hundreds and thousands of them made by us, in the past forty years, breaking in that way, as we certainly should, had there been such a case, for they are all warranted, which none of the wrought are.

Yours Truly
Fisher & Morris [Editor's note: It's Norris]

Trenton, N.J., June 25, 1888

This article shows just how aggressive Fisher & Norris was in promoting their anvils, although some of the statements made were only partly true. For example, the lumping of all foreign anvils into one category, being made of the "poorest material." Peter Wright and Mousehole used the very best reworked wrought-iron they could obtain. The article by M.T. Richardson on the manufacture of *Fisher* anvils was also slightly biased, to say the least.[36] This is understandable, considering that Fisher & Norris was the only American anvil manufacturer advertising in his magazine, *The Blacksmith & Wheelwright,* at the time.

Over the years *Fisher* anvils have proved that they do not stand up as well as the good wrought anvils, especially when abused. However, any high-quality anvil, including the *Fisher*, when used properly, will last almost indefinitely.

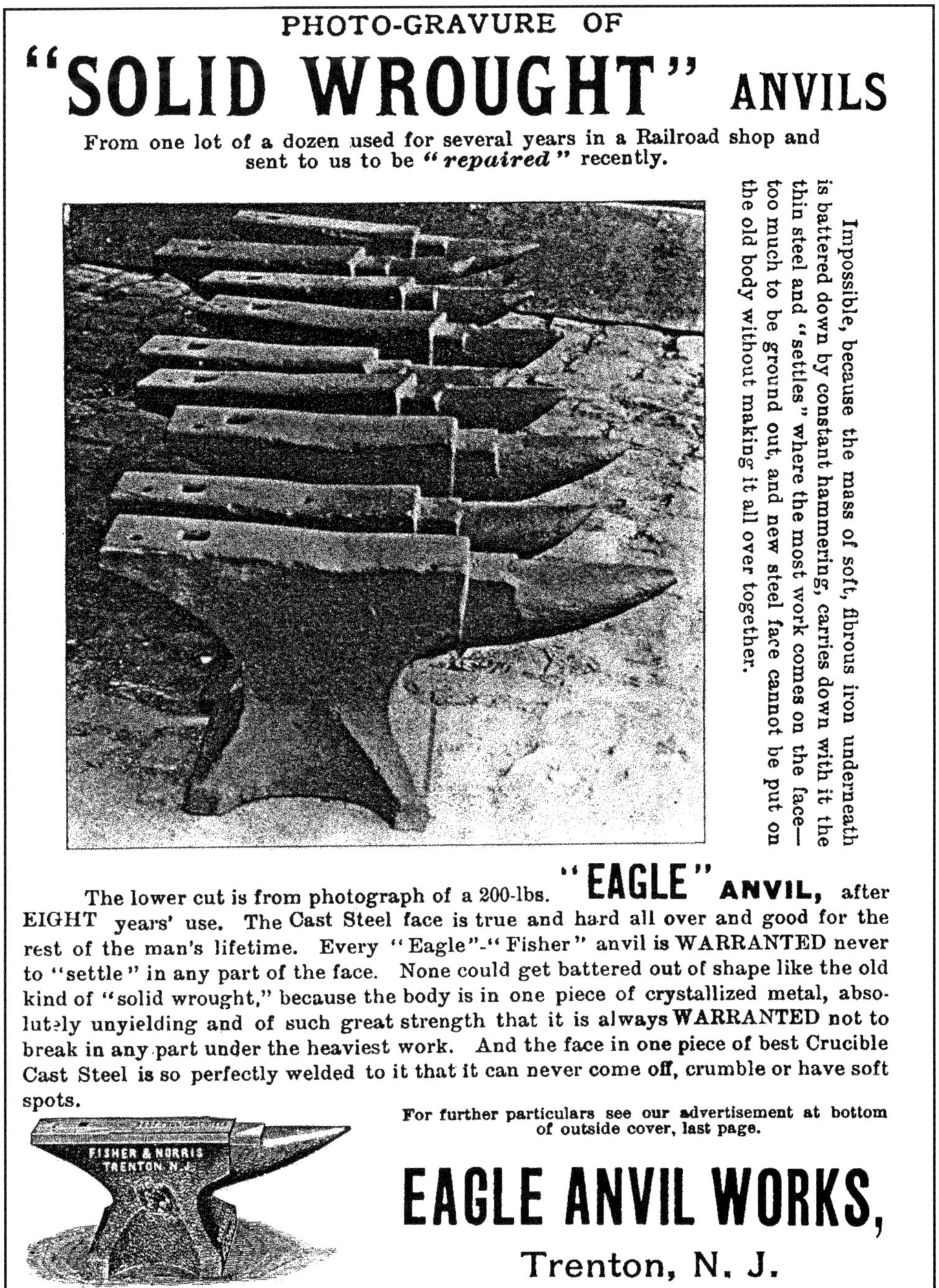

From The American Blacksmith and Wheelwright, March, 1897. It appears that these anvils may be Peter Wrights. (From Mr. Gill Fahrenwald)

The article by Fisher & Norris made the statement that wrought-anvil makers used the "poorest" iron. I found this to be true at one time.[37] From an article found by John Catchings in a London book dated 1842, comes this statement.[38]

Anvils

> *Anvils are made of scrap iron, of common iron, and even of cast* iron, *and their relative value and prices are in the order of this enumeration.... Anvils, even of the second description, when used for such purposes as that last mentioned, are not infrequently seen to crack across their entire substance; though for general purposes and for exportation, they are in very large demand. (The full artical can be found in Appendix H).*

Notice that the writer mentions the anvils made of "common iron." This lowest grade of iron is used in anvils for export, meaning among others, the United States. This would suggest that some English anvil makers were making two grades of wrought anvils. This is what Fisher & Norris was probably referring to in their letter.[39] This is also why so many of the later English anvil manufacturers put "Best Scrap" on their anvils. Note the label in Peter Wright's ad dated May, 1893.

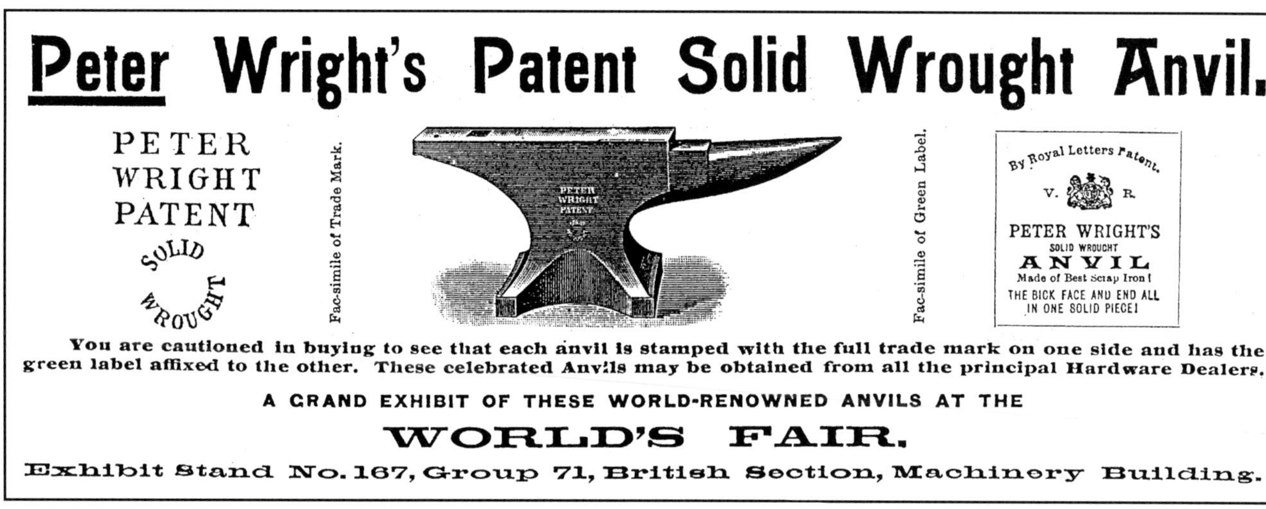

From The Blacksmith and Wheelwright, May, 1893.

In support of the *Eagle* anvil, I am including the end of an article that I reproduced earlier in this book.[40]

> *...Better send the old anvil to Arthur J. O'Leary in Chicago for fixing up, or buy one of the 'Eagle' anvils that are warranted never to have the face come off or 'settle' or the horn to break, and would cost less money. I have worked on one of them, mostly in forging steel, for eighteen years and it is as good and true and hard as in the first.*
>
> *Christian Schmidt*
> *Trenton, N.J., December 10, 1897*

As an aside to the Fisher & Norris story, in the July 1952 issue of the *National Geographic* magazine, there appears a photograph of a man putting the finishing touches on some *Eagle* anvils. Part of the write-up that went along with the photo states that the old water wheel used to power the factory in times past was still in the building.

To close the story on the *Fisher & Norris* anvils are the following copies of the Fisher patents of May 13,1884 and June 28, 1887.

UNITED STATES PATENT OFFICE

Clark Fisher, of Trenton, New Jersey.

ANVIL

Specification forming part of Letters Patent No. 298,567, dated May 13, 1884.

Application Filed December 20, 1883. (No model.)

To all whom it may concern:

Be it known that I, Clark Fisher, of Trenton, New Jersey, have invented certain Improvements in the Manufacture and Construction of Anvils, of which the following is a specification.

My invention relates to anvils of which the body is of cast metal and the face and horn covered with steel welded thereto.

Heretofore all anvils of the above character have been made by placing in the bottom of the mold first the steel plate which forms the face, and then the steel piece which forms the working surface of the horn, the latter being placed with its broad end lapping partially over and resting upon the interior surface of the horn end of the face-plate, substantially as represented in Figure 1. When the mold is filled, the welding is accomplished by the sweeping of the molten iron over the interior or upwardly-projecting surfaces of the steel face-plate and horn-piece, as described in Reissuc No. 227 of the original patent granted to Mark Fisher and William Martin on October 16, 1847. Under the above conditions it is clear that the shoulder or depth of the horn-surface below that of the face of the completed anvil will be equal to the thickness of the steel used for the face-plate, and that the overlapped portions of the face-plate and horn-piece will not be welded, because their overlapping surfaces simply rest upon each other in the manner represented in Fig. 1, and prevent the access of molten metal between them.

In Fig. 2, is represented an anvil made by the method of which Fig. 1 is designed as an illustration, and the fact that there is no welding is indicated by a slight opening between the pen-lines. The thin unwelded part of the face-plate of anvils of the above construction is liable to crack off upon exposure to the hammer of the blacksmith. For many uses it is also desirable that the shoulder between the face and horn should be deeper than the thickness of the steel of the face-plate. Difficulties, however, in otherwise obtaining a perfect weld throughout the entire under surface of the face require that the steel face-plate should be thin, and this necessarily makes a low shoulder. Unless, on the other hand, the two plates are arranged to lap upon each other, and thus form a close joint at the overlap, the molten iron escapes and flows out of the mold instead of filling it and forming the anvil.

Generally stated, my invention is carried into practice by exposing the entire interior surface of the face-plate to the action of the molten iron without having any portion of it in contact with the horn steel, which is accomplished by leaving a space between the overlapping portions of the two plates vertically sufficient to admit the cast metal to flow between them, whereby a more perfect weld of the contiguous extremities of these two pieces is obtained, which, as above explained, could not heretofore be secured on account of the exclusion of the molten welding metal from between the contact surfaces to the entire distance or extent of the overlap.

In the manufacture of an anvil embodying my present improvement the molten metal is prevented from running through the open space between the plates by a dam either of metal or sand core, or by a kindred suitable device, which can easily be arranged at the proper point within the mold by any practical molder.

In Fig. 3 of the drawings is represented an anvil constructed according to my invention.

All of the figures of the drawings are longitudinal vertical central sectional elevations, any further description of which is unnecessary, except to say that A designates the body of the anvil, B designates the face-plate, and C the horn-piece.

The advantages of my improvement reside not only in the securing of a more perfect weld throughout the extent of the face-plate, but in securing a greater depth of shoulder, as at S in Fig. 3, and in preventing the liability of fracture in the front portion of the face-plate.

It is proper to add that the face-plates may, if desired, be made of wrought-iron instead of steel.

Having thus described my invention, I claim and desire to secure by Letters Patent -

An anvil having a steel face-plate and a steel or wrought -iron plate covering the horn, the union being effected by casting the iron body on the plates, in which the adjacent extremities of the horn-piece and face-plate are one above the other, and out of contact with each other, and in which the cast metal extends under the entire surface of the face-plate, and between the end thereof nearest to the horn and the horn piece or plate, substantially as and for the purposes set forth.

In testimony whereof I have hereunto signed my name this 11th day of December, A. D. 1883.

CLARK FISHER

In the presence of-

John Jolley, Jr.,
F. Norman Dixon.

Courtesy of the Smithsonain Institution.

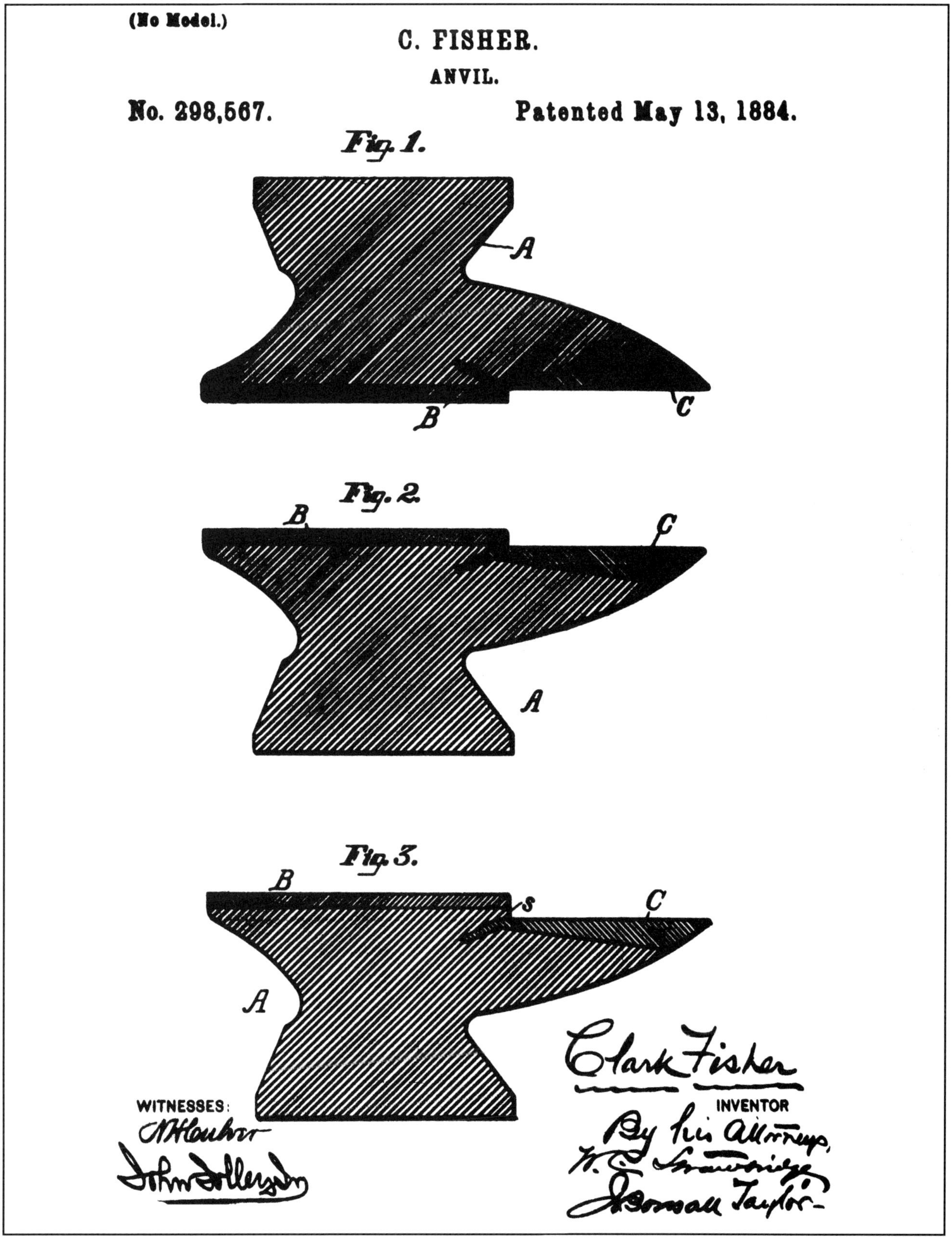
(No Model.)
C. FISHER.
ANVIL.
No. 298,567.
Patented May 13, 1884.
Fig. 1.
A
B
C
Fig. 2.
B
C
A
Fig. 3.
B
s
C
A
WITNESSES:
Clark Fisher
INVENTOR
By his Attorneys,

UNITED STATES PATENT OFFICE

Clark Fisher, of Trenton, New Jersey.

ANVIL.

Specification forming part of Letters Patent No. 365,678, dated June 28, 1887.

Application filed March 24, 1887. Serial No. 232,347. (No model.)

To all whom it may concern:

Be it known that I, Clark Fisher, of Trenton, in the county of Mercer and the State of New Jersey, have invented a new and useful Improvement in Anvils, which invention or improvement is fully set forth and illustrated in the following specification and accompanying drawings.

The object of this invention is to facilitate the welding of steel faces to cast-iron anvils, to facilitate also the hardening and tempering of said faces after welding, to make a stronger and more durable-anvil, and to cheapen the product as a whole or complete article of manufacture.

The complete invention consists of a cast-iron anvil having welded thereto a steel face, of the form hereinafter fully described, and set forth in the claims.

In the accompanying drawings, Figure 1 shows the anvil, in perspective, complete. Fig. 2 shows a geometrical cross section of the steel face of the anvil detached and upon an enlarged scale.

In said figures the several parts are indicated by reference-letters marked thereon, and used in explanation and description, as follows:

All blacksmiths' anvils are constructed of either a cast-iron or wrought-iron body, upon which is welded the face or working surface of steel. The process of welding steel to cast-iron is necessarily different from that of welding steel to wrought iron; but heretofore the steel faces have always been made of uniform thickness for welding to the bodies of all blacksmiths' anvils.

From the brittle nature of cast-iron, additional strength must be derived from the steel overlying it and welded to it, both in the projecting and overhanging portion, called the "tail," x, of the anvil, and also along both edges, where the severe work is likely to come and liable to cause crumbling or breaking out, more than in the central longitudinal portion of the face; but excessive thickness of steel face over the whole surface of the cast iron renders the process of welding to the cast iron both difficult and uncertain; and, likewise, in the subsequent process of reheating and sudden cooling, to give the steel surface the required hardness and temper, there is danger of failure, on account of the difference of expansion and contraction of the two united metals from these sudden changes of temperature when the steel is in one large and uniformly-thick mass. Commercially considered, also, where there is the same thickness of the expensive steel on the central portions of the anvil as on its edges, there occurs in manufacture a waste of material. My invention therefore obviates all these objections by reenforcing the faces of the anvils along said edges lengthwise by welding to the body of the anvil A the plates y, of the general shape or form clearly shown in cross section in Fig. 2, said plates being channeled, so as to be deeper or thicker along their under sides, a b, than at the other portions. The horn of the anvil is indicated by the letter Z. The interior sides of the flanges of the plates y are preferably beveled, as shown in cross section, to facilitate the flow of the cast metal in the process of welding.

Having thus fully described my said improvements of my invention, I claim-

1. A steel face-plate for a blacksmith's anvil having a horizontal transverse section of greater thickness at and about its edges than in its central or remaining portions, said increased thickness being on the lower or welding surface, substantially as and for the purposes set forth.

2. As a new article of manufacture, a blacksmith's anvil having a steel face, as y, welded thereon, and so proportioned as to give the anvil greater thickness and strength of steel along its edges than in its central and remaining working surface of face, substantially as and for the purposes set forth.

CLARK FISHER.

Witnesses:
Chas. W. Edwards,
Edward C. Stover.

Courtesy of the Smithsonian Institution.

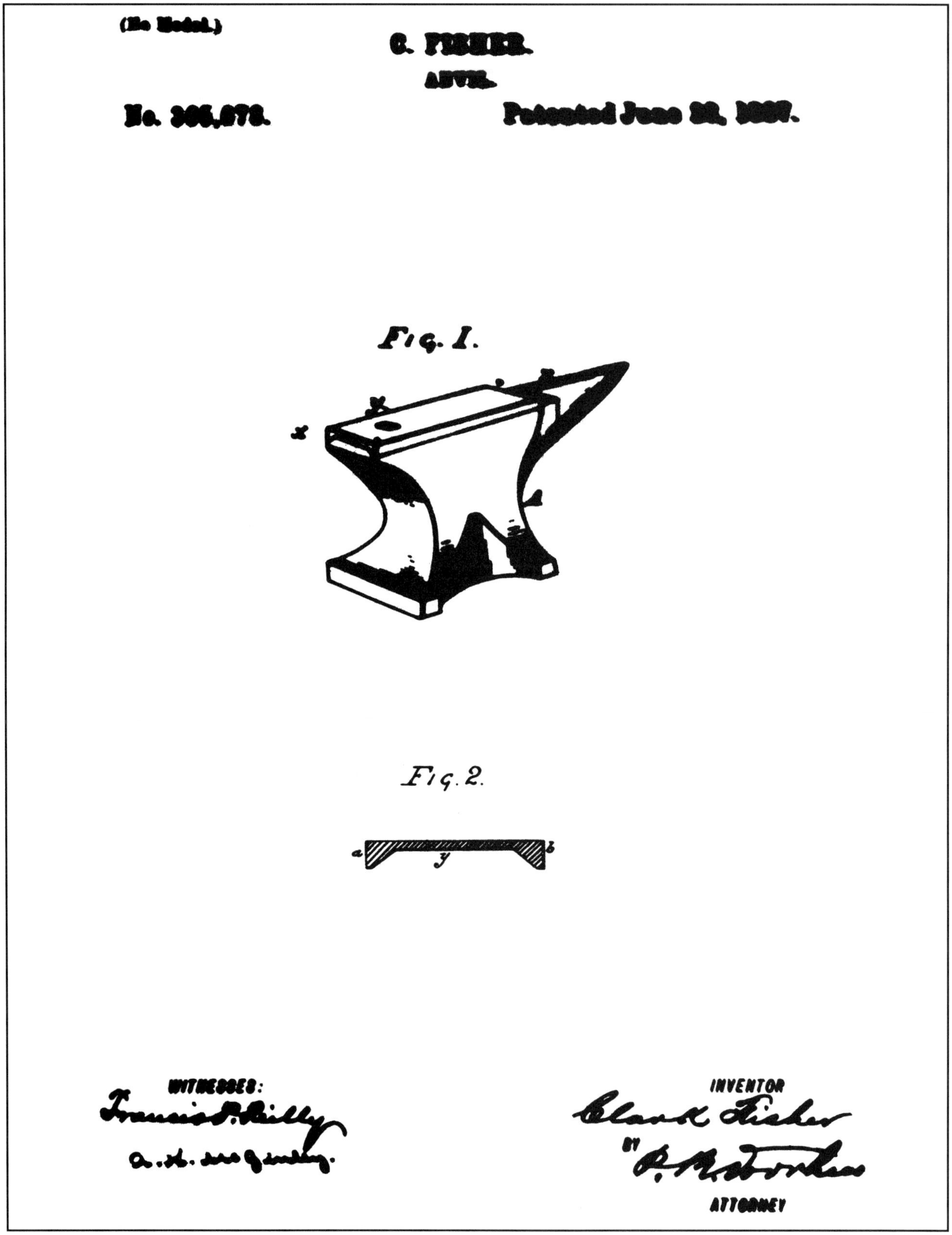
Fig. 1.
Fig. 2.
WITNESSES:
INVENTOR
Clark Fisher
BY
ATTORNEY

The following Fisher & Norris advertisements are typical of most ads that the company produced, from the oldest to the latest. Most of them are self explanatory.

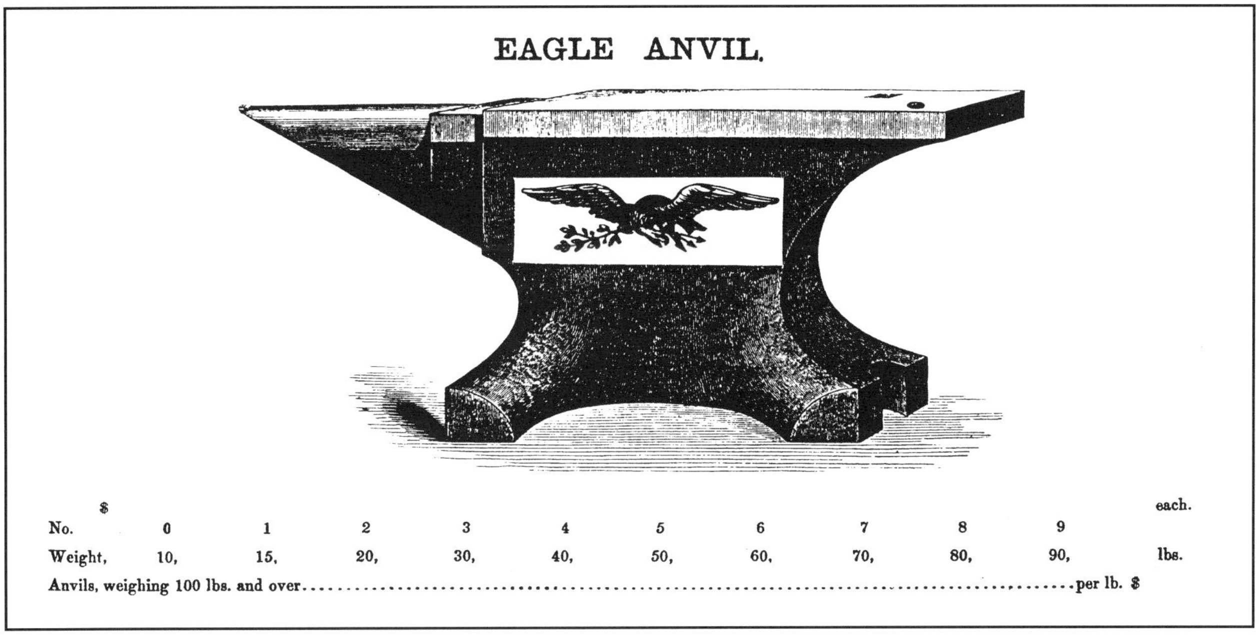

The oldest ad I have discovered is from the 1865, Russell and Erwin Mfg. Co. catalog, New Britain, Connecticut

The Blacksmith and Wheelwright, December, 1882 and May, 1883. This is the oldest ad I have from a trade journal. (from Mr. Hunter Pilkinton)

THE PATENT "EAGLE"-ANVILS.

Fig. 723.

Sizes,	000	00	0	1	2	3	4	5	6	7	8	9
Weight,	½	4	10	15	20	30	40	50	60	70	80	90 lbs.

Larger Sizes from 100 to 800 lbs.

These Tools are fully warranted. The "Peter Wright," and all anvils (except the EAGLE,) are made with a body and horn of fibrous wrought iron, and only for the face steel plates are welded on. This soft iron gradually settles, carrying the face with it, and the Anvil becomes hollowed on the face. Shear steel, also, is used, but it can never be made to take good temper like cast steel. The EAGLE ANVIL body is made of gun metal—crystallized iron—which neither settles nor breaks, and not only the face, but the horn, also, is of the Best Tool Cast Steel, welded on so perfectly by the process of its original and only inventor, the late MR MARK FISHER, that it is warranted never to come off. The face planed to a true surface, is made so hard that the hammer makes no impression on it, and it always remains so. The steel horn is tough and unhardened, and will not break or bend.

Cooke & Company catalog, 1883, New York, New York.

EAGLE ANVIL WORKS,

Trenton, N. J.

The Fisher Double Screw Leg Vise.—Warranted stronger grip than any other Vise. Always parallel and cannot be broken.

Send for Circulars.

The **Eagle Anvil.** — Best Cast Steel Face & Steel Horn. Better than any English Anvil. Fully warranted and lower price.

American Machinist Weekly, August, 29, 1885.

1843. The Eagle Anvil Works, 1893.

FIFTY YEARS.

Horseshoe, 150 lbs.

U. S. Patents.

15 inches long and 4 inches wide.
Solid Steel Horn, 10½ inches long with Side Clip 1¼ inches projection.
Cutter Hole, ⅞ of an inch square.

The above cut is an exact copy of a photograph taken directly from the Anvil itself; it is, therefore, a correct representation of its actual shape, showing the relative proportions of all parts, Face, Horn, &c.

☞Established 1843.

Every Anvil made by these works is *Fully Warranted. None* of the *foreign* made Anvils are ever warranted.

More than 200 different shapes and Weights of Anvils are made by us.

ANVILS weighing 60 lbs. to 800 lbs., 9 cents per lb.

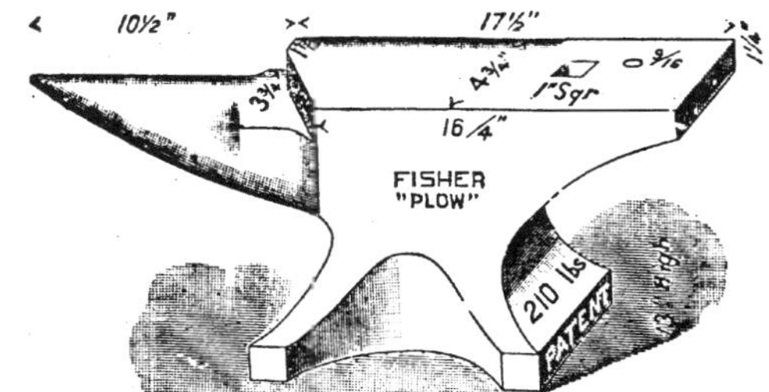

This, our latest Patent, has the **"Fisher" Double-Thick Edge Face of Best Tool Cast Steel** warranted to be in one piece, perfectly welded and never to come loose or break off. It will never "settle" or get out of shape as do the faces of all the foreign-made anvils after a few years of use. It is planed to a true straight surface and given a fine temper, hard and uniform. **The Solid Steel Horn** with Side Clip is long and slim, the best shape for horseshoers' work and it will not break or bend. It is an inch below the face, giving that depth of shoulder which most blacksmiths like. The Cutter Holes are always true and square with the face and have the edges champhered all around to fit the tools. The tail of the anvil is also long and thin, being only **Seven-Eighths** of an inch thick at the end. We have also provided a perfect arrangement for bolting the Anvil to the block by the heavy Lug at each end of the Anvil base, which will be found much more convenient and effective than the old way of spikes and straps.

This pattern of Anvil has been gotten up expressly for horseshoers by the original and only **Eagle Anvil Works** in this country. It is altogether better than the Peter Wright Anvils, or any of the other imported Anvils, is cheaper, and is the only one that is **Fully Warranted** against failure.

Should an Anvil made by us prove imperfect upon using it we will replace it by another without charge. There is no other make of Anvil that gives such a guarantee.

☞Every one of these Anvils has our trade mark of the Horseshoe, also **"Fisher"** stamped in.

The "FISHER" Plow Makers' Anvil,

MADE ONLY BY THE

EAGLE ANVIL WORKS.

Solid Best Cast Steel Face, welded on in one piece and **WARRANTED** never to come off. It is planed to a true surface, tempered to uniform hardness, and *it always remains so.*

The Steel Horn is tough and unhardened, and will not break or bend. It is set low (2 inches below the face), so as to be out of the way, but is long and slim, and of proper shape for general work.

The end of the face is cut off at an angle for greater convenience in Plow work, such as practical experience has shown to be the best. Any variation from this shape can be made to order when desired. While especially for Plow work, it is at the same time every way right for all the work of general blacksmithing. This, like all the anvils made by the "Eagle Anvil Works," is **WARRANTED** never to "settle" in the face, but always keeps its true surface, not only because it is perfectly welded but also because the body under it is of unyielding Crystallized iron and cannot give way under the steel face, as the so-called "wrought iron" ones always do.

The Eagle Anvil Works is the only one in the world that **FULLY WARRANTS** every Anvil sold. Should any imperfection be discovered after being put in use, such imperfect Anvil is immediately replaced with a new one without cost to the customer.

N. B.—None genuine without our "Eagle" Trade Mark and "Fisher" stamped in the side and end of each anvil.

Price of the "FISHER" Plow Anvil, $18.00 Cash with order.

Purchasers are cautioned against buying any of the worthless **"Cast"** Anvils that have been sent out by various parties in cheap imitations of the only genuine and original **"Eagle"** Anvil, which can be made by us alone, as they have been forthe past **FIFTY YEARS.**

☞ Where our EAGLE ANVILS and VISES are not kept by Dealers, we will allow liberal discount to any purchaser sending us the amount with his order, and will forward by lowest freight. Write for Circular of Description and Cash Discount.

Address **FISHER & NORRIS, TRENTON, NEW JERSEY.**

The Blacksmith and Wheelwright, May, 1893. (From Mr. Larry Wood)

FISHER & NORRIS' EAGLE ANVILS

Horseshoers'

The EAGLE ANVIL body is made of gun-metal—crystallized iron—which neither settles nor breaks. The face is in one piece of BEST TOOL CAST STEEL, welded on so perfectly by the process of its original and only inventor, the late Mr. MARK FISHER, that it is warranted never to come off. It is planed to a true surface, tempered with uniform hardness, and it always remains so. The steel horn is tough and unhardened, and will not break or bend. The complaint of deafness, so often occasioned by the "ringing" of the Anvil, is avoided by using the EAGLE ANVIL, which is warranted NOT TO RING.

N. B.—All EAGLE Anvils are now made with the latest Fisher Patent DOUBLE-THICK STEEL on both edges of the face, whole length.

SMALL ANVILS (MINIMS)

No.	Weight, pounds.	Face. Length, inches.	Face. Width, inches.	Cutter-Hole. Square inches.	Horn. Length, inches.	Each.
1	8	5¾	1¾	⅜	3½	$3.50
2	20	6¼	2¼	½	4	4.00
3	30	7¼	2¼	½	4	4.30
4	40	8¾	3	⅝	5¼	4.50
5	50	9¼	3	⅝	6½	5.00
6	60	10	3	⅝	6½	5.50
7	70	10½	3¼	⅝	6½	6.00
8	80	10½	3¼	⅝	7½	7.00
9	90	11	3¼	⅝	7½	8.00

BLACKSMITHS'

Weight, pounds.	Face. Length, inches.	Face. Width, inches.	Cutter-Hole. Square inches.	Horn. Length, inches.	Per pound
100	12	3½	¾	8½	$0.
110	12¾	3½	¾	8½	
120	12¾	3½	¾	8½	
130	13½	3¾	¾	8½	
140	14	4	⅞	8½	
150	15	4	⅞	10	
160	15	4¼	1	10	
170	15	4¼	1	10	
180	15½	4¼	1	10	
200	16½	4¾	1⅛	11½	
225	16½	4¾	1⅛	11½	
250	17¼	5¼	1¼	11½	
275	17¾	5¼	1¼	11½	
300	19	5¼	1¼	11½	
HORSESHOERS'					
150	14½	4	⅞	10	
With Side Clip 1¼ in. long.					

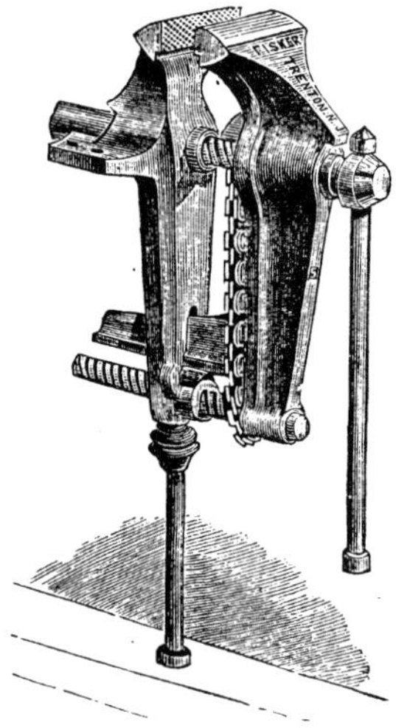

FISHER & NORRIS' DOUBLE SCREW PARALLEL LEG VISES

No.	Weight about	Jaw	Screws	Lever	opens	E…
*2	65 lbs.	4½ × 1 in.	1⅛ in.	13 in.	5½ in.	$10.
3	90 "	5¼ × 1⅛ "	1¼ "	16 "	6½ "	16.
4	120 "	6¼ × 1¼ "	1½ "	19 "	7½ "	20.
5	150 "	7 × 1½ "	1¾ "	24 "	9 "	27.
6	160 "	8 × 1½ "	1¾ "	26 "	10 "	30.

* NOTE.—No. 2 Leg unnecessary, therefore omitted.

The Jaws are of best Tool Cast Steel, welded on, file cut and properly hardened. The Screws are forged of the best Refined Iron, and work in solid cut thread boxes.

The lower screw maintains the parallel position of the two jaws, by having exact [illegible]tion with the upper working screw, through the connecting chain which regulates it.

The chain is very accurately made of steeled links and rivets, and, having no strain [illegible] the work upon it, is therefore as durable as all the other parts.

Russell and Erwin Mfg. Co. catalog, 1899, New Britain, Conn. Note that the anvils are guaranteed not to ring. (From Mr. John Catchings)

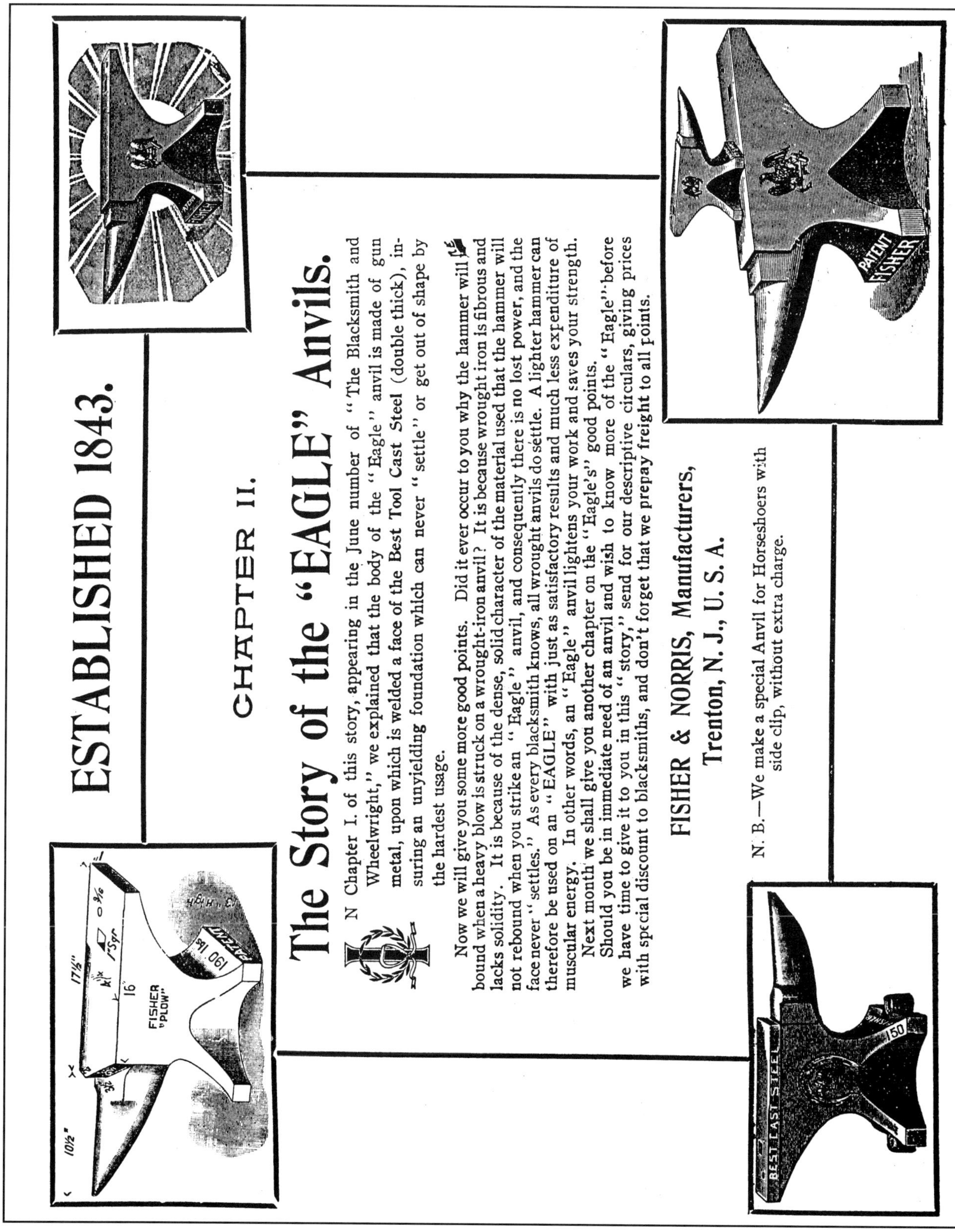

The Blacksmith and Wheelwright, July, 1901. Notice the plowmaker's anvil which was first advertised in early 1894.

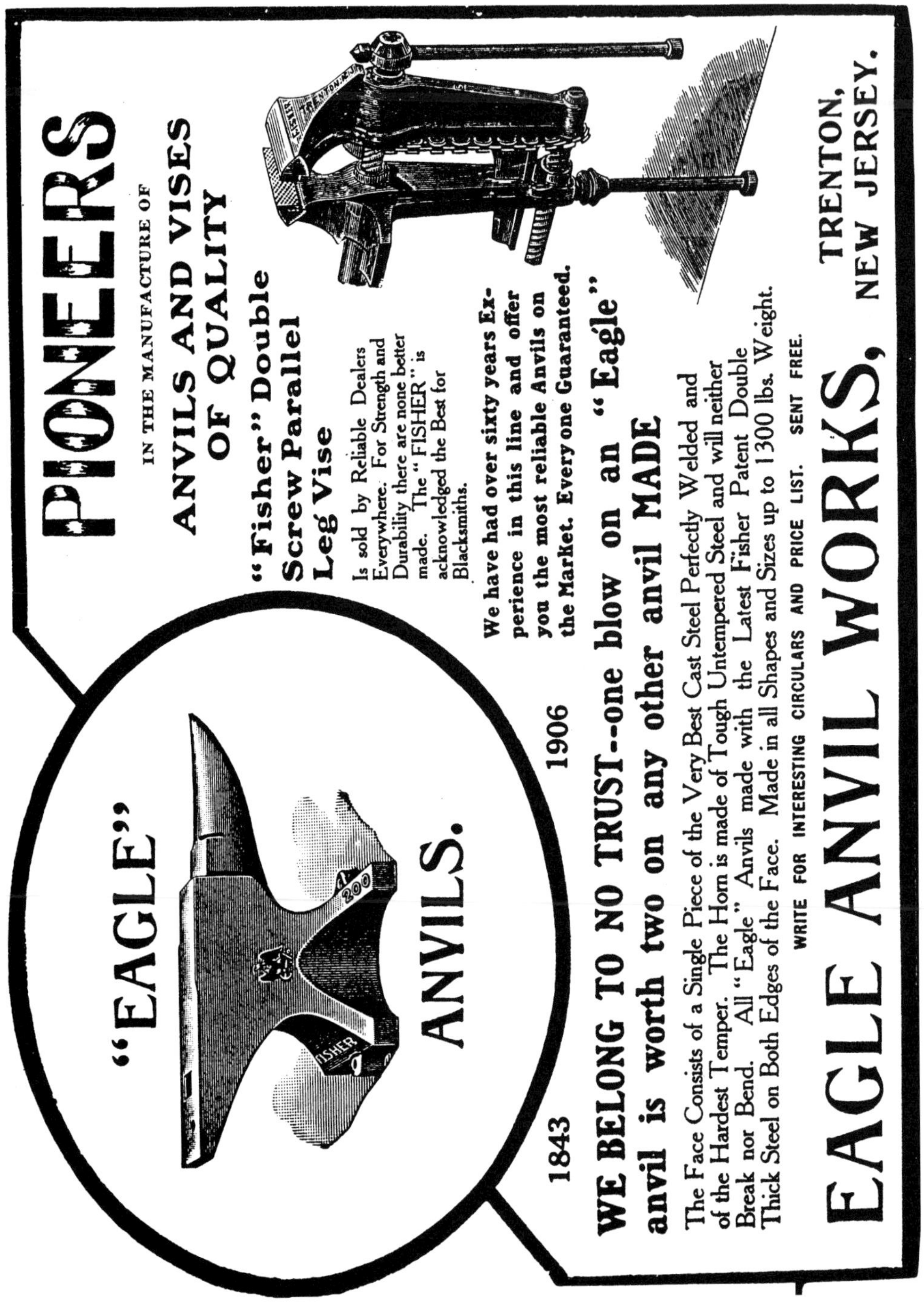

The American Blacksmith, July, 1906. Notice that the anvils are made up to 1300 pounds.

The American Blacksmith, July, 1907. Notice that under the anvil engraving it says that the company will make any "Special" anvils to order. That would require a special pattern.

The Blacksmith and Wheelwright, October, 1909. This ad gives a good idea of their product line at that time. They even came out with a bench vise. (From Mr. Roy Price)

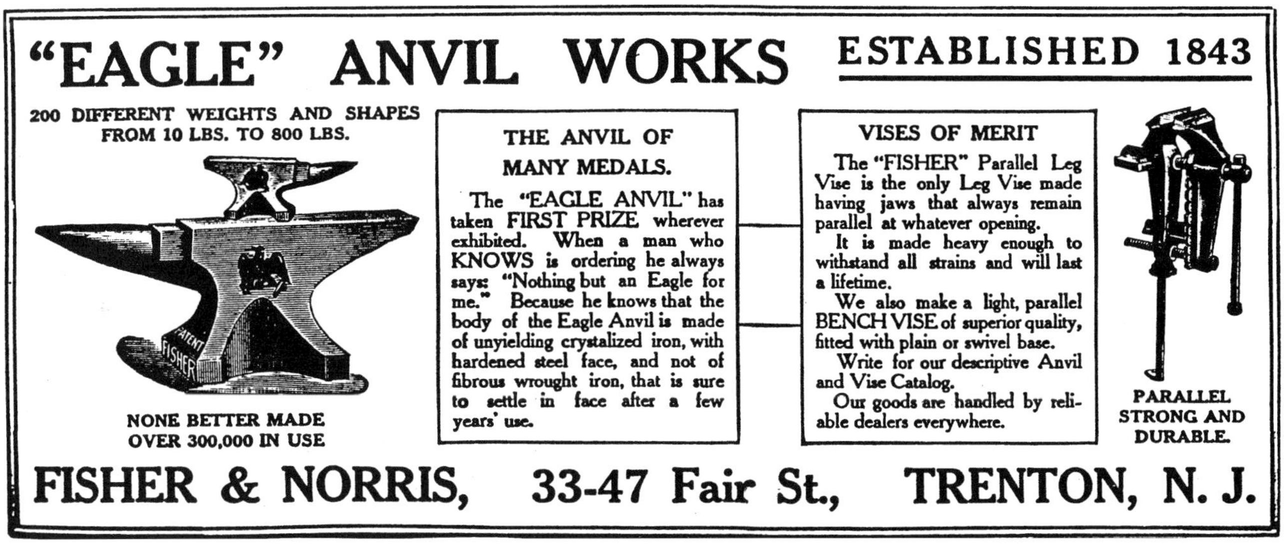

The American Blacksmith, January, 1910.

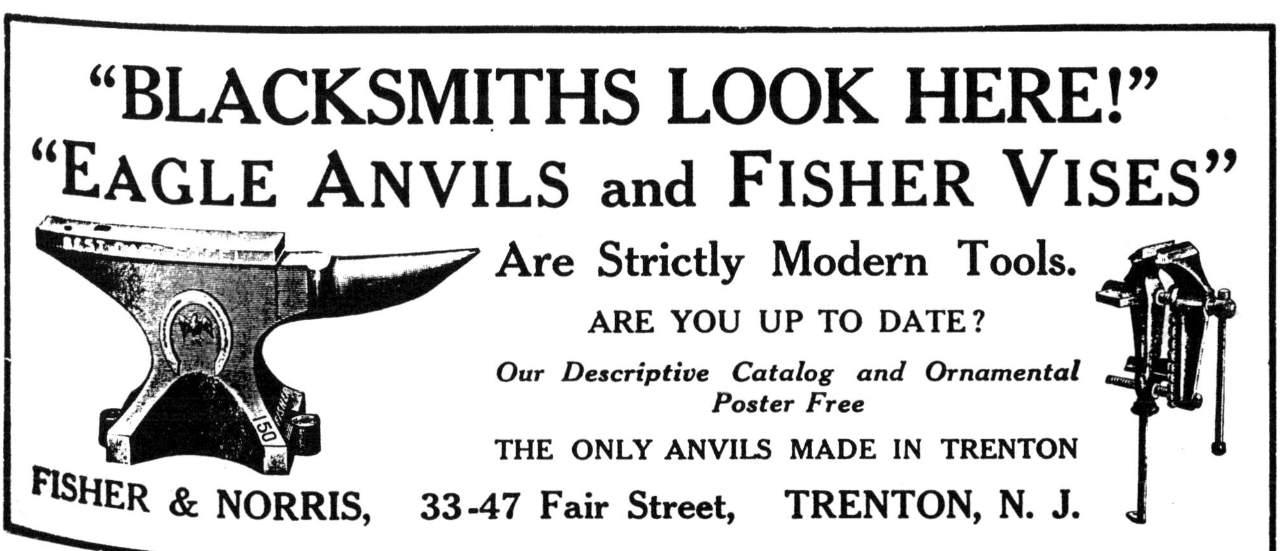

The American Blacksmith, March, 1911. Notice that they are the only anvil makers in Trenton, New Jersey. This is referring to the Trenton anvil made in Columbus, Ohio.

The American Blacksmith, February, 1912.

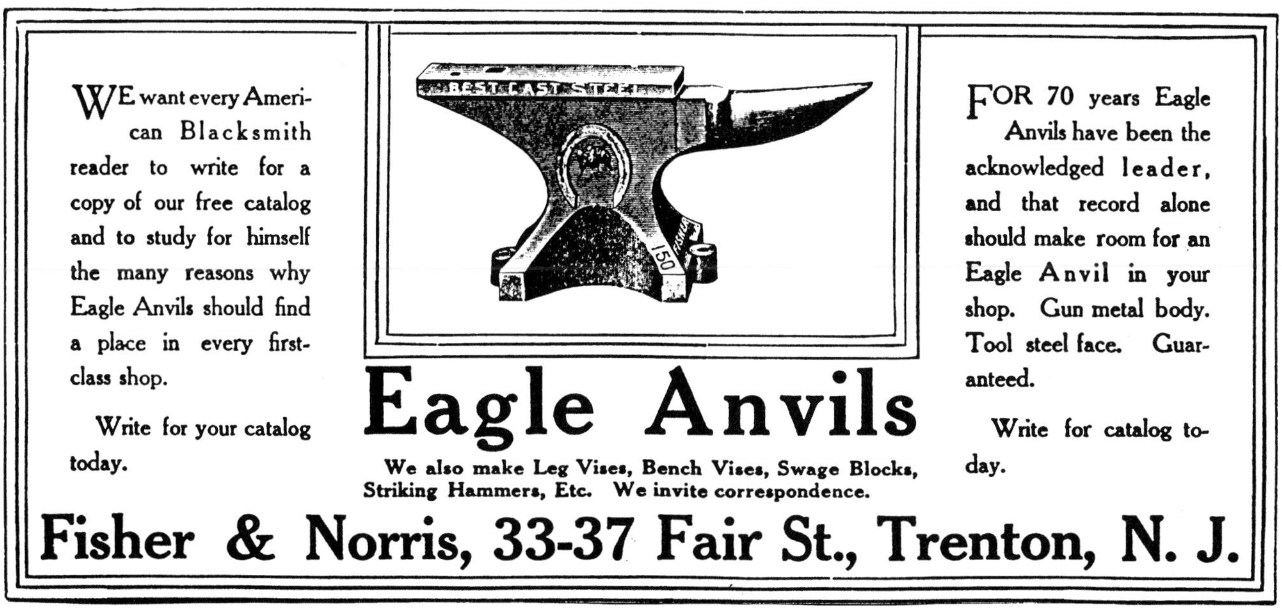

The American Blacksmith, March, 1913.

The Blacksmith and Wheelwright, July, 1916. The government bought many of Fisher & Norris's anvils.

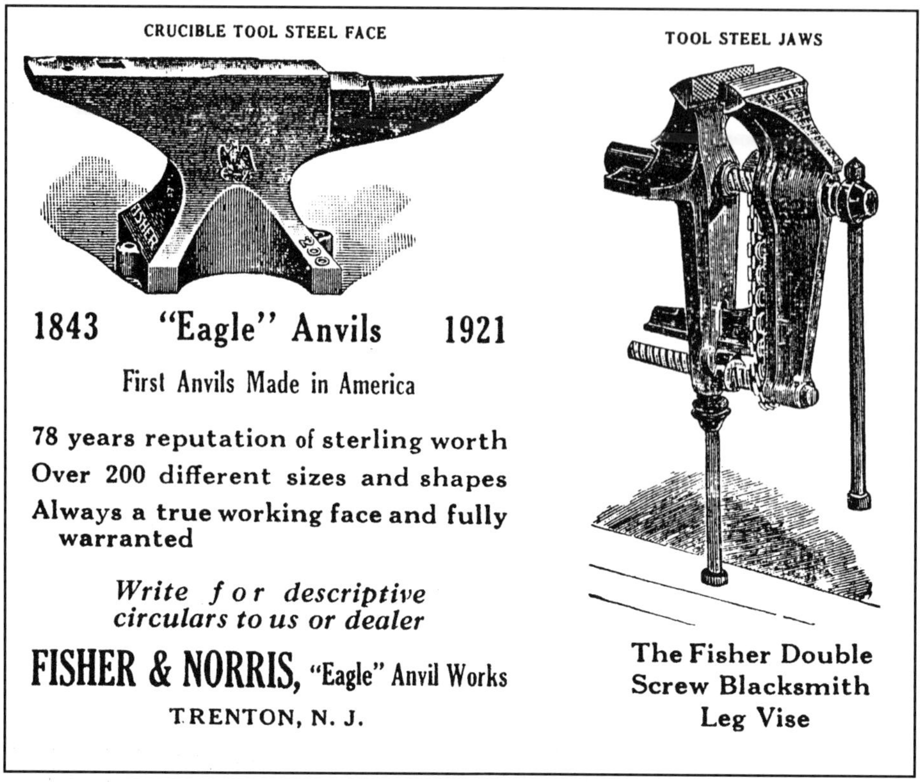

Blacksmith and Wheelwright and Tractor Repair Journal, February, 1922.

Thomas Register, 1941.
(From Mr. Gill Fahrenwald)

Eagle Blacksmiths' Anvils

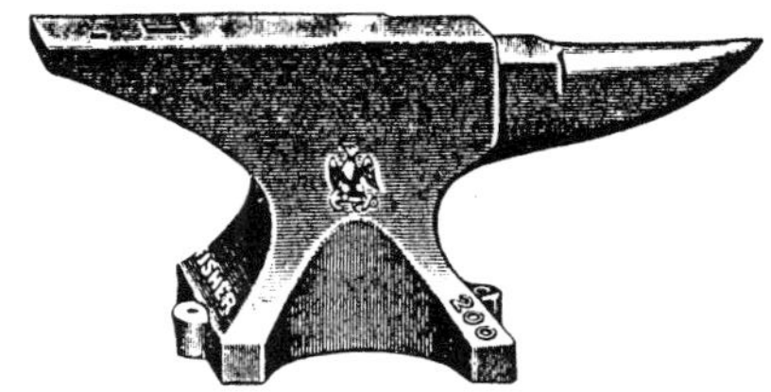

The face is one piece of high grade tool steel, welded to the body, accurately ground and tempered with edges drawn. Body of solid piece of unyielding gun iron of a special analysis. Produces necessary hammer rebound when struck on face. Horn is a solid piece of tough untempered forged steel.

Size Lb.	Face, Inches: Length	Width	Hardie Sq. Hole	Pritchel Hole	Length Horn In.	Height In.
40	9¼	3	5/8	7/16	6¼	7¼
50	10	3	5/8	7/16	6½	8¼
60	10½	3	5/8	7/16	6½	8½
70	11	3¼	3/4	1/2	6½	8¾
80	11½	3¼	3/4	1/2	7½	9⅝
90	11½	3¼	3/4	1/2	7½	10
100	12¾	3½	3/4	9/16	9	10¼
120	13½	3½	3/4	9/16	9	10½
130	14	4	3/4	9/16	9	11
140	15	4	7/8	9/16	10	11
150	15¾	4	7/8	9/16	10¼	11½
160	16	4¼	1	9/16	10¼	11½
180	16½	4¼	1	9/16	10¼	13¼
200	17¼	4¾	1⅛	5/8	11½	12¼
225	18	4¾	1⅛	5/8	11½	12½
250	18½	5¼	1¼	5/8	12½	13¼
275	19¾	5¼	1¼	5/8	12½	14¼
300	19¾	5¼	1¼	5/8	12½	14½
350	21	6	1⅜	3/4	13¼	14¼
400	22	6	1⅜	3/4	13¼	15½
450	22	6	1⅜	3/4	13¼	15¾
500	23	6½	1½	13/16	14½	16
600	24¼	6¾	1½	13/16	14½	17
700	25½	7¼	1⅝	13/16	14½	17¾
800	26¼	7¼	1⅝	13/16	16¾	18
1000	27¼	7¾	...	...	16¾	20

Extras

Size.................pounds	80–1000	70	60	50	40
Add Per Pound.............	.....	$.01	.02	.04	.06

Marshall Supply and Equipment Co. catalog, 1942, Tulsa, Ok. (From Mr. Hunter Pilkinton)

Eagle Blacksmiths' Anvils

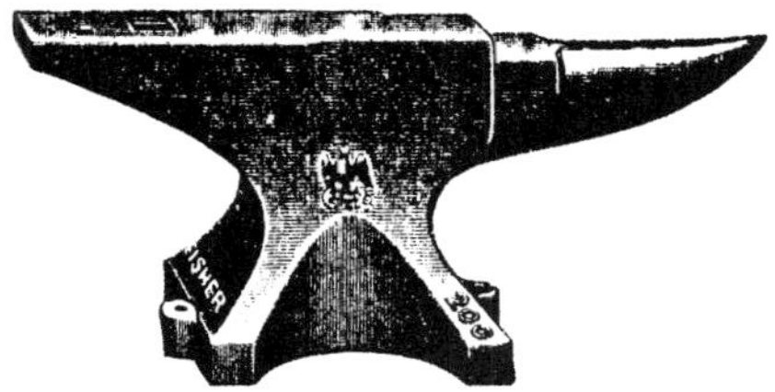

Face is one piece of high grade tool steel, welded to body, accurately ground, tempered with edges drawn. Body is a solid piece of unyielding gun iron. Horn is a solid piece of untempered forged steel.

Size Lb.	Face, Inches: Length	Width	Hardie Sq. Hole	Pritchel Hole	Length Horn In.	Height In.
50	10	3	5/8	7/16	6½	8¼
80	11½	3¼	3/4	1/2	7½	9⅝
100	12¾	3½	3/4	9/16	9	10¼
120	13½	3½	3/4	9/16	9	10½
150	15¾	4	7/8	9/16	10¼	11½
160	16	4¼	1	9/16	10¼	11½
180	16½	4¼	1	9/16	10¼	13¼
200	17¼	4¾	1⅛	5/8	11½	12¼
225	18	4¾	1⅛	5/8	11½	13¼
250	18½	5¼	1¼	5/8	12½	13¼
300	19¾	5¼	1¼	5/8	12½	14½
350	21	6	1⅜	3/4	13¼	14¼
400	22	6	1⅜	3/4	13¼	15½
450	22	6	1⅜	3/4	13¼	15¾
500	23	6½	1½	13/16	14½	16
600	24¼	6¾	1½	13/16	14½	17
700	25½	7¼	1⅝	13/16	14½	17¾

Extras

Size..............................pounds	80–700	50
Add per Pound Over Base..................	......	$.08

Carey Machinery and Supply catalog, 1955, Baltimore, Md.

Fisher & Norris brochure, 1973, Trenton, New Jersey. This brochure was mailed on request. (Mr. Floyd Daniel)

FISHER & NORRIS, INC.

DIVISION OF CROSSLEY MACHINE CO.
EAGLE ANVIL WORKS
(ESTABLISHED 1843)

MANUFACTURERS OF THE

FISHER

PATENT "EAGLE" ANVILS AND DOUBLE SCREW BLACKSMITH VISES, ETC.

301 MONMOUTH STREET
TRENTON, N. J. 08607

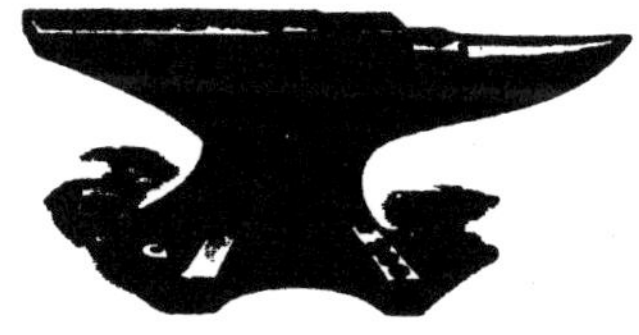

MANUFACTURERS OF THE FIRST ANVIL MADE IN AMERICA

CABLE ADDRESS
"ANVIL" TRENTON

DISTRIBUTOR'S PRICE LIST

EFFECTIVE JANUARY 2, 1973

STANDARD BLACKSMITH ANVILS	STANDARD ANVIL STAND
100-lb. — $100.00 ea.	$ 75.00 ea.
150-lb. — $150.00 ea.	$ 75.00 ea.
200-lb. — $200.00 ea.	$ 85.00 ea.
250-lb. — $250.00 ea.	$ 90.00 ea.

SPECIAL BLACKSMITH'S ANVILS (MADE TO ORDER)	SPECIAL ANVIL STAND (MADE TO ORDER)
50-lb. — $ 75.00 ea.	(Not Available)
70-lb. — $105.00 ea.	(Not Available)
80-lb. — $120.00 ea.	(Not Available)
120-lb. — $180.00 ea.	$ 85.00 ea.
160-lb. — $240.00 ea.	$ 85.00 ea.
180-lb. — $270.00 ea.	$100.00 ea.
225-lb. — $337.50 ea.	$100.00 ea.
300-lb. — $375.00 ea.	$100.00 ea.
350-lb. — $437.50 ea.	$105.00 ea.
400-lb. — $500.00 ea.	$105.00 ea.
450-lb. — $562.50 ea.	$105.00 ea.
500-lb. — $625.00 ea.	$125.00 ea.
600-lb. — $750.00 ea.	(Not Available)
700-lb. — $875.00 ea.	(Not Available)

HORSESHOER'S ANVILS (125-lb. Size Only) $150.00 ea.

ALL SAWMAKERS ANVILS .... $1.25 per lb.

VISES:

NO. 4 Double Screw Parallel Leg Vise	$300.00
NO. 5 Double Screw Parallel Leg Vice	$325.00
NO. 6 Double Screw Parallel Leg Vice	$350.00

Prices subject to change without notice.

Terms of payment: Net 30 Days, subject to suitable credit references.

All prices are F.O.B. Trenton, New Jersey

All contracts and agreements are subject to strikes, accidents, and other causes beyond our control.

Any tax imposed by any present or future law on the sale of the article covered hereby shall be added to the amount to be paid therefore.

Price list, January2, 1973. (From Mr. Floyd Daniel)

FISHER & NORRIS
"EAGLE" ANVIL WORKS

EAGLE ANVILS

The Guaranteed Kind
Fully Warranted

First-Class Blacksmiths, Machinists, Toolmakers and craftsmen have used EAGLE Anvils in this country and abroad for more than a century; EAGLE Anvils have no peer! The superior production techniques and high-grade materials used in their manufacture have kept pace with the demands of modern industry for quality tools which give continued superior performance.

HORN - solid piece, tough, untempered, forged steel, guaranteed not to bend nor break.

FACE - one piece, crucible tool steel, ground and tempered edges, perfectly welded to the body. Will not settle

BODY - solid piece gun iron alloy, warranted not to settle nor break. Will produce satisfactory hammer rebound without unnecessary ringing.

DIMENSIONS - STOCK ANVILS

Wt. Lbs.	FACE				HORN
	Length Inches	Width Inches	Hardie Hole Sq. in.	Pritchel Hole	Length Inches
50	10	3	⅝	7/16	6½
80	11½	3¼	¾	½	7½
100	12¾	3½	¾	9/16	9
120	13½	3½	¾	9/16	9
150	15¾	4	⅞	9/16	10¼
160	16	4¼	1	9/16	10¼
180	16½	4¼	1	9/16	10¼
200	17¼	4¾	1⅛	⅝	11½
225	18	4¾	1⅛	⅝	11½
250	18½	5¼	1¼	⅝	12½
300	19¾	5¼	1¼	⅝	12½
350	21	6	1⅜	¾	13¼
400	22	6	1⅜	¾	13¼
450	22	6	1⅜	¾	13¼
500	23	6½	1½	13/16	14½
600	24¼	6¾	1½	13/16	14½
700	25½	7¼	1⅝	13/16	14½

Empire Machinery and Supply Corp. catalog, 1984, Norfolk, Va. This is probably the last advertisement to appear for Fisher anvils. A cousin of mine is the president of Empire Machinery and I asked him if he still had any Fisher anvils in stock. There weren't any. As with most catalogs a product which is no longer available is often listed for sometime if the catalog is only produced every few years.

The "Fisher" Horseshoer's Anvil, with Detachable Vise.

Weight of anvil, 170 pounds; length of face, 16½ inches; width of face, 4 inches; length of horn, 10½ inches; width of horn, 3 inches; width of face at end next horn, 3 inches; cutter hole, 1 inch square; clip on the horn is 1¼ inches projection.

This has also the "*Fisher*" *double-thick-edge face of best tool cast steel*, warranted to be in one piece, perfectly welded, and never to come loose or break off. It will never "settle" or get out of shape as do the faces of all the foreign-made anvils after a few years of use. It is planed to a true, straight surface and given a fine temper, hard and uniform. The *solid steel horn* with side clip is long and slim, the best shape for horseshoers' work, and it will not break or bend. It is an inch below the face, giving that depth of shoulder which most blacksmiths like. The cutter hole is always true and square with the face, and is chamfered all around to fit the tools. The tail of the anvil is long and thin, being only *seven-eighths* of an inch thick at the end. A perfect arrangement for bolting the anvil to the block is provided by the heavy lug at each end.

This pattern of anvil has been gotten up expressly for horseshoers. It is altogether better than the Peter Wright anvils or any of the other imported anvils, is cheaper and is the only one that is *fully warranted* against failure in any way.

Should an anvil made by us prove imperfect upon using, we will replace it by another without charge. There is no other make of anvil that gives such a guarantee.

The hardened steel jaw hinges, with a blind joint in special socket at the base, and is closed by means of a bolt with slotted head from the other side. This bolt pases through the body of the anvil, its extremity holding the jaw by the thumb screw. The lever pivots with a knife edge on that side of the anvil, its upper end entering the bolt's slotted head, and extending downward it terminates in a treadle of convenient height for the foot. A plain spring on the inner side of the jaw throws it open when the foot is off the treadle, leaving it ready to insert the shoe for turning calks of toe or heel. The whole arrangement is attached to the anvil itself without any connection on the block. The jaw, bolt, treadle and thumb screw, only four pieces in all, can be detached instantly by simply unscrewing the thumb nut, leaving the anvil free for any work required, as shown in the small cut. They can be put back again ready for vise work in *one minute*, thus saving the time and money of a separate vise.

About 5 inches of the front edge is beveled off and rounded for drawing down work. Also the face for a couple of inches in length is reduced to 3 inches, sometimes a convenience for narrow work on the face.

43147 Price of anvil, with vise, complete. $20.00

Montgomery Ward and Company catalog No. 57, 1895. (From Mr. Keith Sommer)

ESTABLISHED 1843

Trenton, N. J.__________19

Bought of FISHER & NORRIS,

MANUFACTURERS OF

THE "EAGLE" ANVILS, THE DOUBLE SCREW PARALLEL VISES, &c.

TERMS F. O. B. Trenton, N. J.. 30 Days Net.
or 2 per cent Cash Discount 10 Days

37 TO 45 FAIR STREET

Your Order

No.

*Shipped this Day, via*__________

*Marked*__________

We hereby guarantee no provision of the Federal Child Labor Act of September 1, 1916 has been violated in the manufacture of the goods covered by the document bearing this imprint.

A Fisher & Norris order form used by the company for many years. Courtesy of the Smithsonian Institution.

Miniature Fisher & Norris advertising anvil.
(Photo by Mr. Gill Fahrenwald)

Fisher & Norris miniature anvil from the othe side.
(Mr. Gill Fahrenwald)

SAMSON

This steel-faced cast anvil was either made for the Van Wagoner & Williams Hardware Company of Cleveland, Ohio, or by them.[41] They were probably made from about 1894 until shortly after the turn of the century. The company ran ads in *The Blacksmith and Wheelwright* journal between June, 1895, and June, 1896. I have been unable to find any further mention of them in this publication after that date.

The Blacksmith and Wheelwright, March, 1896.

The last ad of which I have knowledge is found in the January 1, 1901, catalog of *Farwell-Ozmun-Kirk & Company*, of St. Paul, Minnesota.

SAMSON.

Tough cast iron body, heavily faced with crucible cast steel. Horn has a solid point, and facing of tough untempered steel, superior finish. Every anvil guaranteed.

		Per Pound.
Weight, pounds....	20 30	$0.20
" "	70 80 90	.15
" "	100 120 140	.12

Farwell-Ozmun-Kirk and Company catalog, January, 1901.

The 1899 catalog of the *Russell & Ervin Manufacturing Company,* New Britain, Connecticut, carried the *Samson* anvil and listed 20 different weights between 10 and 250 pounds.

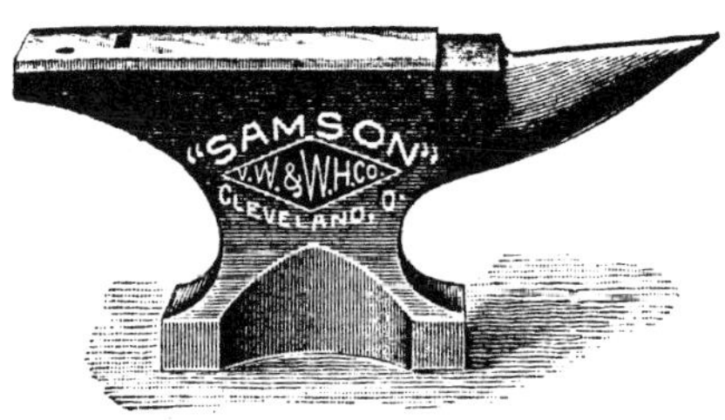

SAMSON ANVILS

10 pounds,	each, $4.50	110 pounds,	each, $16.50
20 "	" 5.00	120 "	" 18.00
30 "	" 5.50	130 "	" 19.50
40 "	" 6.25	140 "	" 21.00
50 "	" 7.50	150 "	" 22.50
60 "	" 9.00	160 "	" 24.00
70 "	" 10.50	175 "	" 26.25
80 "	" 12.00	200 "	" 30.00
90 "	" 13.50	225 "	" 33.75
100 "	" 15.00	250 "	" 37.50

Russell and Erwin Mfg. Co. catalog, 1899, New Britain, Conn. (From Mr. John Catchings)

I have yet to discover one of these anvils to date (1995), but a reliable friend said that he had seen one a few years ago in an antique shop.[42] He stated that it was in very poor condition, but the trademark was very plain to see. He later found a miniature paper weight anvil with the same trademark. He bought the miniature, being a collector of such, and sent me a photo of it. The trademark on the miniature matches those in the advertisements exactly, as well as on the old beat-up *Samson.* The trademark is all there is to go by when attempting to identify this anvil.

Miniature Samson advertising anvil. This is how the trademark would appear on the full size anvil.
(Photo by Mr. Gill Fahrenwald)

SOUTHERN CRESCENT

This cast-iron steel-faced anvil was made by the Southern Skein & Foundry Company of Chattanooga, Tennessee. I have no knowledge at this time as to when the company first started manufacturing anvils, or when they went out of business. They are listed in a 1925 catalog for *Keith-Simmons Company Inc.*, of Nashville, Tennessee, and in a *Kester Machinery* catalog of Winston-Salem, North Carolina, as well as in a manufacturing directory for 1932. The *Kester Machinery Company* catalog shows eleven different sizes from 5 to 250 pounds.

I have recorded several *Southern Crescent* anvils and have found them to look very much like a *Vulcan* anvil except for poorer finishing. The only way to tell the difference is by the trademark. Generally the trademark is not produced very well on the *Crescent* and what you usually see is a "Crescent Moon" with a bar passing through the center, similar to what is shown.

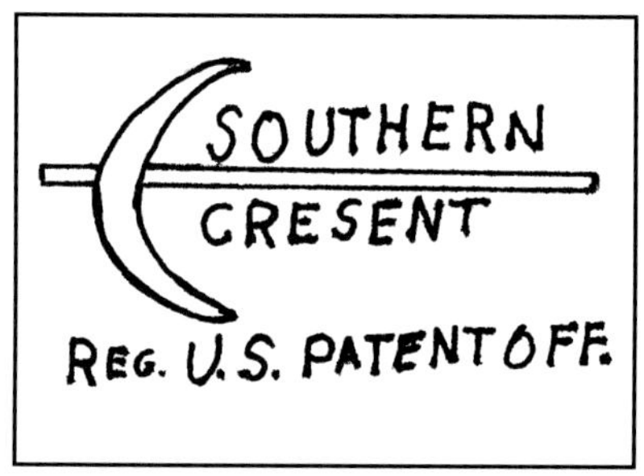

On the anvils that I have recorded, I have found the weight on the side with the trademark under the horn, below the waist. I have done my best to locate additional information about this company and its anvils by phone, but without success.

SOUTHERN CRESCENT

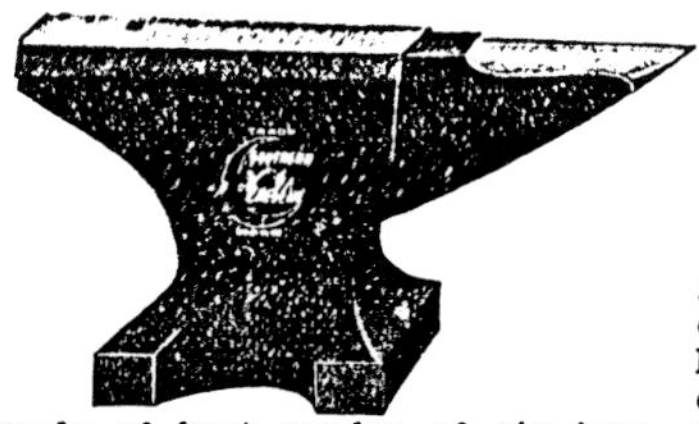

Steel Face

The face is one solid piece of tool steel, *tempered, highly polished* and warranted to be true. The horn is covered with, and its extremity made entirely of tough untempered steel. Body of anvil made of best grades of pig iron, and being much more solid than wrought iron, the work being forged receives the full force of the blow. Face and horn warranted to be thoroughly welded to the body of the anvil and not separate.

Nos.	7	8	9	10	12	15
Wt. each, lbs...	70	80	90	100	120	150
Each	$15.00	16.00	18.00	20.00	24.00	30.00

Open stock.

Keith Simmons Co.,Inc. catalog, 1925, Nashville, Tenn. (From Mr. Hunter Pilkinton)

Southern Crescent Steel Face Anvils

Face is one solid piece of high grade steel, tempered and highly polished. Horn is covered with, and its extremity made entirely of tough untempered steel. Body of best grade pig iron. Face and horn thoroughly welded to body of anvil and will not separate. Polished parts lacquered to prevent rusting and covered with tape.

Size No.	Each	Overall, In. Lgth.	Overall, In. Ht.	Face, In. Lgth.	Face, In. Width	Lgth. Horn In.	Size Hardie Hole In.	Size Pritchel Hole In.	Wt. Lb.
00	$11.60	7	3¾	4	1½	3	½	¼	5
1	17.50	9½	5	5¼	2¼	4¼	½	⅜	15
3	19.50	12¼	6¼	7¼	2¾	5½	½	⅜	30
5	23.00	14¾	7½	8⅝	3	6	⅝	7/16	50
7	29.00	18	8¼	11	3¼	7¾	¾	7/16	70
10	36.25	20¼	9½	12½	3½	7¾	¾	½	100
13	43.50	21⅛	10⅞	13½	3¾	8⅛	1	9/16	130
15	49.50	22⅝	11¼	14½	4	8⅛	1	9/16	150
17½	58.00	24¾	11½	15⅛	4½	9¾	1	9/16	175
20	65.00	25¾	12¼	16	4½	9¾	1⅛	9/16	200
25	81.20	27¼	13	17½	4½	9¾	1⅛	9/16	250

Kester Machinery Co. catalog, 1956, Winston-Salem, N.C. (From Mr. John Catchings)

100-pound Southern Cresent anvil. (R.A.P. photo)

STAR, AMERICAN

144-pound American Star anvil. (R.A.P. photo)

There is a question as to who manufactured this anvil. One advertisement dated in the 1850s states that "Atterbury & Company located at the junction of Assanpick Creek and Railroad St., with offices opposite the State House, Trenton, New Jersey," manufactured these anvils.[43] Another ad of about the same time period from the "Commissioner of Patents" states that "Russell & Co., W.H. Bosnell...C.G. Mc-Chesney" manufactured the *American Star* anvil.[44] This ad suggests that Charles Peters, the person who took out both patents, was the "Superintendent of the American Anvil Factory of Trenton, New Jersey." The first ad calls the foundry or factory the "Anvil Works." There is probably a simple explanation for this confusion, but I have yet to discover it.

Regardless of who owned the foundry, the fact remains that Charles Peters of Trenton, New Jersey, took out two patents on "Improvements" to the *American Star* anvil. The first was dated May 4, 1852, and the second dated February 8, 1853. In other words, here was a direct competitor to Fisher & Norris operating in the same city manufacturing the same type of anvil at the same time. It is very possible and quite likely that Charles Peters had worked for Fisher & Norris before leaving them, and with acquired backing, started the new "Anvil Works." I know from other research that this situation happened with several other anvil manufacturers.

The earliest advertisement that I found for any American anvil made after the 1850s was from the *Russell Erwin Manufacturing Company* catalog of 1865. It lists only the *Eagle* and the *Star* anvils. The last ad mentioning the *Star* anvil was in the 1870 *J.M. Warren & Company,* catalog of Troy, New York. A very interesting ad of 1903 or 1904 in the *C.M. McClung & Company* catalog of Knoxville, Tennessee, mentions the *Eagle* and *Star* as an anvil and then shows an illustration of a Fisher *Eagle.* This last ad suggests that, in all likelihood, Fisher & Norris bought out the Star Anvil Works. This probably happened sometime in the 1860s or 1870s. Fisher & Norris may have continued to manufacture the *Star* anvil on a limited basis, or the ad in the *C.M. McClung & Company* catalog was in error, picked up as copy from an older catalog.

American Star Anvil.
With Tempering Cavity.

$						per lb. each.
No.	0	1	2	3	4	
Weight,	10	15	20	30	40	lbs.
$						per lb. each.
No.	5	6	7	8	9	
Weight,	50	60	70	80	90	lbs.

Anvils, weighing 100 lbs. and over......................per lb. $

Russell & Erwin Mfg. Co. catalog, 1865, New Britain, Conn.

BLACKSMITHS' TOOLS, ETC.

(CONTINUED.)

STAR ANVIL.

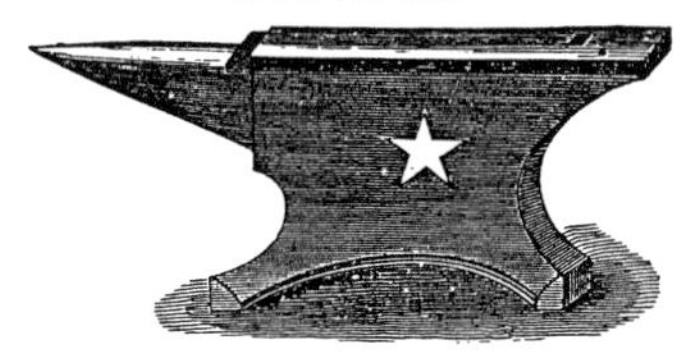

Cast Steel Face,..$ per lb.

J.M. Warren and Co. catalog, 1870, Troy, New York. (From Mr. Hunter Pilkinton)

Eagle and Star.

Double Thick Edges. Face of Best Tool Steel. Solid Steel Horn. Warranted.

Weights	60	70	80	90	100	110	120
Eagle, per pound, ...	......	......	......	......	......	......	......
Star, "	......	......	......	......	......	......	......

C. M. Mc Clung and Co. catalog, 1903-04, Knoxville, Tenn. (From Mr. John Catchings)

I am of the opinion that the *American Star* was manufactured from approximately the early 1850s to the late 1860s, or for about twenty years. I have recorded six of these anvils (1995) and have heard of others.

The first time that I came across one of these *Star* anvils it did not register with me that this was an important anvil manufacturer. The trademark appears on the side in the form of a five-pointed star in relief, with no other markings. I therefore thought it was just another "no-namer." However, when I looked under the base I found a 1 7/8-inch round hole going up 8 3/4 inches into the main body. This was so unusual for a cast-iron anvil that I recorded it. It also had handling holes in both ends of the waist, again unusual. I was hoping to be able to identify the maker some day, which was soon in coming. While searching through the patent files in the Smithsonian Institution I came across a patent calling for a large cavity in the base of an anvil. The patent referred to this as a "tempering cavity." I then remembered that the 1865 ad which mentioned the *Star* had also mentioned a "tempering cavity." Putting two and two together I realized that I had previously recorded this very anvil. The additional information that I uncovered later only confirmed my previous discovery.

This first-recorded *American Star* anvil weighs 143 pounds, is 21 1/2 inches long, with a 12 7/8 x 4-inch face and is 11 1/8 inches high. This means that the "tempering cavity" comes within 2 3/8 inches of the top of the face. The other *Star* anvils that I have recorded show similar proportions, in reference to the cavity size. The company must have sold a large number of these anvils for so many to have survived until now, or maybe I have just been lucky in locating them.

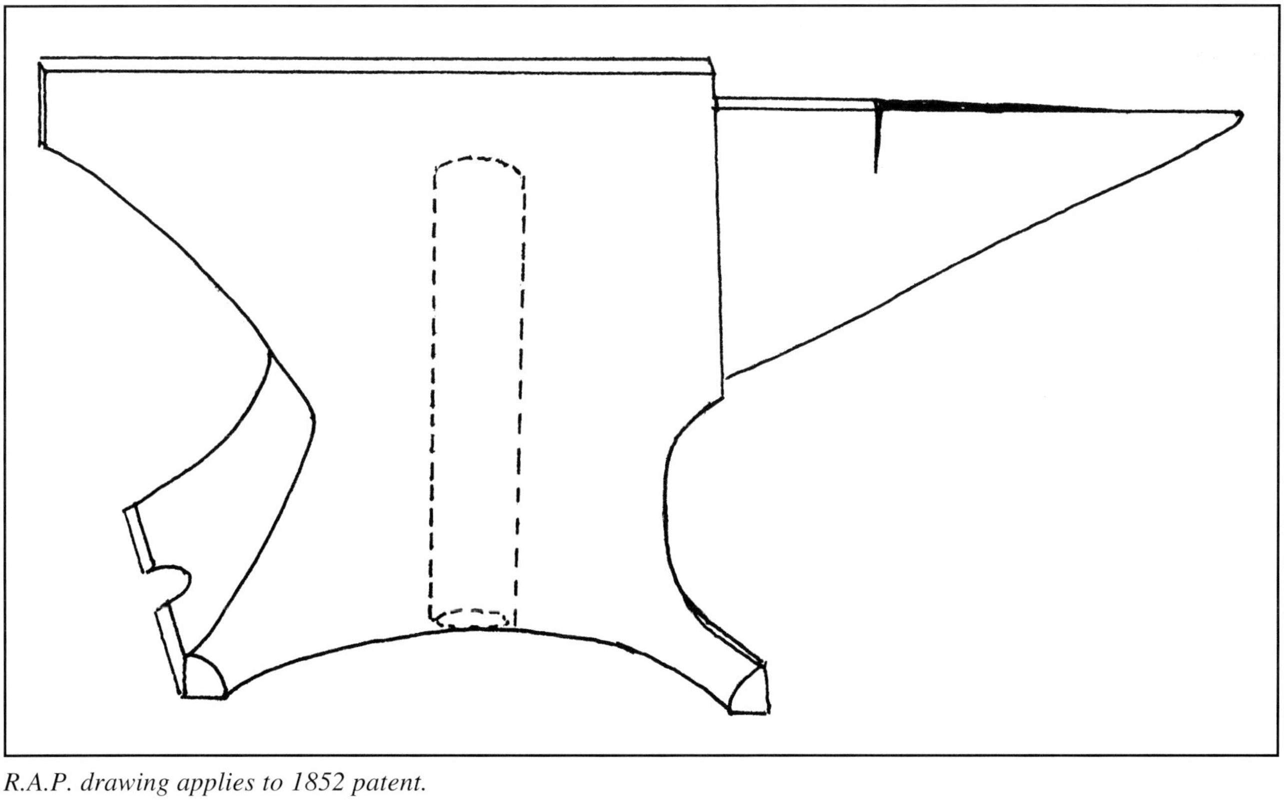

R.A.P. drawing applies to 1852 patent.

UNITED STATES PATENT OFFICE

Charles Peters, of Trenton, New Jersey, and William Fetter, of Bucks County, Pennsylvania.
ANVIL

Specification of Letters Patent, No. 8925, dated May 4, 1852

To all whom it may concern:

Be it known that we, Charles Peters, of the city of Trenton, in the state of New Jersey, and William Fetter, of the county of Bucks, in the State of Pennsylvania, have invented a new and Improved Mode of Preventing Anvils with Steel Faces from Cracking in the process of Tempering.

The nature of our invention consists in providing a cavity in any part of the body of the anvil; which cavity may vary in size or in shape, and have its aperture in the bottom, sides, or any part of the anvil convenient for cooling the interior thereof while the face is undergoing the process of hardening: - the purpose of said cavity being to admit a current of water, oil or other hardening fluid or mixture while the face is subjected to the hardening process so as the more expeditiously to lower the temperature of the interior of the anvil and prevent checks and flaws which frequently occur in solid anvils when the hardening is done altogether from the outside.

To enable others skilled in the art to make and use our invention we will proceed to describe its construction and operation.

We construct our anvils, with steel faces, in any of the known modes, making a cavity in the body of the anvil of any size or form convenient for letting in water or other fluid into the same. As a matter of convenience we usually make the aperture of said cavity in the bottom of the anvil as is represented in the accompanying drawing. We then subject the face of the anvil to the process of tempering in any of the usual modes, and at the same time we cause water or any other cooling fluid to flow into the cavity above described, so that, by this means, the whole of the anvil cools together and is not subject to those flaws and cracks which occur so frequently in the common mode of manufacture where the face and body of the anvil are in such different states of temperature.

What we claim as our invention and desire to secure by Letters Patent is -

A cavity in the body of anvils for the purpose of cooling the same, by the introduction of water or other fluid into the said cavity, while the faces of the said anvils are undergoing the process of tempering.

Chas. Peters.
Wm. Fetter

Witnesses:
B. A. Ward
M. B-----?

Courtesy of the Smithsonian Institution

THE AMERICAN STAR ANVIL.

Tempering Cavity,
Patented May 4, 1852

Imp'd mode of welding
Patented February 8, 1853

FACED WITH CAST-STEEL IN ONE CONTINUOUS PIECE

The subscribers, manufacturers of the Improved Cast-steel Faced Anvil, would call the attention of merchants, manufacturers, and others, to the following remarks of the Commissioner of Patents:

Report of 1852-3 ---- Page 380.

"A simple improvement in Anvils bids fair to obviate an important practical difficulty in their construction. This difficulty has its origin in the heat retained for a long time in the immense mass behind, or rather below, the centre of the steel face, in the process of hardening, which heat prevents the rapid cooling of the steel face, and generally leaves a soft spot near its centre. By forming the body of the Anvil with a cavity of some size, extending from its bottom nearly to its face, a portion of the metal at centre is dispensed with, and facility for the introduction of a stream of cold water into the centre of the mass, and almost upon the bottom of the face, is afforded during the process of hardening. The centre of the mass is therefore cooled almost as rapidly as its exterior, and a sound and equally hard face is, in consequence, a matter of easy attainment."

The tempering cavity not only insures a uniform temper to the face, but relieves the Anvil from that strain to which all castings are subject when heated and suddenly chilled by the application of cold water to the surface only. This *strain* or *tension* in the solid Anvils (after tempering) is so great that a few smart blows will often break them.

In forming the tempering cavity, about five pounds of iron to the hundred is dispensed with; the purchaser gets the same length and width of face, and *five per cent. less iron,* than in any other Anvil of similar size. The Star Anvil, being hollow, has a degree of elasticity, and gives a rebound to the hammer, the want of which is one of the objections to the solid cast Anvil.

These Anvils, for strength, durability, and finish, are considered unsurpassed, having been fully tested in the largest shops in the country.

We warrant our Anvils sound in every respect, the face put on in one piece, and perfectly welded to the body, the metals of the best quality, and the temper uniform.

Charles Peters, Patentee and Superintendent.

ROSSELL & CO.
(W. H. Rossell....C. G. McChesney)
American Anvil Factory, Trenton, N.J.

Depot in Philadelphia, 29 Commerce -street.
A. B. Shipley, agent.

Agency in New York, 31 John-street.
B. Corning & Co. Agents.

An Advertisement of Rossell & Co., Trenton, New Jersey. Courtesy of the Smithsonian Institution

UNITED STATES PATENT OFFICE

Charles Peters, of Trenton, New Jersey.
IMPROVEMENT IN MOLDS FOR UNITING STEEL TO CAST-IRON

Specifications forming part of Letters Patent No.9,578, dated February 8, 1853.

To all whom it may concern:

Be it known that I Charles Peters, of the city of Trenton, in the county of Mercer and State of New Jersey, have invented a new and Improved Case to Molds in Which Steel or Wrought-Iron is to be Welded to Cast-Iron; And I hereby declare that the following is a full and exact description thereof.

The nature of my invention consists in forming a solid basis of iron, brick, or other hard material to the mold, with an aperture therein of the shape of the steel or wrought-iron sought to be welded, and thus by means of the said aperture subjecting the steel or wrought-iron while in the mold to the fire until the same is heated to a welding-heat.

To enable others skilled in the art to make use of my invention, I will proceed to describe its construction and operation, reference being had to the accompanying drawing.

My description and drawing relate to the casting of anvils with steel or wrought-iron faces, from which the application of my invention to other articles will be readily understood.

I form a base for my mold of iron, fire-brick, or other solid material the shape of the flask required to mold the anvil. I then form an aperture in the said base the shape of the steel or wrought-iron face of the anvil, but somewhat smaller, so that the steel or wrought-iron will cover the aperture and be exposed on the under side of the case. I also have a rib surrounding the case of the thickness of the steel or wrought-iron face, so as to form a recess around said steel or wrought-iron. I have also set-screws and loops to retain the steel or wrought-iron in its place. I then daub the recess with molding-sand, so as to prevent the molten iron from passing around the edges of the steel or wrought-iron face and the screws and recess from getting heated. Upon this base thus formed I then build my mold, precisely as molders are accustomed to do, upon the ground. I then place the mold over a furnace level with the ground, so that the solid part of the base shall be on the ground and nothing but the steel or wrought-iron face shall be exposed through the said aperture to the fire. Having obtained a welding-heat I slide the mold off the fire onto a bed of sand, to prevent the steel or wrought-iron from burning. The molten iron is then poured into the mold in any of the modes practiced by founders. I thus dispense with the heating of the steel or wrought-iron face while out of the mold, and also the formation of a new base for each casting.

What I claim as my invention, and desire to secure by Letters Patent, is -

The use of a solid base to molds in which steel or wrought-iron is to be welded to cast-iron, with an aperture in the same, so that by means thereof the said steel or wrought-iron can be subjected to the heat of the furnace while in the mold.

Chas. Peters

Witnesses:

Saml. Evans, A. W. Beasley.

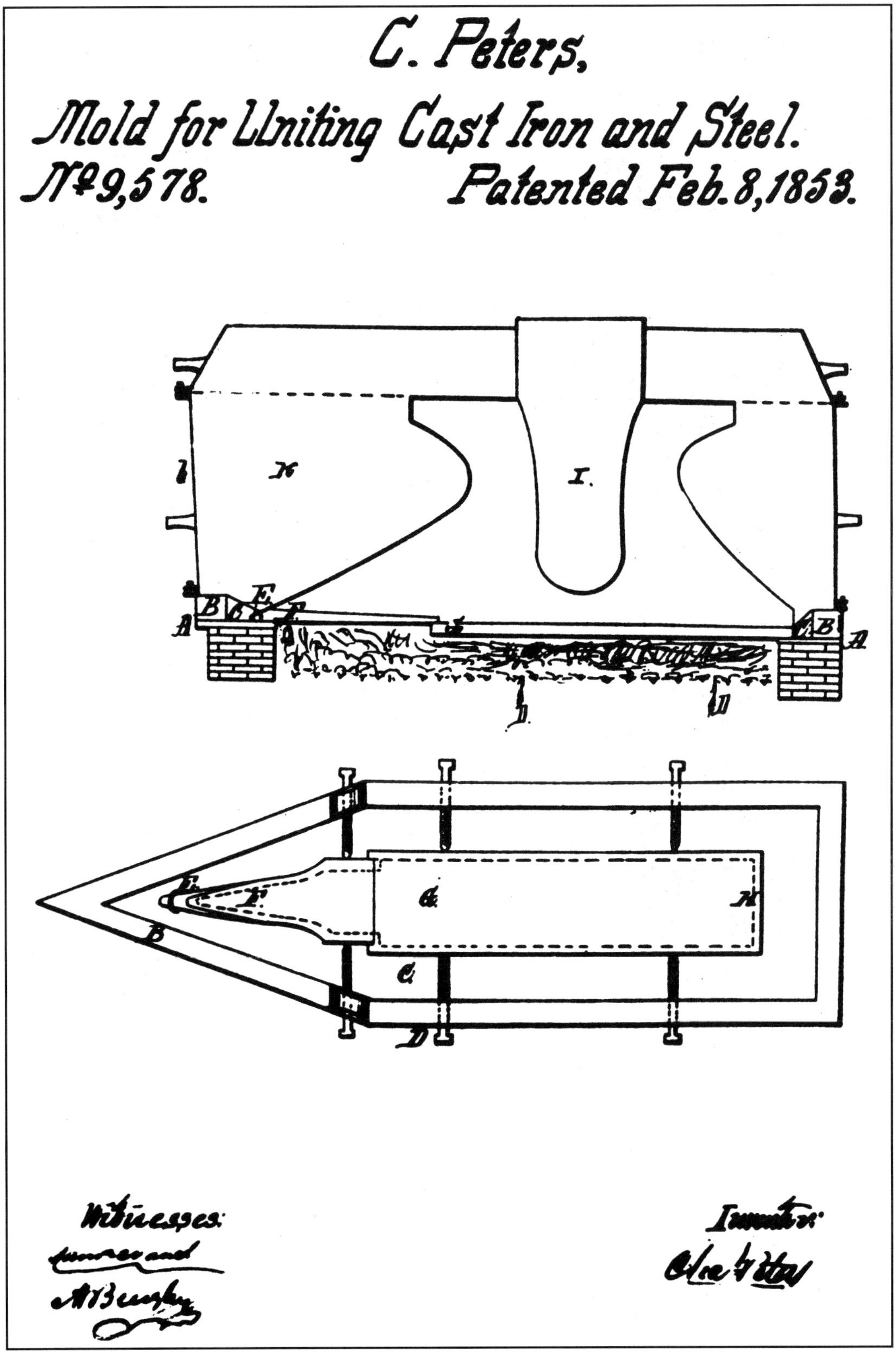

Patent and drawing courtesy of the Smithsonian Institution.

VULCAN

The following information was included as part of an ad for the *Vulcan* anvil, in the 1895 *Charles A. Strelinger & Company* catalog, Detroit, Michigan.[45] Although this early ad illustrates the *Vulcan* anvil the text applies to all cast-iron steel-faced anvils.

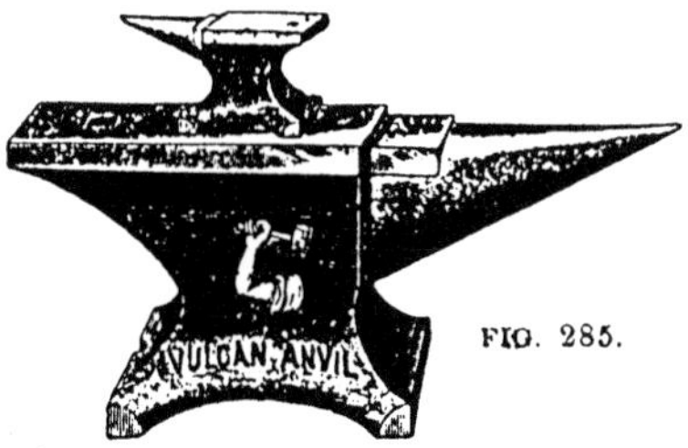

FIG. 285.

Cast iron anvils with tool steel face and horn, are coming into more general use; they are much less in price, and answer for two-thirds of the purposes for which anvils are used.

A point of advantage claimed by one maker of cast anvils is, that they are free from the noisy clang and ring of the wrought iron anvil, but your average blacksmith would'nt know how to do a job on an anvil that didn't ring—doubtless a question of habit and poetic sentiment.

No.	Each.	Wt.	No.	Each.	Wt.
0	$2.20	10 lbs.	8	$6.40	80 lbs.
2	3.20	20 "	9	7.20	90 "
4	4.20	40 "	10	8.00	100 "
5	4.80	50 "	12	9.68	120 "
6	5.20	60 "	14	11.20	140 "
7	5.80	70 "	15	12.00	150 "

The Charles A. Strelinger Co. catalog, Detroit, Mi.

The *Vulcan* steel-faced cast-iron anvil was made by the Illinois Iron & Bolt Company of Carpentersville, Illinois, from about 1875 until about 1969.[46] In the mid-1980s I attempted to obtain information on this company but without success. Therefore, most of my information has come from advertisements and the anvils themselves. Fortunately these anvils are very plentiful, even today.

The Illinois Iron & Bolt Company manufactured a variety of products as well as anvils, as the name of the company would indicate. They produced wagon tire shrinkers, tire benders, screw jacks, front and rear car axles, post drills, fullers and swages, and twyere irons (fire pots), to mention a few.[47]

I would not rate the *Vulcan* anvil as being of the same quality as the Fisher & Norris *Eagle* or the wrought anvils, and slightly higher quality than the strictly cast-iron anvils. Many were bought by farms, schools, and smaller shops, places that did not depend on making a living using an anvil. There appears to be as many or more *Vulcan* anvils in use today as there are Fisher & Norris *Eagles,* though they certainly did not manufacture nearly as many. One reason for the *Vulcan*'s success was that more hardware catalogs advertised the brand than any other single make of anvil. They were also cheaper than the other steel-faced cast-iron anvils. It is hard to pick up an old catalog that does not offer the *Vulcan.* Sears Roebuck and Montgomery Ward carried them continually, in spite of the fact that they changed from brand to brand on the wrought-iron anvils they offered. However, I have never found an advertisement for the *Vulcan* anvil in any blacksmith and related journals. I did find a short write-up of the Illinois Iron & Bolt company under "Blacksmith's Tools" in the May, 1884, issue of *The Blacksmith & Wheelwright* journal. " This company makes Thimble Skeins as well as a great variety of Blacksmith's tools...."[48]

Sometime in the 1960s the Illinois Iron & Bolt Company bought the American Skein & Foundry Company (see *Badger*) and had the *Vulcan* manufactured by them.[49] Both companies went out of business in the late 1960s or early 1970s.

VULCAN ANVILS

DEALER PRICE LIST — **EFFECTIVE JANUARY 1, 1966**

Trade Name Reg.

FACE AND HORN SOLID STEEL

BODY of BEST QUALITY GRAY CAST IRON

The face of the Vulcan Anvil is of solid steel, warranted to be thoroughly welded to the cast iron body of the anvil . . . Ground, tempered and polished to a smooth finish. The horn is covered with -- and its point is made entirely of — tough untempered steel. The Vulcan is a high grade anvil and has been used successfully for many years.

VULCAN ANVIL SPECIFICATIONS

Size No.	Dealer Price	Weight lbs.	Length Overall	Length of Face	Width of Face	Height Overall	Length of Horn	Size of Hardie Hole	Size of Pritchel Hole
00	**$ 14.75**	5	7⅛"	4"	1½"	3⅞"	3⅛"	⅜"	¼"
1	**20.00**	15	9½"	5 3/16"	2¼"	5"	4 5/16"	½"	⅜"
3	**27.50**	30	12"	7¼"	2¾"	6¼"	4¾"	½"	⅜"
5	**35.00**	50	14¾"	8⅝"	3"	7½"	6⅛"	½"	⅜"
7	**41.00**	70	18½"	11"	3¼"	8⅛"	7½"	½"	⅜"
10	**50.00**	100	20¼"	12⅝"	3½"	9¼"	7⅝"	¾"	½"
13	**56.50**	130	21⅛"	13½"	3¾"	10⅞"	7⅝"	¾"	9/16"
15	**64.50**	150	22⅝"	14½"	4"	11½"	8⅛"	⅞"	9/16"
17½	**81.00**	175	24¾"	15⅛"	4½"	11½"	9⅝"	⅞"	9/16"
20	**87.50**	200	25¾"	16¼"	4½"	12¼"	9½"	1¼"	9/16"
25	**100.00**	250	27¼"	17½"	4½"	13"	9¾"	1¼"	9/16"

F.O.B. Factory, Carpentersville, Ill. Terms 2% - 10 days, Net 30

Prices subject to change without notice.

Manufactured By

AMERICAN SKEIN & FOUNDRY COMPANY

DIVISION OF

ILLINOIS IRON & BOLT CO.

Carpentersville, Illinois

Price list, January 1, 1966. You will notice that the Illinois Iron & Bolt Co. had purchased the American Skein & Foundry Co. by that time. (From Mr. Hunter Pilkinton)

ANVILS IN AMERICA

On most *Vulcan* anvils the trademark is the "Arm" holding a "Hammer," similar to the Arm & Hammer Baking Soda trademark. Some of the earlier anvils had the initials "I.I.& B. Co." located on the top of the foot, front or rear, but these anvils are rare. The trademark itself changed three times during its manufacture, but only slightly. The first was a circle down near the base between the feet, as follows:

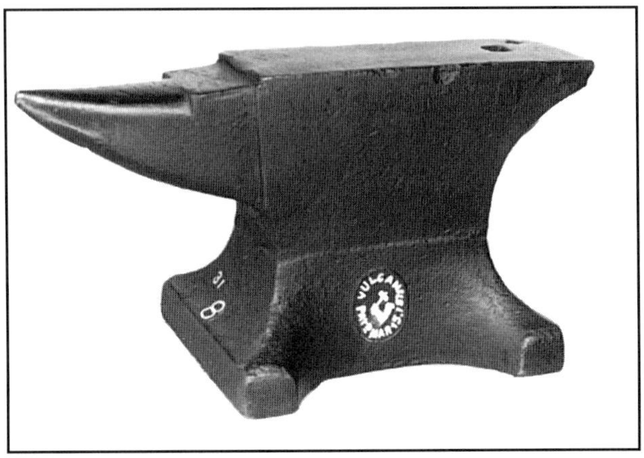

80-pound early Vulcan anvil. The first style trademark. It reads: "Vulcan, Pat. Mar. 15, 1875 or 76." Notice how blocky it is. (R.A.P. photo)

100-pound. The second style trademark. It reads: "Vulcan, Brand." (R.A.P. photo)

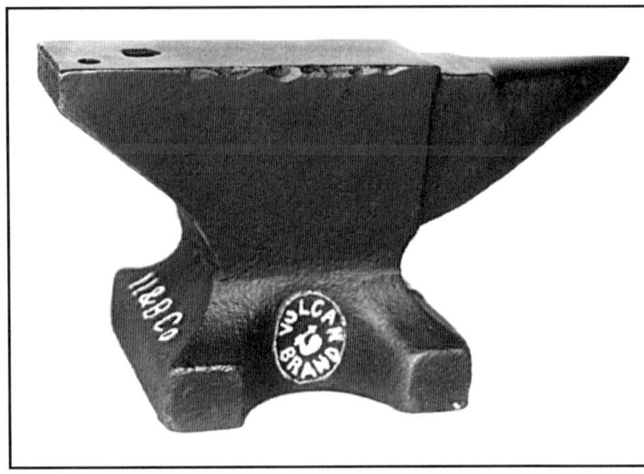

50-pound Vulcan anvil with the second trademark and the "II & B Co." Trademark on it as well. (R.A.P. photo)

140-pound Vulcan anvil with the third and final trademark change. It reads: "Vulcan, Reg. U. S. Pat. Off." In Every case the last digit of the weight is left off of these anvils. (R.A.P. photo)

The patent date could be 1875. Sometime in the 1880s or '90s, the company changed their anvil style from the *Old English* which was very blocky, to the *Modern* style, changing the trademark during this time as well. In place of the patent date, the word "Brand" was substituted making it read *Vulcan Brand*, with the "Arm & Hammer" logo in the center. The circle also became more oval in shape from top to bottom. The last change to the trademark came in the 20th century, replacing the "Brand" with "Reg. US Pat OFF." The anvil itself did not change physically.

Apparently the company manufactured only the smith's pattern anvil, and did not get into the variety of other anvils that Fisher & Norris offered.

An advertisement in 1930 shows 22 anvils of different sizes ranging in weight from 5 to 250 pounds. The last ad I have seen is from 1966 and shows 11 different sizes, from 5 to 250 pounds. The oldest ad that I have come across is dated 1895, and shows 20 different sizes, again from 5 to 250 pounds. I have personally recorded a 10-pound and a 250-pound *Vulcan,* and many sizes in between.

Vulcan Anvils

The horn is covered with, and its extremity made entirely of tough, untempered steel. The body is charcoal iron, and the face is covered with one solid piece of tool steel welded to body.

No.	Approx. Wt., Lbs.	Price Each	No.	Approx. Wt., Lbs.	Price Each	No.	Approx. Wt., Lbs.	Price Each
00	5	$2.00	7	70	$7.25	15	150	$15.00
0	10	2.75	8	80	8.00	16	160	16.00
1	15	3.25	9	90	9.00	17½	175	17.50
2	20	4.00	10	100	10.00	20	200	20.00
3	30	4.50	11	110	11.00	22½	225	22.50
4	40	5.25	12	120	12.00	25	250	25.00
5	50	6.00	13	130	13.00	...	...	......
6	60	6.50	14	140	14.00	...	...	......

The Charles A. Strelinger Co. catalog, 1930, Detroit, Mi.

ANVILS.

VULCAN.

Number, .	00	0	1	2	3	4
Weight, lbs.,	5	10	15	20	30	40
Each, .	$2 00	2 75	3 25	4 00	4 50	5 25

Number, .	6	7	8	9	10	11
Weight, lbs.,	60	70	80	90	100	110
Each, .	$6 50	7 25	8 00	9 00	10 00	11 00

Number, .	13	14	15	17½	20
Weight, lbs.,	130	140	150	175	200
Each, . .	$13 00	14 00	15 00	17 50	20 00

The above Anvils are cast iron with a steel face.

Paddock-Hawly Iron Co. catalog, 1895, St. Louis, Mo. (From Mr. Hunter Pilkinton)

ANVILS IN AMERICA

It is interesting to note the change in the prices on these anvils, from 1895 to 1966. *Vulcan* price list using a 100 pound anvil as the example from various catalogs. Most of these seem to be wholesale prices:

1895 $10.00 Charles A. Strelinger Co.
1896 $10.00 Canton Hardware
1898 $ 7.30 Sears Roebuck
1901 $10.80 Montgomery Ward
1903 $10.00 Pittsburgh Gage & Supply Co.
1906 $10.00 Machinists Supply Co.
1910 $10.00 Kelly, Maus & Co.
1920 $10.00 Vulcan Supply Co.
1920 $20.00 Geo. B. Carpenter & Supply Co.
1920 $20.00 Schwabacher Hardware Co.
1925 $15.00 Bostwich - Braun Co.
1925 $15.00 H. Channon Co.
1925 $10.00 South Bend Supply
1926 $10.50 Montgomery Ward
1924 $21.30 Marshall - Wells Co.
1928 $10.00 Minneapolis Iron Store
1927 $10.00 Taylor Iron-Works & Supply Co.
1928 $18.80 Marshall - Wells Co.
1930 $16.00 Janney, Semple, Hill & Co.
1930 $20.00 Charles A. Strelinger Co.
1932 $13.50 Hilbard, Spencer, Bartlett & Co.
1933 $11.24 McMaster - Carr Supply Co.
1934 $29.25 Belknap Co.
1935 $16.00 South Bend Supply
1936 $16.00 H. Channon Co.
1937 $16.00 W. Bingham Co.
1937 $11.20 McMaster - Carr Supply Co.
1940 $16.00 Belknap Co.
1941 $38.00 W. Bingham Co.
1954 $36.25 Barrett - Christie Co.
1955 $41.50 Bostwich - Braun Co.
1956 $36.25 Mossman Yarnell Co.
1961 $148.50 Belknap Co.
1966 $50.00 Illinois Iron & Bolt Co. (wholesale)

It is important to remember that the *Vulcan* anvil is not in any way related to the *Arm & Hammer* anvil, although they both used the "Arm & Hammer" as part of their trademark (see *Arm & Hammer*). The only thing they have in common is that they are both anvils. Many people become confused due to the similarity in these trademarks. The following selected ads and photos are explained individually, and the ads are not usually representative of the *Vulcan*'s appearance.

VULCAN.

Cast Iron, Steel Face.

No.	0	1	2	4	6	8	10	11	12	14
Weight......pounds,	10	15	20	40	60	80	100	110	120	140
Each	$2 75	3 25	4 00	5 25	6 50	8 00	10 00	11 00	12 00	14 00

The Canton Hardware Co. catalog, 1896, Canton, Ohio.

VULCAN ANVILS

We would call the attention of blacksmiths and others to the Vulcan Anvil, now sold by us.

The face of this Anvil is one solid piece of English tool steel, thoroughly welded to the body of the anvil by a patented process. It is then accurately ground and tempered, so it will stand the most severe test.

The Horn is covered with, and its extremity made entirely of, tough, untempered cast steel.

The body of the anvil is made of the best Lake Superior Charcoal Iron, and, being much more solid than wrought iron, the work being forged receives the full force of the blow. The face of the anvil will remain perfectly true after years of service, and it is warranted to be thoroughly welded and not to separate.

PRICE LIST.

Number.	Weighing about.	Price.	Number.	Weighing about.	Price.
00	5 lbs.	$2 00	8	80 lbs.	$8 00
0	10 "	2 75	9	90 "	9 00
1	15 "	3 25	10	100 "	10 00
2	20 "	4 00	11	110 "	11 00
3	30 "	4 50	12	120 "	12 00
4	40 "	5 25	13	130 "	13 00
5	50 "	6 00	14	140 "	14 00
6	60 "	6 50	15	150 "	15 00
7	70 "	7 25	--	------	-----

Machinist Supply Co. catalog, 1903-1906, Chicago, Ill.

Vulcan Anvils

Cast Iron Base—Tool Steel Face

These anvils are cheaper than our Bullock warranted anvils, but for occasional use will give satisfactory service.

The body or base is made of Lake Superior charcoal iron of superior mixture—no scrap used.

The face is one solid piece of tcol steel thoroughly welded to body by special process.

Face is ground tempered and polished.

Horn is covered with and its extremity made entirely of tough untempered steel.

Number	Approximate Weight, Pounds	Size of Face, Inches	Hardy Hole	Pritche Hole	Purpose	Price Each
3	30	2¾ x 7½	½	⅜	Very light repair work......................	$ 6.75
4	40	3 x 7¾	⅝	⅜		8.00
5	50	3 x 8¾	⅝	7/16		9.00
6	60	3¼ x 10½	¾	7/16		9.75
8	80	3¼ x 11½	¾	7/16		12.00
10	100	3½ x 12½	⅞	½	Farmers and contractors use, light blacksmith-	15.00
12	120	3¼ x 13	1	½		18.00
15	150	4 x 14	1	½	ing, etc....................................	22.50

H. Channon Co. catalog, 1925, Chicago, Ill.

94 ILLINOIS IRON & BOLT CO.

Vulcan Anvil

We call the attention of blacksmiths and others to the **Vulcan Anvil,** the face of which is one solid piece of English tool steel, thoroughly welded to the body of the anvil by a patented process. It is then accurately ground and tempered, so it will stand the most severe test.

The horn is covered with, and its extremity made entirely of tough, untempered steel.

The body of the anvil is made of the best **Lake Superior Charcoal Iron,** and, being much more solid than wrought iron, the work being forged receives the full force of the blow. The face of the anvil will remain perfectly true after years of service, and it is warranted to be thoroughly welded and not to separate.

PRICE LIST

NO.			NO.		
00,	weighing about 5 lbs.	$ 2.00	11,	weighing about 110 lbs.	$11.00
0,	" " 10 "	2.75	12,	" " 120 "	12.00
1,	" " 15 "	3.25	13,	" " 130 "	13.00
2,	" " 20 "	4.00	14,	" " 140 "	14.00
3,	" " 30 "	4.50	15,	" " 150 "	15.00
4,	" " 40 "	5.25	16,	" " 160 "	16.00
5,	" " 50 "	6.00	17½,	" " 175 "	17.50
6,	" " 60 "	6.50	20,	" " 200 "	20.00
7,	" " 70 "	7.25	22½	" " 225 "	22.50
8,	" " 80 "	8.00	25,	" " 250 "	25.00
9,	" " 90 "	9.00	30,	" " 300 "	30.00
10,	" " 100 "	10.00			

Illinois Iron & Bolt Co. brochure, 1920s or '30s. This is the only listing I have found offering a 300-pound Vulcan anvil. (From Mr. James Moon)

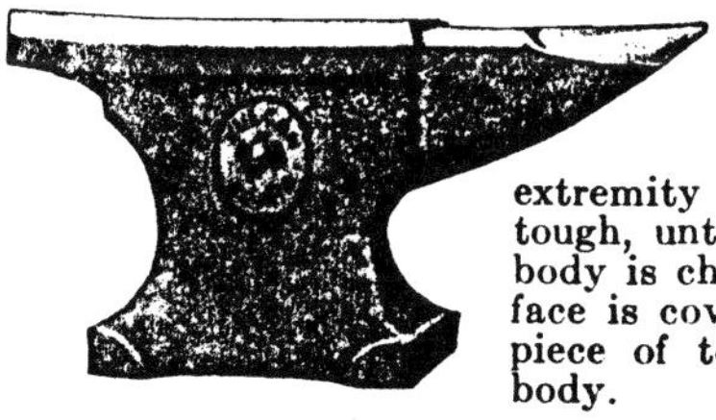

Vulcan Anvils

The horn is covered with, and its extremity made entirely of tough, untempered steel. The body is charcoal iron, and the face is covered with one solid piece of tool steel welded to body.

No.	Approx. Wt., Lbs.	Each	No.	Approx. Wt., Lbs.	Each	No.	Approx. Wt., Lbs.	Each
00	5	$6.00	7	70	$13.00	15	150	$26.00
0	10	7.00	8	80	14.00	16	160	28.00
1	15	7.50	9	90	15.00	17½	175	31.00
2	20	8.00	10	100	16.00	20	200	35.00
3	30	9.00	11	110	18.00	22½	225	38.00
4	40	10.00	12	120	20.00	25	250	40.00
5	50	11.00	13	130	22.00	..	...	.....
6	60	12.00	14	140	24.00	..	...	.....

The South Bend Supply Co. catalog, 1935, South Bend, In.

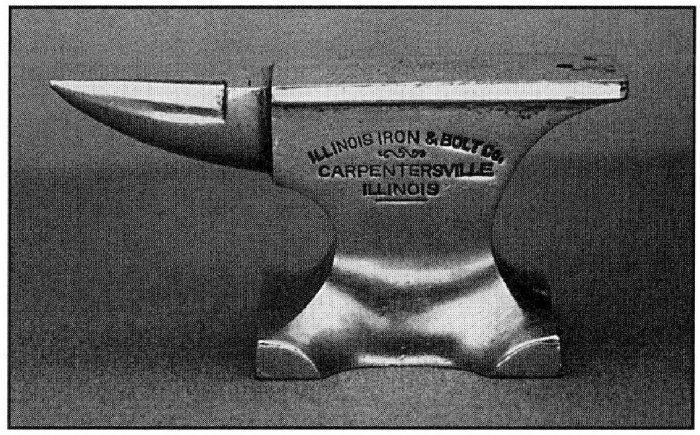

Miniature Vulcan advertising anvil. (Mr. Gill Fahrenwald photo)

Vulcan Anvils

The face is one piece of high grade steel — thoroughly welded to the body of the anvil. It is then accurately ground and tempered.

Horn is covered with—and its extremity made entirely of—tough, untempered steel. The body is made of the best quality grey cast iron.

Size No.	Each	Wt., Lbs.	Length Overall, Ins.	Length of Face, Ins.	Width of Face, Ins.	Height Overall, Ins.	Hardie Hole, Ins.
00	11.60	5	7	4	1½	3¾	½
1	17.50	15	9½	5¼	2¼	5	½
3	19.50	30	12¼	7¼	2¾	6¼	½
5	23.00	50	14¾	8⅝	3	7½	⅝
7	29.00	70	18	11	3¼	8¼	¾
10	36.25	100	20¼	12½	3½	9½	¾
13	43.50	130	21⅛	13½	3¾	10⅞	1
15	49.50	150	22⅝	14½	4	11¼	1
17½	58.00	175	24¾	15⅛	4½	11½	1
20	65.00	200	25¾	16	4½	12¼	1⅛
25	81.20	250	27¼	17½	4½	13	1⅛

Barrett-Christie Co. catalog, 1954.

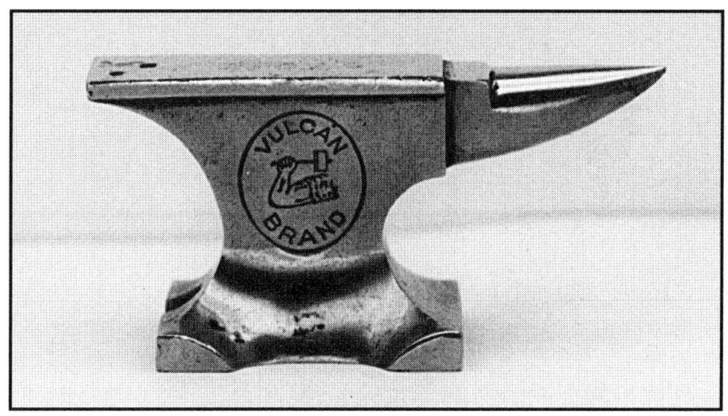

Reverse side of the Vulcan miniature anvil. (Mr. Gill Fahrenwald photo)

CAST-STEEL ANVILS

The cast-steel anvil was unknown until the very last part of the 19th century, with the development of the "open hearth" and "electric furnace." Then production of cast-steel anvils became a reality. The Swedes were most likely the first to produce these anvils, from "open hearth" steel. It was picked up in the United States by one major manufacturer shortly after the start of the 20th century.

Other steel foundries produced anvils at a later date, but very few of these examples are to be found today. As an example of this later type, in the 1920s Sears Roebuck offered a cast-steel anvil named the *Fulton* which was named after the street that ran in front of Sears main offices in Chicago, Illinois. It was produced by a now-unknown steel foundry especially for Sears and I consider it a very good anvil. After a year or more, Sears dropped this cast-steel anvil and went back to selling wrought-iron anvils in its place. The foundry and factory that made the *Fulton* probably went on to manufacture anvils under another name for other clients.

Sears of course did not make the anvil, they just sold them with their trademark shown on the side. In fact, one foundry cast the anvils, another factory probably machined the anvils, and yet still another company or heat treat shop hardened and tempered the anvil, making up to three companies involved in its manufacture. So who actually manufactured the *Fulton* anvil? Not Sears, they just sold them. Not the foundry, they just cast the anvil to order. Not the heat treaters, they just hardened and tempered the anvils. It was even possible that yet another concern ground and polished the hardened surfaces and painted them, though this was most likely performed by the shop machining them. Therefore, with but a couple of exceptions almost all solid cast-steel anvils were manufactured by a number of unknown foundries and the name on the anvils only applied to the seller. It is therefore possible to attribute names to anvils based only on who sold them.

Anvils manufactured after the 1960s for the most part came from smaller Mom & Pop type operations, who finish the anvils after they were received from a foundry. That is not to say that these are of inferior quality. On the contrary, some of these small operations produced cast-steel anvils as good as and even better than any of those ever manufactured.

COLUMBIAN

The Columbian Hardware Company, of Cleveland, Ohio, is still in business, but under another name, and since 1925 the only anvil they have manufactured is a bench or craft anvil weighing just under 9 pounds.

BENCH ANVILS
COLUMBIAN

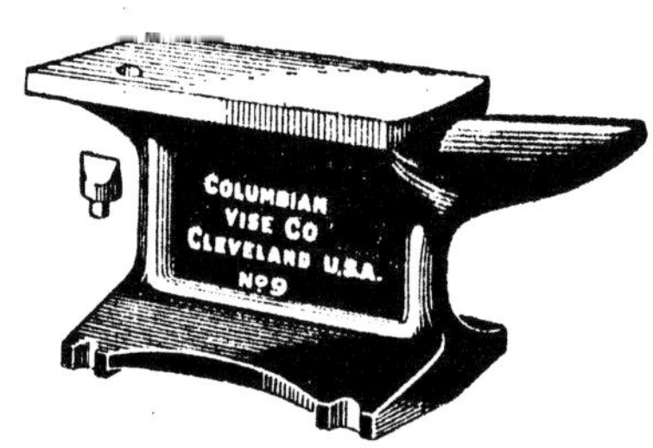

A splendid tool for a wide range of work, from industrial and home shop benches to dentists' and jewelers' work tables. Made of shock-resisting "Kromiron" castings, with a broad, tough face. Equipped with a steel cut-off tool, carefully heat-treated and hardened. Face—3x6 inches, horn—3 inches, base 4x5½ inches. Weight 9 lbs.

No. 9A .... Each $3.90

The Bostwick-Braun Co. catalog, 1955, Cleveland, Ohio.

There is no one with the company today (1990) who knows for certain when the company was founded, or when they became known as The Columbian Hardware Company. My guess would be that the company was founded in the early 1890s and started producing anvils about 1903 or 1904. Their brochure for 1907 states on the cover, "The Columbian Hardware Company, world's largest maker of Vises and Anvils." [50] They may have made more vises and anvils in combination than any other company, but they certainly did not manufacture more anvils than many other anvil manufacturers. They existed as an unincorporated company until 1913, and therefore there are no previous records listed with the Secretary of State's Office of Ohio.[51] In May of 1926, they incorporated as the Columbian Vise & Manufacturing Company, by which time they had ceased production of large anvils.[52] In December of 1972 they merged with the Cincinnati and Warren Tool Companies, to form The Warren Group.[53]

The name "Columbia" was very popular in the 1890s, with the Columbian Exposition of 1892 in Chicago. Therefore, the company probably picked the name because it stood out and was well known. The name would not have been worth much had they not produced an excellent cast-steel anvil. In all likelihood, they were the first cast steel anvil makers in the United States and definitely the most successful.

The Columbian Hardware Company manufactured smith's anvils up until 1922 or 1923, after which they imported cast-steel anvils from *Söderfors* of Sweden until 1926 or 1927.[54] If my dates are correct, and I suggest that they are, this company made anvils for about 20 years. What is significant about these few years is the fact that there are still so many *Columbian* anvils to be found today, most of them still in use.

I have recorded five of these *Söderfors / Columbian* anvils, dated 1923, two in 1924, 1925 and 1926. These anvils have the full *Söderfors* trademark, then above it is stamped "Columbian" (see Söderfors). The *Columbian* anvils manufactured by The Columbian Hardware Company have no markings, with the exception of a trademark; an inverted recessed pyramid with a raised "C" in the center. Usually there is a raised letter on the other side or in some cases somewhere else. This letter is a foundry or pattern mark, and as far as I can tell has no relationship to the date of manufacture or weight. However, the 100-pound anvils generally have an "M" in relief on them. I have recorded 30 *Columbian* anvils to date (1994), and maybe after recording about 100 or more the letters will show a pattern and have meaning.

The oldest ad I have obtained is from the August, 1905, *The American Blacksmith* journal and the most recent from the January, 1919, issue of *The Blacksmith & Wheelwright* journal.

I have also obtained two additional ads in hardware catalogs dated 1920 and 1925, sent to me by Mr. Hunter Pilkinton. However, as I have stated before, hardware companies often continue to show a product in their catalogs long after the product's demise.

The company seems to have manufactured only blacksmith pattern anvils, and some European patterns for export, four of which I have recorded. The anvils were manufactured in sizes ranging from 10 to 800 pounds.[55]

The American Blacksmith August, 1905.

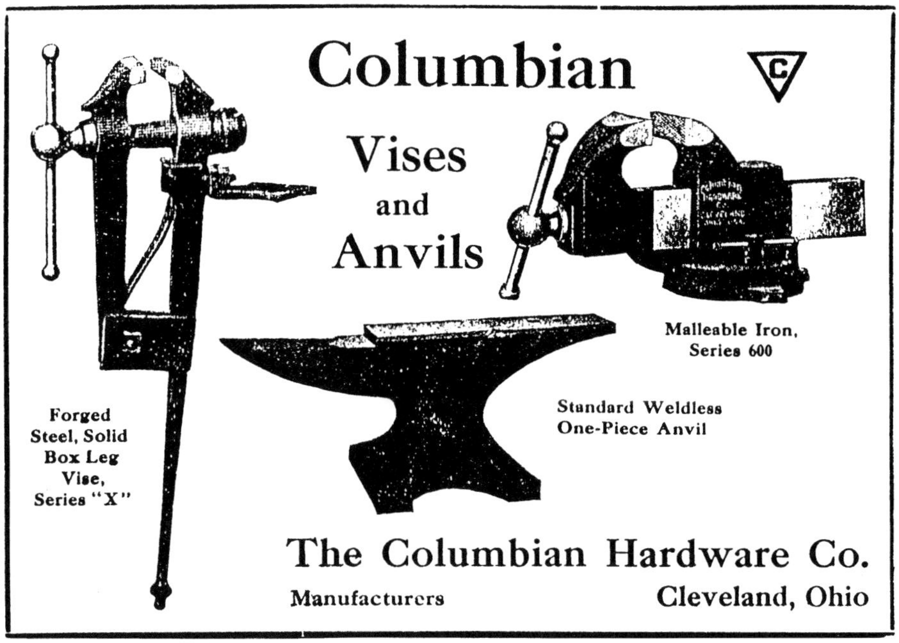

The Blacksmith and Wheelwright, January, 1919.

COLUMBIAN ANVIL

ALL STEEL

Fig. 5290

Scientific Design. The width of base is equal to the height, insuring the rigidity necessary for securely fastening to block. Large face surface, milled before tempering, and ground after tempering, so as to insure a perfect finish and rectangular edges.

80 to 425 lbs	Base..........	per lb.	Market
426 to 625 lbs...........	per lb.	extra	$0.00½
626 to 800 lbs...........	"	"	.01
70 to 79 lbs...........	"	"	.00½
60 to 69 lbs...........	"	"	.01
50 to 59 lbs...........	"	"	.02
40 to 49 lbs...........	"	"	.03
30 to 39 lbs...........	"	"	.05
20 to 29 lbs...........	"	"	.08
10 to 19 lbs...........	"	"	.15

George B. Carpenter Co. catalog 110, 1920, Chicago, Ill.

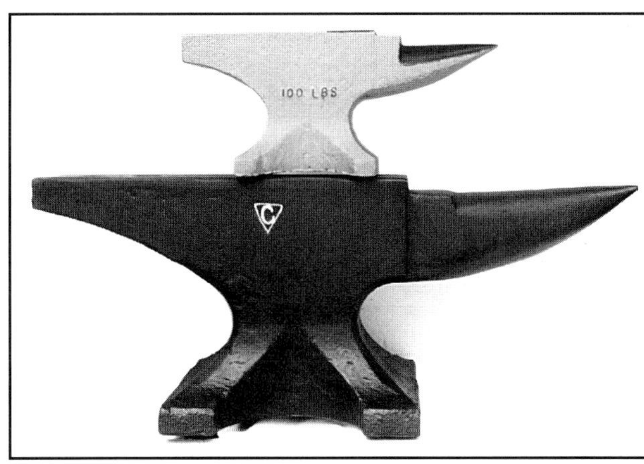

511-pound Columbian anvil. Note the trademark and compare with the 100-pound anvil. (R.A.P. photo)

The largest *Columbian* anvil I have recorded weighs 511 pounds and the smallest 33 pounds. Reproduced below is a part of the company's 1907 brochure, courtesy of the Columbian Vise & Manufacturing Company showing types and sizes of anvils offered.

COLUMBIAN HARDWARE

In Keeping with

20th Century Ideas

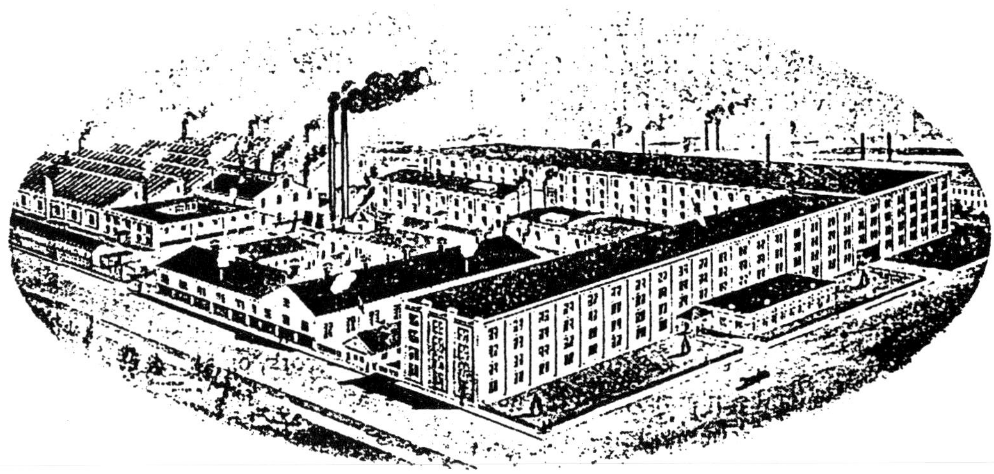

THE COLUMBIAN HARDWARE CO.

CLEVELAND, OHIO, U. S. A.

PRODUCERS OF ART HARDWARE

1907

BRANCHES

NEW YORK
168 Church St.

SAN FRANCISCO

CHICAGO
45-47 E. Lake St.

LONDON

PARIS

SYDNEY

Complete Stock Carried at New York and Chicago

Inside cover of the 1907 brochure. Courtesy of the Columbian Vise & Mfg. Co.

The Columbian Hardware Company *Manufacturers*

Columbian All-Steel

Anvils

Standard Pattern

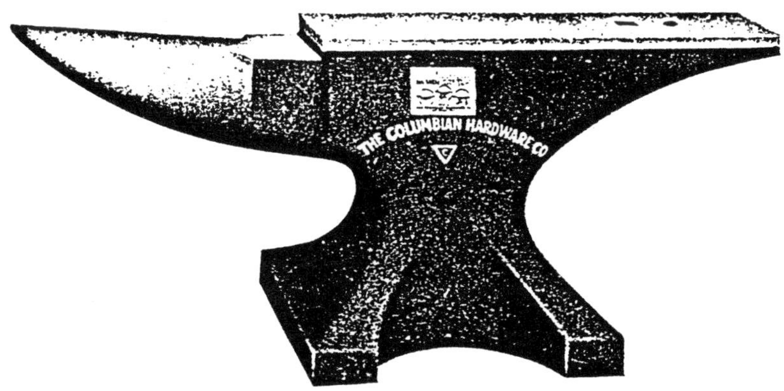

All Columbian Anvils are made in one solid piece of crucible steel of such analysis as assures tough hard working surface.

The true mechanic recognizes that the anvil must be made from the best quality of steel to produce the necessary ring and rebound.

Scientific Design. The width of base is always equal to the height, thereby lowering the center of gravity, and insuring the rigidity necessary for securely fastening to block. The horn is long and rounded throughout its full length, and the tail is long. Large face surface, which is milled before tempering, and ground after tempering, so as to insure a perfect finish and rectangular edges.

Sizes

80 to 425 lbs.	Base
426 to 625 lbs.	½c per lb. extra net
626 to 800 lbs.	1c " " " "
70 to 79 lbs.	½c " " " "
60 to 69 lbs.	1c " " " "
50 to 59 lbs.	2c " " " "
40 to 49 lbs.	3c " " " "
30 to 39 lbs.	5c " " " "
20 to 29 lbs.	8c " " " "
10 to 19 lbs.	15c " " " "

From the 1907 brochure.

The Columbian Hardware Company *Manufacturers*

Blacksmiths' Anvil Dimensions

Order by Number

Lbs.	Face		Horn	Blight			Square Hole	Round Hole	
Size of Anvil	Length	Width	Length	Length	Width	Depth Below Face	Size of Square Hole	Size of Round Hole	Height of Anvil
10	6 1/2	1 3/4	3 1/4	13/16	1 1/2	1/4	3/8	1/4	4 1/16
20	8	2 1/4	4	1	2	21/64	1/2	5/16	5
30	9 1/4	2 1/2	4 5/8	1 5/32	2 1/4	3/8	9/16	3/8	5 25/32
40	10 1/4	2 3/4	5 1/8	1 9/32	2 1/2	27/64	5/8	7/16	6 13/32
50	11	3	5 1/2	1 3/8	2 11/16	29/64	3/4	7/16	6 7/8
60	12	3	6	1 1/2	2 11/16	1/2	3/4	7/16	7 1/2
70	12 1/2	3 1/4	6 1/4	1 9/16	2 15/16	1/2	3/4	7/16	7 13/16
80	13 3/8	3 1/4	6 11/16	1 11/16	2 15/16	17/32	3/4	7/16	8 3/8
90	13 3/4	3 1/2	6 7/8	1 23/32	3 1/8	9/16	7/8	7/16	8 9/16
100	14 1/2	3 1/2	7 1/4	1 13/16	3 1/8	9/16	7/8	7/16	9 1/8
110	14 3/4	3 3/4	7 3/8	1 27/32	3 3/8	9/16	7/8	7/16	9 3/16
120	15 3/8	3 3/4	7 11/16	1 15/16	3 3/8	21/32	1	1/2	9 5/8
130	15 7/8	3 3/4	7 15/16	2	3 3/8	21/32	1	1/2	9 15/16
140	16	4	8	2	3 1/2	21/32	1	1/2	10
150	16 1/2	4	8 1/4	2 1/16	3 1/2	11/16	1	1/2	10 5/16
160	17	4	8 1/2	2 1/8	3 1/2	11/16	1	1/2	10 5/8
175	17 1/4	4 1/4	8 5/8	2 5/32	3 3/4	11/16	1	1/2	10 3/4
200	18	4 1/2	9	2 1/4	4	3/4	1 1/8	5/8	11 1/4
225	19	4 1/2	9 1/2	2 3/8	4	13/16	1 1/8	5/8	11 7/8
250	19 1/2	4 3/4	9 3/4	2 7/16	4 1/4	13/16	1 1/4	3/4	12 3/16
275	19 3/4	5	9 7/8	2 15/32	4 1/2	13/16	1 1/4	3/4	12 3/8
300	20 1/2	5	10 1/4	2 9/16	4 1/2	13/16	1 1/4	3/4	12 13/16
325	20 5/8	5 1/4	10 5/16	2 9/16	4 3/4	7/8	1 3/8	7/8	12 7/8
350	20 3/4	5 1/2	10 3/8	2 19/32	5	7/8	1 3/8	7/8	13
375	21 1/4	5 1/2	10 5/8	2 21/32	5	7/8	1 3/8	7/8	13 1/4
400	21 1/2	5 3/4	10 3/4	2 11/16	5 1/4	7/8	1 3/8	7/8	13 7/16
425	22 1/4	5 3/4	11 1/8	2 25/32	5 1/4	15/16	1 3/8	7/8	13 7/8
450	22 1/2	6	11 1/4	2 13/16	5 1/2	15/16	1 3/8	7/8	14 1/16
475	23	6	11 1/2	2 7/8	5 1/2	15/16	1 1/2	1	14 3/8
500	24	6	12	3	5 1/2	1	1 1/2	1	15
525	24 1/4	6 1/4	12 1/8	3 1/32	5 1/2	1	1 1/2	1	15 1/8
550	24 1/2	6 1/4	12 1/4	3 1/16	5 1/2	1	1 1/2	1	15 5/16
575	24 3/4	6 1/2	12 3/8	3 3/32	5 3/4	1	1 1/2	1	15 7/16
600	25 1/2	6 1/2	12 3/4	3 3/16	5 3/4	1	1 1/2	1	15 15/16

Dimensions of the standard blacksmith's anvil from the 1907 brochure.

The Columbian Hardware Company *Manufacturers*

Columbian All-Steel

Double Horn Anvil

Series "S. A."

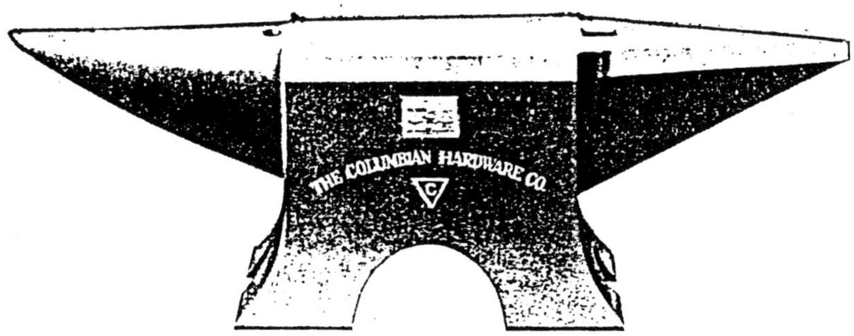

Order by Number

Weight		Total Length Inches	Length of Horn Inches	Width of Face Inches	Length of Face Inches	Height Inches
Kilog.	Pounds					
9	20	13	4⅛	3	4¾	4½
15	33	16¼	5¼	3¼	5¾	4¾
20	44	18⅜	5¾	3 9/16	6⅞	5⅜
25	55	19⅞	6¼	3⅞	7⅜	5¾
30	66	21	6⅝	4⅛	7¾	6 3/16
35	77	21⅞	7	4⅜	7⅞	6½
40	88	22¾	7⅜	4 9/16	8	6⅞
45	99	23⅝	7 11/16	4¾	8¼	7 3/16
50	110	24½	8	4 15/16	8½	7½
55	121	25¼	8¼	5⅛	8¾	7¾
60	132	26	8½	5⅜	9	8
65	143	26⅝	8 11/16	5½	9¼	8⅛
70	154	27¼	8 15/16	5⅝	9⅜	8½
90	198	29½	10	6¼	9½	9
100	220	30	10¼	6⅞	9½	10¼
125	275	31½	10⅝	7½	10¼	10⅝
150	330	33	11	7¾	11	11
185	407	33¾	11½	8⅛	11⅜	11⅜

Anvils for export from the 1907 brochure.

The Columbian Hardware Company *Manufacturers*

Columbian All-Steel

Double Horn Anvil

Series "F"

Order by Number

Weight		Total Length Inches	Length of Horn Inches	Width of Face Inches	Length of Face Inches	Height Inches
Kilog.	Pounds					
40	88	20½	6⅝	4	7	7¾
50	110	21⅝	7	4⅜	7⅝	8¼
100	220	29	9⅝	6	9¾	10½
140	310	31½	10½	6⅜	10½	10⅞
180	397	34½	11½	6¾	11½	12⅛

Anvils for export from the 1907 brochure.

The brochure states that their anvils were cast using "Crucible Steel," which was most likely steel from an open hearth furnace. H. Bradley Smith's book, *Blacksmith's and Farrier's Tools at Shelburne Museum*, quotes the chairman of the company as saying that the anvils were manufactured "from electric furnace cast steel."[56] However, I doubt that this was true, at least for the first few years of the company's history. Mr. Smith states in another part of his book that the "First commercial production of electric furnace steel in the United States [was] between 1902 and 1905."[57] "In 1906, steel was produced by the electric arc process, in Syracuse, New York, for the first time in this country."[58] The company's claim to have used "Crucible Steel" is also unlikely, as its cost would have been prohibitive and was generally used instead in making cutlery and other implements needing to hold a very sharp edge. The *Columbian* anvil sold in the market place for about the same price as the top-quality wrought-iron anvils of the time. This was due, most likely, to the quality tool steel they used in the process. Though expensive, the cost of cast-tool steel was offset by the labor savings in the cast versus wrought-iron anvil manufacture.
The majority of the *Columbian* anvils that I have recorded have weighed over 200 pounds, which suggests that industry bought a large share of them. The *Columbian* anvils have held up well over the years, as most that I have recorded are in very good condition. One area of weakness that does occur in cast anvils is the thin heel, which is very likely to break if subjected to hard use. I have not recorded a *Columbian* anvil with a cracked or broken part. Maybe the anvils that broke were junked, as they would have been very difficult to repair.

Before 1926, the Columbian Hardware Company's bench or craft anvil was drop-forged in closed dies and weighed 10 pounds. I don't know whether this is the 10-pound anvil offered in the brochure or not, because it is very different in style from that of the smith's pattern. The trademark on the early bench or craft anvil is as follows, and its "ring" will give you an earache. The bench anvil made from 1926 on is of a cast iron.

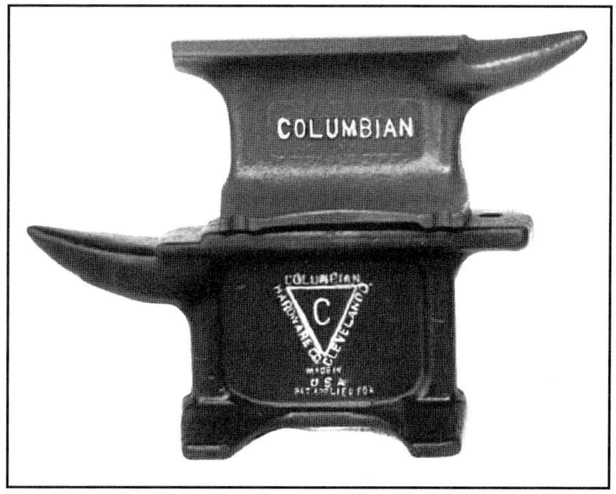

The early 10-pound craft anvil is on the bottom and the later 8-pound craft anvil is on the top. (R.A.P. photo)

Note: If there is a "W" in the triangle, the anvil is a "West."

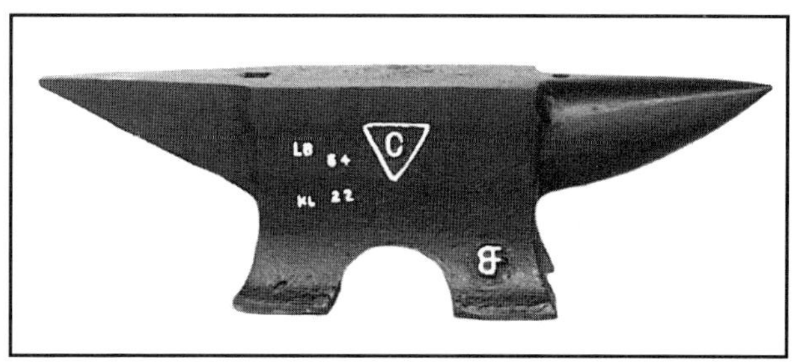

54-pound Columbian export style anvil. See brochure. (R.A.P. photo)

Various advertisements and photos of *Columbian* anvils follow. The *Columbian* anvil was not advertised very often in the trade journals.

The American Blacksmith, June, 1907.

The American Blacksmith, April, 1913.

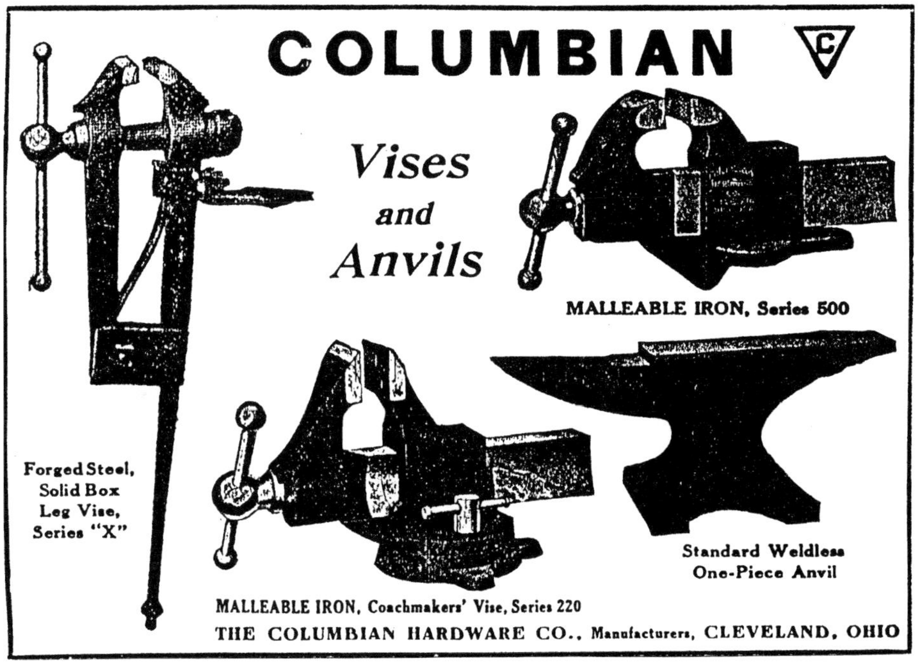

The Blacksmith and Wheelwright, July, 1918.

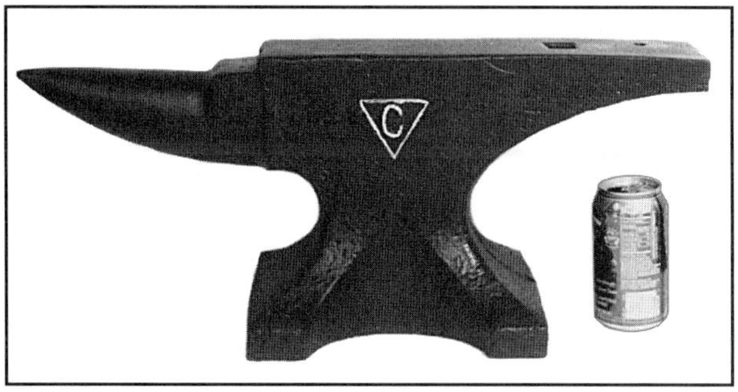

100-pound Columbian anvil. (R.A.P. photo)

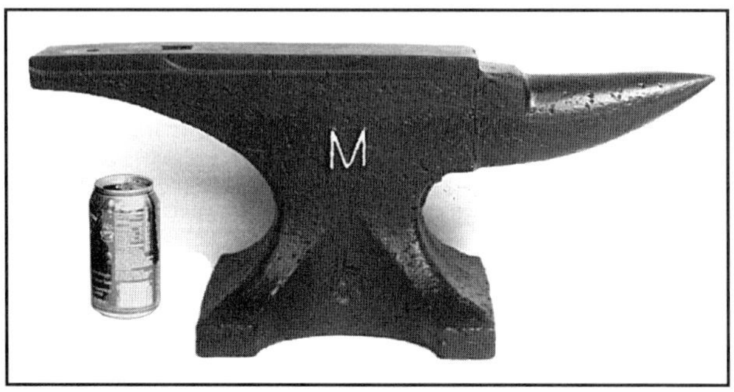

100-pound Columbian anvil from the other side. (R.A.P. photo)

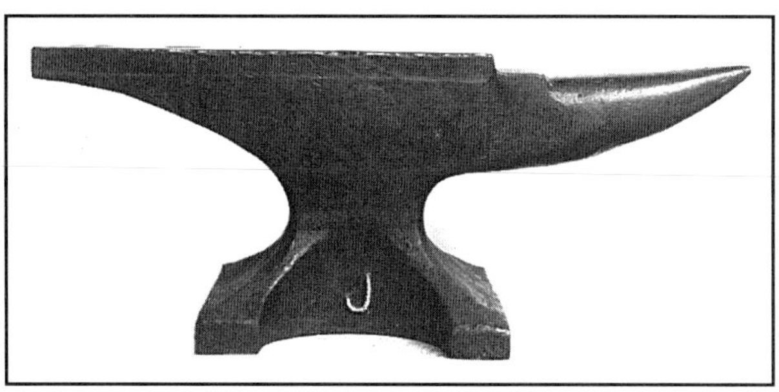

200-pound Columbian anvil,opposite the trademark side. The "J" is most likely a foundry pattern mark. (R.A.P. photo)

Following are three different advertising miniature anvils issued by the Columbian Hardware Co. (From Mr. Gill Fahrenwald collection)

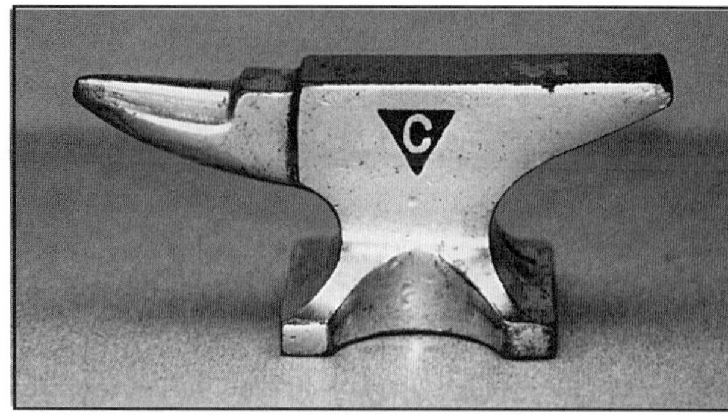

Trademark side. (Mr. Gill Fahrenwald photo)

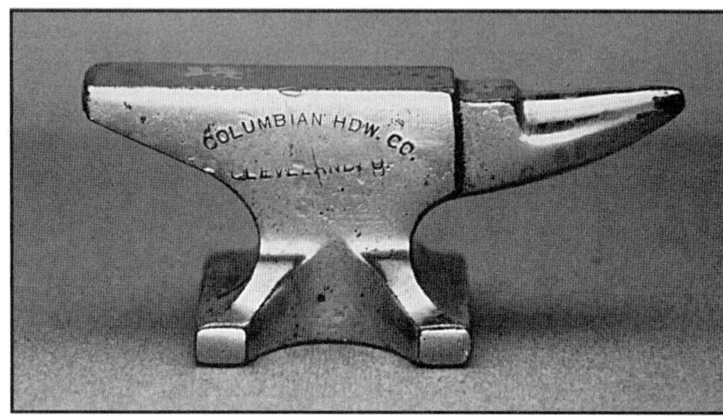

The other side of the same miniature. (Mr. Gill Fahrenwald photo)

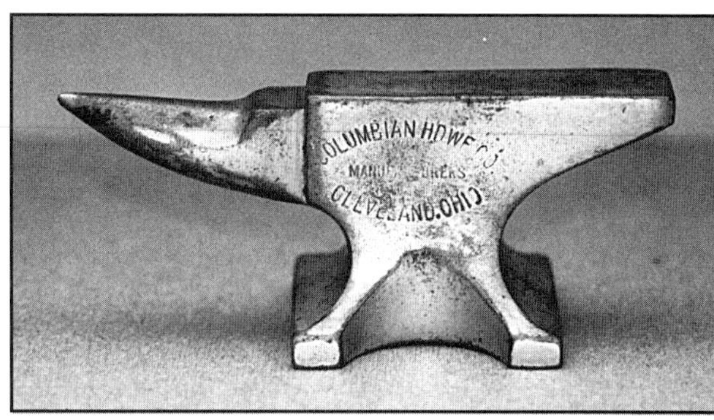

There was no trademark on the other side of this miniature (Mr. Gill Fahrenwald. photo)

Trademark side. (Mr. Gill Fahrenwald photo)

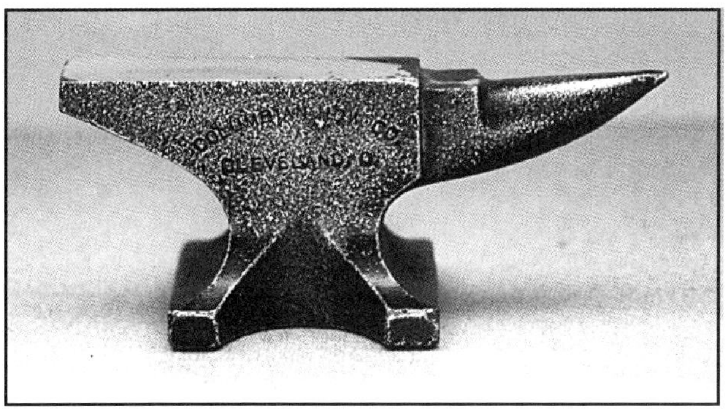

The other side of the same miniature. (Mr. Gill Fahrenwald photo)

VARIOUS MAKES

The following names and trademarks may be found on the following cast-steel anvils:

FULTON

The name is cast in relief on the side. These anvils were made for Sears Roebuck and Company in the 1920s. Producer unknown.

101-pound Fulton anvil. (R.A.P. photo)

OMSTEEL

I do not know what trademark appears on this anvil or what the anvil looked like. Manufactured by Omaha Steel of Nebraska. This information came from The Automobile Trade Directory, February, 1926.

ROCKWOOD

These anvils are marked *Rockwood* or *Rockwood U.S.A.*, maker unknown. They weigh about 65 pounds. The face does not appear to be finished or heat-treated.

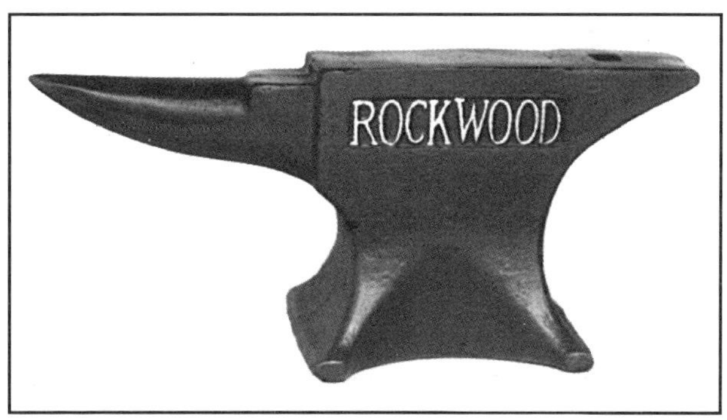

65-pound Rockwood anvil. (R.A.P. photo)

ANVILS IN AMERICA

REV-O-NOC

These anvils were manufactured by either the Columbian or West companies for the Hilbrand, Spencer, Bartlett & Company of Chicago, Illinois, as early as 1913. *Rev-o-noc* was one of their trademarks and was found on a tag attached to the anvil. The manufacturer's trademark is also located on the anvil. The anvil that I recorded was made by the West Steel Casting Company of Cleveland, Ohio.

REVONOC ANVILS.

Special Ingot Steel, One-piece Construction Carefully Hardened and Tempered Face and Horn; Milled, Ground and Polished Face; Body Coated with Pure Asphaltum Paint; Given a Sledge and File Test Before Shipping.

Fully Warranted.

Nos.	R50	R60	
Dimensions of face	11×3	12×3	in.
Size of hardie hole	¾	¾	in.
Size of pritchel hole	7/16	7/16	in.
W't about	50	60	lbs.
Per lb	$0.25	.23	
	AE	AC	
Nos.	R70	R80	
Dimensions of face	12½×3¼	13⅜×3¼	in.
Size of hardie hole	¾	¾	in.
Size of pritchel hole	7/16	7/16	in.
W't about	70	80	lbs.
Per lb	$0.21	.19	
	AB	AA	
Nos.	R90	R100	
Dimensions of face	13¾×3½	14½×3½	in.
Size of hardie hole	⅞	⅞	in.
Size of pritchel hole	7/16	7/16	in.
W't about	90	100	lbs.
Per lb	$0.19	.19	
	AA	AA	
Nos.	R110	R120	
Dimensions of face	14¾×3¾	15⅜×3¾	in.
Size of hardie hole	⅞	1	in.
Size of pritchel hole	7/16	½	in.
W't about	110	120	lbs.
Per lb	$0.19	.19	
	AA	AA	
Nos.	R135	R150	
Dimensions of face	16×4	16½×4	in.
Size of hardie hole	1	1	in.
Size of pricthel hole	½	½	in.
W't about	135	150	lbs.
Per lb	$0.19	.19	
	AA	AA	
Nos.	R175	R200	
Dimensions of face	17¼×4¼	18×4½	in.
Size of hardie hole	1	1⅛	in.
Size of pritchel hole	½	⅝	in.
W't about	175	200	lbs.
Per lb	$0.19	.19	
	AA	AA	

Rev-o-noc anvils have a greater surface area for their weight than any other on the market, owing to the correct distribution of the metal. For illustration: a Rev-o-noc anvil 3½ inch by 14½ inch surface weighs 100 pounds, and frequently less; in order to obtain this same surface in a wrought anvil of other makes, the user is often compelled to buy 115 pounds or more of metal.

The width of the base is always equal to the height, so they will not tip nor break loose from their fastenings by side blows.

The long horn is round throughout its full length, a perfect ring can be forged and welded at any point.

The long tail admits of bending small perfect U shapes, and leaves a large surface for welding.

Ad from the Hibbard, Spencer, Bartlett & Co. catalog, 1913, Chicago, Ill.

SKAGIT

These anvils were manufactured by the Continental Emsco Company of Sedro, Washington. They were offered in two sizes, 45 and 120 pounds. I have one of these anvils in my possession. The trademark is only a recessed rectangle on the side with no mention of a company name or logo. The company is no longer in business. Gill Fahrenwald sent me this clipping from a Washington state newspaper dated November 12,1993.

> *Sedro Wolley- Sea Dog Corp. of Everett, which bought the old Skagit Steel plant here, has no tenants lined up right now, but says several are interested and the regional economy makes the purchase a good buy. ....The plant, founded as Skagit Steel and Ironworks in 1910, has been abandoned since 1990. It employed more than 1,000 people at its peak in the 1960s, when it manufactured forest- harvesting equipment.*

And I might add manufactured anvils. Another anvil gone, victim of the Spotted Owl.

45-pound Skagit anvil. (R.A.P. photo)

SUPERIOR STEEL

This steel foundry was located in Benton Harbor, Michigan, and manufactured anvils under its own trademark as well as for other concerns. A 20-pound anvil I have recorded has the following trademark, plus the company's name showing on the side. The company went out of business in the 1970s.

20-pound Superior Steel anvil. From Mr. Howard Busse collection. (R.A.P. photo)

VANADIUM STEEL

Sears Roebuck and Montgomery Ward both sold these anvils under the *Vanadium Steel* trademark in the late 1930s and early 1940s, prior to World War II. There was a Vanadium Steel Corporation, which manufactured tools, and they are the ones most likely to have manufactured these anvils. They are listed in the catalogs in sizes 70, 100, 125 and 150 pounds. A 100-pound anvil is listed in the 1941-42, Fall & Winter *Montgomery Ward* catalog for the great sum of $17.95. The trademark, *Vanadium Steel* is cast in bold relief on the side with the weight listed below.

Catalog No.	Wt., Lbs.	Face, In.	Hardie Hole	Prt'l Hole	Each
99 HM 5293	70	3⅛x12⅜	⅞ In.	½	$9.95
99 HM 5294	100	3¼x12⅞	⅞ In.	½	14.45
99 HM 5295	125	3¼x15⅜	⅞ In.	9/16	17.95
99 HM 5296	150	4 x16	1 In.	9/16	21.75

Sears Roebuck, Spring-Summer catalog, 1938. (From Mr. Hunter Pilkinton)

100-pound vanadium steel anvil. (R.A.P. photo)

WEST

The West Steel Casting Company of Cleveland, Ohio, manufactured anvils to-order for other concerns, such as the Hilbrand, Spencer, Bartlett Hardware Company, that I have mentioned before. They also manufactured anvils under their own name and trademark, which on the smaller anvils is an inverted pyramid, with a raised "W" in the center.

The largest anvils I have seen, weighing 750 pounds, have *West* and the weight,"750," cast in very large relief on the side of the anvil. The *Steel* journal of 1932 states that these anvils were manufactured from vanadium steel from an electric furnace, which makes a very good anvil.[59] Up to this time I have been unable to obtain any literature listing their different sizes, but I have recorded anvils weighing 100, 150 and 750 pounds, and I believe there are anvil sizes in between.

W

100 -pound

WEST
STEEL

150-pound

WEST

750-pound

West anvil made for "Rev-O-Noc." (R.A.P. photo)

750-pound West anvil. (Mr. Norm Davis photo)

Tapping vanadium steel from an electric furnace, and an anvil cast in one piece from such steel. Americans with improved technique in manufacture, and higher tariff on anvils, are beginning to hope for a better market in this country. From Steel magazine, November 24, 1932. (From Mr. Bill Hill)

YOST

The Yost Manufacturing Company of Meadville, Pennsylvania, was primarily a vise manufacturer and I believe they are still in business today (1995) under another name. The 1923 *Gibson Company Hardware* catalog of Indianapolis, Indiana, shows *Yost* anvils in 100-, 125- and 150-pound sizes, at 40 cents a pound. That is about the same price that forged anvils were selling for at the time. I believe they used the same trademark on their anvils as they did on their vises, a circle with a large "Y" in the center.

Y

From Machinery journal, March, 1922.

Blacksmith's Anvil

	Size Pounds	Length Face	Length Blight	Width Face	Width Blight	Hght. Anvil	Lgth. Horn
No. G116	100	14¼"	3"	3¼"	1⅞"	9"	7"
No. G117	125	15"	3¼"	3⅝"	2"	9½"	7½"
No. G118	150	16⅛"	3½"	4"	2"	10¼"	8"

Price 40c per pound.

The Gibson Co. catalog, 1923, Indianapolis, In.

MODERN CAST ANVILS

The following United States anvil manufacturers have produced cast anvils since about 1960. With a few exceptions, these anvils are of the farrier's pattern. These patterns can change from time to time, as the farrier's pattern is still evolving to meet the demands of the modern farrier's trade. The production of these anvils is small and most of these concerns can be considered "Mom & Pop" type businesses. Therefore the only way to find out if these companies are still in business is to contact them. The ones still in business are the only anvil makers in the United States. As of June 1997 the following companies were still making anvils although there may be others unknown to me.

ACROSS THE ANVIL

Across the Anvil is the same as J.H.M.. Manufacturing. This is the older name for the company and anvils. Their advertisement in the September, 1987, *Anvil* farrier's journal lists four different weights, one of which was a smith's pattern anvil weighing 260 pounds. The basic trademarks are as shown: *Across the Anvil* and *Bruce Daniels*. Bruce Daniels was president of the American Farriers Association and designed the anvil bearing his name.

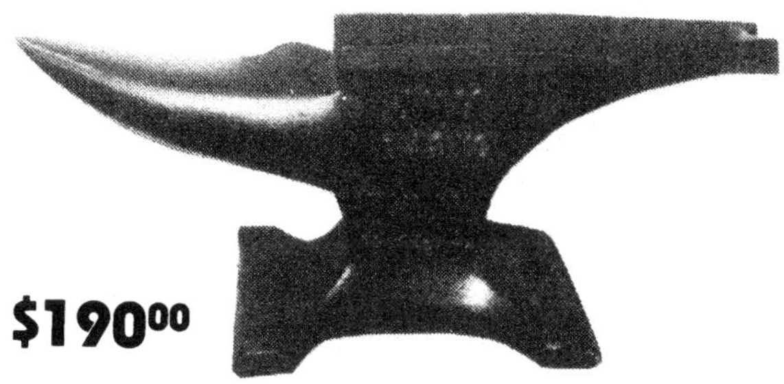

$190.00

The "BASIC" Anvil

Weight: 70 lbs.
Horn: 9"
Height: 9"
Base: 10½" x 8¼"
Face: 3¾" (tapered to a 3" heel) x 12"
Hardie Hole: 1"
Clip Horn
Pritchel Hole: ½"
Turning Cams in Heel

$325.00

The "JOURNEYMAN" Anvil

Weight: 124 lbs.
Face: 4" x 17" (tapered to a 3" heel)
Horn: 12"
Height: 11"
Hardie Hole: 1"

First of two-part ad from Anvil magazine, September, 1987 (see J.H.M. Mfg.).

Second part of an ad from Anvil Magazine, September, 1987 (see J.H.M. Mfg.).

A. P. TOOL CO.

A.P. Tool Company; 201 Porter Road, Conroe, Texas, 77301. Ph. # (409) 756-7114. This company produced at least seven sizes and styles of farrier's anvils. They are marked as they appear in the brochure below. A.P. Tool company bought the G.E. pattern anvil and named it the *Special*; I assume they marked them the same way as mentioned in the brochure. These anvils are made of heat-treated steel. I was told they started producing anvils in the early 1970s.

BIG TEX ANVIL

Undoubtly the best designed and made anvil to be had anywhere. Made of the finest steel alloy (chrome, nickel, Molydinum, carbon) the foundry has to offer. Heat treated to 48 Rockwell C hardness. Face is perfectly ground. The horn has been perfected so any size shoe can be worked with ease. Heel thickness of 5/8" allows complete utility.

SPECIFICATIONS:
All figures approximate.

Weight:	116 lbs.	Height:	9½"
Overall Length:	31"	Base:	9" x 14"
Face:	3¼" x 16½"	Pritchel Holes:	½"
Horn:	4¾" x 14½"	Hardy Hole:	1"

LITTLE TEX ANVIL

For a light weight anvil designed expressively for the farrier, this anvil cannot be beat. It has the same steel alloy as the Big Tex, heat treated to 48 Rockwell C. Face is machine ground; the horn is hand finished. An ideal travel anvil with total utility.

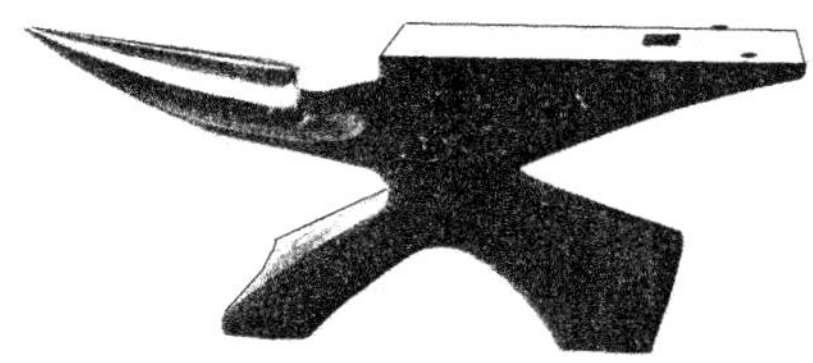

SPECIFICATIONS:
All figures approximate.

Weight:	70 lbs.	Height:	7½"
Overall Length:	22½"	Base:	7" x 12"
Face:	3¼" x 11½"	Pritchel Holes:	½"
Horn:	3½" x 11"	Hardy Hole:	1"

Double Clip Blocks (right & left)

RANGER ANVIL

A one of its kind design. This anvil is designed for the hot shoer, cold shoer, western horse shoer, show (gaiter & walking horse) shoer, track shoer, (both running horses as well as harness horses). All facets of desired design have been incorporated into this anvil. Made of the same steel alloy as the Big Tex, heat treated to 48 Rockwell C, machine ground face, hand finished horn.

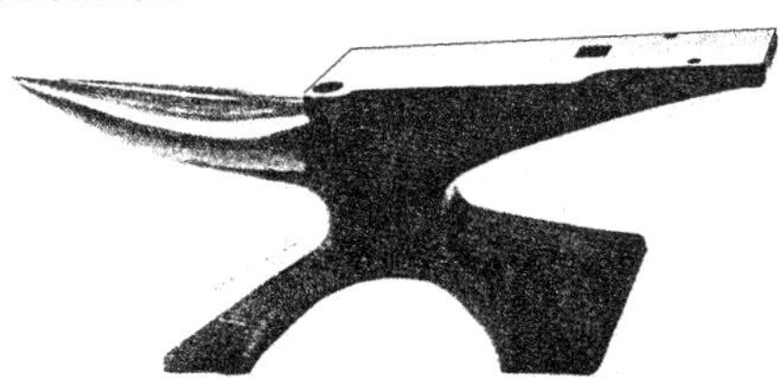

SPECIFICATIONS:
All figures approximate.

Weight:	103 lbs.	Height:	8½"
Overall Length:	24½"	Base:	9" x 14"
Face:	4¼" x 9" x 3¼" x 5"	Pritchel Holes:	½"
	(overall length of face 14")	Heel Hooking Hole:	1" rd
Horn:	3¾" x 10½"	Hardy Hole:	1"

RANGER A ANVIL

Undoubtly the finest medium weight anvil to be found anywhere. This anvil is designed for the farrier wanting to have enough weight for all types of iron work, but not too heavy for transport. The same steel alloy as the Big Tex, heat treated to 48 Rockwell C hardness. The face is machine ground, horn hand finished.

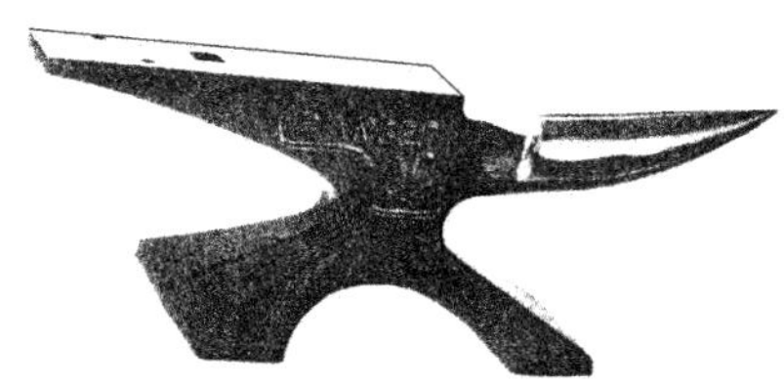

SPECIFICATIONS:
All figures approximate.

Weight:	99 lbs.	Height:	8½"
Overall Length:	26½"	Base:	9" x 14"
Face:	3¼" x 14"	Pritchel Holes:	½"
Horn:	3¾" x 12½"	Hardy Hole:	1"

AP "SPECIAL" FARRIERS ANVIL
DESIGN NO. 1

This anvil is constructed of the same high quality steel alloy as the other AP anvils and the entire anvil is heat treated to 48 Rockwell C hardness. Both the top and bottom of the anvil are machined to a flat surface.

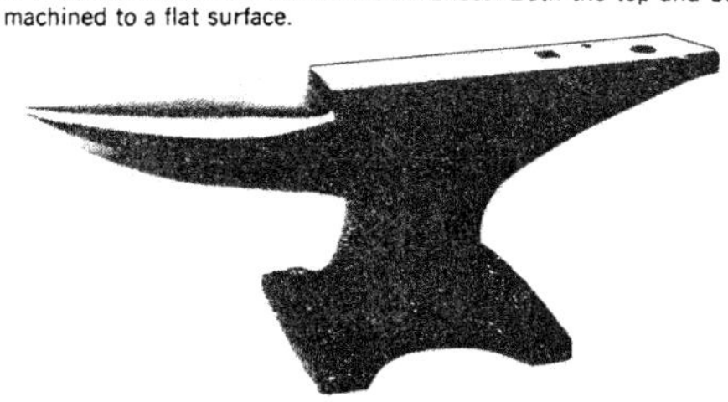

SPECIFICATIONS:
All figures approximate.

Weight:	120 lbs.	Base:	10¼" x 9"
Overall Length:	28"	Pritchel Holes:	½"
Face:	3¼" x 16"	Heel Hooking Hole:	1" Chamfered
Horn:	12"	Hardie Hole:	7/8"
Height:	11"		

AP "SPECIAL" FARRIERS ANVIL
DESIGN NO. 2

This anvil is constructed of the same high quality steel alloy as the other AP anvils and the entire anvil is heat treated to 48 Rockwell C hardness. Both the top and bottom of the anvil are machined to a flat surface.

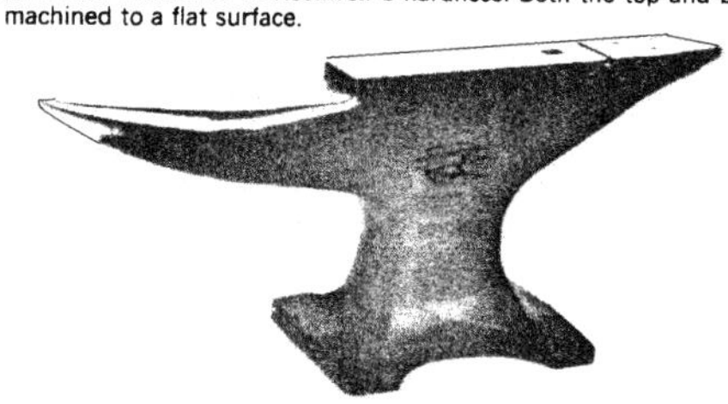

SPECIFICATIONS:
All figures approximate.

Weight:	120 lbs.	Height:	11"
Overall Length:	28"	Base:	10¼" x 9"
Face:	3¼" x 16" with a Tapered Heel	Pritchel Holes:	½"
Horn:	12"	Hardie Hole:	7/8"

From the company's older brochure.

AP ANVILS

The Finest Quality Anvil Made Anywhere

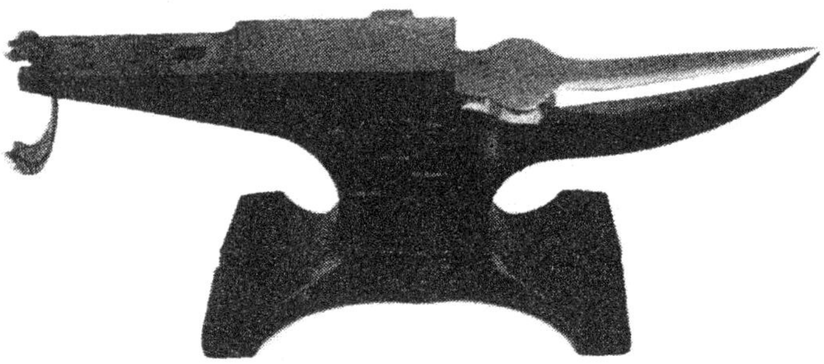

RANGER W ANVIL

• TURNING CAMS • EXTRA WIDE FACE
• SQUARE TOES • DOUBLE CLIP HORNS
• SOLID BASE

SPECIFICATIONS
All figures approximate

Weight:	100 lbs.	Height:	8⅝"
Overall Length:	25⅝"	Base:	9" x 13¾"
Face: 4¼" x 7½" x 3¼" x 6½"		Pritchel Holes:	½"
Horn:	11½"	Hardy Hole:	1"

LITTLE RANGER ANVIL

• TURNING CAMS • EXTRA WIDE FACE
• DOUBLE CLIP HORNS • SOLID BASE

SPECIFICATIONS
All figures approximate

Weight:	62 lbs.	Height:	8"
Overall Length:	19¼"	Base:	7⅞" x 11"
Face:	4" x 6" x 3" x 4¼"	Pritchel Holes:	½"
Horn:	8¾"	Hardy Hole:	1"

From their latest catalog, 1997. Part one of brochure.

AP ANVILS

The Finest Quality Anvil Made Anywhere

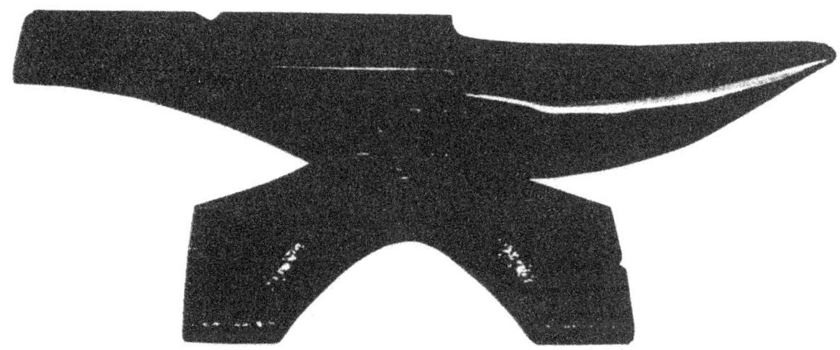

BABY TEX

The Baby Tex is a 22 lb. platers anvil. It is made of the same high quality steel as the other AP anvils. It has been heat treated a longer life and greater durability. The Baby Tex is equipped with two shoe working grooves on either side of the heel hooking hole.

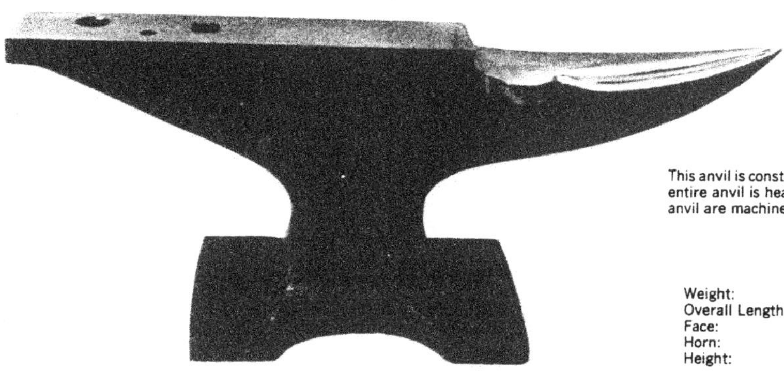

AP "SPECIAL" FARRIERS ANVIL
DESIGN NO. 1

This anvil is constructed of the same high quality steel alloy as the other AP anvils and the entire anvil is heat treated to 48 Rockwell C hardness. Both the top and bottom of the anvil are machined to a flat surface.

SPECIFICATIONS:
All figures approximate.

Weight:	120 lbs.	Base:	10¼" x 9"
Overall Length:	28"	Pritchel Holes:	½"
Face:	3¼" x 16"	Heel Hooking Hole:	1" Chamfered
Horn:	12"	Hardie Hole:	7/8"
Height:	11"		

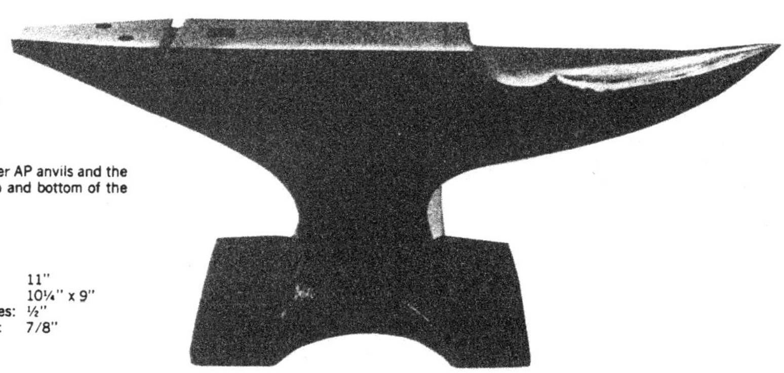

AP "SPECIAL" FARRIERS ANVIL
DESIGN NO. 2
TAPERED HEEL

This anvil is constructed of the same high quality steel alloy as the other AP anvils and the entire anvil is heat treated to 48 Rockwell C hardness. Both the top and bottom of the anvil are machined to a flat surface.

SPECIFICATIONS:
All figures approximate.

Weight:	120 lbs.	Height:	11"
Overall Length:	28"	Base:	10¼" x 9"
Face:	3¼" x 16" with a Tapered Heel	Pritchel Holes:	½"
Horn:	12"	Hardie Hole:	7/8"

From their latest catalog, 1997. Part two of brochure.

CLIFF CARROLL'S

Cliff Carroll's; Horseshoer's Supplies, Inc., 261 W. Perry Park Ave., Larkspur, Colorado, 80118. Phone # (303) 681-2643. This farrier's pattern anvil weighing 70 pounds has two round protrusions close together, coming out the side of the anvil body. They are used in bending horseshoes, and are called "turning cams." I am unsure of how the trademark appears, but assume the company's name and address is cast on the side of the anvil. These anvils may be marked "HSI."

1997 photo of Cliff Carroll's anvils, 35-,70-, and 125-pound. These are made of heat-treated ductile iron.

EMERSON HORSESHOE SUPPLY

Emerson Horseshoe Supply; 2301 Vanceville Road, Bossier City, Louisiana, 71111. Phone # (318) 742-5991. The company started manufacturing anvils in 1982. They are made of 4140 heat-treated steel. Emerson Horseshoe Supply is on the side of the anvils. The ad shows three farrier patterns and a smith's pattern.

ANVILS DESIGNED BY DILTON EMERSON

ANVIL #101 - $325
100 LB. FLAT HORN DESIGN

FACE - 12 1/8" x 3 1/4"	BASE - 10" x 11 3/4"
HORN - 11"	HARDIE HOLE - 1"
HEIGHT - 8 1/2"	PRITCHEL HOLE - 5/8"

ANVIL #102 - $325
100 LB. ROUND HORN DESIGN

FACE - 14" x 3 1/2"	BASE - 14" x 8 3/4"
HORN - 13"	HARDIE HOLE - 1"
HEIGHT - 8 1/2"	PRITCHEL HOLE - 5/8"

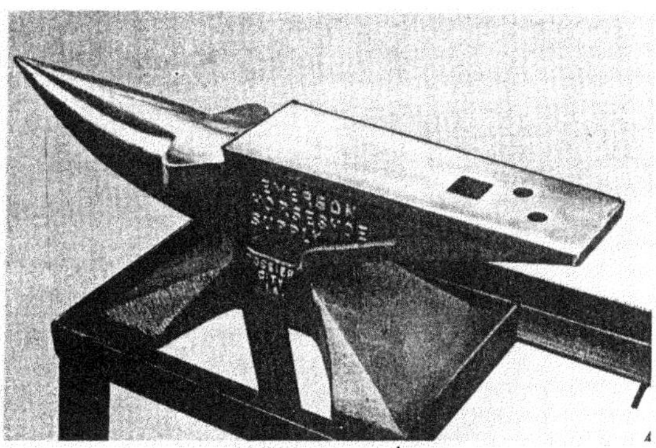

ANVIL #103 - $200
60 LB. FLAT HORN DESIGN

FACE - 12 1/8" x 3 1/4"	BASE - 10" x 11 3/4"
HORN - 11"	HARDIE HOLE - 1"
HEIGHT - 8 1/2"	PRITCHEL HOLE - 5/8"

ALL 3 ANVIL DESIGNS
CAN HAVE NAME AND ADDRESS FOR A SMALL FEE

ANVILS MADE OF 41-40-C HEAT TREATED STEEL

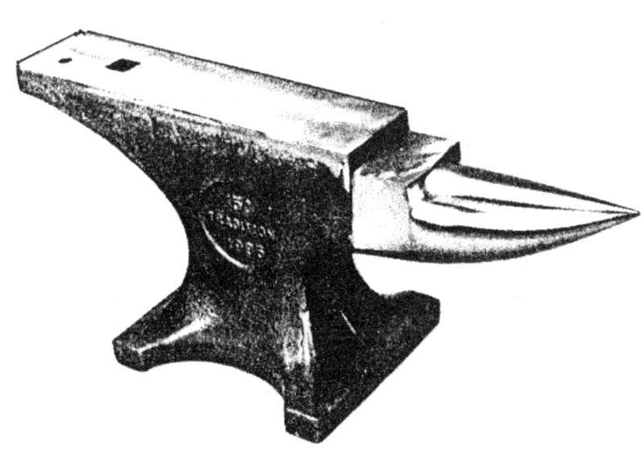

158# TRADITION ANVIL - $485

FACE - 16 3/4" x 4 3/16"	BASE - 10 3/8" x 9 7/16"	HORN - 10 3/16"
HARDIE HOLE - 1" SQUARE	HEIGHT - 11 3/8"	PRITCHEL HOLE - 9/16" ROUND

Page 28

1997 company flier.

DELTA

Delta Horseshoe Company; Remco Van Linden, California. Ph. # (916) 624-7181. These anvils have an aluminum base and is possibly the first of its kind. The company produces three farrier patterns with *Future 1, Future 2* and *Future 3* cast in relief which is in a rectangular depression on the side. *Future 1* weighs 85 pounds, *Future 2* and *Future 3* weigh 98 pounds.

"Future 2". Anvil magazine, December, 1993.

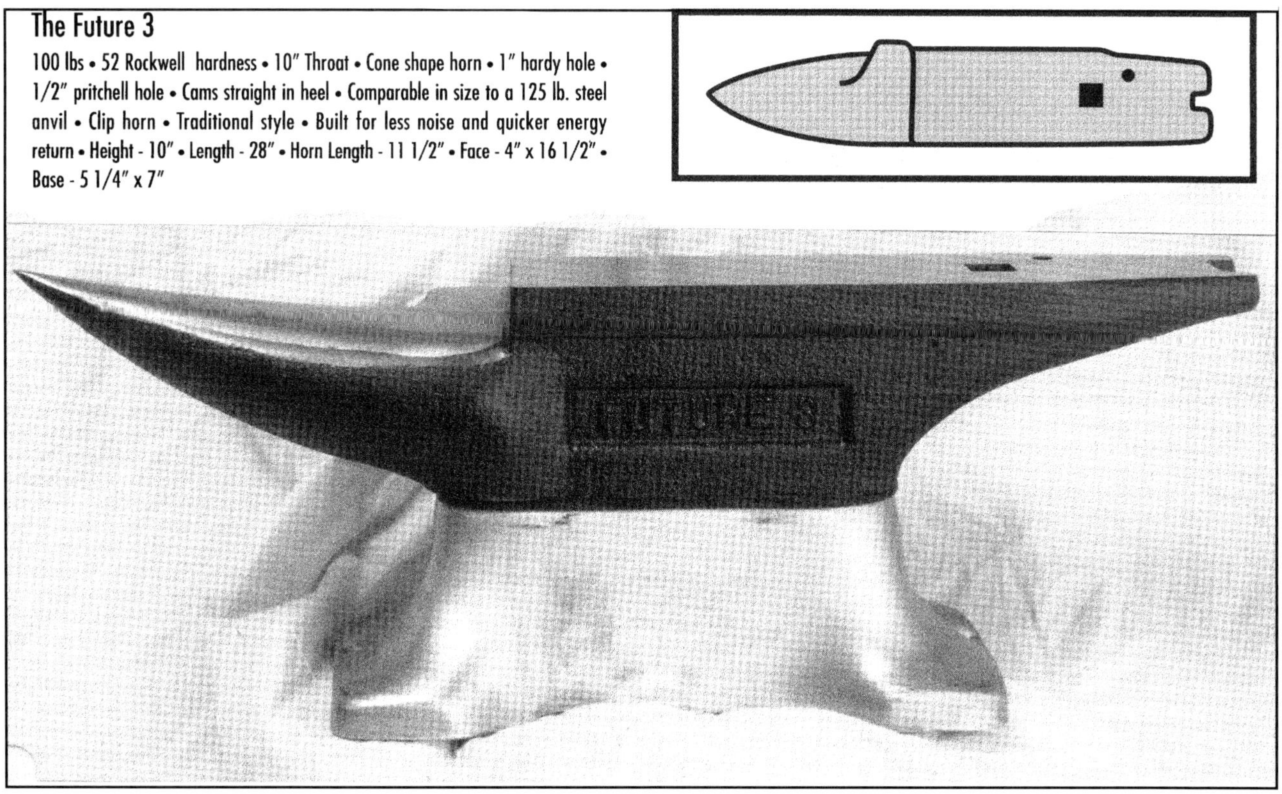

"Future 3". 1997 flier.

J.H.M. MANUFACTURING

J.H.M. Manufacturing; P.O. Box 93, Plaster, Texas, 76485. Ph. # (817) 594-6710. These anvils have J.H.M. cast in relief on the side. The company started producing anvils in 1984. They make six sizes of anvils (see Aross The anvil). These anvils are made of a ductile iron composite and heat-treated.

From Anvil magazine. Date unknown (1980s).

LAUREL MACHINE & FOUNDRY

Laurel Machine & Foundry Company; P.O. Box 1049, 810 Front St. Laurel, Mississippi, Ph. # (601)-428-0541. These anvils are cast from chrome-molybdenum steel. The design was developed November, 1993. The first anvil was poured August, 1994. As of September 27, 1996 forty-eight anvils have been made. It appears to be an excellent anvil. As you can see the anvils are of the smith's pattern.

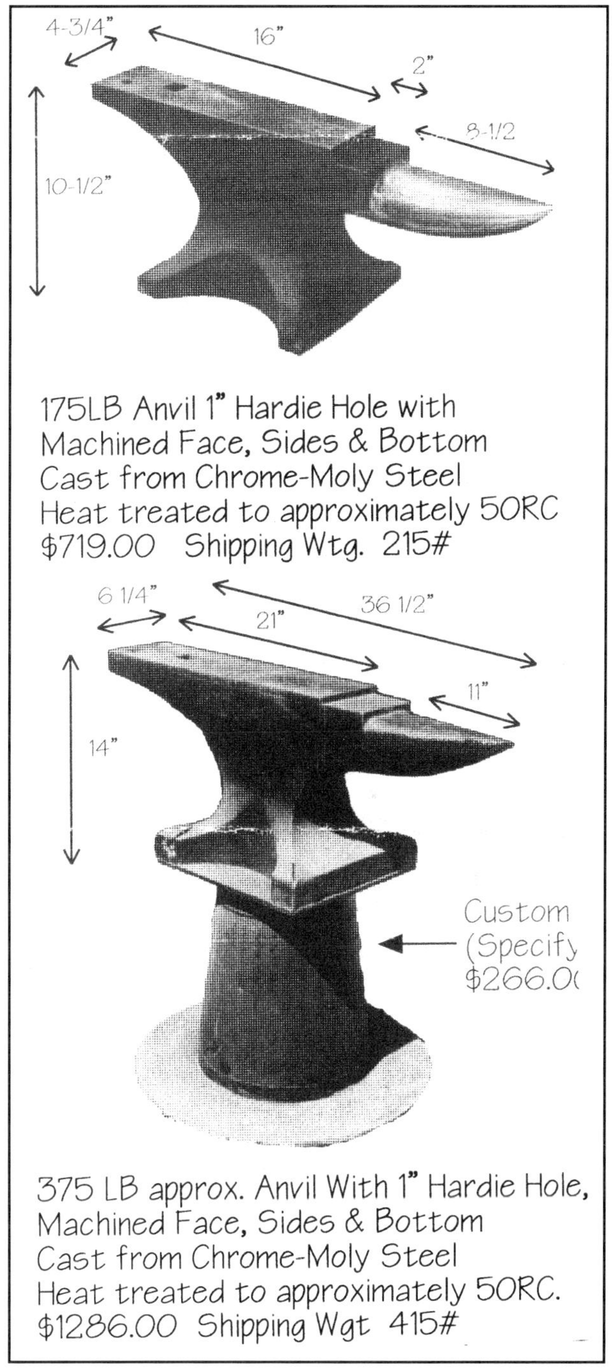

Laurel Machine and Foundry Co. flier, 1997.

MANKEL

Mankel Blacksmith Shop; 7836 Cannonsburg Road, P.O. Box 29, Cannonsburg, Michigan, 49317. Ph. # (616) 874-6955. Their 1994-1995 catalog lists four sizes of anvils, including a 130-pound smith's pattern. The anvils are cast of 8630 steel by a Michigan foundry and the Mankel Blacksmith Shop, machines, tempers, grinds and finishes them. The farrier patterns come in 70-, 115- and 165-pound sizes. In the past the shop has also produced a 95-pound farrier's and a 125-pound smith's pattern anvil.

The first anvils were produced in 1971, with the 95-pound farrier's pattern. By December of 1984, 3,100 anvils had been sold. The word *Mankel* is cast on the side of the anvil body. Older anvils had "Mankel" on both sides, and the address on just one side. Some of the first anvils also have a circle "S" on them, which stood for Superior Steel Foundry of Benton Harbor, Michigan. All Mankel anvils are painted black.

MANKEL'S

All Mankel Anvils are made from 8630 steel
Cast here in Michigan
Machined and tempered at Mankel Blacksmith Shop

MANKEL HORSE-SHOERS' ANVIL - 115#

Well-balanced
horse-shoers'
Pattern Anvil

The all-new middleweight anvil from Mankel has it all wrapped up in one perfectly balanced package. Your favorite features of a lighter shoer's anvil combined with the solid durability and base of heavier shop models. Check the list:

- •Inch square hardy hole
- •One or two clip horns
- •Two ½" pritchel holes
- •10½" horn, 4¼" swell
- •9 x 10" base
- •Tempered to Rockwell "C"
- •Shipped motor freight
- •Overall 25 x 10" with 3¼" x 14¾" face

•Anvils can be ordered with Tapered Heel or Turning Cams.

From the 1997 Mankel catalog. Part 1.

MANKEL'S

MANKEL LIGHT WEIGHT ANVIL - 70#

8630 - Cast Steel - Tempered to a 48 Rockwell "C" hardness
Size - 19" long, 8½" high
Base - 7¾" x 11"
Weight - 68 lbs.
Face - 3⅛" x 10"
Horn - 3¾" x 9", 3¾" swell to horn
Hardie Hole - 1 inch
Clip Horn - 1¾" x 1"
Two Pritchel Holes - 7/16" diameter
Shipped U.P.S.

Part 2.

MANKEL SHOP ANVIL - 165#

8630 - Cast Steel - Tempered to a 48 Rockwell "C" hardness
Size - 26⅜" long x 10⅝" high
Base - 11½" x 9¾"
Weight - 165 lbs.
Face - 3¾" x 15¼"
Horn - 4½" x 11", 4½" swell to horn
Hardie Hole - 1"
Clip Horn - 2" x 1"
Two Pritchel Holes - ½" diameter
Shipped Motor Freight

Part 3.

MANKEL BLACKSMITH (OR LONDON) PATTERN - 130#

8630 - Cast Steel - Tempered to a 48 Rockwell "C" hardness
Size - 25" long x 11" high
Base - 10¼" x 9½"
Weight - 130 lbs.
Face 4" x 15½"
Horn 4" swell x 10" long
Hardie Hole - 1"
Two Pritchel Holes - ½" diameter
Shipped Motor Freight
NO CLIP HORN

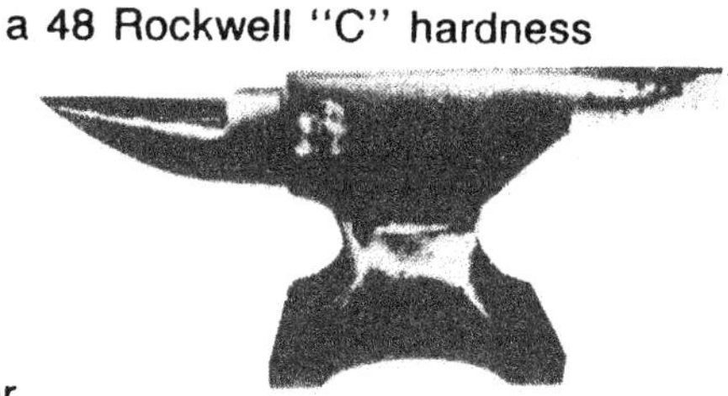

Part 4.

125-pound Mankel smith's pattern anvil. 25 of these were made. (R./A.P. photo)

95-pound Mankel farrier's model. (R.A.P. photo)

N.C.

N.C. Tool Company; 8568 Hunt Road, Pleasant Garden, North Carolina, 27313. Ph.# (910) 674-5654 and (800) 446-6498. This company produced at least two farrier's pattern anvils of about 70 pounds. The first has an inverted "horseshoe" trademark cast on the side, with *N.C.* shown above. The second anvil has *Short Sugar* cast on the side. They now produce four farrier's pattern anvils.

The following is from their 1997 catalog.
NC Anvils by Donald Jones:

** Developed in1977*
** Developed for our students at North Carolina School of Horseshoeing*
** Made of ductile iron, a recycled steel product*

* *Face is hardened to 46 to 48 rockwell*
* *Popular because horseshoers do not need excessive weight for fitting shoes*
* *Easy to lift and easy to haul on a mobile shoeing rig*
* *All anvils made after 1985 have the cast date on its side*
* *4 different models - NC, Short Sugar, Big Face and Cavalry*
* *Over 25,000 sold*

Miniature advertising N C anvil. (Mr. Gill Fahrenwald photo)

Miniature N C anvil from the other side. (Mr. Gill Fahrenwald photo)

NC Anvil

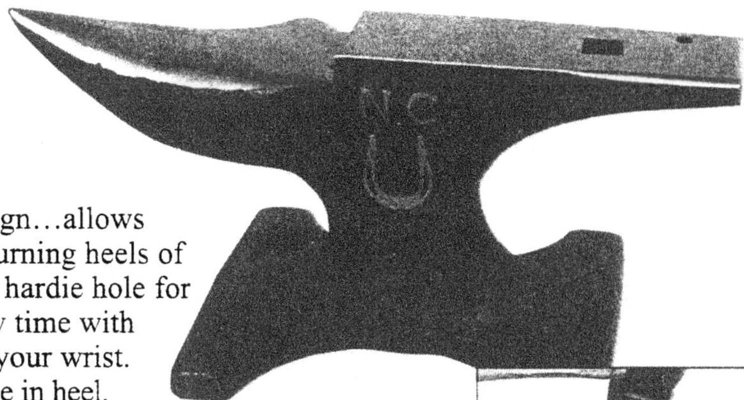

Thin heel design...allows hooking and turning heels of shoes through hardie hole for better fit every time with less stress on your wrist.

- 1" hardie hole in heel.
- Pritchel hole in heel.
- Fits NC/BigFace Anvil Stand.

Specifications:
Weight: 70 lbs.
Face: 3-1/4" x 10-1/2"
Horn: 4-1/2" x 9-1/2"
Height: 8-3/4"
Base: 11" x 9"

$195.00
(plus shipping)
Item #0200

$215.00
With Punch Slot
(plus shipping)
Item #0202

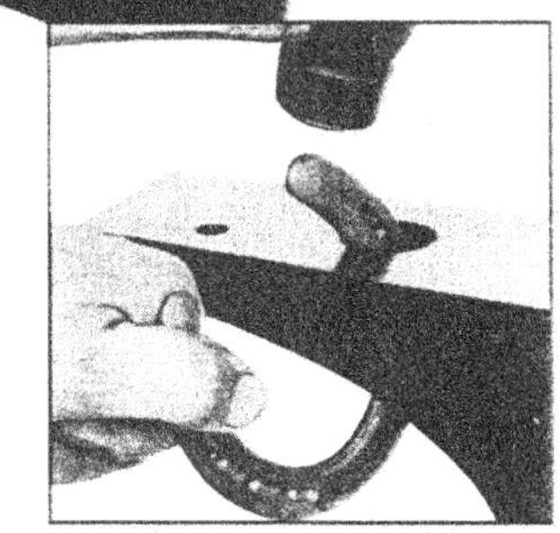

Turn heels easily...
Thin heel design allows hooking heels on all NC anvil models.

Short Sugar Anvil

Perfectly Balanced.

- Thin heel design.
- Square Cliphorn
- 1-1/4" chamfered, round turning hole in heel.
- 1" hardie hole in horn.
- Pritchel hole in heel.
- Fits Short Sugar Anvil stands.

Specifications:
Weight: 70 lbs.
Face: 3-1/4" x 11-1/2"
Horn: 4" x 8"
Height: 9"
Base: 11" x 8-1/2"

$202.00
(plus shipping)
Item #0204

With Punch Slot
$222.00
(plus shipping)
Item #0206

From 1997 N C catalog. Part 1.

Anvils

Quality alloyed castings...
all our anvils are precision milled and heat treated...
Quality assured for many years of dependable service.

BigFace Anvil

Our Most Innovative Anvil.
Fully featured to work shoes, hot or cold, easier than ever.

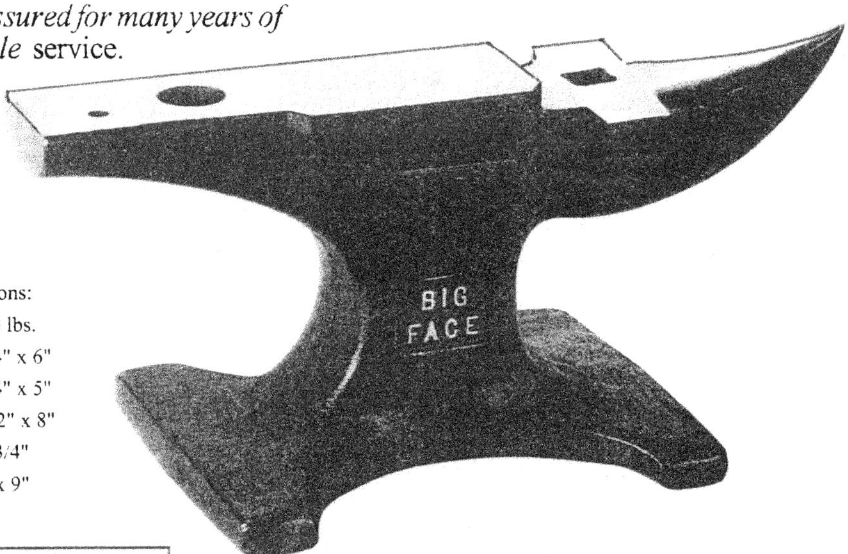

Specifications:
Weight: 70 lbs.
Face: 3-3/4" x 6"
Heel: 3-1/4" x 5"
Horn: 3-1/2" x 8"
Height: 8-3/4"
Base: 11" x 9"

Optional for all anvils:
¼" punch slot milled across face...easily punch hand-mades.

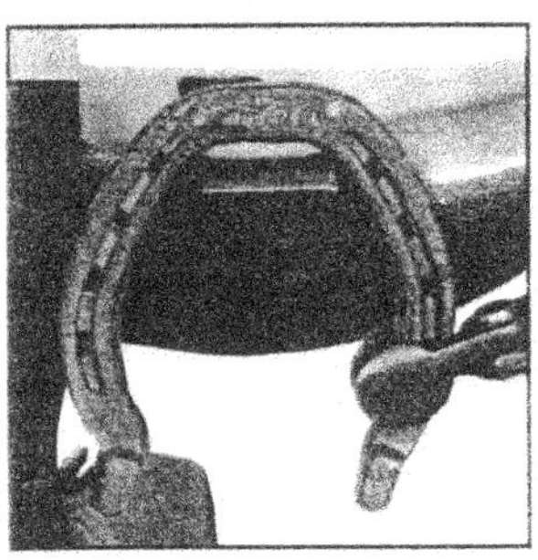

Square cliphorn on BigFace & Short Sugar models make squaring toes a snap.

- 3-3/4" wide face (max.)... great for leveling shoes.
- Opposing square & round cliphorns...for making great clips and square toes. Also convenient for left handed shoers.
- Thin heel design.
- 1-1/4" chamfered, round turning hole in heel.
- 1" hardie hole in horn.
- Pritchel hole in heel.
- Fits NC/BigFace Anvil Stand.

$225.00
(plus shipping)
Item #0207

$245.00
With Punch Slot
(plus shipping)
Item #0208

From 1997 N C catalog. Part 2.

Cavalry® Anvil

Balanced Alloyed Casting

$385.00
(plus shipping)
Item #0209

$405.00
with punch slot
(plus shipping)
Item #0209A

- 4-5/8" x 7" Wide face
- 1-1/2" Square clip horn
- 1-1/4" Round turning hole in heel
- 1" Hardie hole in horn
- 3/8" Pritchel hole in heel
- May be ordered with round clip horn at no extra charge
- Fits NC/Big Face Anvil Stands

A heavy weight in efficiency and dependability...
yet an easy traveler.

Top View

Specifications:
Weight: 112 lbs.
Face: 4-5/8" x 7"
Heel: 3-5/8" x 6"
Horn: 4" x 9-1/2"
Height: 11"
Base: 11" x 9"

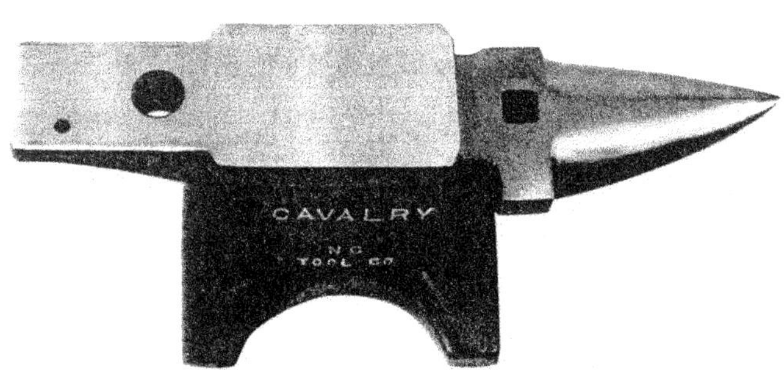

From 1997 N C catalog. Part 3.

The following companies produced anvils between about 1955 to the early 1990s.

CENTAUR FORGE, LTD.

Centaur Forge; Ltd., P.O. Box 240-V, 117 N. Spring St., Burlington, Wisconsin, 53105. Ph. # (414) 763-9175. They sold a farrier's anvil in 90- and 125-pound weights, which was manufactured by Kohlswa of Sweden to Centaur's specifications and with Centaur's trademark. Centaur no longer carries the *Kohlswa* anvils(1997). Centaur is probably the largest blacksmith supply company in the United States. Centaur now carries a number of anvil brands both foreign and domestic (1997).

125-pound Centaur anvil. (R.A.P. photo)

G.E. FORGE & TOOL

G.E. Forge & Tool sold it's anvil patterns to the A.P. Tool Company as stated earlier. The farrier's anvil manufactured by G.E. would have *G.E.* cast in relief on the side of the anvil.

JIM LINZY

Jim Linzy farrier's pattern anvils were manufactured in 50-, 70-, 125- and 135-pound weights. They have a distinctive turned-up nose on the horn. The trademark is a diamond, with the combined letters JL in the center. The trademark appears as follows:

From Anvil magazine, December, 1986.

MULTI PRODUCTS COMPANY

This company produced farrier's pattern anvils from 1955 to 1985. They were cast of chrome-molybdenum steel in weights of 85, 125 and 135 pounds. The trademark is probably *Multi Products* cast in relief on the side of the anvils.

RED BARN

This company made at least two farrier's pattern anvils: 65 and 125 pounds. The words *Red Barn* are the trademark cast in relief on the side of the anvils. I was told that they were cast of 4130 steel. The company made anvils from the middle 1970s until the late 1980s. I was also told that the anvils were mostly for local sales.

From Anvil magazine, 1984.

SOUTHERN STEEL & SUPPLY CO.

Southern Steel & Supply Company, 12350 Los Nietos Road, Santa Fe, California. This company's address is used as a trademark on their farrier's pattern anvil. A *SSS* may also be used. I have seen one of these anvils in use, but have not located any ads for them. They probably no longer produce an anvil.

The following names might be found on farrier's cast anvils: *Hoover, Thomas, Valley and Enders.* I have seen them in person, seen advertisements for them, or viewed a photo of these anvils. In 1975, Enders address was Original Enders, Guttenberg, Iowa, 52052, according to an ad. They manufactured a 125-pound farrier's pattern anvil and carried six different-sized smith's pattern anvils in stock at one time.

Enders Tool Company, Albert Lea, Minnesota, make 10-and 25-pound anvils. Most likely ductile or cast-iron anvils. A.P. Tool Company is one of their dealers.

MISCELLANEOUS CAST ANVILS

Many foundries produced cast-iron and cast-steel anvils, with and without names. Some of them are of good quality and some are not. One particular anvil comes to mind, because I know its origin.

The Clark Equipment Company of Buchanan, Michigan, had a steel foundry in that city. They have since moved. They manufactured cast-steel axles, among other things. When some steel was left over from casting a job it was sometimes poured into molds made up by some of the workers. One such mold was for an anvil pattern that someone in the shop produced. Just about every workman in the foundry had one of these anvils, and these anvils eventually were spread around the local area. In time, they have spread even further, winding up in yard sales and flea markets, etc. Over time, who knows where they will wind up? Some collector will stumble upon one of these and think he has located a new anvil brand. This particular anvil is unmarked, as are most of such locally made anvils and were originally not meant for resale.

50-pound Clark Equipment Co. anvil. (R.A.P.)

CHAPTER 5
American Wrought Anvils

Frederick C. Budden. Co-founder of Hay-Budden Manufacturing Co.
(Courtesy of Mrs. Florence Budden Sage)

ANVILS IN AMERICA

Undoubtedly a number of attempts were made to produce wrought anvils in the United States before the 1890s. This is shown by the patent of 1844 (see Appendix F) and has been mentioned in Chapter 2. However, any success Americans enjoyed in making wrought anvils before the 1890s was short-lived. The Americans either could not compete with the English anvils or went out of business because of impractical methods such as those shown in the 1844 patent. To date I have not recorded an American manufactured wrought anvil made before the 1890s.

Here is the account of one attempt to produce wrought anvils in America in the 1860s.[1]

> *In the late fifties or early sixties of the last Century there came to New York an Englishman by the name of James Case, a smith who had been in the employ of Peter Wright at Dudley, England. For several years Case was employed at the Novelty Iron Works, 'Allen Stillman' Dry Dock, New York. On their going out of business, Mr. Case started in for himself, as a repairer and maker of vises and anvils, on East 15th St., New York. In 1864 he got out his first batch of two-piece American wrought anvils, a half dozen all told. Mr. Case might have done a prosperous business and made himself independent, but it is said that, with him, religious fervor took the place of business and he, (with the golden opportunities of a virgin field) failed and died poor.*

I have not located anything else about Mr. Case or his anvils. It is a shame that he was not successful, but I must say it is far better to die poor with a *hope* then to die rich without it.

It is very possible that one or more of the *James Case* anvils still exists, but if his name is not on the anvil it will be impossible to tell who made it.

A number of concerns repaired and refaced wrought anvils in the 19th century before any American wrought anvils were produced. Facing and hardening the face of an anvil is the most difficult part of the process and it is even more difficult to reface and harden a used anvil, so one wonders why these concerns did not manufacture anvils, as well. It might well have been due to the fact that *Peter Wright, Mousehole,* and *Wilkinson* so dominated the market that James Case and the other potential anvil makers had little chance to introduce new wrought anvils. These concerns seem to have been successful in their undertakings, however. One such company was Smith and O'Leary of Chicago, Illinois, a company I will write about later. As you will see from the following article and in the sections in this book on *Hay-Budden* and *Trenton*, it was not easy for a new maker of wrought anvils to establish themselves.

The following article published in *The American Blacksmith* was named the *American Anvil.*[2] The *U.S.* anvil referred to was made by Hay-Budden for a large Chicago hardware house, S.D. Kimbark, a well-established concern who had been selling imported German-made anvils under their trademark. Hay-Budden must have had a contract with Kimbark from their very beginning because Kimbark dropped the German-made anvils and started buying *Hay-Buddens* shortly after the Hay-Budden Manufacturing Company went in business, thus establishing Hay-Budden as a successful anvil maker. Hay-Budden made some of the early anvils under their own trademark, but most of the early ones that I have located were made under the *U.S.* trademark for the smith pattern anvil and *S.D. Kimbark* for the farrier's pattern. Kimbark was also the inventer of the farrier pattern.

166 -pound U. S. anvil. Made by Hay-Budden ca. 1892. (R.A.P. photo)

128-pound S. D. Kimbark anvil. Made by Hay-Budden ca. 1893. (R.A.P. photo)

The extent of the anvil trade in this country and method of manufacture are probably not generally known by our people.

Previous to 1891 the wrought-iron anvils used in this country were imported from England, Germany, and Sweden. For the ten years preceding 1892 the average importation per year was 830 tons. The highest record was 959 tons in 1888. Several attempts to manufacture anvils in this country are on record, but the real birth of this industry, now so firmly established, dates from 1891.

Its growth in 1892 and a part of 1893 was rapid. In 1894 the importations were only 380 tons and in 1895 estimated at 500 tons, while the consumption for each of these years averaged about 800 tons. This shows a satisfactory progress for the American manufacturers.

[Author's note: To my knowledge there was only one American manufacturer of wrought anvils in 1894 and if the anvil's average weight was 200 pounds, American smiths were buying an average of 8000 anvils a year at that time.]

ANVILS IN AMERICA

It is judged that the consumption for the present year will reach at least 800 tons, and it is believed that 600 tons of this amount will be produced at home. We have been speaking of wrought anvils only. It must be remembered that cast-iron anvils with steel faces have been successfully manufactured in this country for many years. Their quality is excellent and will answer well for many purposes.

Formerly wrought-iron anvils were built up of six pieces welded to a central core- four corner pieces, the heel piece and the horn piece. These six parts were welded to the core and the whole then hammered and shaped. The steel face was then welded on and tempered and ground. This method was generally pursued in the manufacture of foreign anvils, but more recently improvements have been adopted by some so that a less number of pieces enter into their construction.

*American manufacturers have discarded Old World methods entirely. We happen to know the methods used by the manufacturers of the anvil known as the 'U.S'. With improved furnaces for heating the blooms, ponderous steam hammers for forging, steel dies for shaping the heel and horn , new processes for tempering and grinding, novel appliances for swinging the anvil, as required in the course of construction, they produce a superior ar*ticle. *This anvil is rightly named 'U.S.' for the capital invested, the material used, the labor employed, are wholly American. In the manufacture of the 'U.S.' anvil the material is selected and piled to form the bloom or mass sufficient in size to produce the entire upper half, including the heel and horn, and a separate one to form the entire lower half. These piles are run into a furnace and when heated to the proper degree become blooms, which are brought under a steam hammer and each shaped and punched and then welded in the center. Next the steel face is welded on, only special American crucible cast steel is used, and the anvil is swung to another hammer for the finishing strokes. Then follows the tempering of the face by a process originating in this plan, which insures the most even and perfect results, with a great saving of time in handling. After this is the testing, inspection, japanning, labeling and bagging, and a warranted anvil is ready for the market.*

It is a pleasure to note the establishment of new industries in our country - the success of one leads to others and thus adds to the general prosperity.[3]

The Blacksmith and Wheelwright, March, 1896.

This is an appropriate introduction to American-made wrought anvils. However, there were only three successful American wrought anvil manufacturers, and they began production when blacksmithing was reaching its "high water mark," which one writer suggests was about 1910.[4] The American anvil makers hardly outlasted their foreign competitors.

There was a fourth wrought anvil manufacturer but it lasted not more than ten years. However, they made high-quality wrought anvils.

The following American wrought anvil makers will be presented alphabetically, not chronologically.

AMERICAN

The least known of the American wrought anvils is the *American,* made by the American Wrought Anvil Company of Brooklyn, New York. Because the concern never incorporated, it is only a guess as to their going in or out of the business of anvil-making. However, various pieces of evidence suggest that they began to do business about 1899 and ceased anvil production in 1910 or 1911.[5]

How I discovered this *American* anvil maker may be of interest. I had recorded a number of anvils with the trademark *American* stamped on the side, with the weight in pounds just below the trademark. The *American* trademark was in the shape of a diamond, thus:

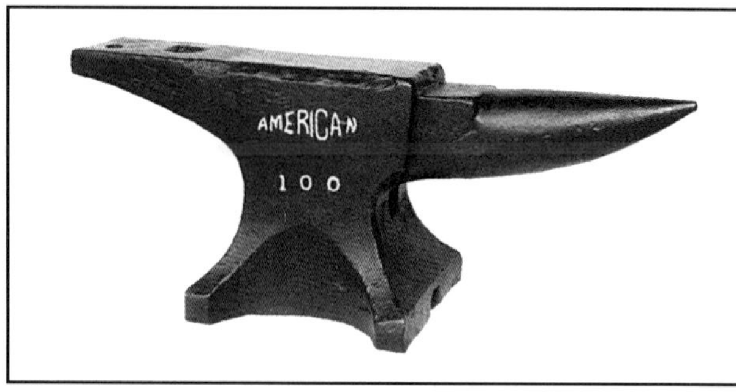

100-pound American anvil. (R.A.P. photo)

The features of this anvil, such as four handling holes and flatness under the base, resembled the very early *Trentons*, but there was no serial number on the front of the foot and the very shape of this *American* anvil did not correspond to the early *Trentons.* I then assumed that the trademark *American* was that of a large hardware company who was importing German- or Swedish-made anvils. The *American* features did not correspond to any of the English anvil makes I knew of, either. The fact that I kept recording numbers of these anvils left me a little bewildered. I could not locate any large hardware company that used the trademark *American,* and who imported large numbers of German or Swedish anvils. A number of trade catalogs mentioned *American* wrought anvils, made in America, along with other known American makes, but these ads seemed to suggest that the *American* was used simply to identify those anvils made in America. I even located a patent for an unnamed anvil, the patentee submitting the application from Brooklyn, New York. However, many applications for patents are submitted but never reach production of the product. I could not identify these *American* anvils with any known American manufacturer or any known importer or imported brand.

The break came a number of years after I had recorded the first *American* anvil. I was in a used book store, browsing through old technical books. I always make a habit of looking in the indexes under Anvil. In *The Automobile Trade Directory* for October, 1909, I found two different addresses of anvil makers in Brooklyn, New York. One was Hay-Budden Manufacturing Company at 254 N. Henry St. (well known to me). The other was the American Wrought Anvil Company at 12-14 Richardson St., an address and manufacturer's name I had never known before.

The name of this anvil manufacturer suggested that they were the makers of the *American* anvil. It also suggested that the patent of Mr. Walter F. Ring that I had previously uncovered was related to the American Wrought Anvil Company and that Mr. Ring may have worked for Hay-Budden before starting his own concern. Mr. Ring, Mr. Hay and Mr. Budden all lived in the same borough of Queens, New York. (It was just too much of a coincidence). Also confirming my suspicions about a connection between the two companies, the patent issued to Mr. James Hay, a year after the Ring patent was issued, refers to the Ring patent. The Hay patent stated that the methods shown in the Ring patent were impractical for producing a one-piece anvil, and that the Hay patented method was practical. How many patentees attack another patentee if they are not connected in some way?

In July, 1994, Mr. Gill Fahrenwald sent me a short article he found in the January, 1898, issue of *The Blacksmith and Wheelwright* journal. The article was dated December 22, 1897, and concerned misuse of second grade or flawed *Hay-Budden* anvils being sold as first-class anvils by some dealers. At the bottom of the article sent to the journal it was signed: "Hay-Budden Manufacturing Company Walter F. Ring, Secretary and Treasurer." Mr. Ring must have had a falling out with Mr. Hay and Mr. Budden, and Mr. Ring started a new company to manufacture anvils for himself. This has to be the American Wrought Anvil Company of Brooklyn, New York (see Appendix M).

This concern either went out of business after Hay-Budden came out with their plateless anvil in 1908 or Hay-Budden bought them out. They may not have gone out of business at all, but ceased making anvils, because of the keen competition in the early-20th century. A forge shop is a very expensive investment and all anvil makers did job shop forging besides producing anvils.

If the name *American* or *American Wrought* is not on the side of an anvil it will be almost impossible to say whether it is an *American* anvil or not, as it closely resembles other American-made anvils. All the *American* anvils I have recorded have four handling holes and are flat under the base like the European anvils and none have a serial number. The smallest one I have recorded weighs 52 pounds, and the largest, 350 pounds. From the appearance of those that I have seen they seem to be a very good anvil.

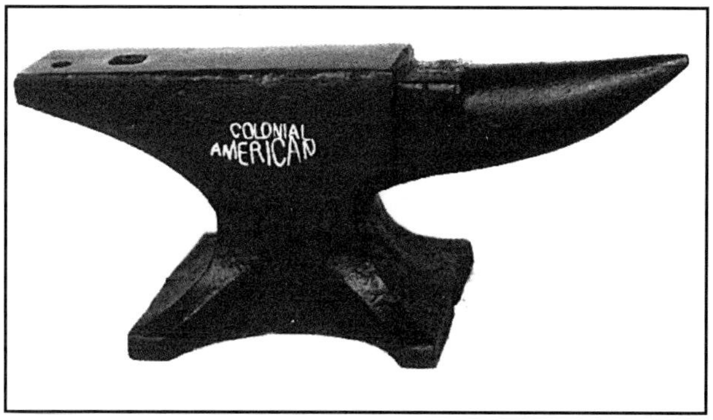

88-pound American anvil. Made for the Schwabacher Hardware Co., "Colonial" being one of their trademarks. (R.A.P. photo)

UNITED STATES PATENT OFFICE[6]

Walter F. Ring, of New York, N. Y.

Method of Forming Anvils.

No. 896, 155. **Specification of Letters Patent** **Patented Aug. 18, 1908**

Application filed January 11, 1907. Serial No. 351,853

To all whom it may concern:

Be it known that I Walter F. Ring, A citizen of the United States, and resident of the borough of Queens, in the city and State of New York, have invented a new and useful Method of Forming Anvils, of which the following is a specification.

My invention relates to the method of making anvils, with the object in view of forming an anvil in an expeditious manner and consists broadly in subjecting a blank held in a die to pressure upon its opposite ends to press the blank into shape with its top and bottom of greater cross section than the narrowest part of the die chamber.

Means for carrying my method into practical effect are represented in the accompanying drawings, in which Figure 1 is a view of the inside face of one half of the die in elevation showing the blank therein and the plunger heads in position to approach each other to press the blank into shape. Fig. 2 is a similar view showing the plungers in their advanced position and the anvil formed. Fig.3 is a transverse section through the die in the plane of the line A-A of Fig. 1, the blank being removed. Fig. 4 is a top plan view of the die with the blank removed, and Fig. 5 is a bottom plan view of the die.

The blank is denoted by 1. It is of a general T-shape and its lower or stem portion is intended to be about the width of the space between the side walls of the die and extends downwardly below the point which the bottom of the formed anvil is to occupy to supply the metal for forming the base of the anvil while the top portion of the blank is composed of a mass of metal sufficient to fill the die between the advanced positions of the plungers.

The die is formed in two sections, denoted, respectively, by 2 and 3, which sections are intended to be firmly held together by suitable means, such for instance as bolts 4 having between them a space of such size and shape as may be required to form the anvil.

The top plunger is denoted by 5 and the bottom plunger by 6. The top plunger is intended to fit with a close sliding fit the upper portion of the die chamber and is provided with a projecting lip 7 for off-setting the horn S of the anvil from the body portion 9.

The base portion of the die chamber is provided with rounded cheek pieces 10 and 11 projecting inwardly from the opposite sides thereof for the purpose of forming a curved recess 12 in each of the opposite sides of the base of the completed anvil.

In operation, the half sections of the die having been fastened firmly together and the blank 1 having been properly heated and placed in position as shown in Fig. 1, the plungers 5 and 6 are caused to approach each other by any well known or approved means not shown herein, and the blank 1 is thereby pressed into the form of the anvil 8, 9, 12. The plungers are then withdrawn and the half sections of the die are separated to permit the removal of the completed anvil. The process is preferably carried out by a single operation by causing the plungers 5 and 6 to move simultaneously toward each other to act upon the blank. During this process, the base of the blank has been pressed outwardly into a form wider than the narrowest part of the die chamber and the top has also been further elongated and formed but the completed anvil is readily removed by separating the half sections of the die.

What I claim is:-

The method of forming an anvil in a single operation consisting in seating a T-shaped blank within a die chamber shaped to form the walls of the anvil and having a contracted portion intermediate of the top and bottom of lesser width and length than the base and head of the completed anvil and simultaneously exerting pressure upon the opposite ends of the seated blank thereby forcing the blank laterally against the walls of the die chamber to form the top and bottom of the anvil of greater length and width than the most contracted portion of the die chamber.

In testimony that I claim the foregoing as my invention. I have signed my name in presence of two witnesses, this 7th day of January 1907.

Witnesses: *Walter F. Ring*

Henry Thieme
F. George Barry

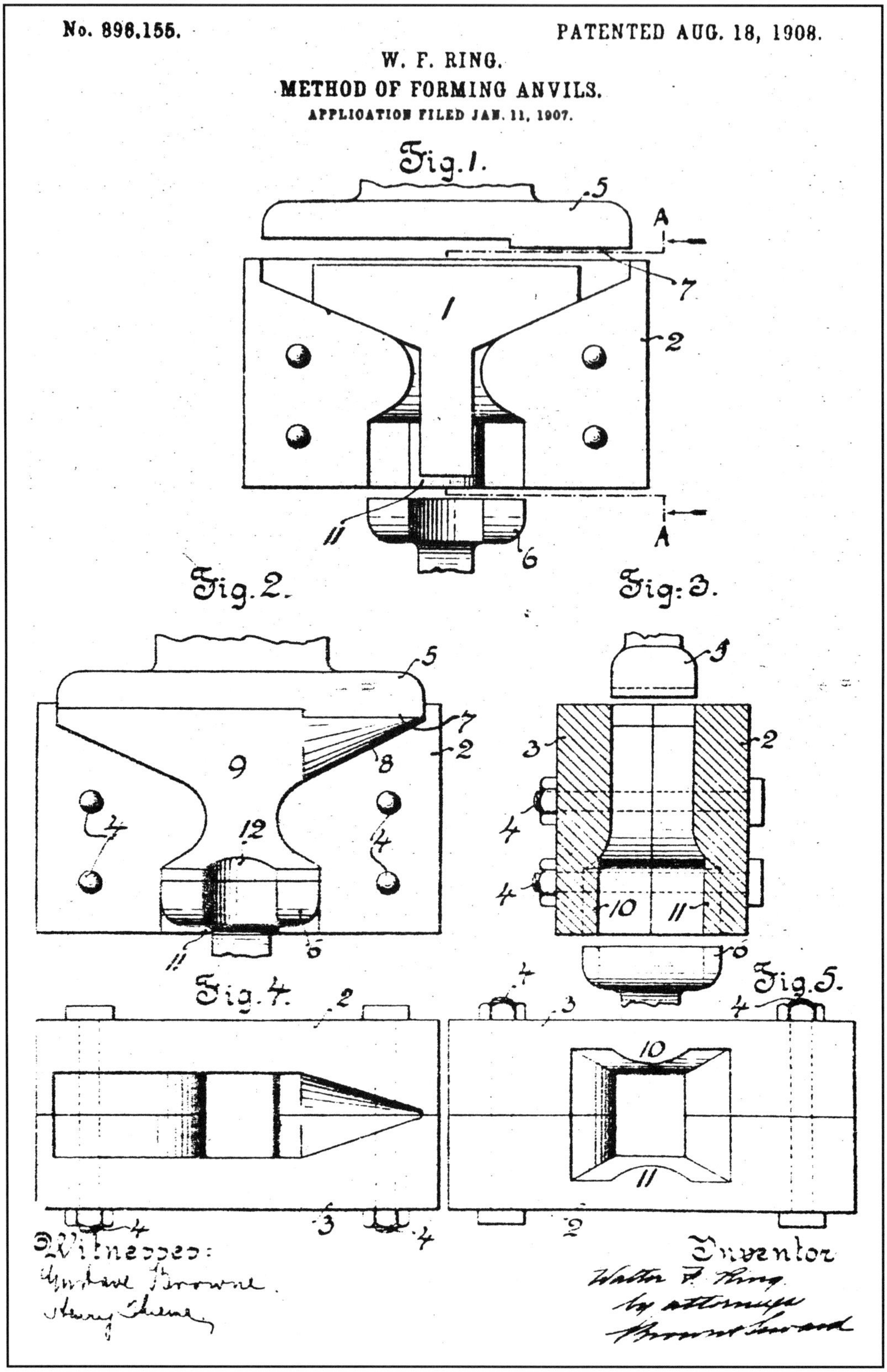

The patent and patent drawing courtesy of the Smithsonian Institution.

ARM & HAMMER

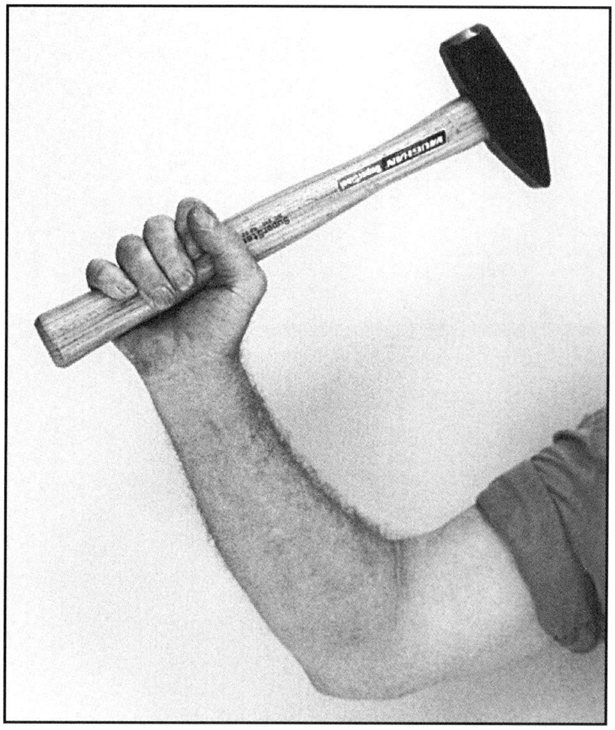

The *Arm and Hammer* anvil was produced by the Columbus Anvil and Forging Company of Columbus, Ohio, between the years 1900 and 1950. When anvil production was dropped in 1950, the company continued to produce forged products until November of 1955 when the company went out of business.[6] From my perspective today the *Arm and Hammer* anvil was undoubtedly one of the finest wrought anvils ever manufactured anywhere. However, despite its apparent quality very few people today who use anvils or collect them know this anvil existed. As of this writing I have not discovered any mention of this company or their anvil in any written material when referring to anvils. The anvil is only represented in advertisements in blacksmith trade journals and hardware catalogs of the past. My reason for considering this anvil one of the best wrought anvils comes from the very good condition of the face plate of the thirty plus *Arm and Hammer* anvils that I have recorded (1991). Percentage-wise, it has fared better than any other make of anvil I have recorded. I don't know why the *Arm and Hammer* has held up so well compared to the other makes, but considering that the vast majority were made before 1920 I can think of no other reason than quality of material and craftsmanship.

The *Arm and Hammer* anvil sold well during the early years of its manufacture, but it was late coming on the market and was competing with high-quality, well established anvil makes for about the same price. Therefore, it never achieved the success of some of the other wrought anvils. The very fact that the anvil, even though made in small numbers after 1920, continued to be produced until 1950 speaks for itself.

There are a number of reasons why it is not well-known among those using anvils. One reason is the fact that only 51,000 to 52,000 were ever made, and at least eighty percent of these were made before 1920. They also seem to be well-distributed around this country from what I have observed. It is hard to say how many were shipped overseas and how many were scrapped. Therefore, they are not encountered very often. Another reason is that they are confused with the *Vulcan* anvil which is more plentiful. When I first started my research on anvils I was under the impression that the *Arm and Hammer* was a better grade anvil made by the same company that made the *Vulcan* (Illinois Iron and Bolt Company). The logo or trademark on both anvils is the arm holding a hammer (like the baking soda logo), except that the *Vulcan* trademark is cast in relief on the side of the anvil and the *Arm and Hammer* logo is stamped into the side. I had assumed that because the *Arm and Hammer* was forged it was the company's top-grade anvil and the *Vulcan*, being cast-iron with a steel face, was their second-grade anvil. I even had a smith tell me he had a *Vulcan* that had a nice ring to it (cast-iron anvils don't ring). When I had a chance to look at the anvil I discovered that it was an *Arm and Hammer* and not a *Vulcan*. All that was still visible on the side of the anvil was the logo of the arm holding the hammer. The only anvil he had ever heard of that had this trademark was the *Vulcan*. Still another reason for the *Arm and Hammer* not being readily recognized is the fact that it is confused with the *Trenton* anvil when the trademark is gone. On the majority of the *Arm and Hammer* anvils that I have recorded the trademark is obliterated or very obscure, because that side of the anvil is worked in normal use, and was a poor stamping to begin with. In shape the *Arm and Hammer* resembles the *Trenton* (a better known anvil) very closely. In fact the base on both are cast, and at times were made by the same foundry, and probably from the same pattern. This leads those who have an *Arm and Hammer* anvil without the trademark to believe that they have a *Trenton*.

The Columbus Anvil and Forging Company was a spinoff of the Columbus Forge and Iron Company (*Trenton*). The story goes that a Mr. Tom Long, one of the anvil makers with the Columbus Forge and Iron Company (he probably made many of the early *Trentons*), had problems getting along with his employers. He finally quit or was fired, but then with local backing he started the Columbus Anvil and Forging Company which was located about two miles south of his former employer. I was told that, while Tom Long was still with the Columbus Forge and Iron Company, he had an argument with the plant superintendent. Thereupon, Tom was fired and ushered out of the gate and told not to come back. Tom then went around to the back of the plant, climbed over the fence and went back to the job he had been doing. The superintendent, upon seeing him again, threw up his hands and said in effect, "If he is this determined to stay--well, I guess I will let him."[8] Apparently, that did not last very long.

At first the Columbus Anvil and Forging Company made only anvils and repaired broken and worn wrought anvils of other makes, as far as I can determine. From the anvils I have recorded, many of these early anvils did not even use the name *Arm and Hammer*, but appeared as follows:

WARRANTED

SOLID **WROUGHT**

The company prospered from the start and soon expanded into general job shop work, forging any item to order that their single frame steam hammers could handle on open dies. (This forge shop continued to use these single frame steam hammers exclusively until the plant was auctioned off in 1955).[9] A machine shop was added next to the forge shop after a few years, thereby allowing the company to produce many finished products that were first forged by them. In later years, one of their main products were crank shafts for punch presses.[10] These were forged and then finished by machining.

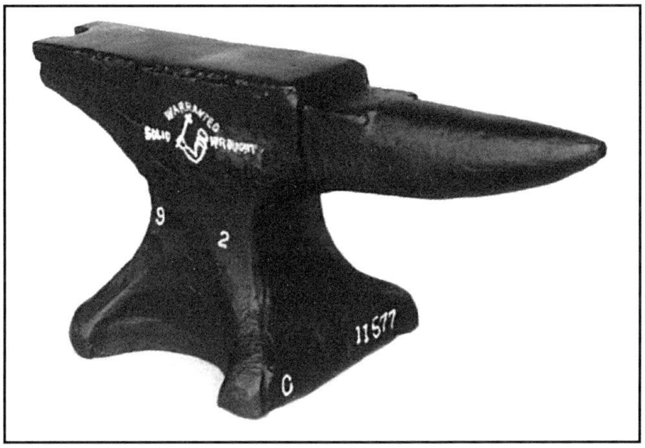

92-pound Arm & Hammer anvil ca. 1910. (R.A.P. photo)

The Columbus Anvil and Forging Company incorporated on January 11, 1900, and continued to prosper until after World War I.[11] By the middle of the 1920s, the company was in financial difficulty, going in and out of business twice, but was finally bought by Mr. J. E. Finneran, Sr. Mr. Finneran was a purchasing agent connected with the Buckeye Steel Casting Company, which might have sold cast-steel bases to the Columbus Anvil and Forging Company, Mr. Finneran managed to get the company back on a sound financial basis.[12] Mr. Tom Long was still with the company when Mr. Finneran bought it, and he was retained as plant superintendent.[13] At this time anvils became a secondary product. The anvil market was soft after World War I, which probably contributed to one of the company's financial problems, as was also true for the Hay-Budden Manufacturing Company.

I was told by Mr. J. E. Finneran, Jr., who assumed the management in 1938 or 1939, that Mr. Earl M. Tilton, President of the Columbus Forge and Iron Company, came to see him in late 1949 or early 1950 on the subject of making anvils. At the time the Columbus Forge and Iron Company was making their *Trenton* anvils of up to 200 pounds in closed dies, and they did not want to keep the open die steam hammers for making the larger anvils, as orders were very few. The *Arm and Hammer* anvils could be made in sizes up to 500 pounds in open dies. Therefore, Mr. Tilton wanted to make a deal. The Columbus Forge and Iron Company would make anvils of 200 pounds or less and the Columbus Anvil and Forging Company would make the ones from 200 to 500 pounds. Mr. Finneran said in reply "You are too late, we discontinued anvil-making a week ago."

I was also told by Mr. Finneran, Jr., that in the late 1920s someone from the Hay-Budden Manufacturing Company came to Columbus to talk to them, and probably also went to see the Columbus Forge and Iron Company. He wanted all the anvil manufacturers in the country to get together in order to send a lobby to Washington, D.C., to try to get a tariff put on imported anvils. The American wrought anvil makers did not worry too much about the competition of the English anvils (mainly *Peter Wright*), because they always cost at least one cent a pound more than the American wrought anvils. It was the Swedish anvil makers with their lower priced, high-quality cast-steel anvils.

Machining a punch press crankshaft. (Photo courtesy of Mr. J. E. Finneran Jr.)

that were taking much of what little there was left of the American anvil market. Apparently, nothing came from the visit, but it does suggest that the Hay-Budden Manufacturing Company was still making or trying to make anvils in the late 1920s (see *Hay-Budden*). I wrote this last sentence in 1991 and as you will see from new information obtained in 1996, I was wrong. Incidentally, the Swedish-American Steel Corporation, the importers of the Swedish anvil *Kohlswa,* was located in Brooklyn, New York, not far from the Hay-Budden Manufacturing Company. Apparently something did come of the Hay-Budden visit, because on June 17, 1930, the tariff was raised on imported anvils from 1 5/8 cents per pound to 3 cents per pound[13] (see Appendix N for the article).

From my recording, the first *Arm and Hammer* anvils were shaped just like the *Trentons.* They were made in three pieces, had four handling holes, and the base was of cast steel. However, the cast-steel base was flat underneath, unlike the *Trenton's* "hour glass" shape. The Columbus Anvil and Forging Company may have had to pay a royalty for using the patented cast-steel base of the Columbus Forge and Iron Company. Some of the *Arm and Hammer* bases were forged, especially on larger anvils. If these early *Arm & Hammer* or *Trentons* do not have their trademark, they are very hard to distinguish from one another or from the *American.*

The oldest *Arm and Hammer* anvil I have recorded has the serial number 162 and weighs 260 pounds. So far I have learned from recording and from Mr. Finneran that the serial numbers were consecutive from the first anvil to the last. The highest serial number I have recorded is 51,799, weighing 180 pounds. I have not recorded an *Arm and Hammer* that did not have a serial number, although some of them were so worn that I could not make out what the numbers were.

By serial number 1?824 (about 1909), the bottom of the base was the same as the later *Trentons.* That is, they had a depressed oval. It had only two handling holes, one in the waist under the horn and the other in the front of the foot under the horn. By at least serial number 21910 (about 1913 or 1914), there were no handling holes at all. The lack of handling holes and the oval under the base remained standard from then on.

Mr. Finneran said that they made anvils weighing from 25 to 500 pounds. Their brochure confirms this. The brochure follows, courtesy of Mr. J. Edward Finneran Jr.

The Columbus Anvil & Forging Co.

MANUFACTURERS OF

"ARM AND HAMMER"
WROUGHT IRON ANVILS

GENERAL FORGINGS OF WROUGHT IRON OR STEEL

++++++

The Columbus Anvil & Forging Co.
115-129 W. FRANKFORT ST.
COLUMBUS, OHIO, U. S. A.

Part A

For centuries wrought iron has been recognized as the best material available for anvil purposes.

Being fibrous, it is very tough, enabling an anvil made from it to withstand the hard usage given such a tool.

Next to a compact, tough body, the working surface is the most important part of an anvil. Only the highest grade crucible steel has proven equal to the service demanded of an anvil face.

ARM & HAMMER brand ANVILS are made from American ingot or wrought iron, welded at the waist, thus making a solid forging, to which is welded a solid plate of crucible cast steel, the best we can obtain, which, welded to the wrought iron, makes a perfect weld.

In view of these facts, it is a pleasure to offer to a prospective customer our ARM & HAMMER brand, knowing there is no anvil on the market, which will give better service or greater satisfaction.

Part B

The above cut, made from a photograph of a piece of the iron from which our anvils are made, illustrates the toughness of the body of our anvils. This piece of iron was bent into this shape while cold. Note the absence of any cracks or flaws.

Other Special Features Are:

1—Horn and heel in correct proportions.

2—Base with sufficient area, or spread, so as not to require strapping down, thus insuring greater stability and absence of rocking or tipping.

3—Special temper of edges to prevent chipping.

4—Face level and true, and of correct proportion for practical work.

5—Hardy and pritchel holes and other mechanical details of correct accuracy.

6—Correct resilience between anvil and hammer, and musical ring.

7—Correct proportions, graceful design, stability on the block and the absence of superfluous weight in the base.

"Arm & Hammer" Anvils Are Fully Guaranteed

Part C

REGULAR BLACKSMITHS' ANVIL

The type of anvil used by the majority of smiths.

FARRIERS' CLIP HORN ANVIL

The type used by smiths the majority of whose work consists in shoeing.

Part D

BASE SIZES, 80 TO 425 LBS.

Advance on Weights other than base sizes and for Farriers' Anvils.

425 to 500	1	cts. per lb. Extra
70 to 79	1	" " "
60 to 69	2	" " "
50 to 59	4	" " "
40 to 49	6	" " "
30 to 39	10	" " "
20 to 29	16	" " "
10 to 19	20	" " "
Farriers' Plain Horn	1	" " "
Farriers' Clip Horn	2	" " "

Our Standard Dimensions of Size of Face to Weight of Anvil.

SIZE OF FACE	WEIGHT	SIZE OF Hardy Hole
2¼ in. x 9½ in.	25 to 30 lbs.	⅝ in.
2¼ " x 10 "	30 to 35 "	⅝ "
2½ " x 10½ "	35 to 40 "	⅝ "
3 " x 11 "	40 to 50 "	¾ "
3¼ " x 11½ "	50 to 60 "	¾ "
3¼ " x 12 "	60 to 70 "	¾ "
3½ " x 12½ "	70 to 80 "	⅞ "
3½ " x 13 "	80 to 90 "	⅞ "
3½ " x 13½ "	90 to 100 "	⅞ "
3¾ " x 14 "	100 to 110 "	1 "
3¾ " x 14½ "	110 to 120 "	1 "
3¾ " x 15 "	120 to 130 "	1 "
4 " x 15½ "	130 to 140 "	1 1/16 "
4 " x 16 "	140 to 150 "	1 1/16 "
4 " x 16½ "	150 to 160 "	1 1/16 "
4¼ " x 17 "	160 to 175 "	1⅛ "
4¼ " x 17½ "	175 to 200 "	1⅛ "
4½ " x 18 "	200 to 225 "	1⅛ "
4¾ " x 19 "	225 to 250 "	1⅛ "
5 " x 20 "	250 to 300 "	1¼ "
5½ " x 22 "	300 to 350 "	1¼ "
5½ " x 24 "	350 to 400 "	1¼ "
6 " x 26 "	400 to 500 "	1⅜ "

Pritchel hole, base sizes, ½ in. to 9/16 in.; other sizes in proportion.

An additional charge will be made on SPECIAL SIZES, and definite specifications required.

Part E

We also make a line of Sawmakers' Anvils, in which we maintain the high quality obtained by using a combination of wrought iron and high-grade steel.

Made in weights from 25 to 250 pounds.

Our Sawmakers' Hammers are not excelled.

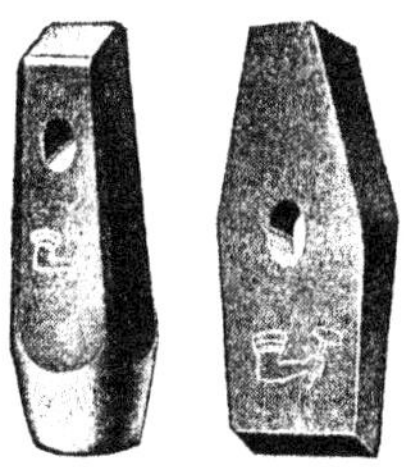

Part F

The smallest *Arm and Hammer* anvil I have recorded weighs 65 pounds and the largest weighs 405 pounds. The anvils from 125-150 pounds sold the best, Mr. Finneran said. From the pictures of their forge hammers you will be able to see that it would not have been practical to forge an anvil weighing much over 500 pounds. Besides, after about 1900 very few anvils over 500 pounds were made by any of the manufacturers.

150-pound Continental pattern anvil. (photo courtesy of Mr. J. E. Finneran Jr.)

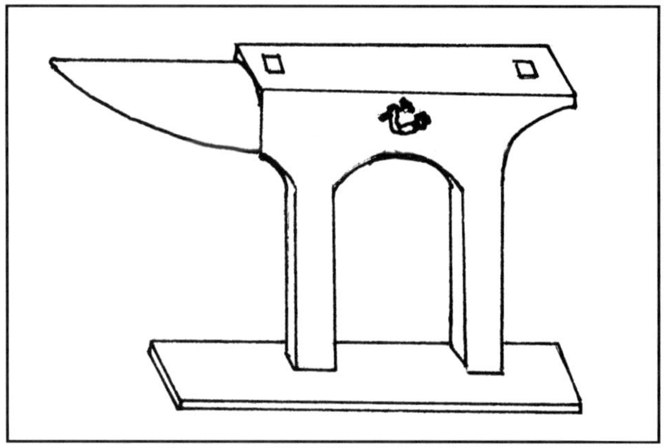

"Try the Arm & Hammer mule anvil-Solid Wrought-Iron -Cast Base. Crucible steel face, made in weights from 205 to 500 lbs. The Columbus Anvil & Forging Co." (Courtesy of Mr. J. E. Finneran Jr.)

To have equipped a blacksmith shop to use very large anvils after 1900 would not have been a wise investment, because the small inexpensive mechanical power hammer began coming on the market at about this time, and it could replace the very large anvils and the smith's helper. If any one reading this should locate an *Arm and Hammer* anvil weighing over 500 pounds, I would like to know about it.

Most of the *Arm and Hammer* anvils were made in the general blacksmithing pattern. I have recorded only one of the farrier's patterns and know of another. The brochure shows sawmaker's anvils from 25 to 250 pounds. I have the original print blocks which show that they advertised a bridge or railroad anvil (they call it a *Mule* anvil) as well as a European double-horned pattern. Included is a picture of one European pattern anvil that they forged.

Large single frame steam hammer at the Columbus Anvil and Forging Co. (Photo courtesy of Mr. J.E. Finneran Jr.)

The Columbus Anvil and Forging Company's anvil repair work was probably fairly extensive. Most of their advertisements emphasize the repair of wrought anvils regardless of condition. The other wrought anvil makers did the same, but did not advertise as such, preferring not to specialize in repair work.

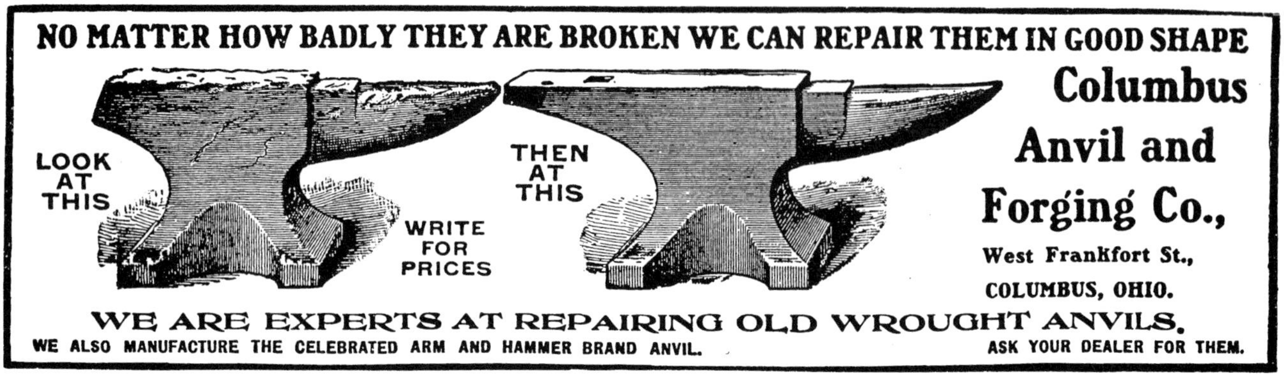

The American Blacksmith, May, 1908.

I know from the records of the makers of the *Trenton* anvil that they would repair any wrought anvil sent to them. There were other forge companies who specialized in anvil repair, although they did not make anvils.

Mr. Finneran said that until the late 1930s, the body of the *Arm and Hammer* was made of one piece of the best wrought iron. At that time they switched to low-carbon steel for the body because wrought iron was becoming too expensive to be profitable. At about the same time they began experiencing considerable difficulty getting the forge weld at the waist to hold. This was attributed to the material that was in the cast-steel base. The cast base was made from reclaimed scrap steel, and bits of impurities such as copper were getting into the scrap and into the castings causing the welds not to take or causing them to break loose in many cases. Therefore, in 1940 the company discontinued forge welding at the waist and replaced it with electric arc welding using consumable electrodes. This solved the problem, but some of the anvil's ring was lost.

Ironically, the change over to electric welding at the waist was hastened on May 21, 1940, when the Columbus Anvil and Forging Company burned to the ground. The forge shop building was completely destroyed as was the furnace used in heating the base and body for forge welding. Only the roof of the machine shop was burned off because there was a concrete wall between the two parts of the building. Mr. Finneran believed the fire was started on the roof of the forge shop by sparks from a nearby train. Believe it or not, the forge shop was back in operation by summer under open skies with their full line of products, plus the few anvils that they were making. Included is a newspaper article about the fire.

TWO-ALARM BLAZE HITS TWO FACTORIES IN CITY; ORIGIN UNDETERMINED

C. & O. Yard Crew Is First to See Fire; Warning Is Given With Blasts On Whistle of Locomotive

Undetermined today was the cause of the two-alarm fire which last evening struck two South Side plants, inflicting damage which may run into tens of thousands of dollars. No official estimate of the loss has been made.

Th*e Columbus Anvil & Forging Co., 117 W. Frankfort-st, was destroyed by flames, which soon spread to the Superior Die Tool & Machine Co. plant, 115 W. Frankfort-st, consuming its roof before firemen could bring the blaze under control.*

Attention of employes was drawn to the fire about 5:25 p. m. by a passing yard crew of the C. & O. Railroad. Engineer Charles W. Reade, 590 S. 18th-st; Fireman H. J. Rea, 325 Clinton-st, and Conductor C. A. Weir, 173 Colombian -av, saw the flames, and Engineer Reade, by prolonged blast from the engine's whistle, directed an employe's attention to the flames.

Origin Undetermined

Although Fire Department said the origin was undetermined the blaze apparently started near the roof at the southwest corner of the Columbus Anvil & Forging Co. plant.

Sixty workmen fled from the two smoke-choked buildings to safety, but in fighting the fire Captain Pearl Cloud of No. 15 engine house was blistered on the face and neck and Fireman Dennis J. Caroll of the same engine house was cut on the right knee. Both received first aid treatment and returned to duty.

J. E. Finneran, 89 S. Dawson-av, is president and general manager of both companies, but he was in New York at the time of fire and will not return to Columbus until this afternoon.

P. J. Hilstein, vice president and engineer of the Superior Die Tool & Machine Co., said rush orders for both plants would be filled with other local companies. The two firms, he said, were operated independently.

Quiz Hits Snag

Fire department investigators were still seeking the probable cause of the fire this afternoon. Fire Chief Edward P. Welch said the investigation has been hampered because the west side, where the blaze apparently started, is so nearly destroyed.

J. E. Finneran Jr., superintendent of the Columbus Anvil & Forging Co., said no decision on rebuilding the plants will be made until his father returns. He said sabotage is not suspected because the firms had no Government orders.

This article was taken from ***The Columbus Citizen*** newspaper dated Wednesday May 22, 1940. *(Courtesy of Mr. J.E. Finneran, Jr.)*

The forge shop back in business. Notice the stack of anvil bases to the left and center of the photo. Two steam hammers can be seen. (Photo courtesy of Mr. J. E. Finneran Jr.)

The forge and machine shop shortly after the fire. (Photo courtesy of Mr. J. E. Finneran Jr.)

The lowest serial numbered anvil I have recorded with the arc-welded waist is 46555, which would place its manufacture in the 1940s. The process of forge welding a steel plate on the body was retained even after the fire and I asked Mr. Finneran why the company had not changed over like all the other companies had done by at least the 1930s, including *Peter Wright.* He said that they were making so few anvils by that time that it was not worth their while to experiment with the plateless anvil. The old method was working fine and besides, they put a better grade of steel on the face than the other makers could by making the face and body of one piece. As far as I am concerned I believe time has proven him correct.

After Mr. Finneran, Sr., bought the company in the 1920s, I am told that all of their anvils were made for the trade only under the *Arm and Hammer* trademark. They did not put the name or trademark of any other company or hardware supply house on their anvils. However, before Mr. Finneran took over I know for certain that the Columbus Anvil and Forging Company made anvils for other companies, putting that company's trademark on the anvils. H. Channon Company, of Chicago, Illinois, was one of these. *Bullock* was this company's logo, and it appeared on the anvils in this manner:

BULLOCK
H. Channon Co.
Chicago

I have recorded five of these Columbus Anvil and Forging Company-made *Bullocks.* On one, serial number 37209, appears with the Channon trademark stamped in the side with what remains of the *Arm and Hammer* trademark under it. Whoever ground the side to take off the *Arm & Hammer* did not complete the job. The Columbus Anvil and Forging

Company also made anvils for Belknap Hardware company of Louisville, Kentucky, marking the anvils as follows:

BELKNAP
BH & M CO.
LOUISVILLE

152-pound Arm & Hammer anvil made for H. Channon Co. (R.A.P. photo)

If they made anvils for any other companies I have not discovered them as yet.

The manufacture of *Arm and Hammer* anvils differed very little from other wrought anvils made after 1900. The details of their manufacture that follows was mostly given to me by Mr. J. E. Finneran, Jr.

The body was made of one piece (not built up) of wrought iron, then later of low-carbon steel. The heel was drawn out first, then the waist and sides were brought to the proper size on flat open dies. Next, special open-horn dies were used to draw out the horn, and the table was created. The face plate was then forge welded on while the body was held in a jig designed for that purpose. Where the face plate joins the body, a demarcation line is very evident, more so than on any other anvil make with a steel face plate. When the body and face were still white-hot (2300-2400 degrees F.), the hardie and pritchel holes were punched out. In later years they were drilled and the hardie hole was broached square before heat treatment. The body and base, after heating to welding temperatures, were brought together under a steam hammer and forge welded at the waist. After 1940 this would be an electric arc weld. The face was then planed flat, the face heated to hardening temperature and placed in a large tub under a large pipe connected to a water tower. The water in this tower was continually circulated so that it maintained a constant temperature the year around. A head of water from the tower was let down on the anvil face with great force, breaking the steam barrier at the face and quenching the face to a very hard temper. There was no draw of the temper, I am told. After cooling, the face was ground on a large surface grinder to remove any warping that might have taken place during the heat treatment. The anvil was now finished except to be weighed and stamped, the serial number and trademark put on, and painted black. In the last years, the trademark was stamped in deep while the anvil was still hot.

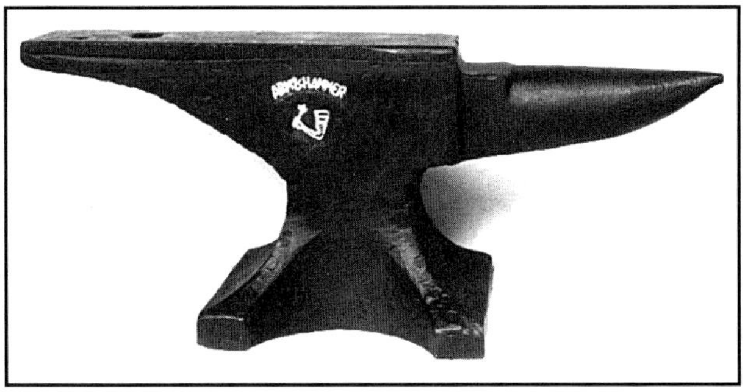

180-pound Arm & Hammer anvil dating from the late 1940s. The latest trademark. (R.A.P. photo)

Identification, as indicated earlier, can be difficult when the trademark is missing. The *Arm and Hammer* anvil can be confused with other anvils, especially the *Trenton*. Here are some guidelines to *Arm and Hammer* anvil identification.

First, all *Arm and Hammers* are wrought; therefore, they should have a nice ring when tapped with a hammer if they are loose on their stand and do not have a loose face plate or inner flaws. One of my blacksmith friends has a very nice *Arm and Hammer* anvil, but he told me it did not ring. When I saw the anvil, its feet were wedged tightly under two pieces of angle iron on a steel stand. He was right, it was almost dead so far as ring was concerned. I asked him if he would remove it from the angle iron and I was sure it would ring. He said he did not think so, whereupon he produced a ten- pound sledge and proceeded to smash the anvil at the waist with several blows until it came loose. Of course, this procedure put a number of dents in the waist. Another blacksmith friend standing nearby said that I should have seen my face when the anvil was struck. He said my face reflected absolute horror. It must be an indication of my care for anvils. Anyway, after the anvil was placed on top of the angle iron it rang nicely with only a tap of the hammer.

Second, look for the trademark on the side when the horn is to your right. If the trademark is visible, the *Arm and Hammer* logo will be there, but the words may vary. It might have *Warranted* over the logo and *Solid Wrought* as indicated before, will be on either side of the logo on the very early anvils. If *Arm & Hammer* is over the logo, there will be either *Solid Wrought* or *Wrought Iron* under the logo.

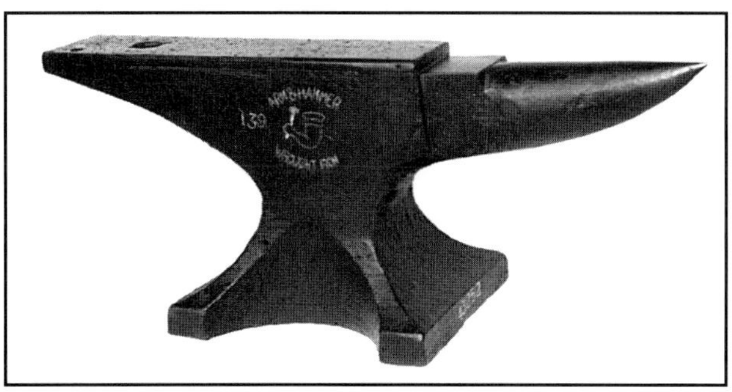

138-pound Arm & Hammer anvil ca. 1930. A typical trademark. (R.A.P. photo)

Also watch for the *Bullock* or *Belknap* trademarks. So far I have not discovered any other anvil maker producing anvils for these two hardware suppliers other than the Columbus Anvil and Forging Company, using these hardware company's trademarks. Both of these companies generally sold anvils under the maker's trademark. Selling anvils under their own trademark probably took place over a short period of time when anvil sales were high.

Generally the weight is stamped in pounds under the trademark, but it may be to one side and slightly under the trademark. I guess it depends on who was doing the weighing and stamping on a particular day. After about 1930 the weight was put on the front of the foot to the right of the serial number.

I have found that the trademark is usually obscure, either from being hammered on, or from rust, paint or heavy grime. Use a wire brush on the area where the markings should be. If the anvil is painted and you don't have a power wire wheel you may have to get paint remover. After brushing rub with a dry cloth. If there are any markings left this will usually bring them out. This procedure applies to all makes of anvils.

Third, look for the cast base. The pattern dividing line will come down the middle of the base front and rear under the horn and heel. Do not make too much of a cast base; this may indicate that it is a *Trenton* anvil as well. Other than that, the very early anvils will have forged bases that are flat beneath. The later ones will have a depressed oval under the base. This can also be true of the *Trenton*. I have not found an *Arm and Hammer* yet that had the *Trenton* "hour glass" shape under the base, although that does not mean the Columbus Anvil and Forging Company did not use it.

Fourth, look for a serial number stamped on the front of the foot under the horn. It is generally in the middle or to the right on anvils before 1930. After 1930, it is to the left. It usually takes a little wire brushing to bring it out. The serial number will never be more than five digits. There may or may not be a foundry mark to the left on the front of the foot. This mark is usually a letter. It is possible that the letters are those of the anvil maker.

Fifth, look at the general shape of the anvil. *Arm and Hammer* anvils generally have a long, thin heel and a narrow waist when viewed from the side. The horn has a nice long, rounded shape relative to the size of the anvil. However, as every wrought anvil is somewhat different, do not depend on the shape for identification alone. It is only an indicator of a particular make. Also look for the distinct separation line between the face plate and the body.

If you are still in doubt of the anvil's make, look under the heel. This is usually a dead giveaway and possibly should be the first identification mark to be looked for on an *Arm and Hammer* anvil. Of all the anvils that I have looked at not one other wrought anvil maker leaves the underside of the heel with such outstanding fullering marks on such a consistent basis as did the Columbus Anvil and Forging Company. If the anvil is smooth under the heel it is probably not an *Arm and Hammer.* These marks are left by the fullering tool while the heel is being finished after having been drawn out under the blows of the steam hammer. The other wrought anvil makers smoothed out these marks more or less; *Peter Wright* did the best job. Since this area of the anvil is almost never used, the Columbus Anvil and Forging Company probably thought it was a waste of time to smooth it out, and it would keep the cost of the anvil down. Everything done to the *Arm and Hammer* was strictly utilitarian.

Roughness under the heel of an Arm & Hammer anvil. (R.A.P. photo)

Last of all, study the pictures and drawings, as they will be a help. If you cannot identify the anvil, do not lose hope. There are some wrought anvils whose maker may never br determined.

Several years ago a notice of an auction caught my attention. An old welding shop was being sold, and the list of tools included an anvil. I wanted to see the anvil before I went to the auction in order to insure that attending the auction would be worth my time. However, I had to go after working all day and when I got to the welding shop, it was closed. I found a dirty window and peered in, hoping to catch a glimpse of the anvil. There it was on its block, with the heel pointing in my line of sight. From this obstructed view, I made a guess that the anvil was probably an *Arm and Hammer*. When I went to the auction I found that I was right. It was an *Arm and Hammer,* and a particular type

I had been looking for. What I had seen through that window was a thin heel and heavy ridges under it.

Following is a partial list of the *Arm and Hammer* anvils I have recorded, arranged by serial number, and the approximate date of their manufacture:

Serial Number	Date
162	1900
4507	1901
15238	1912
16824	1913
21910	1914
22096	1915-16
22805	1915-16
25828	1916-17
26810	1916-17
28529	1917-18
30011	1917-18
31706	1917-18
32546	1918
32908	1918
36774	1919
37209	1922
37319	1922-23
38020	1922-23
38048	1922-23
38156	1922-23
38911	1922-23
43252	1935
443??	1938
46555	1941
49880	1944-45
51799	1948-49

When the Columbus Anvil and Forging Company was auctioned off on November 8, 1955, it had five forging steam hammers; two Erie single frame hammers each with a 4,000 pound rating, and three McDougall & Potter single frame hammers each with a 1,500 pound rating.[14] These five forging hammers produced all their forged products and anvils.

I am including part of a letter that I received from Mr. J.E. Finneran Jr., the last owner of the Columbus Anvil and Forging Company, written August 21, 1991. Mr. Finneran has since passed away. The letter speaks for itself.

> *There were a number of reasons why we went out of the anvil & forging business. One of the more important ones was that our only anvil maker was getting up in years and none of the younger fellows wanted to become an anvil maker. The same was true of all of our hammersmiths. We had one young man who was making a good hammersmith. We had another who was willing but although a high school graduate could not do simple arithmetic. He did not know what a division sign was.*
>
> *We also had the problems facing most small businesses. TOO MUCH OVERHEAD. When I joined the company in the late 1920s, we had one girl who did all the office work; typing, billing, shipping, payroll, bookkeeping etc. When we went out of business, we had 2 girls, a half-time CPA, and another fellow who did nothing but payroll and fringe benefits etc. We had not yet gotten into pensions etc. But that was probably next. Heaven only knows what fringe benefits we would be into today. Baby sitting etc.*
>
> *There was an article in The Wall Street Journal a few years back. It told that 20-30 years ago there were app. 3000 forge shops & foundries in existence. Now there are only a handful and most of those are owned by large corporations who expect to operate them at a loss but they need the products.*
>
> *As far as I know The Buckeye Steel Casting Co. never made any anvil bases. They made their money in the 1920's*

making railroad castings. That was their hey day along with the railroads. They have been going downhill ever since. They are now part of a local conglomerate-Worthington Industries.

One problem we had when we went to electric welding was that we lost most of the ring, but that could not be helped. There were too many advantages otherwise.

Ed Finneran.

Some *Arm and Hammer* prices by date:

1925 - .31 / lb.
1929 - .46 / lb.
1932 - .25 / lb.|depression
1937 - .25 / lb.| years
1941 - .60 / lb.
1947 - 48 - .40/lb. Sligo Inc.

An explanation will be given where necessary with the following advertisements and photos. The company did not advertise in the trade journals continuously. The selection presented here is typical of the ads.

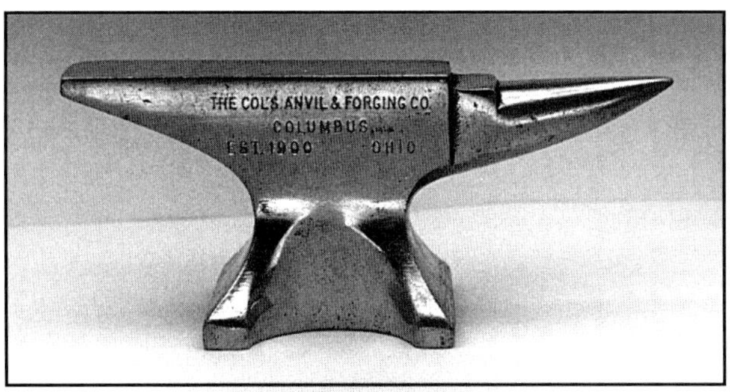

Aluminum miniature advertising anvil.
(Mr Gill Fahrenwald photo)

BEFORE BUYING

A New Anvil, write us.
It will pay you

THE COLUMBUS ANVIL & FORGING CO.
COLUMBUS, OHIO

The American Blacksmith, February, 1913.

Arm & Hammer Wrought Iron Blacksmiths' Anvils

Forged from wrought iron, welded at the waist, to which is welded a solid plate of crucible tool steel.

Pritchel hole, base size, ½ to 9/16 inch; other sizes in proportion.

An additional charge will be made for special sizes, and definite specifications required.

Standard Dimensions

Size of Face Inches	Weight Pounds	Size of Hardy Hole Inches	Size of Face Inches	Weight Pounds	Size of Hardy Hole Inches
2¼x 9½	25 to 30	⅝	4 x15½	130 to 140	1 1/16
2¼x10	30 to 35	⅝	4 x16	140 to 150	1 1/16
2½x10½	35 to 40	⅝	4 x16½	150 to 160	1 1/16
3 x11	40 to 50	¾	4¼x17	160 to 175	1⅛
3¼x11½	50 to 60	¾	4¼x17½	175 to 200	1⅛
3¼x12	60 to 70	¾	4½x18	200 to 225	1⅛
3½x12½	70 to 80	⅞	4¾x19	225 to 250	1⅛
3½x13	80 to 90	⅞	5 x20	250 to 300	1¼
3½x13½	90 to 100	⅞	5½x22	300 to 350	1¼
3¾x14	100 to 110	1	5½x24	350 to 400	1¼
3¾x14½	110 to 120	1	6 x26	400 to 500	1⅜
3¾x15	120 to 130	1	....	....	...

Advance on Weights Other Than Base Sizes

Base Sizes, 80 to 425 Pounds

Weight Pounds	Extra Over Base Per Lb.	Weight Pounds	Extra Over Base per Lb.
425 to 500	$.01	40 to 49	$.06
70 to 79	.01	30 to 39	.10
60 to 69	.02	20 to 29	.16
50 to 59	.04	10 to 19	.30

H. Channon Co. catalog, 1908, Chicago, Il.

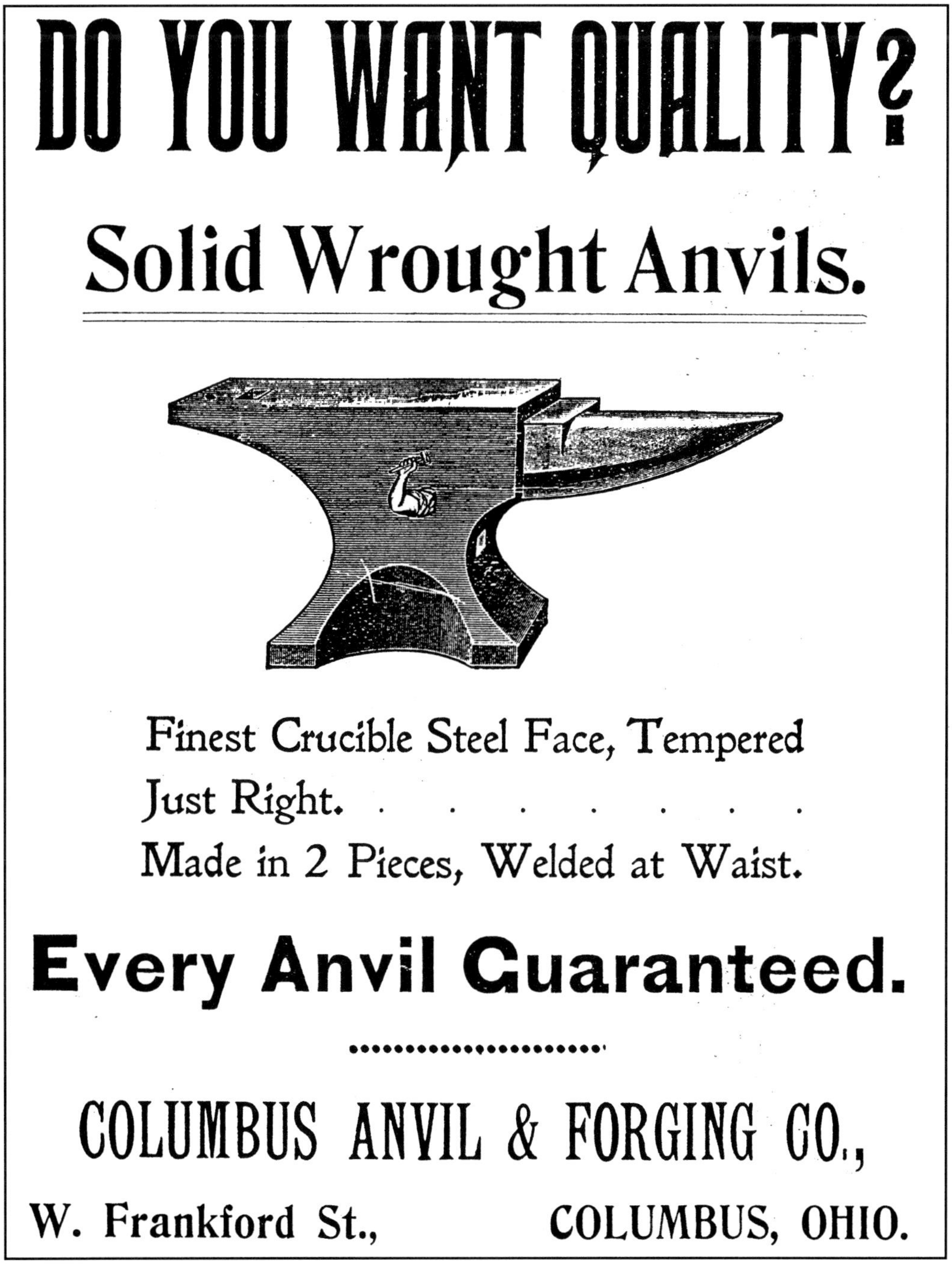

The Blacksmith and Wheelwright, March, 1901. The first ad to appear in the journal.

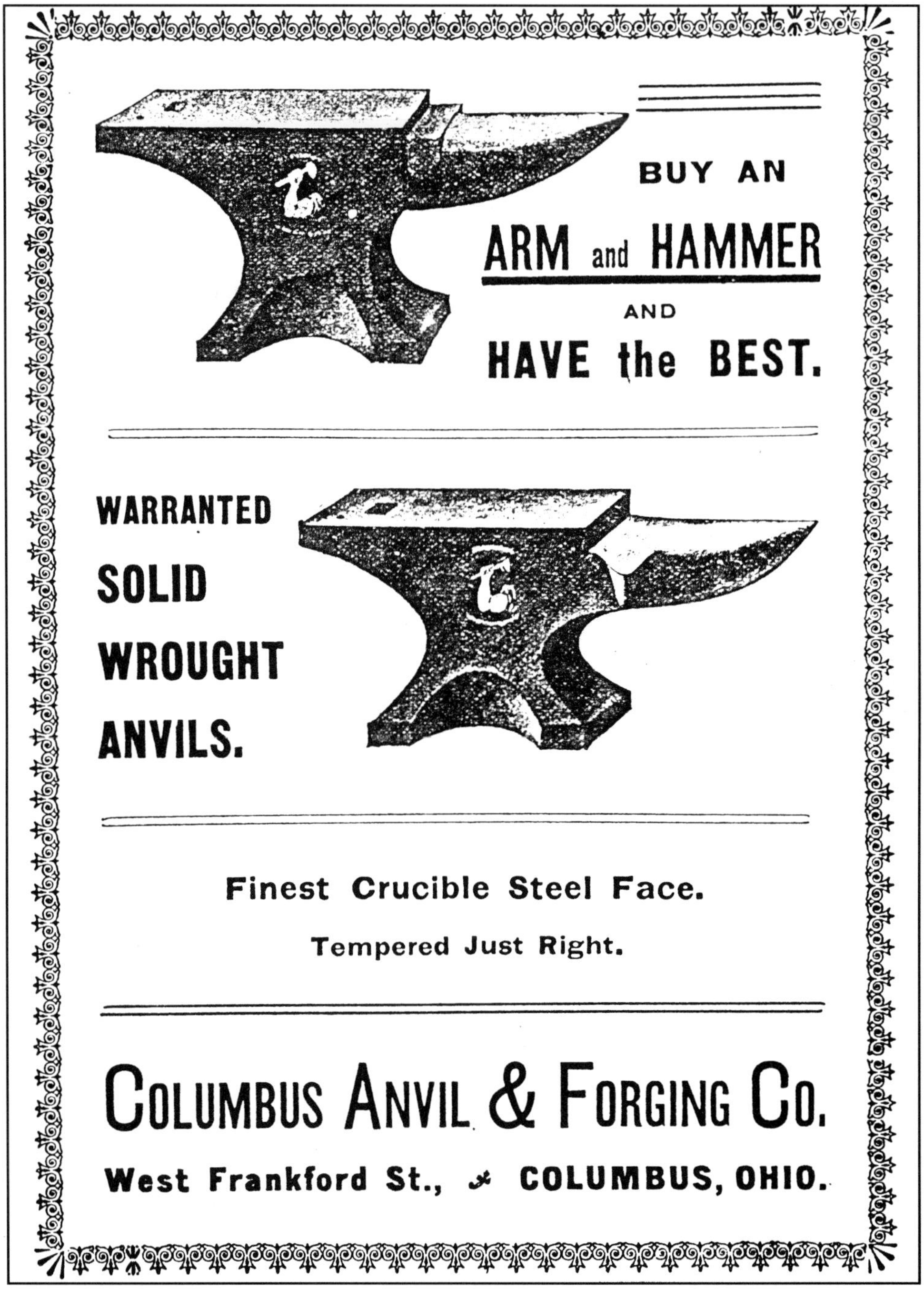

The Blacksmith and Wheelwright, April, 1901. Their second ad to appear in this journal.

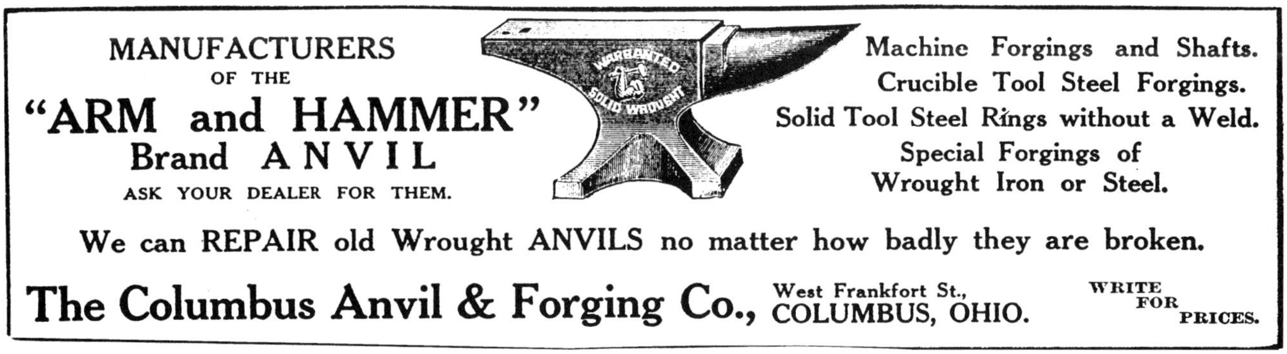

The American Blacksmith, January, 1909. Notice that they were making other items beside anvils.

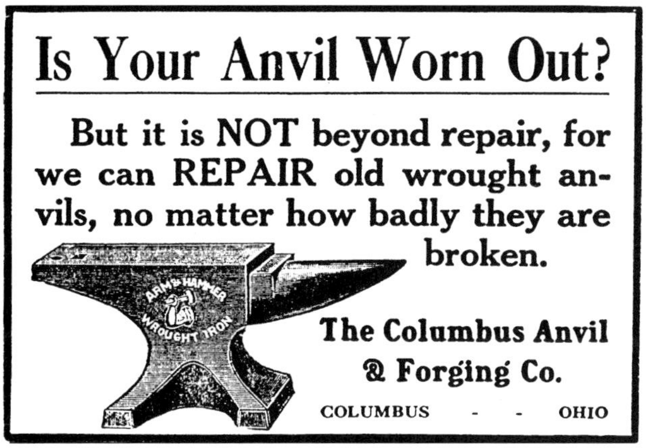

The American Blacksmith, March, 1914.

The American Blacksmith, December, 1914.

The Blacksmith and Wheelwright, December, 1917. The same ad as that of 1908 with a slight change.

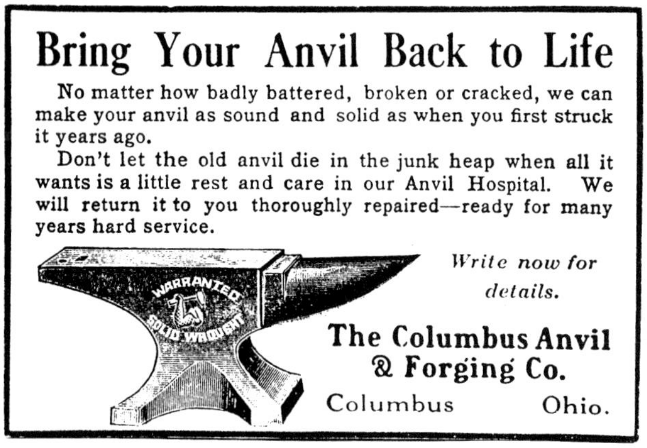

The American Blacksmith, November, 1915.

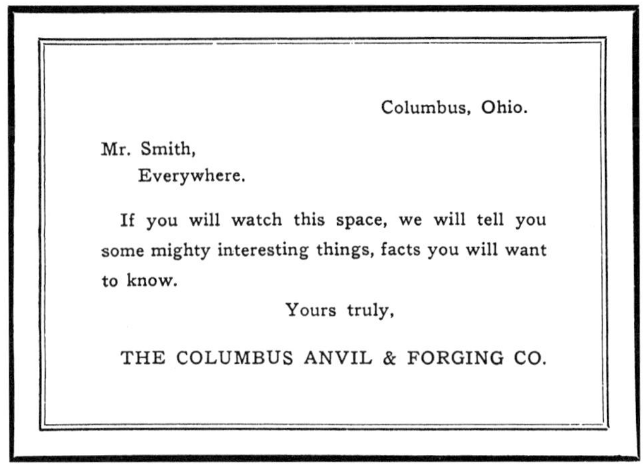

The Blacksmith and Wheelwright, June, 1919.

The Blacksmith and Wheelwright, January, 1925. Though in financial trouble they continued to advertise.

The Blacksmith and Wheelwright, June, 1926. (From Mr. Keith Sommer)

Mill Supply Buyer's Guide catalog, 1928, Chicago, Il.

BLACKSMITHS' ANVILS

Arm & Hammer brand Anvils are made from American ingot or wrough iron, welded at the waist, thus making a solid forging, to which is welded a solid plate of crucible cast steel, the best obtainable which, welded to the wrought iron, makes a perfect weld.

Special Features

1—Horn and heel in correct proportions.

2—Base with sufficient area, or spread, so as not to require strapping down, thus insuring greater stability and absence of rocking or tipping.

3—Special temper of edges to prevent chipping.

4—Face level and true, and of correct proportion for practical work.

5—Hardy and pritchel holes and other mechanical details of correct accuracy.

6—Correct resilience between anvil and hammer, and musical ring.

7—Correct proportions, graceful design, stability on the block and the absence of superfluous weight in the base.

Base Sizes, 80 to 425 pounds............... Per Pound **$0.40**

Advance on Weights, other than base sizes and for Farriers' Anvils

425 to 500 lbs., Advance over Base	Per Pound	$0.00½
70 to 79 lbs., Advance over Base	Per Pound	.00½
60 to 69 lbs., Advance over Base	Per Pound	.01
50 to 59 lbs., Advance over Base	Per Pound	.02
40 to 49 lbs., Advance over Base	Per Pound	.03
30 to 39 lbs., Advance over Base	Per Pound	.05
20 to 29 lbs., Advance over Base	Per Pound	.08
10 to 19 lbs., Advance over Base	Per Pound	.15

Standard Dimensions of Size of Face to Weight of Anvil

Size of Face	Weight	Size of Hardy Hole
2¼ in. x 9½ in.	25 to 30 lbs.	⅝ in.
2¼ in. x 10 in.	30 to 35 lbs.	⅝ in.
2½ in. x 10½ in.	35 to 40 lbs.	⅝ in.
3 in. x 11 in.	40 to 50 lbs.	¾ in.
3¼ in. x 11½ in.	50 to 60 lbs.	¾ in.
3¼ in. x 12 in.	60 to 70 lbs.	¾ in.
3½ in. x 12½ in.	70 to 80 lbs.	⅞ in.
3½ in. x 13 in.	80 to 90 lbs.	⅞ in.
3½ in. x 13½ in.	90 to 100 lbs.	⅞ in.
3¾ in. x 14 in.	100 to 110 lbs.	1 in.
3¾ in. x 14½ in.	110 to 120 lbs.	1 in.
3¾ in. x 15 in.	120 to 130 lbs.	1 in.
4 in. x 15½ in.	130 to 140 lbs.	1 1/16 in.
4 in. x 16 in.	140 to 150 lbs.	1 1/16 in.
4 in. x 16½ in.	150 to 160 lbs.	1 1/16 in.
4¼ in. x 17 in.	160 to 175 lbs.	1⅛ in.
4¼ in. x 17½ in.	175 to 200 lbs.	1⅛ in.
4½ in. x 18 in.	200 to 225 lbs.	1⅛ in.
4¾ in. x 19 in.	225 to 250 lbs.	1⅛ in.
5 in. x 20 in.	250 to 300 lbs.	1¼ in.
5½ in. x 22 in.	300 to 350 lbs.	1¼ in.
5½ in. x 24 in.	350 to 400 lbs.	1¼ in.
6 in. x 26 in.	400 to 500 lbs.	1⅜ in.

Pritchel hole, base sizes, ½ in. to [illegible] in.; other sizes in proportion

An additional charge will be made on SPECIAL SIZES, and definite specifications required.

Buhl and Sons Co. catalog, 1929, Detroit, Mi. (From Mr. John Catchings)

Arm & Hammer Wrought Iron Blacksmiths' Anvils

Forged from wrought iron, welded at the waist, to which is welded a solid plate of crucible tool steel.

Pritchel hole, base size, ½ to 9/16 inch; other sizes in proportion.

An additional charge will be made for special sizes, and definite specifications required.

Size of Face Inches	Weight Pounds	Size of Hardy Hole Inches	Price per Pound
2¼x 9½	25 to 30	⅝	....
2¼x10	30 " 35	⅝	....
2½x10½	35 " 40	⅝	....
3 x11	40 " 50	¾	....
3¼x11½	50 " 60	¾	....
3¼x12	60 " 70	¾	....
3½x12½	70 " 80	⅞	....
3½x13	80 " 90	⅞	....
3½x13½	90 " 100	⅞	....
3¾x14	100 " 110	1	....
3¾x14½	110 " 120	1	....
3¾x15	120 " 130	1	....
4 x15½	130 " 140	1 1/16	....
4 x16	140 " 150	1 1/16	....
4 x16½	150 " 160	1 1/16	....
4¼x17	160 " 175	1⅛	....
4¼x17½	175 " 200	1⅛	....
4½x18	200 " 225	1⅛	....
4¾x19	225 " 250	1⅛	....
5 x20	250 " 300	1¼	....
5½x22	300 " 350	1¼	....
5½x24	350 " 400	1¼	....
6 x26	400 " 500	1⅜	....

C. A. Turner, Inc. catalog, 1937, Pittsburgh, Pa.

STANDARD REGULAR BLACKSMITH

Made from very low carbon steel to which is welded a solid plate of electric furnace high carbon steel, which when welded to the low carbon steep top makes a perfect weld. The top is then electric welded to a cast base at the waist.

A low carbon steel, as close to wrought iron as we can obtain, is perfect material for a top as it will absorb quite a bit of shock and is also soft enough that the Blacksmith can shape the horn to his requirement for his job.

STANDARD ANVIL SPECIFICATIONS

Weight Lbs.	Size of Face In.	Size of Hardy Hole In.	Each
100	3¾x14	1	$60.00
125	3¾x14½	1	75.00
150	4 x16	1 1/16	95.00
200	4¼x17½	1⅛	125.00
250	4¾x19	1⅛	155.00
300	5 x20	1¼	195.00

The W. Bingham Co. catalog, 1941, Cleveland, Oh. By this time low carbon steel had been substituted for wrought iron for the body of the anvil.

The Columbus Anvil and Forging Co. building ca. 1935. (Courtesy Mr. J. E. Finneran Jr.)

The following article is from a Columbus, Ohio newspaper ca. 1950. There were several photos with the article, but they were not good enough to be reproduced. The "Columbus Bolt & Forging Co., 291 Marconi-blvd.", referred to is a mistake by the reporter who wrote the article. He probably got the Columbus Bolt & Forging Company mixed up with the Columbus Forge & Iron Company makers of the *Trenton* anvil, who were the only other maker of forged anvils in the United States in 1950. (Article courtesy of Mr. Mike McGee.)

Many a Village Smith Got His Anvil From Local Forge

Columbus Firm Still Makes Pieces

by Ernest Leogrande

An old-time industry- the making of anvils- still is being carried on in Columbus, but it may soon join the crank telephone and the horse-drawn buggy.

The Columbus Anvil & Forging Co. at 117 W. Frankfort-st has been making anvils since 1900 when the company was founded.

The market for the iron-shaping blocks has been dwindling, and the company's president, Ed Finneran Jr., says it soon will stop making them.

The manufacture of other iron and steel pieces will keep the forge working full-time at the company's furnaces and hammers.

Mr. Finneran, who is 36, says he knows of only two places in the country where the old-style forged anvils are made.

One is in his shop and the other is at the Columbus Bolt & Forging Co., 291 Marconi-blvd. There is one hammer in Mr. Finneran's shop. It is used only for putting the "horns," or pointed ends, on anvils.

A forged anvil is the only kind a real blacksmith will work on," Mr. Finneran said.

Blacksmithing never was practiced much in the old Columbus Bolt [Anvil] & Forging, its present owner said, probably not for more than a year after it was incorporated on Jan. 13, 1900.

A fire leveled the plant in 1940, and little more than the office records was saved. It was soon rebuilt, however.

Mr. Finneran took over the presidency when his father, James Edward Finneran Sr., died in 1948.

Previously Mr. Finneran Sr. had been purchasing agent for the Buckeye Steel Casting Co. He took control of Anvil and Forging in 1924 when it fell into the hands of receivers.

The plant appears small from the front, but the buildings are low and spread out. It has a forge shop with four hammers and furnaces; a sawroom where the billets, or bars of metal are stored and cut; a boiler-room; a rough-turn room with four lathes, where unfinished forgings are turned out; offices and shipping room.

In one part of the buildings is the Superior Die & Tool Co., which is a separate company.

The Plant's furnaces, boilers lighting and all its other equipment have kept step with the times.

Such marvels as fluorescent lighting, Dictaphones and oil-fired furnaces would have made the Anvil & Forging's first customers rub their eyes in disbelief.

Today modern charts in the president's office keep constant check on the progress of each order.

There was nothing so fancy in 1900. One man who can prove it is James H. Bateman, engineer and fireman.

Jim, as he is called, has been with the firm since Sept. 1, 1900. He has seen men and equipment come and go. The steam boiler over which he stands guard is placed beside an old-style one, no longer used.

Jim admits that the fabulous half-century which he has seen pass was one of change. But he says that some things have endured.

Just as steel is alloyed to give it durability, some old-time industries have lasted by changing to suit the times.

This photo was probably taken at the Columbus Anvil and Forging Company in the 1930s. It is one-half of the composite used to produce the painting "The Last Anvil Makers." (Courtesy of Mr. J.E. Finneran Jr.)

HAY-BUDDEN

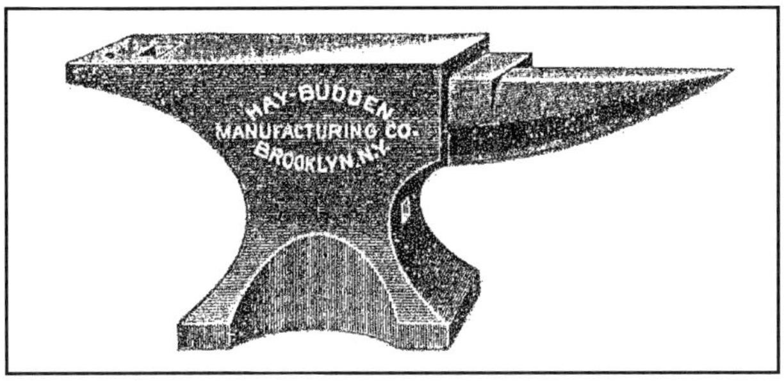

The *Hay-Budden* anvil is the best-known of the American-made wrought anvils. Only *Peter Wright* (English) is as well-known. The *Hay-Budden* was produced by the Hay-Budden Manufacturing Company of Brooklyn, New York. They started making anvils sometime in the late 1880s and ceased production about 1925 or 1926. When I first became interested in anvils I asked a demonstrator working at an anvil what make anvil he was using. The only names of anvils that I had heard of up to that time were *Fisher, Vulcan,* and *Peter Wright.* In a quick reply he said, a *"Hay-Budden."* I thought he must be joking with me because what he said sounded more like *"Hay-Bud"* to me. I could not imagine someone naming an anvil *Hay-Bud.* It did not take the smith long to straighten out my thinking. He then pointed at the name on the side, *Hay-Budden.* The name, incidentally, comes from the last names of the two partners who started the company--James Hay, President and Treasurer and Frederick C. Budden, Vice President. Hay was a Scotsman and Budden was an Englishman, both blacksmiths.[15]

In fact, today the *Hay-Budden* anvil is one of the most sought-after anvils by both professional and hobby smiths. To own a *Hay-Budden* was to have the Rolls Royce of anvils. This was so around the turn of the 20th century and many think it so today (1991). Here is a statement dated 1911 that I came across while looking for information about anvils,[16]

> *...There is a great difference in the make and also the quality of anvils. The author has forged tools on almost every make of anvil manufactured in Great Britain or the United States, but the kind that has given the most satisfactory results, and which can be relied upon, is known as the Hay-Budden. Manufactured by the Hay-Budden Manufacturing Company, Brooklyn, New York, U.S.A., this make of anvil is fully guaranteed. They are made from the very best material and by expert workmen, and the face is perfectly hardened. There are no soft spots, neither do they sag down or get hollowing in the face, as in a great many other makes, including the Peter Wright. This is no hearsay, neither is the author favoring any particular manufacturer, but this advice is founded on experience which is for the reader's benefit, as in order to do good work a good anvil is necessary, which is perfectly free from all hollows and soft spots. . ..*

Many smiths must have taken his advice because in approximately 35 years of making anvils the Hay-Budden Manufacturing Company sold more wrought anvils in this country than all other American wrought anvil makers combined: well over 300,000. I have no idea whether the author ever used an American-made *Trenton* or an *Arm and Hammer* anvil, as both were fairly new to the market when this paragraph was written.

It had bothered me for a while to think that a company such as Hay-Budden should lose out in the anvil market while the other American wrought anvil makers continued to produce anvils until well after World War II.

Then, I made contact with Mrs. Warren Budden Sage (granddaughter of Frederick C. Budden, Sr.) and I began to understand what probably happened. Mrs. Sage is also my source for information on any incidents relative to the company's operation other than where credit is given.

One of the reasons for the company's demise was that anvils were a major part of their product line and after World War I the anvil market declined rapidly. To my knowledge no other anvil makers in this country depended so much on the sale of anvils, except maybe *Fisher* (not a wrought anvil). A blank check from the company indicates that they also specialized in steel rings without a weld joint and die and steel forgings, but it suggests that anvils were their main product.

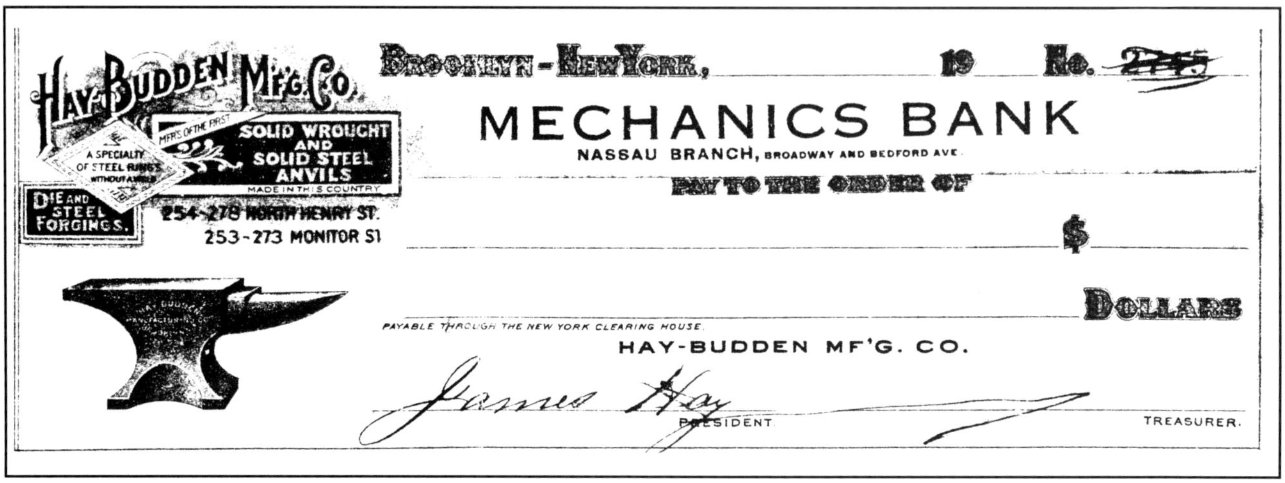

Blank check courtesy of Mrs. Florence Budden Sage.

Manufacturer's Record, April 7, 1899. A southern industrial newspaper, Baltimore, Md. Other products of Hay-Budden.

Second, what probably hurt the company most was a breakup within the company at a critical time. Internal strife has often destroyed many well-established firms. One of the partners, Mr. Hay, apparently wanted to retire and leave the business and in order to do so sold off his share of the equipment about 1925, after which the company never recovered. Advertisements in blacksmith publications indicate that they were making or still selling anvils in 1925. I was told that the company also tried to get into the auto engine block business at about this time. What really happened to the company was known only to the management of that time, but with the beginning of the Depression in 1929 the end came. The Hay-Budden Manufacturing Company was dissolved by proclamation on December 15, 1930.[17]

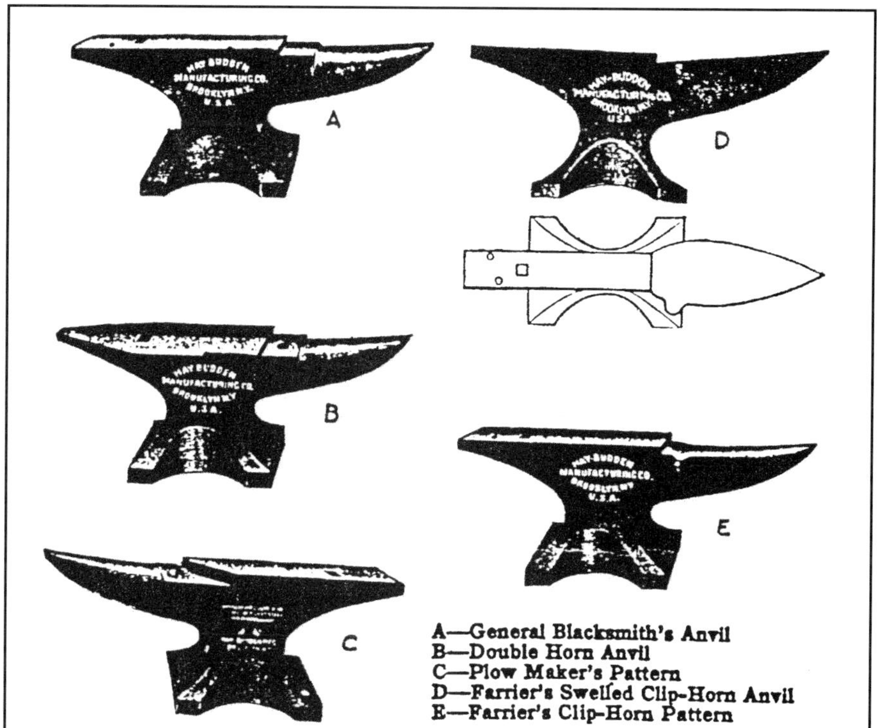

SOME OF THE VARIOUS ANVIL PATTERNS MADE BY HAY-BUDDEN CO.

The American Blacksmith, October, 1914.

What is known about the early beginnings of the company is contained in part of an article appearing in an October, 1914, issue of *The American Blacksmith* journal.[18]

> *Later, or in 1886, two enterprising young men, namely James Hay and Frederick C. Budden, both blacksmiths (Hay a Scotsman, Budden an Englishman) started in to make anvils, on Front Street, Brooklyn. These young men, under the firm name of Hay-Budden Mfg. Co., have done well. Since they commenced to do business in the early days of their anvil-making, they were beset with many difficulties, not least of which was the ingrown American prejudice against home products. Even now this prejudice in many respects remains as a survival of the times when it was thought that anything really worth having had to come from abroad. It is, however, gradually dying a natural death and everything American, with the exception of some sons-in-law, is becoming good enough for Americans.*
>
> *From the very beginning, the Hay-Budden Mfg. Co. demonstrated the fact that they*

could make anvils as well, if not in some respects better, than the old-established anvil makers of England. Needless to say, their process of manufacture is essentially that of Peter Wright, but with some improvements, one of which is the welding on of the steel faces, regardless of size, in one piece. Some years ago they still further improved their anvils by making the top, from the waist up, of one solid piece of high carbon open hearth steel. The advantage of this is at once obvious. Besides improving old methods, and devising new mechanical appliances for the manufacture of anvils, Messrs. Hay & Budden have been able to make many minor improvements on their anvils.

The difference between the standard American anvils and other makes may be noted by comparing the different groups of anvils shown in this article.

In a hardware catalog of 1895 of Charles A. Strelinger & Company, I found the following interesting item.[19]

In wrought iron anvils, the Wright, Wilkinson and Armitage 'Mousehole' are the best known; they are all English made. We believe there is but one American concern, and their output is small. Have been informed (by the house who sells them) that they are equal to the best English.

This was written before 1895 and could only be referring to the Hay-Budden Manufacturing Company, because as their advertisements maintain, they were the first successful wrought anvil maker in this country.

In the book, *Blacksmith's and Farrier's Tools at Shelburne Museum*,[20] Mr. Smith mentions a *Hay-Budden* brochure of 1905. Mr. Smith says that according to the brochure the company started manufacturing in 1890. (I don't think this is a contradiction to the 1886 date, which I will explain later). Smith also says the brochure shows the factory in 1890, 1895 and 1905 with the smoke stacks of the buildings increasing from one in 1890 to six in 1905.

I found the following statement in *The Blacksmith & Wheelwright* journal of November, 1894:[21]

The Hay-Budden Mfg. Co. has started to work in their new plant. With their improved machinery and new methods, they will be able to turn out the celebrated Hay-Budden anvils with even better satisfaction than in the past.

It seems that each time they made an improvement to their manufacturing capacity they also moved. They moved three times in Brooklyn, New York, from Front Street to 253-273 Monitor Street to 254-278 North Henry Street. The blank check shows that they maintained the Monitor Street as well as the North Henry Street address. Their advertisements use only the North Henry Street address, which undoubtedly was their main plant and final location.

Mrs. Sage says that some of the correspondence between her grandfather and relations in England indicate that the company experienced a number of fires including an especially bad one at the turn of the 20th century.

Frederick C. Budden, Sr., was born in 1856 in Southhampton, England. He and James Hay were about the same age, I am told.[22] That would have put both men in their early thirties when they commenced manufacturing. Frederick C. Budden, Jr. (Mrs. Sage's father and vice president of the company after his father), was born in the United States in 1884 and his two older sisters were born in England, so Mr. Budden, Sr., must have come to this country in the early 1880s.[10] With this information it can be presumed that Mr. Budden, Sr., would have been about twenty-five to twenty-eight years old when he arrived in America. That is a very young age to have become a master blacksmith and especially an anvil maker. I am therefore speculating that he must have served a blacksmith's apprenticeship at a very young age and learned quickly. Most likely, he and /or Mr. Hay served their apprenticeship with Peter Wright and Company, because there were no wrought anvil makers in this country and the early *Hay-Budden* anvils closely resemble the *Peter Wrights.*

I have been unable to find out anything about James Hay's background except that he was a blacksmith, was President-Treasurer of the company, and their anvil patent of 1909 was taken out in his name.

Unfortunately Mr. Budden, Sr., died in 1916. I do not know whether it is coincidental or not, but beginning at about this time the company's trade journal advertisement began to get smaller until in 1923-1925, it was a very small insert on an inside page of *The American Blacksmith and Motor Shop* journal.[24]

Blacksmith and Wheelwright, January, 1925. This appears to have been the company's last ad in this journal. There was an ad in every previous issue since the company started advertising. There was no ad in the June, 1926 issue. (From Mr. Keith Sommer)

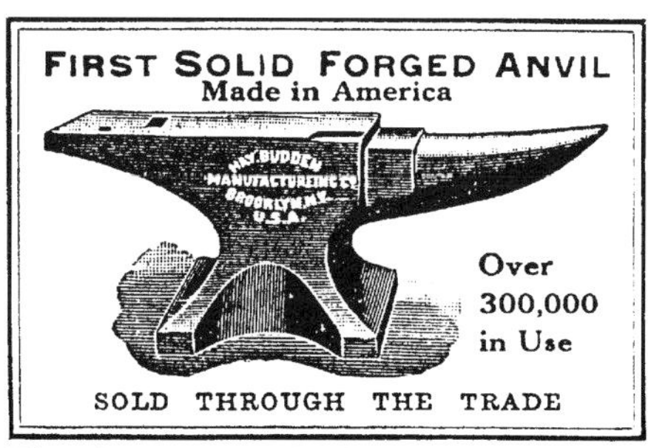

The American Blacksmith and Motor Shop, December, 1923.

If the company's advertisements are correct, between October, 1913, and December, 1923, they sold over 100,000 anvils, which would be some of the best years in their business history. One would believe that with such prosperity, the advertisements would have grown in size. Before 1916 the *Hay-Budden* advertisements would cover about one third of the outside back cover of *The American Blacksmith and Wheelwright.* It is peculiar that at the same time that Hay-Budden's advertisements were becoming smaller the

Columbus Forge and Iron Company's *(Trenton)* advertisements were increasing in size though anvils were a smaller part of their total product line. In any event smiths were deprived of a quality anvil all too soon.

No records survive of the company's operation to my knowledge; therefore, all the data I will be giving will come from the company's advertising and from their anvils which I have recorded or have personal knowledge of.

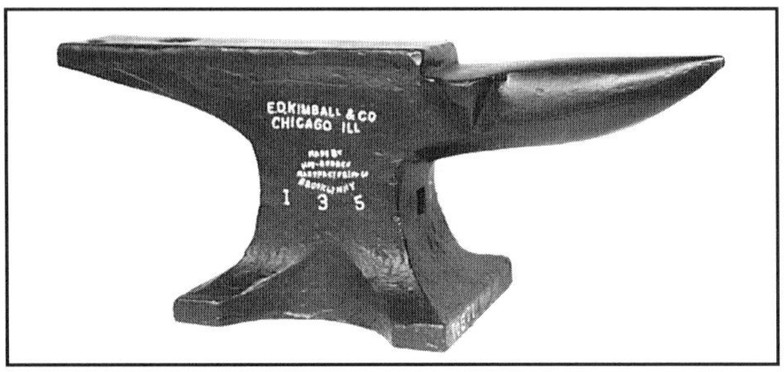

135-pound Hay-Budden anvil made for E. D. Kimball and Co., Chicago, Il. dated ca. 1898. Note the trademark just above the weight. (R.A.P. photo)

The Hay-Budden Manufacturing Company made anvils for many large and also smaller hardware companies. These anvils were marked with the hardware company's logo or trademark and not Hay-Budden's, although I discovered several brands with a small Hay-Budden trademark below the hardware company's. However, there are certain characteristics that distinguish a *Hay-Budden* from other makes regardless of whose name is on it. I will discuss these distinctions later.

S. D. Kimbark of Chicago was one such hardware company and was probably very much responsible for Hay-Budden's early success. The following quote is about this company from *The Blacksmith and Wheelwright* journal of April, 1898.[25]

> *Previous to the year 1882, there was not known in the trade a distinctive horseshoer's anvil. Out of the largest stocks, it was difficult to select an anvil suited to the work of turning and fitting shoes [S.D. Kimbark was the originator of the farrier's anvil]. For ten years, it [the farrier's anvil] was made in Germany, but since 1892, it has been made in this country, and all its special features have been constantly retained...*

144-pound Hay-Budden anvil made for S. D. Kimbark, Chicago, Il. dated ca. 1893. (R.A.P. photo)

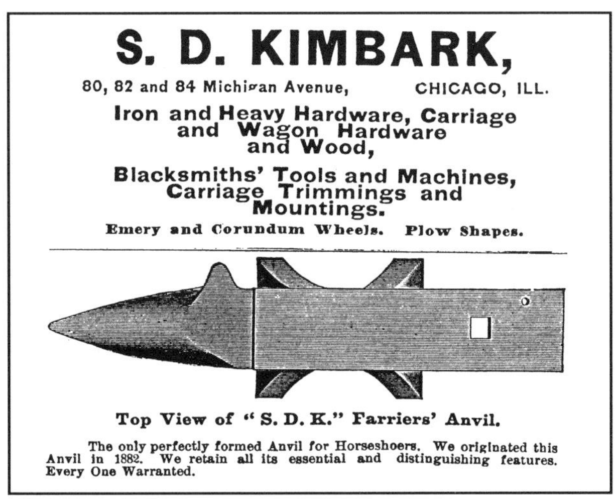

The Blacksmith and Wheelwright, July, 1898.

I have three of these Kimbark anvils which I have dated 1892 or 1893. The distinctive *Hay-Budden* characteristics leave no doubt that these anvils were made by the Hay-Budden Manufacturing Company.

Therefore in 1892, S. D. Kimbark, one of the largest blacksmith supply houses in the country, gave Hay-Budden Manufacturing Company a very large order. That probably accounts in part for their expansion and why they incorporated October 4, 1893.[26] The first farrier's anvils made in this country came from the Hay-Budden Manufacturing Company.

Following are some of the trademarks for hardware companies for whom Hay-Budden Manufacturing Company. made anvils:

Acme, S/N 14?58?, (*ca.* 1908); one of Sears, Roebuck & Company's trademarks.

American Wrought, Horse Shoe Brand, (Montgomery Ward catalog, 1895) S/N 932?, (*ca.*1894).

B. I. C., Omaha. S/N 43818, (1907).

Columbia, H. Channon Company, Chicago, Illinois. (1908) The ad indicates by its content that it was made by Hay-Budden.

E.D. Kimball & Co., Chicago. S/N 14218, (*ca.* 1894).

Hercules, S/N 61539, (*ca.* 1901); unknown (a western hardware company).

Holliday, S/N 13867, (*ca.* 1895); name of an Indianapolis, Indiana hardware company.

Hudson, (Charles Williams Store, N.Y.) S/N 200777, (*ca.*1913); (This is the highest serial number that I have recorded with a trademark other than Hay-Budden's).

J.E. Pilcher, Blackjack, St. Louis, S/N 48023, (*ca.* 1900); Simmons Hardware Company, St Louis, Missouri.

John Henry, S/N 4637, (*ca.* 1894); Dubuque, Iowa, hardware company.

Johnson, Hardware, This name was below the *Hay-Budden* trademark.

Keen Cutter, Blackjack, St. Louis, S/N 210796, (*ca.* 1914); Simmons Hardware Company, St. Louis, Missouri.

Keen Kutter, Blackjack, St.Louis, S/N ?, St. Louis, Missouri.

King Hardware, S/N 58???, (*ca.* 1901); name of an Atlanta, Georgia, hardware company.

Lakeside, S/N 143428, (*ca.* 1908); one of Montgomery Ward's trademarks.

R & H; *S.F. Our Best*, a San Francisco hardware company. Hay-Budden trademark on the other side.

Raola, H & I Co., S/N 10288, (*ca.* 1894) ; Indianapolis, Indiana, hardware company.

S.D. Kimbark, S/N 3373, (*ca.* 1893); Chicago, Illinois.

SLIGO, S/N 20969, (*ca.*1895); St. Louis, Missouri.

S W I C O, 150 lbs. S/N 5214, (*ca.*1893).

U.S. (S.D. Kimbark) S/N 2306, (*ca.* 1892); Chicago, Illinois.

166 -pound Hay-Budden anvil made for S.D. Kimbark, Chicago, Il. (R.A.P. photo)

I am sure they made anvils for other companies also, but these are the brand names I have recorded as of 1996.

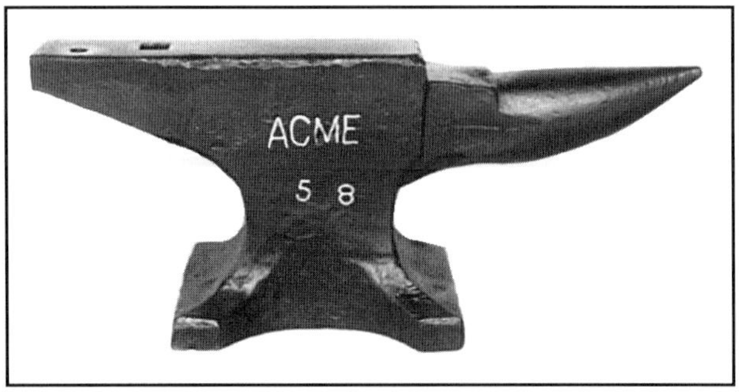

58-pound Hay-Budden Acme anvil. (R.A.P. photo)

84-pound Hay-Budden anvil made for Montgomery Ward. (R.A.P. photo)

Wrought Anvils.

43150 The American Wrought "Horse Shoe" brand Anvil, *fully warranted*, made of the best American wrought iron, and faced with the best crucible steel, are guaranteed superior to all other makes, to be sound, free from flaws and to have faces hard and true. Try one, if not satisfactory return at our expense. Any weight from 85 lbs. and upward. Per pound..........................$0.11

43151 The American Wrought Horse Shoe Brand Anvil up to 85 lbs. Per pound..........$0.11¼

43152 PeterWright's Blacksmiths' Wrought Iron Anvil, steel face. Price, per pound, from 85 lbs. and upwards...$0.11¼

43153 PeterWrights Celebrated Blacksmiths' Wrought Iron Anvil, steel face. Price per pound, from 60 to 85 pounds........$0.12¼
Price per pound from 50 to 60 pounds...........13
Cannot always furnish exact weights wanted in anvils. Make second or third choice in weights when ordering

Cast Iron Anvils.

43154 With steel face, is only intended for light work and is not guaranteed, as we do not take back or exchange if they prove defective.

No.........	2	3	4	5	6	7	8	9
Weight....	20	30	40	50	60	70	80	90 lbs.
Each....	$2.70	3.04	3.54	4.05	4.39	4.82	5.40	6.08

Montgomery Ward & Co. catalog, 1894-95. (From Mr. Hunter Pilkinton)

COLUMBIA ANVILS

Solid Wrought. Crucible Steel Face. Warranted.

Our Columbia Anvils are made of all wrought iron doubly refined and consequently are **Solid Forgings.**

The Crucible Steel on face, which is **the very best** that can be bought for the purpose, is welded to the wrought iron. They are the best anvils on the market, none excepted.

General Blacksmiths' Pattern

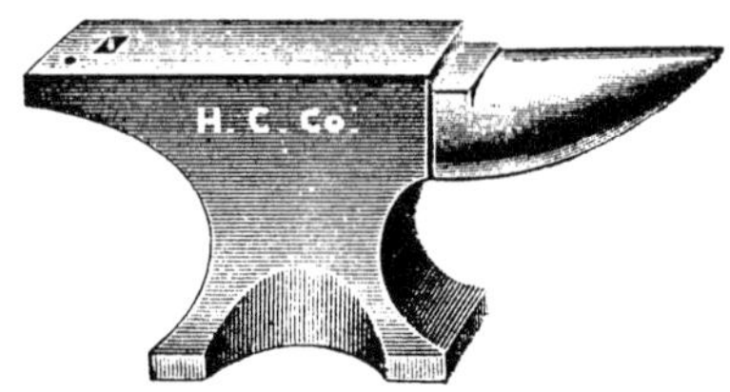

PRICES:

80 to 400 lbs.		base per lb......
70 to 79	½	cts. per lb. Extra
60 to 69	1	" " "
50 to 59	2	" " "
40 to 49	3	" " "
30 to 39	5	" " "
20 to 29	8	" " "
10 to 19	15	" " "

Size of Face to Weight of our Columbia Anvils.

Face	Weight	Hardy Hole
3½ in. x 12½ in.	80 to 90 lbs.	⅞ in.
3½ " x 13½ "	100 to 110 "	⅞ "
3¾ " x 15 "	125 to 135 "	1 "
4 " x 15½ "	140 to 145 "	1$^{1}/_{16}$ "
4 " x 16 "	150 to 160 "	1$^{1}/_{16}$ "
4¼ " x 16½ "	170 to 180 "	1⅛ "
4¼ " x 17 "	190 to 200 "	1⅛ "
4½ " x 18 "	250 "	1⅛ "
5 " x 20 "	300 "	1¼ "
5½ " x 22 "	350 "	1⅜ "
6 " x 24 "	400 "	1⅜ "

Punch hole is ½ inch diameter in anvils with 4 inch face and smaller. 9-16 inch in larger sizes.

COLUMBIA FARRIERS' CLIP HORN ANVILS

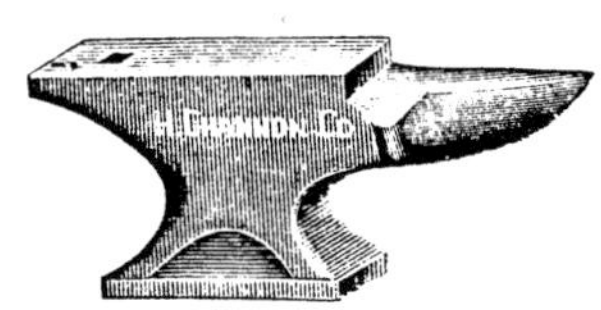

Made from 80 to 400 lbs. Generally with 3½ and 3¾ face. Drop on step from ⅞ to 1 inch.
Add 1 cent per lb. to Blacksmith Pattern prices.

H. Channon Company catalog, 1908, Chicago, Il.

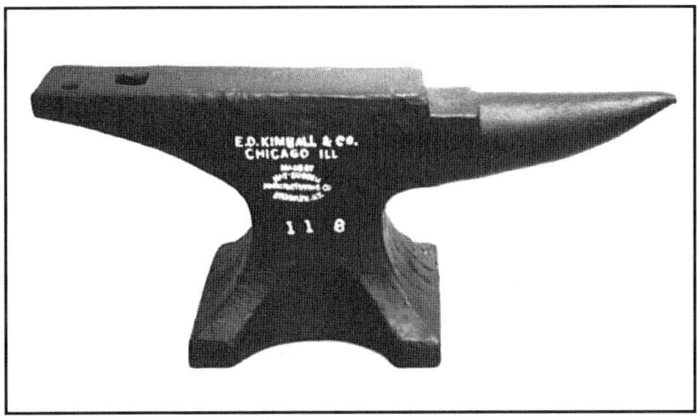

118-pound Hay-Budden anvil made for E. D. Kimball & Co., Chicago, Il. (R.A.P. photo)

123-pound Hay-Budden anvil made for Holliday Hardware Co. Indianapolis, In. (R.A.P. photo)

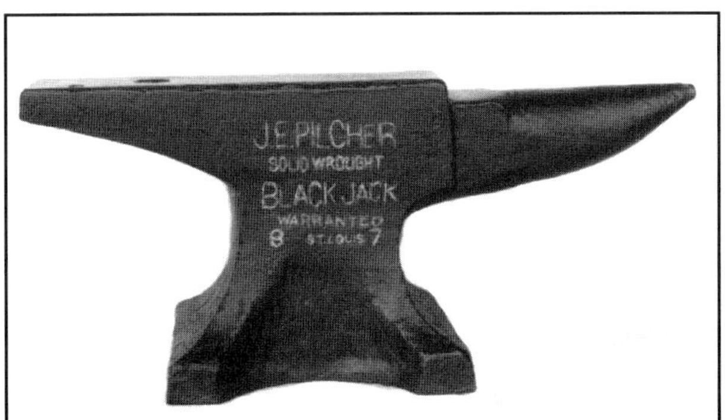

87-pound Hay-Budden anvil made for Simmons Hardware Co., St. Louis, Mo. (R.A.P. photo)

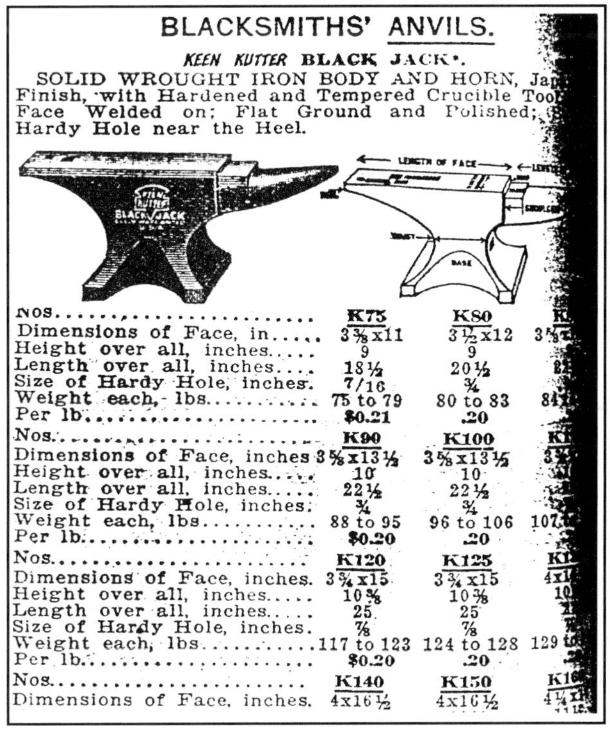

BLACKSMITHS' ANVILS.

KEEN KUTTER **BLACK JACK***.

SOLID WROUGHT IRON BODY AND HORN, Jap Finish, with Hardened and Tempered Crucible Tool Face Welded on; Flat Ground and Polished; Hardy Hole near the Heel.

Nos.	K75	K80
Dimensions of Face, in.	3⅜x11	3½x12
Height over all, inches	9	9
Length over all, inches	18½	20½
Size of Hardy Hole, inches	7/16	¾
Weight each, lbs	75 to 79	80 to 83
Per lb	$0.21	.20
Nos.	**K90**	**K100**
Dimensions of Face, inches	3⅝x13½	3⅝x13½
Height over all, inches	10	10
Length over all, inches	22½	22½
Size of Hardy Hole, inches	¾	¾
Weight each, lbs	88 to 95	96 to 106
Per lb	$0.20	.20
Nos.	**K120**	**K125**
Dimensions of Face, inches	3¾x15	3¾x15
Height over all, inches	10⅜	10⅜
Length over all, inches	25	25
Size of Hardy Hole, inches	⅞	⅞
Weight each, lbs	117 to 123	124 to 128
Per lb	$0.20	.20
Nos.	**K140**	**K150**
Dimensions of Face, inches	4x16½	4x16½

Simmons Hardware Co. catalog, 1913, St. Louis, Mo. (From Mr. David Poppke)

126-pound Hay-Budden anvil made for Simmons Hardware Co., St. Louis, Mo. (R.A.P. photo)

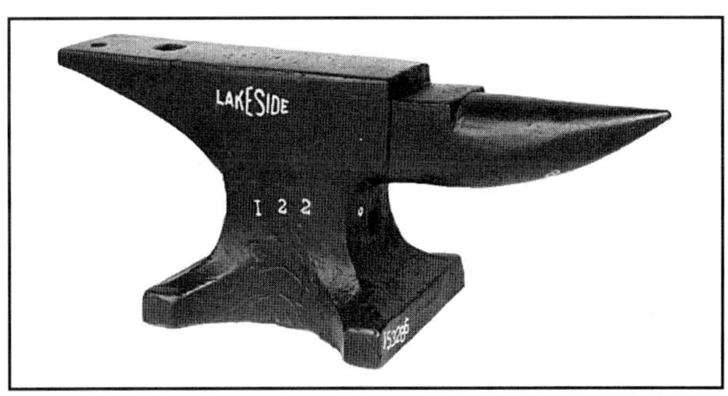

122-pound Hay-Budden anvil made for Montgomery Ward & Co. (R.A.P. photo)

The majority of anvils made by Hay-Budden Manufacturing Company for various hardware companies were produced around the turn of the 20th century or before. After the Hay-Budden name became synonymous with quality, these hardware companies sold anvils under the maker's name rather than their own, with the exception of Sears Roebuck and Company and Montgomery Ward and Company. The Hay-Budden trademark had become one of excellence and there was less advantage to using the hardware company trademarks.

The *Hay-Budden* general purpose or smith's pattern anvil went through two minor and one major body style change.

The first style, and I think the best proportioned, looked very much like the other better wrought anvils made at that time, especially *Peter Wright*, except that the *Hay-Budden* had a thinner waist than the *Peter Wright*.

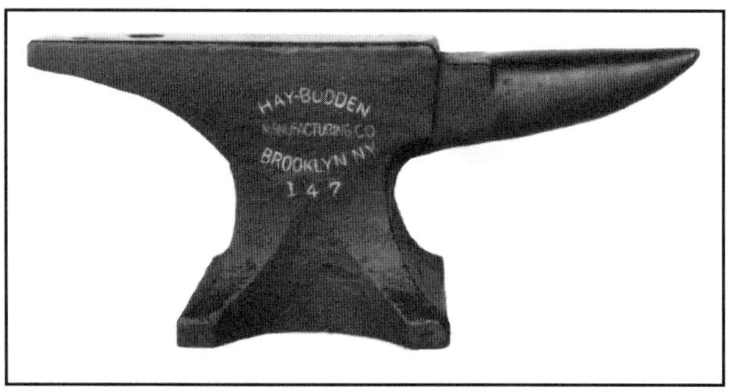

147-pound early Hay-Budden anvil, ca. 1895. (R.A.P. photo)

The second style had the upper body thickened toward the anvil's face which gives it a top heavy appearance when viewed from the side. This style must have been made for several years, considering the number to be found.

81-pound Hay-Budden anvil of the second style, made ca. 1900 for the King Hardware Co., Atlanta, Ga. (R.A.P. photo taken at Mr. Floyd Daniel's)

The third and major style change came about in 1908 when the method of making the top half of the anvil was completely changed. Instead of welding a tool-steel face plate on to the wrought iron body, the entire anvil from the waist up was made of one piece of open hearth tool steel. In my view this last change, as far as the style of the anvil is concerned, was the least aesthetic of all the general purpose anvils they made. The waist is almost straight instead of a continuous curve and the horn starts out straight from the body then makes a sharp angle to the point, when the anvil is viewed from the side.

128-pound Hay-Budden anvil of the third style, made ca. 1919. (R.A.P. photo)

In 1908 the Hay-Budden Manufacturing Company started experimenting with a process in which the entire anvil could be forged from one piece of steel or the upper half of one piece. The following article appeared in *The American Blacksmith* journal for October, 1909.[27]

A NEW PROCESS IN ANVIL MANUFACTURING

> *There are today three distinct classes of Anvils on the market, the Forged anvil, the Cast Iron anvil, and the Cast Steel Anvil. Most of the Wrought Anvils have heretofore been made by welding a solid forged iron top to a forged bottom, and welding a steel face plate to the same. This is the method that has been used by the leading Wrought Anvil manufacturers for years, but the Hay-Budden Manufacturing Co. have, during the past year, made up most of their anvils in a manner which does away with the welding on of the steel face plate.*
>
> *After considerable experimenting they have succeeded in obtaining for the entire top of their Anvil, a steel of such analysis that will take a satisfactory temper and yet be sufficiently strong and tough to stand the usage for which the tool is intended.*
>
> *This steel top is forged in such a manner as to impart additional strength and toughness to the material. By this method they overcome what has been a weakness in the Wrought Anvil, that of loose faces, for by this process it is obvious that there cannot be any loose faces. By a process of their own the steel top's welded to the forged base in such a manner as to insure a lasting union at that point.*
>
> *An earlier announcement of this step forward in their method of manufacture had not been made for the reason that they desired a number of these Anvils to be placed in actual use and a working test made, and they report that in the past 14 months over 10,000 of these New Process Anvils have been shipped out with entirely satisfactory results. Their advertisement will be found on the rear cover.*

I suggest that this new process was stimulated at least in part by three influences. First, other new anvil manufacturers were coming into the market with less expensive methods of making high quality wrought anvils. One of these was a direct competitor in this country, the Columbus Forge and Iron Company *(Trenton)*. They lowered the price of high-quality anvils. Second, the Swedish makers had turned to a solid cast-steel anvil of good quality and more of these were coming into the country from several Swedish manufacturers. The Columbian Hardware Company of Cleveland, Ohio, also had started making a one-piece anvil of cast steel.

Although the material used in a solid cast-steel anvil is more expensive, the expense is more than offset by the reduction of the labor involved in making the anvil with a steel face plate. Wrought anvil making was labor-intensive, using skilled craftsmen. Although cast anvils are not considered as high in quality as wrought anvils, time has proven that these cast-steel anvils have stood up as well as the wrought ones. Most of the Swedish and *Columbian* anvils I have recorded have been in good to excellent condition although they are not near as numerous as the wrought anvils. They also sold for a little less money than wrought anvils in most cases.

Third, in 1907 a Mr. Walter Ring (American Wrought Anvil Company) filed an application with the U. S. Patent Office for a method to forge a one-piece anvil. This patent was granted August 18, 1908 (see American). In October of 1908, Mr. James Hay filed an application for their new process with the U.S. Patent Office and it was granted on May 26, 1909. Here is the interesting part. In the Hay patent, James Hay refers to the older Ring patent, suggesting that Ring's method was impractical, expensive and in all probability would not work, whereas, the Hay patent would. I have not read a great number of patents, but I have read all those I could find that dealt with anvils and anvil-making and I have yet to discover a patent in which one patentee attacked another. This would indicate to me that there was very close competition. As far as I know, neither proceeded to produce a one-piece forged anvil that was commercially successful as stated in the patents. Both patents required too many expensive dies and too much power. If anvils of only one or two-sizes were to be made, the method might have been successful, but hundreds of different patterns and sizes were required for the market. However, Hay-Budden could not let a local competitor get a jump on them. I am told by Mr. John Lemoine that he has a *Hay-Budden* anvil forged of one piece of steel. Mr. Lemoine sent me a photo of this anvil of 100 pounds, but it was not good enough to reproduce for this book.

UNITED STATES PATENT OFFICE[28]

James Hay, Of New York, N. Y.

MANUFACTURE OF ANVILS

No. 922, 475 **Specification of Letters Patent** **Patented May 25, 1909.**

Application filed October 12, 1908. Serial No. 457,222

To all whom it may concern:

Be it known that I, James Hay, a citizen of the United States, residing at Richmond Hill, in the borough and county of Queens, city and State of New York, have invented certain new and useful Improvements in the Manufacture of Anvils, of which the following is a specification.

The common practice of making anvils is to make them in two or more pieces welded together. For example, a common type consists of a steel top-plate hardened to a high temper, a horn and body of wrought iron, and a base of wrought iron all welded together. According to this invention substantially the same relative hardness and toughness of the parts is preserved, while a great economy is effected by forming the body, together with the part corresponding to the top-plate, and preferably also the base, in a single intregal mass of steel. The hardness of the top is then secured by heating and quenching. The shape of the anvil is preferably secured by pressing a block to substantially or approximately the desired shape in a press mold, the block being of course heated to a suitable temperature.

Other features of improvement are referred to in detail hereinafter.

The accompanying drawings illustrate the method.

Figure 1 is a cross-section through the mold, indicating the plungers and anvil in elevation. Fig. 2 is an end view of the rear plunger. Fig. 3 is a section at right angles to Fig. 1, showing the plungers in end elevation. Fig. 4 is an end view of the forward plunger. Fig. 5 is a side elevation of the finished anvil.

Referring to the apparatus illustrated, the die is formed of a pair of sides A -A which are separable from each other along the longitudinal plane passing through the longitudinal axis of the head and the corresponding axis of the base of the anvil, as indicated by the dotted line B -B of Fig. 1. The rear part of the anvil is formed by a plunger C reciprocated by hydraulic or other suitable means, and the front part is formed by a corresponding reciprocating plunger C. Any suitable mechanism may be utilized for securing the relative movements of the plungers and the die, and the opening and closing of the die. *The plunger C forms the horn and the forward faces of the body and base of the anvil. The rear end of the top of the anvil and the rear faces of the body and base are formed by the plunger C. The sides of the mold A form the longitudinal sides of the anvil, and the top curved face of the horn being provided for the latter purpose with projections indicated in dotted lines at A in Fig. 3. The A corresponds with the extreme lateral edge of the horn, as shown in Fig. 1. This arrangement permits the easy entrance and withdrawal of the forward plunger C.*

The block is of steel of a composition best adapted to admit of the forging operation involved and at the same time to permit a subsequent hardening of the top. For example, steel of 0.70 per cent. carbon will be found suitable.

By the process described the entire anvil can be produced in a small fraction of the time necessary for its production by previous methods, and the subsequent operations hereinafter referred to involve only trifling expense. Upon the completed anvil there is a savings of from 40 to 50 percent, in cost, and there is an improvement in the anvil in its entire freedom from joints.

The finished anvil is shown in Fig. 5. The body E and base F are entirely completed in the mold (except for the removal of fins caused by the division lines of the mold). The top G of the anvil is hardened, preferably after withdrawing from the mold, by heating and quenching in water or oil in any usual or suitable way. The point H of the horn J is preferably also drawn out and hardened somewhat after removal from the mold.

Where it is desired to have a base of iron, the process above described may be applied to the making of the body only, the latter being then attached to the base by welding or any other desired manner.

This invention is designed particularly as an improvement upon the invention shown in the Patent of Ring No.896,155 of August 18, 1908, in which it is proposed to clamp a T-shaped blank between walls corresponding to the sides and ends of the anvil and to form the latter by means of plungers pressing upon the top and bottom of the anvil. In attempting to make anvils of large size by the Ring process the area of work engaged by the ends of the plungers, and especially the plunger engaging the top of the anvil is very great so that the total pressure required for the operation is enormous. Regardless of the size of the anvil to be made, the making of the T-shaped blank from which the anvil is afterward made is an item of heavy expense, heavier in fact than that of making the anvil from the blank. By arranging the plungers as in my invention so that they engage the ends of the work, the fixed mould being arranged to engage the sides and top and bottom, the plungers require to have a longer stroke but the area of work which they engage is very much less than at the sides or top or bottom of the anvil, and consequently the total pressure required for the plunger is within a practicable limit even for larger size anvils. Furthermore, this arrangement has the great advantage from the point of view of economy of operating with blanks which are simple blocks of steel cut from billets of appropriate cross-section, thus avoiding the necessity of shaping an intermediate blank as in the prior process above referred to.

What I claim is:

The method of manufacture of an anvil which consists in providing a mold corresponding with the shape of the top, bottom and longitudinal sides of the anvil, introducing into said mold a block of steel and forcing said block into the shape of the mold and forming the opposite end faces of the anvil by an endwise pressure from suitably shaped members engaging the opposite end faces of the work.

JAMES HAY

Witnesses:

Domingo A. Usina,

Theodore T. Snell.

Patent and patent drawing Courtesy of the Smithsonian Institution.

The American Blacksmith, November, 1909. The first ad for the solid top anvil.

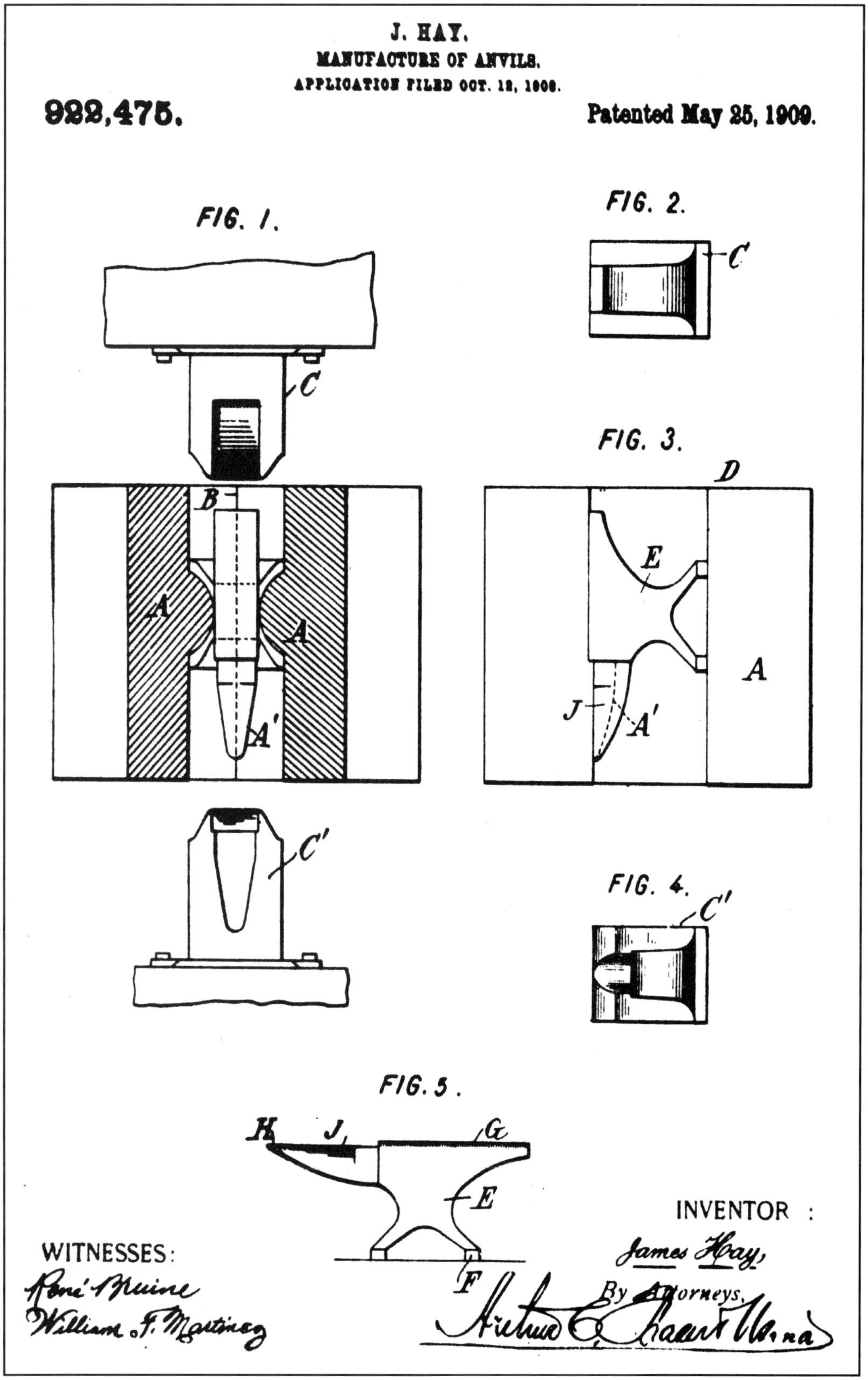
J. HAY.
MANUFACTURE OF ANVILS.
APPLICATION FILED OCT. 12, 1908.
922,475.
Patented May 25, 1909.
FIG. 1.
FIG. 2.
FIG. 3.
FIG. 4.
FIG. 5.
WITNESSES:
INVENTOR:
James Hay,
By Attorneys,

What came from the Hay patent was a two-piece anvil. The base was forged from wrought iron or low- carbon steel and blocked to size in a die. The top (body and face) was forged on open dies and then the top and base were forge welded together, thus producing the first plateless wrought anvil. With this process, Hay-Budden Manufacturing Company most likely put the American Wrought Anvil Company out of the anvil business, or they bought them out. The shape from the side of these new *Hay-Budden* anvils made after 1908 is very distinctive from all other anvil makes and if you are familiar with anvils will recognize one at a glance.

Identification: The *Hay-Budden* anvil is the easiest to identify except for their earlier ones. The base of the early ones is the most revealing characteristic because in most cases, the trademark or name is very poor or only a few letters are still visible even after cleaning. It may also have a different trademark or name. The forged base of the early ones was made up of slabs of wrought iron plate or scrap forge welded together. When wire-brushed or sometimes without cleaning, the lines where these plates join is visible in most cases and follows the contour of the base when viewed from the side. The bottom of the base is essentially flat with a slight rim about one-half inch wide extending down about one-sixteenth of an inch around the perimeter. This rim keeps the center part from touching the mounting block, thereby making the anvil stable and less likely to rock on its mounting when in use. To my knowledge no other anvil maker made the bottoms of the base like this. I have found a few of these older anvils with the rim almost worn away.

These early *Hay-Budden* anvils also have a serial number on the front of the foot under the horn. Corrosion tends to take place at this point, so it is often hard to make out the numbers even after cleaning. I don't believe the company started putting serial numbers on the anvils until after they incorporated. I will speak more about serial numbers later.

The weight of the anvil in pounds (they never used the English weight system) was stamped on the same side as the trademark. The weight is sometimes hard to find because the numbers were usually stamped in lightly or they were obliterated with the use of the anvil.

The line where the face plate joins the body is easy to discern on the *Hay-Budden* although this line is not as clear as it is on the other American wrought anvils. American manufacturers did not try to hide this line like the English makers did, especially Peter Wright. Any *Hay-Budden* anvil with a plate can be said to have been made before 1909 unless it was a later model that has been refaced.

Most of the early *Hay-Budden* anvils have three handling holes, one in the bottom and one on either end of the waist under the horn and heel. The one under the heel usually tends to be smaller than the other two. I have not found an anvil that I have identified as a *Hay-Budden* to have a hole in the front of the foot under the horn like the *Peter Wright*. However, anvils over 300 pounds usually had a handling hole in the front of the foot under the heel with or without one in the waist under the heel. There can be exceptions to the number of handling holes, considering that these anvils were essentially custom-made.

Most of the *Hay-Budden* anvils made before 1909 have a number stamped on the front of the waist under the horn to the left or right of the handling hole. This number was sometimes stamped on both sides of the hole. The numbers ranged from 0 to 9 and were usually stamped in quite deep. This number was probably an inspector's code mark. The earliest anvils I have inspected do not have these numbers. This number on the waist is a sure identification mark when all other markings are missing. The nice part about the waist mark is that it is located where the anvil is never used; only rust, paint and crud can obscure it. With a little wire brushing, it can almost always be found if there is a number there.

The *Hay-Budden* anvils made after 1908 are marked similarly to the earlier models. They use the same trademark, weight markings and continue the serial numbers. It is rare to find one without the inspector's number on the front of the waist. To my knowledge, only the Hay-Budden Manufacturing Company used this number on the waist.

These later plateless anvils had, in addition, numbers stamped on the opposite side of the body from the trademark and weight. The ones I have recorded have these numbers up close to the face and toward the heel. These numbers are usually two or three digits and are random; that is, not consecutive in relationship to the serial numbers or to the time of manufacture. Here are a few of them taken from anvils I have recorded: **05, 101, 68, 92, 094, 089** and **046**. I believe that they are steel heat numbers, indicating a steel mill run of a particular lot of steel. Should a group of anvils made from steel with the same heat numbers prove to be defective, the Hay-Budden Manufacturing Company could trace the steel used back to the particular steel mill or supplier. I suggest that this is true because I have recorded two anvils with the same number, **094**, and the serial numbers are close, **A31487** and **A31587**. On a few of the *Hay-Budden* anvils, both old and new models, I have found other letters or numbers put on at the factory. I found the letters **"BB"** stamped to the right of the serial number and have found numbers of one digit stamped under the horn. I have no explanation for them. As I said before, wrought anvils are essentially custom-made and markings could have been put on at the factory for any number of reasons, but these are the exception and can not be used for identification. Some of the Swedish anvils have letters and numbers in various places all over the anvil body and base, put on at the factory. I have also found letters and parts of names stamped here and there on *Hay-Budden* anvils. Factory markings are generally put on hot, except the weight, and these other markings which were put on cold with hand stamps are usually not very straight and have rather sharp edges. They are put on either by anvil owners for inventory purposes or to identify ownership. Some marks seem to have been stamped just to test out a metal stamp.

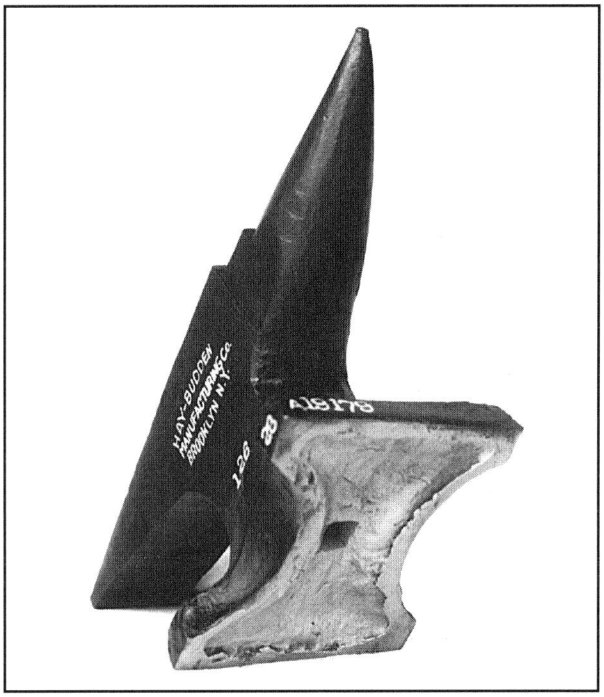

Under base. (R.A.P. photo)

The base of those anvils manufactured after 1908 appear to be made of one piece of wrought iron or low- carbon steel. The underside of the base is hollowed out as much as two inches deep toward the center, with about a one-half-inch-wide flat rim. This depression lightened the base, ensured the anvil stood steady, and was a savings in metal without any loss of quality. To my knowledge and according to their advertisements, these bases were forged. Because anvils sold by the pound in those days, the loss of weight would also be a saving to the customer if the price per pound remained fairly constant.

After 1908, most of the *Hay-Budden* anvils had only two handling holes, one under the base and one under the horn in the waist. A small one (57 lbs.) that I recorded had only one, and it was in the waist under the horn. The larger ones (300+ lbs.), besides the two holes, usually had one in the face of the foot under the heel and another under the heel in the waist like the older models did. Some of the last ones made may not have had any handling holes at all, but I have never recorded or heard of any of these. The other American wrought anvil makers dropped the handling holes by the 1920s or before, using clamps and tongs to manipulate the hot anvils.

I suggest that sometime between 1915 and 1917, the serial number system in use was dropped and Hay-Budden Manufacturing Company started all over again from number one, but prefixed the numbers with an "A". The last anvil I have recorded under the old system is serial number **218321** and the lowest one I have recorded under the new system is serial number **A1661**. This change may have taken place in 1916 when Mr. Budden, Sr., died and the management changed somewhat. One source has informed me that the old system stopped at 250,000, but I cannot verify this. The same source also told me that a "B" was prefixed to the serial number in very late models, but again, I cannot substantiate this information at this writing.[29] There is one other difference in those anvils with the "A" prefix. The number on the front of the waist under the horn on all I have recorded is always a "4". Sometimes the "4" is upside down. Maybe by that time the company was using only one inspector.

Hay-Budden's advertisements indicate that they made anvils weighing from 10 to 800 pounds. The smallest I have recorded weighed 26 pounds and the largest weighed 685 pounds. The small one was made after 1908, but I could not see the serial number because of red paint. The large one was made about 1906. I hope that a 800-pound anvil is still out there somewhere waiting to be recorded, and possibly a 900-pound one. Their most popular-size anvil was approximately 200 pounds.

LIST OF ANVIL SIZES AND WEIGHTS[30]

Weight Lbs.	Face Inches	Horn Inches	Hardie Hole	Pritchel Hole
10	2 x 6	3 1/2	1/2	5 /16
20	2 1/4 x 7	4	5/8	3/8
30	21/2 x 8	5	5/8	3/8
40	2 3/4 x 9	6	3/8	7/16
50	3 x 9 1/2	6 1/2	3/4	7/16
60	3 1/4 x 10	7	3/4	7/16
70	3 3/8 x 11	7 1/2	3/4	7/16
80	3 1/2 x 12	8 1/2	3/4	7/16
100	3 5/8 x 13 1/2	9	3/4	7/16

Weight Lbs.	Face Inches	Horn Inches	Hardie Hole	Pritchel Hole
125	3 3/4 x 15	10	7/8	9/16
150	4 x 16 1/2	10 1/2	7/8	9/16
175	4 1/4 x 17	11	1	9/16
200	4 1/2 x 18	11 1/2	1 1/8	11/16
250	4 3/4 x 20	12 1/2	1 1/8	11/16
300	5 x 21	13	1 1/4	11/16
350	5 1/2 x 22	13 1/2	1 1/4	11/16
400	6 x 23	14 1/21	1 3/8	11/16
450	6 1/2 x 24	15 1/2	1 3/8	11/16
500	6 3/4 x 25	16	1 1/2	13/16
600	7 x 26	17	1 1/2	13/16
700	7 1/2 x 27	18	1 3/4	7/8
800	8 x 28	19	1 3/4	7/8

Hay-Budden Manufacturing Company also either made or had a miniature paper weight anvil made weighing about a pound, as did the other major manufacturers. The trademark on it is just like the one on the large anvils. I know of a number of these as of this writing (1996).

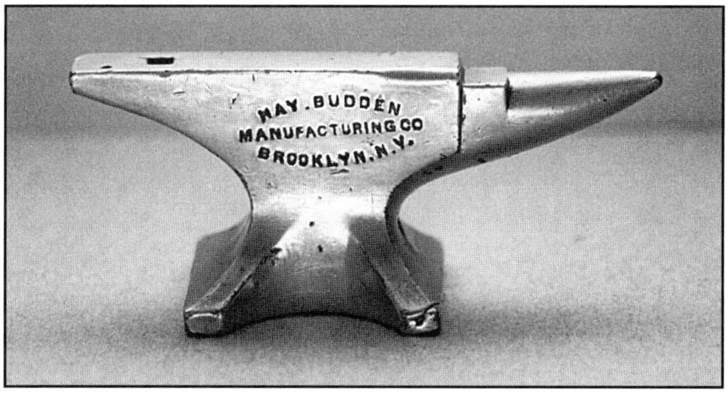

Miniature advertising Hay-Budden anvil.
(Mr. Gill Fahrenwald photo)

The Hay-Budden Mannufacturing Company made mostly general blacksmith and farrier's pattern anvils. Their farrier's anvil dominated the market before 1925. They advertised plowmaker's, double-horned (European style), hornless, sawmaker's, instrument maker's, chainmaker's, axe maker's and cooper's anvils or any pattern the customer desired, "made on short notice."[31] The only *Hay-Budden* patterns I have seen are the general purpose, plowmaker's and farrier's anvils. I have a cooper's anvil that looks like the one they made, but I cannot find any writing on it. To my knowledge, Hay-Budden is the only American anvil manufacturer who advertised a cooper's anvil (a cooper makes wood barrels and such).

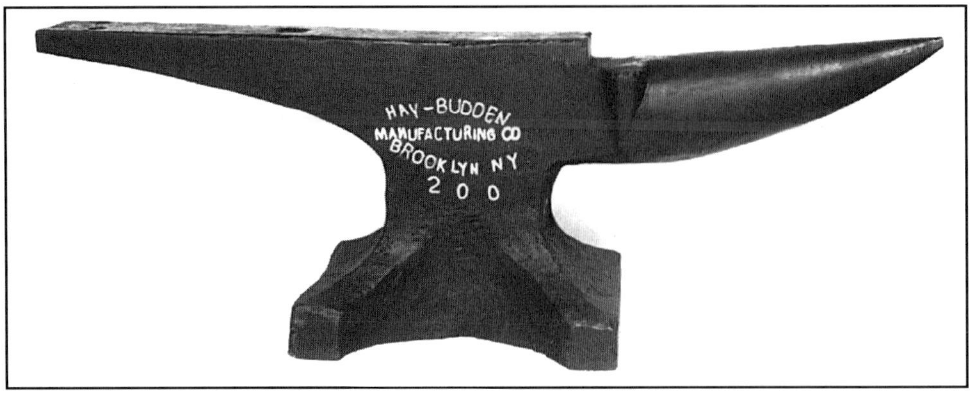

200-pound Hay-Budden farrier's pattern anvil. Mr. Mike McGee's anvil. (R.A.P. photo)

160-pound Hay-Budden plowmaker's pattern anvil. Mr. Ken Wilson's anvil. (R.A.P. photo)

I have no knowledge of Hay-Budden Manufacturing Company ever having made a blacksmith's vise, although many of the other anvil makers did.

Fortunately, Hay-Budden's advertising gives some indication of the number of anvils they made and when they made them. They used a system similar to McDonald's restaurants. The first I have found in *The Blacksmith and Wheelwright* journal is July, 1901, saying that over 60,000 anvils were in use. In September of 1902, over 70,000 and in December, over 75,000. After 1910 the increments listed in the ads jumped by 50,000. They list making over 100,000 in 1905, over 150,000 in 1909, over 200,000 in 1913, over 250,000 in 1920 and over 300,000 by July, 1922. With this information and the serial number of the anvils themselves, I can arrive at the approximate date of an anvil's manufacture. The lowest serial number I have recorded is **2127**, which I assigned to the year 1893 and the highest of the old system I have recorded is **218321**. I assigned the last one the date of 1914, knowing that was when this anvil was purchased. I have approximated the date of the anvils with the "A" prefix primarily because I have not recorded enough of them (1991). It is fairly reasonable to assume that they were all made after 1916.

The Blacksmith and Wheelwright, July, 1901. From the back cover of the journal.

The following list of anvil serial numbers gives the approximate date within two years of their manufacture. With future evidence, the dates may be changed.

Serial Number	Date	Serial Number	Date
1-3,000	1892	160,001-170,000	1909
3,001-9,000	1893	170,001-180,000	1910
9,001-16,000	1894	180,001-190,000	1911
16,001-23,000	1895	190,001-200,000	1912
23,001-30,000	1896	200,001-210,000	1913
30,001-37,000	1897	210,001-220,000	1914
37,001-42,000	1898	220,001-230,000	1915
42,001-50,000	1899	230,001-243,000	1916
50,001-60,000	1900	243,001-250,000	1917
60,001-75,000	1901	A1-A12,000	1918
75,001-82,000	1902	A12,001-A20,000	1919
82,001-89,000	1903	A20,001-A30,000	1920
89,001-96,000	1904	A30,001-A40,000	1921
96,001- 109,000	1905	A40,001-A50,000	1922
109,001-125,000	1906	A50,001-A54,000	1923
125,001-150,000	1907	A54,001-A56,000	1924
150,001-160,000	1908	A56,001-A57,000	1925

With the serial number missing, any *Hay-Budden* with a face plate can be placed before 1909. Possibly a few made after 1908 were faced, but they would be extremely rare.

The Hay-Budden Manufacturing Company averaged approximately 9,000 anvils per year from their beginning until about 1926. The advertisements show that between the years 1907 and 1908, they could produce about 25,000 anvils a year, if necessary. That is 80 anvils per day for a six-day week. It may be interesting to note that the Mousehole Forge could produce about 30 anvils per day at peak capacity, using only two water-powered tilt hammers.[32]

It is hard to imagine the activity that must have gone on at the factory when you consider that these anvils weigh between 10 and 800 pounds and the various patterns were made one at a time. The slow period of 1903 to 1904 probably indicates the building of the new factory on North Henry Street after the fire. Their normal output of all sizes and patterns of anvils appears to be about 10,000 per year. That is still about 33 anvils per day for a six-day week. Undoubtedly, their main product was anvils. I am told that many of the factory workmen were of Polish descent or immigrants. Maybe that accounts for the quality and high output. I am also told that there were labor management relation problems from time to time.[33]

One author claims that Mr. Hay and Mr. Budden began business in 1886.[34] Another says that the *Hay-Budden* catalog of 1905 states that they began manufacturing in 1890.[35] I account for the four-year difference as follows. The 1886 date was given by a magazine journalist in 1914 who either visited the factory or corresponded with the company and they gave him the 1886 date as the beginning of their anvil making, which was probably in a smaller forge shop. The few anvils that they produced were probably so well-liked that they began expanding their anvil production. By 1890, they most likely considered themselves a successful anvil producer. Most companies and product lines fail in the first couple of years, but if after a few years they are successful, they will list the starting date of the venture when mentioning the company's history to someone. By 1890 Hay-Budden was manufacturing in a small factory and that would be the likely date to appear in an advertising brochure. It must also be remembered that they did not incorporate until 1893.

I have no way of knowing whether they started using serial numbers in 1886, 1890 or when they incorporated in 1893, but my best guess is 1893 as stated. The lowest serial number anvil I have recorded from a *Hay-Budden* is 2127. It was an early farrier's pattern anvil, and it was not until 1892 that S.D. Kimbark had Hay-Budden Manufacturing Company start making farrier's pattern anvils. S.D. Kimbark, as mentioned earlier, invented the farrier's anvil, not Hay-Budden. Therefore, I assumed a 1893-1894 manufacture date for anvil number 2127. By 1895 I use the serial number 16,000, which I consider to be approximately the number of anvils they produced from early 1893 through 1895. If someone should find a very early type in very good condition (I'm sure there are some) with no serial numbers you can be assured that it was made before 1893. These would be very rare.

In February, 1991, I was informed that Hay-Budden Manufacturing Company operated a steel foundry as well as a forge shop, but I cannot confirm this.[36] This may account for the attempt to make engine blocks.

In 1926 the Hay-Budden Manufacturing Company ordered a number of anvils from the Columbus Forge and Iron Company. Most likely they were filling some final orders.[37] The following is part of an article from a Hay-Budden advertisement.[38]

> *For general blacksmith work, anvils of this pattern weighing less than 100 pounds have the faces ground flat and all edges square. Anvils over 100 pounds have the faces ground slightly oval, and the front forward edge slightly rounded. Hardie hole countersunk in anvils 200 pounds and over.*

The following selected advertisements and material will be explained with each entry. Between at least November 1894 and January 1925 there was an advertisement for *Hay-Budden* anvils in every issue of *The Blacksmith and Wheelwright* journal. Most of *The Blacksmith and Wheelwright* ads for *Hay-Budden* anvils were on the back outside cover and do not reproduce well because the cover is orange in color. The ads presented are typical.

The Blacksmith and Wheelwright, November, 1894. This is possibly the first ad to appear in this journal. There was no ad in the April, 1894 issue. (From Mr. Hunter Pilkinton)

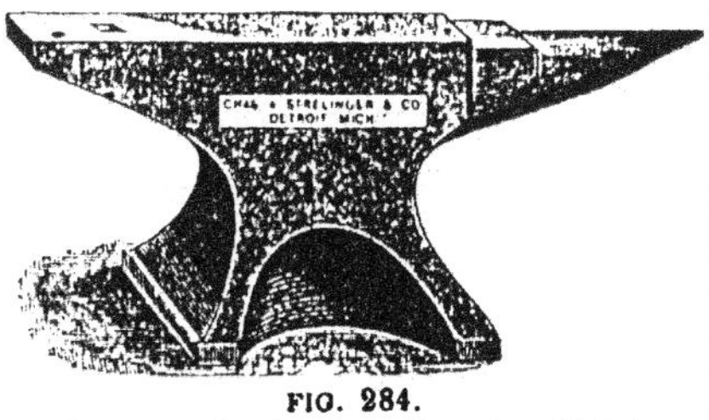

ANVILS.

FIG. 284.

In wrought iron anvils, the Wright, Wilkinson and Armitage "Mousehole" are the best known; they are all English made. We believe there is but one American concern, and their output is small. Have been informed (by the house who sells them) that they are equal to the best English.

The price has been in the neighborhood of twelve cents per pound, varying from one-half to one cent more or less for a great many years.

Charles A. Strelinger catalog, 1995, Detroit, Mi. This must be the Hay-Budden Mfg. Co.

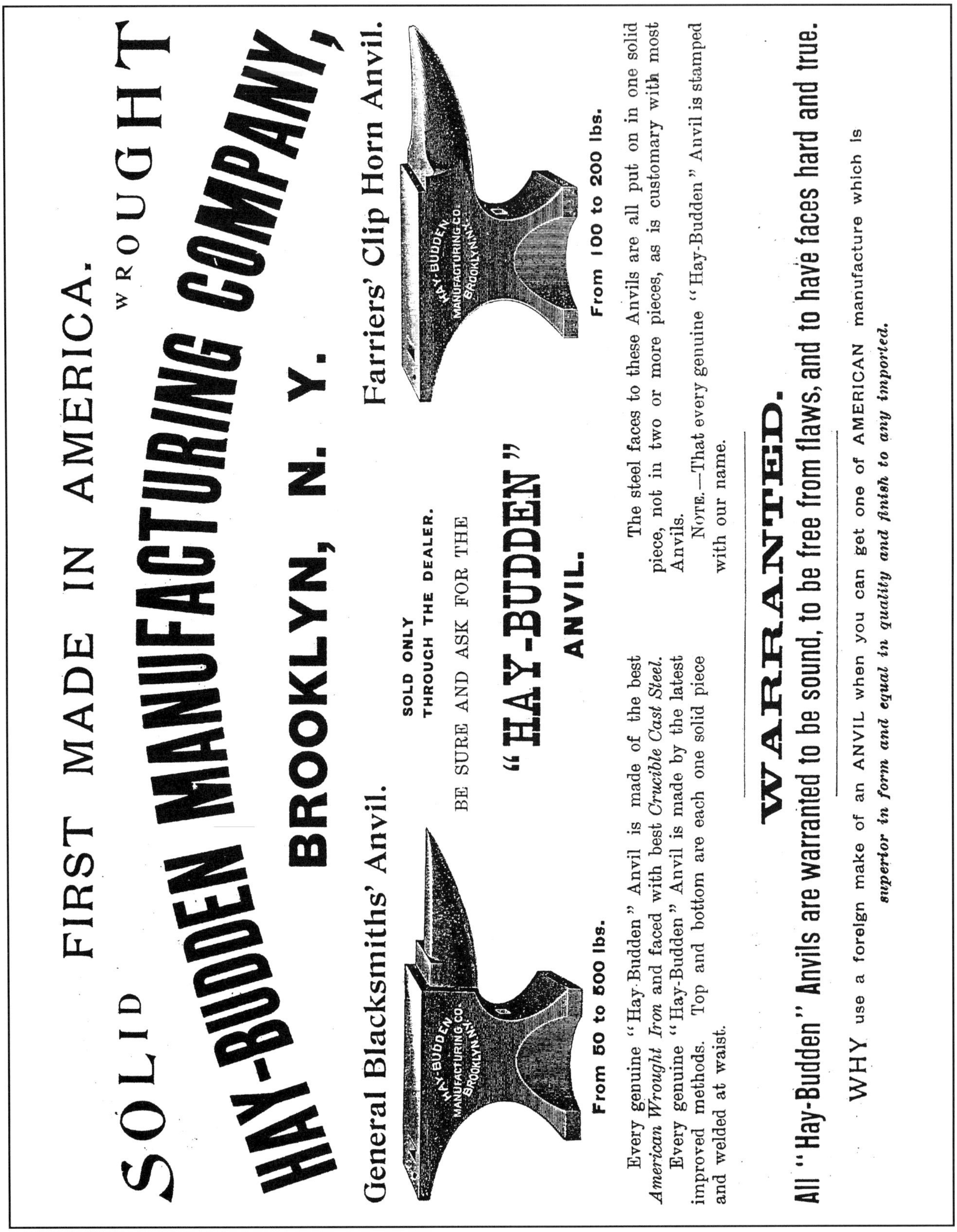

The Blacksmith and Wheelwright, March, 1896.

HAY BUDDEN CO.'S BEST

American Wrought Anvils.

We carry in stock a great variety of sizes from 80 to 300 lbs.

The Charles C. Louis Co. catalog, 1896 or 1897. (From Mr. Hunter Pilkinton)

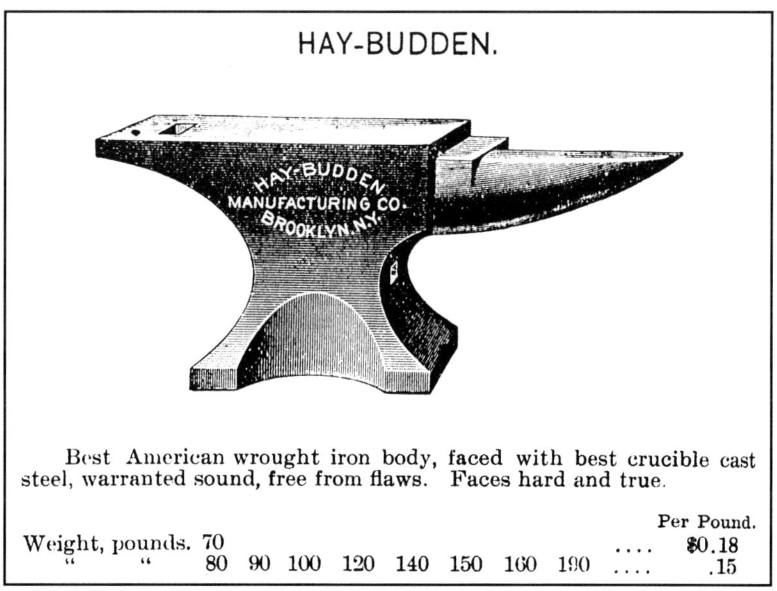

HAY-BUDDEN.

Best American wrought iron body, faced with best crucible cast steel, warranted sound, free from flaws. Faces hard and true.

		Per Pound.
Weight, pounds.	70	$0.18
" "	80 90 100 120 140 150 160 190	.15

Farwell-Ozmun-Kirk and Co. catalog, January, 1901.

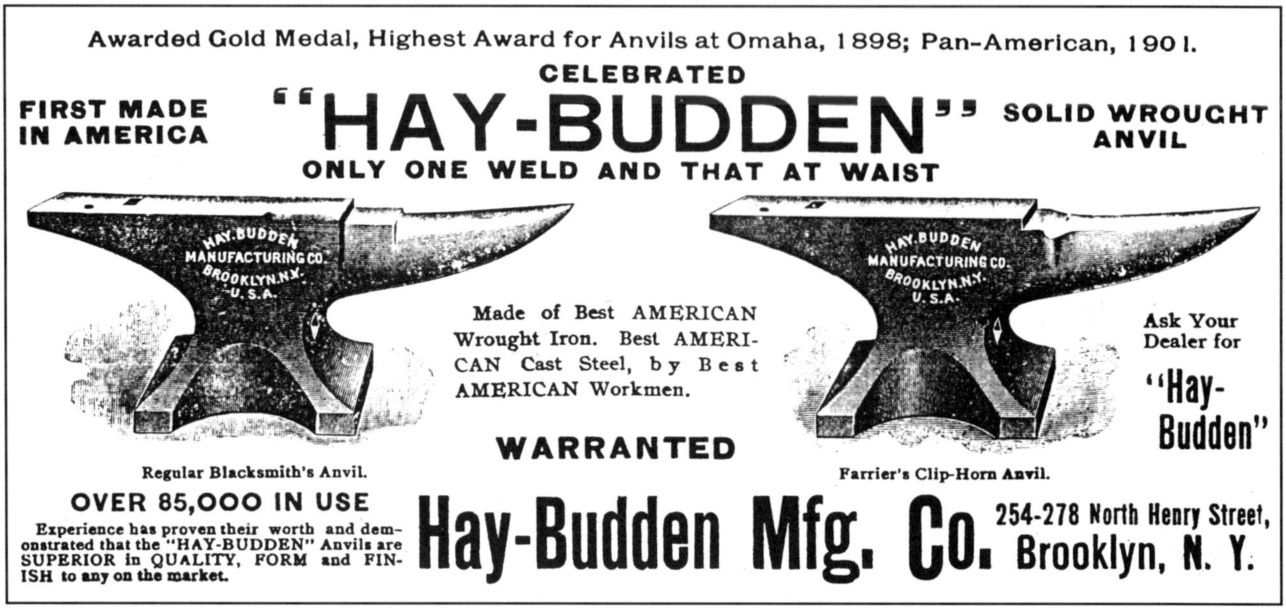

The Horse Shoer's Journal, January, 1904.

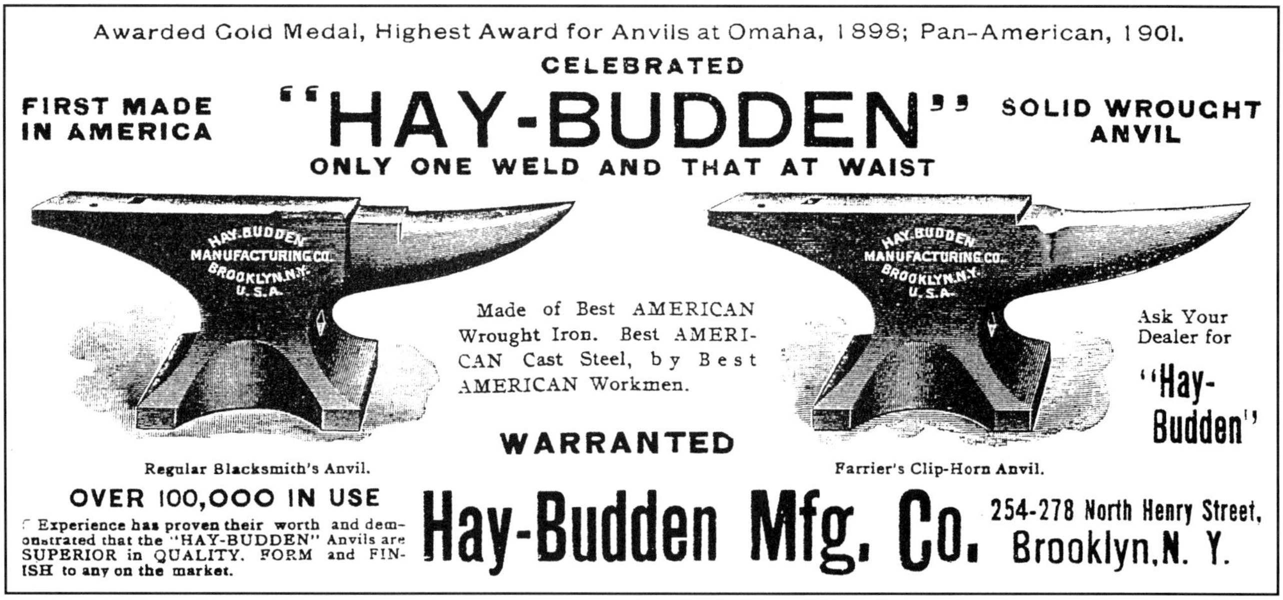

The Horse Shoer's Journal, September, 1905. (From Mr. Gill Fahrenwald)

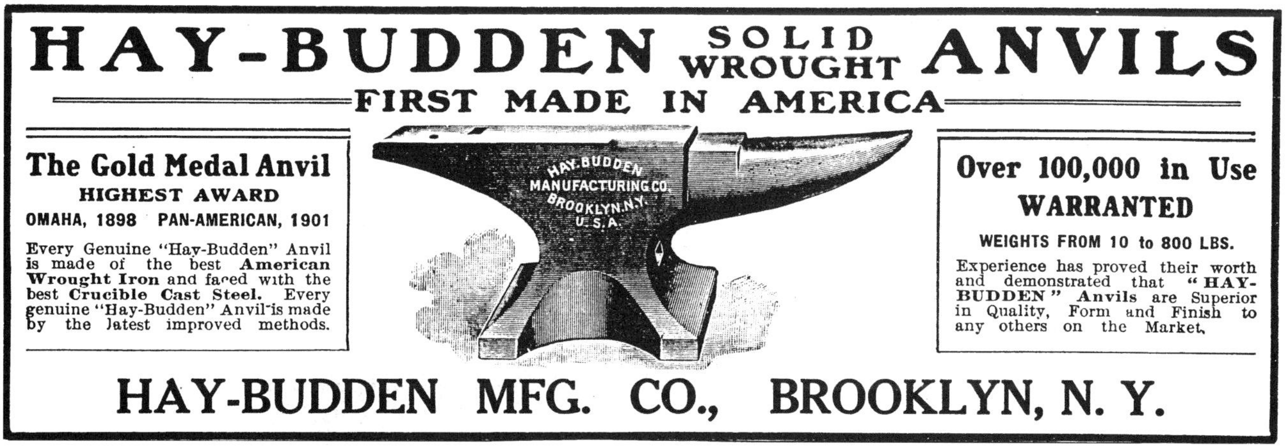

The American Blacksmith, September, 1908.

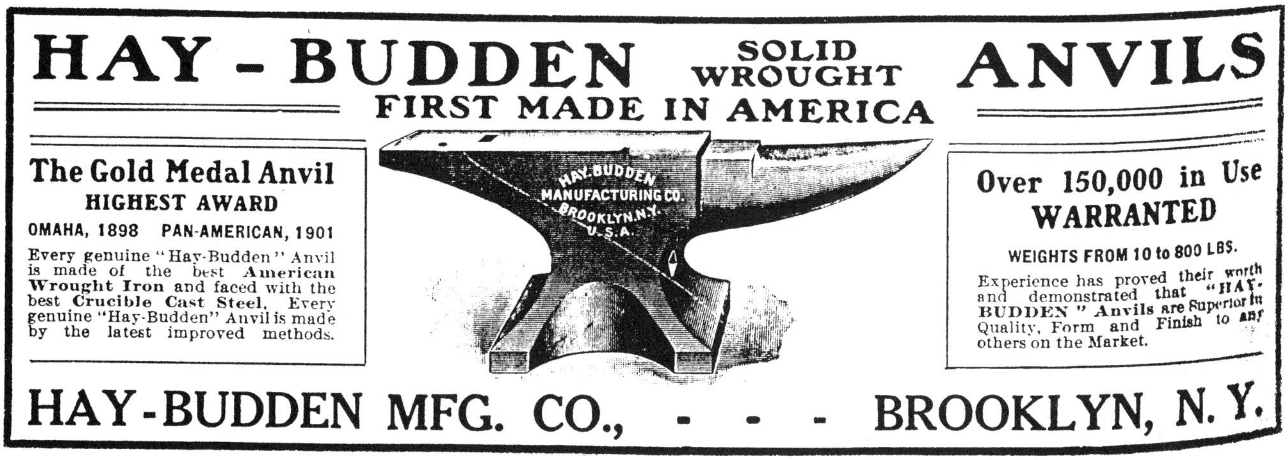

The American Blacksmith, October, 1909.

The Crow Bar Journal, January, 1916, Minneapolis, Minnesota.

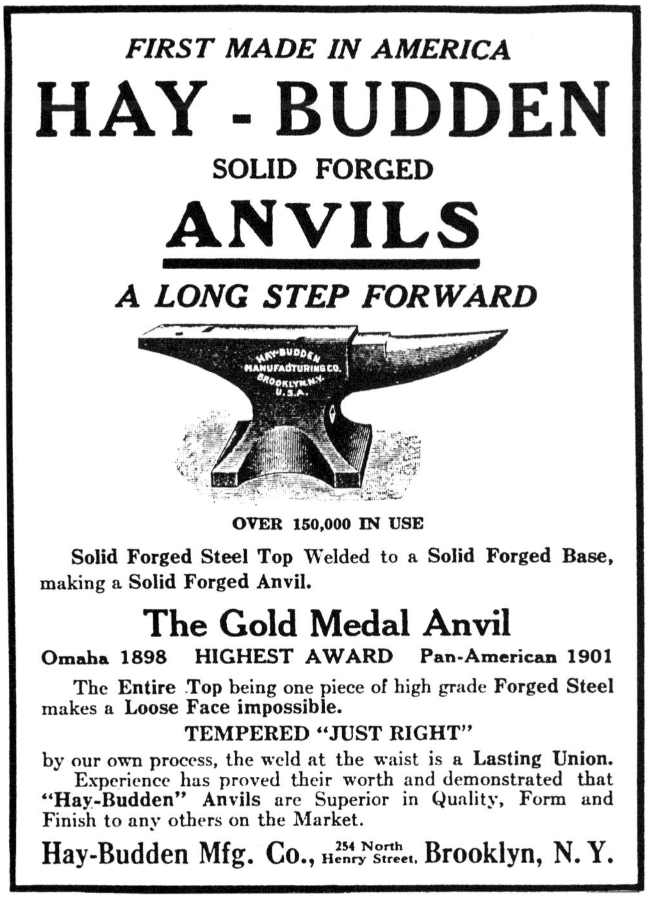

The American Blacksmith, June, 1912.

The American Blacksmith, March, 1914 through October, 1919. This journal became The American Blacksmith, Auto and Tractor Shop.

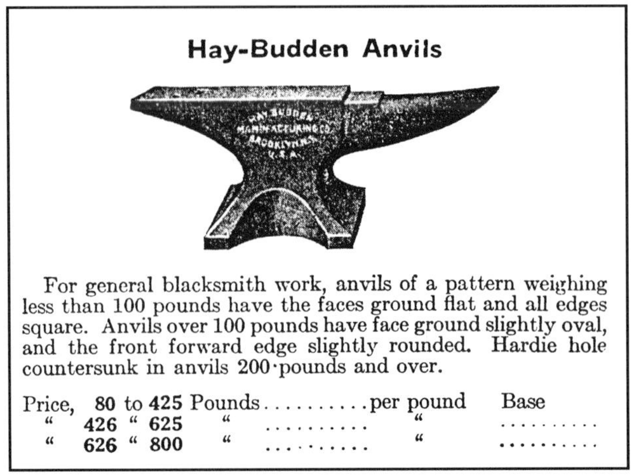

Hay-Budden Anvils

For general blacksmith work, anvils of a pattern weighing less than 100 pounds have the faces ground flat and all edges square. Anvils over 100 pounds have face ground slightly oval, and the front forward edge slightly rounded. Hardie hole countersunk in anvils 200-pounds and over.

Price,	80 to 425 Pounds	.... per pound	Base
"	426 " 625 "	.... "	....
"	626 " 800 "	.... "	....

The South Bend Supply Co. catalog, 1925, South Bend, In.

The American Blacksmith, Auto and Tractor Shop, December, 1919.

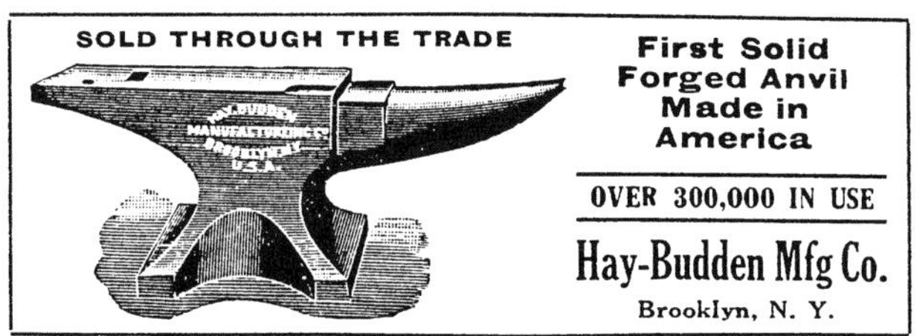

The Horse Shoer's Journal Sept. 1922. (From Mr. Ken Wilson.)

ANVILS

PETER WRIGHT

First quality English Wrought steel.

85 to 250 lbs.............................Per Pound, $ 0.16

HAY-BUDDEN

First quality wrought steel.

85 to 250 lbs.............................Per Pound, $ 0.14

FARMERS' CAST

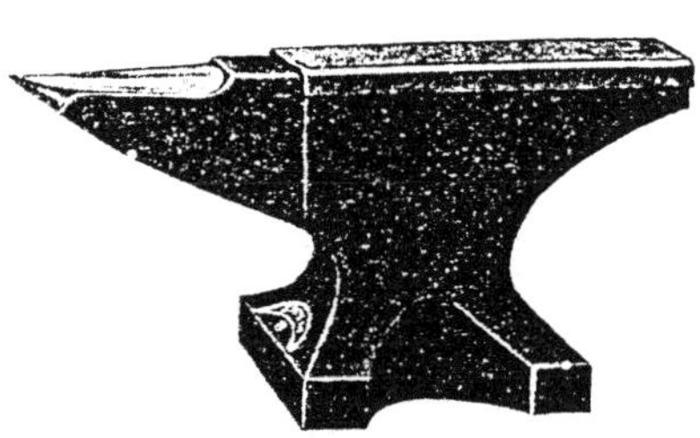

Cast iron. Chilled face.

100 lbs..Each, $ 7.50

H. G. Lipscomb and Co. catalog, 1913, Nashville, Tenn. Notice the differences in anvil prices. (Mr. Hunter Pilkinton)

Solid Wrought Anvils.—The attention of our readers is directed to the advertisement of the Hay-Budden Manufacturing Co., 254 North Henry street, Brooklyn, N. Y., on our last cover page. They manufacture solid wrought anvils. Every Hay-Budden anvil they say is made from the best American wrought iron, and faced with the best crucible cast steel. These anvils are sold only through dealers, but you can obtain a descriptive catalogue giving full particulars by writing to the manufacturers as above and mentioning "The Blacksmith and Wheelwright."

The Blacksmith and Wheelwright, September, 1900.

The following article and drawings are from *The Iron Age* journal, May 11, 1899. The Hay-Budden Manufacturing Company had a booklet made out of the article and used it for advertising purposes. To me there appear to be a number of glaring errors in the article as well as in the drawings. You would have thought that the Hay-Budden Company would have corrected the errors before publishing the booklet. Could I be wrong?

A copy of the booklet was sent to me courtesy of Mr. John Catchings. The photographs of the drawings were made by Mr. Keith Retterer.

Hay-Budden Manufacturing Co.

BROOKLYN, N. Y.

Manufacturers of the Celebrated

"Hay-Budden"

Solid Wrought Anvils.

FIRST MADE IN AMERICA

◆ ◆ ◆

Why USE A FOREIGN ANVIL, when you can get one of Domestic manufacture which is superior in Form, Quality and Finish to any imported ?

OUR Anvils ARE

FULLY WARRANTED.

REPRINTED FROM
THE IRON AGE,
MAY 11, 1899.

Front cover of booklet.

BE SURE AND ASK FOR THE

"Hay-Budden" Anvil

SOLD ONLY THROUGH
THE DEALER.

Received Diploma and Gold Medal,
Highest Award for Anvils at Omaha Exposition, 1898

Back cover of booklet.

GENERAL BLACKSMITHS' ANVIL.

From 25 to 800 lbs.

We have carefully studied all the advantages and disadvantages of the various makes of anvils on the market, and we here present an anvil which is as near perfect as it is possible to be. The size of face, length and fullness of horn—depth of drop have all been considered, and we defy a better shaped anvil for general blacksmithing.

FARRIERS' CLIP HORN ANVIL.

From 80 to 400 lbs.

We call special attention to our **Farriers' Clip Horn Anvils.** These anvils are the result of years of careful study on our part of the Farrier's wants and we are now able to present an anvil that is perfection

We also make narrow face Farriers' Anvils without clip horn.

SAWMAKERS' ANVILS AND BLOCKS.

Anvils of Special Dimensions and Shape Made at Short Notice.

NOTE—That every genuine "Hay-Budden" Anvil is stamped with our name.

Inside left cover of booklet.

THE CELEBRATED

"Hay-Budden" Solid Wrought Anvils

Have Been Manufactured in the United States for the Past Ten Years.

No means have been spared to turn out a thoroughly first-class anvil. Our efforts were not only to produce anvils equal to others on the market, but, if possible, to make them superior. Our results have been crowned with success.

After extensive experimenting we are pleased to state that we have produced a steel for the faces of our anvils which will take a harder temper and be less liable to chip than any on the market.

Every anvil is thoroughly tested before it leaves our works; and the blacksmith who wants a strictly first-class anvil can make no mistake in purchasing a

"Hay-Budden."

SOLD ONLY THROUGH THE DEALERS.

"America for Americans."
"Home Goods for Home Consumers."

Inside right cover of booklet.

Making a Solid Wrought Iron Steel Faced Anvil

Although the ordinary anvil is being gradually displaced by steam and drop hammers, it is more than doubtful if it will ever be entirely superseded by any other device or appliance upon which to pound and shape things. It is essentially an adjunct upon which the skill of the blacksmith depends in the closest degree, and without which his vocation would be seriously hampered. Its form has not been changed materially in the memory of man, the only alterations being such as have been required by special operations or the whims of the user.

The blacksmith's anvil must as prerequisites embody certain features. The body and horn should be able to withstand the severest usage, while the face should be of such material as to successfully resist the blows of a hard faced hammer, to resist chipping or flaking, and to be unaffected by the heat imparted by the article operated upon. It is essentially a rough and ready tool, in the use of which but little discretion is exercised, and which, if it fails from any cause, will undoubtedly be emphatically condemned.

The tone of an anvil, upon which more or less stress has always been laid, may or may not mean that it is a superior article. While all anvils could not enter an "anvil chorus," it yet remains a fact that a poor anvil may have a rich and perfect tone and at the same time be weak in its structure. On the other hand it may be put down as being almost axiomatic that the good anvil is invariably of good tone. Judging, therefore, its physical qualities by its musical characteristics is not always a safe and sure guide. The old blacksmith who has pounded for years upon the same anvil probably appreciates its ring more than any other feature, and to him it means a harmonious instrument of high grade.

Solid Wrought Steel Faced Anvil

The wrought solid body steel faced anvil built by the Hay-Budden Mfg. Company of Brooklyn, whose works a representative of The Iron Age recently had the privilege of visiting, is made essentially of three parts: the wrought iron base, a wrought iron body and a steel face of special composition. In the making of an anvil these three parts are manipulated separately and brought as near as possible to their final form, when they are united by welding, the anvil brought to its true form, tempered and finished. The method of manufacture and material employed in all the parts insure the integrity of the finished article.

It is very evident that the selection of material is of prime importance, since it controls the quality of the output. Skill in the manipulation insures good results as far as workmanship is concerned, but without perfectly adapted material to start with the utmost skill and experience would be thrown away. The best material may be, of course, spoiled in the handling, but a quality of the highest grade is necessary to perfect results. Wrought iron is used for the body of the anvil, for the reason that it is less liable to be injured by the successive heatings and hammerings to which it is subjected than steel, and for this reason the result can be more surely relied upon.

In watching the various operations one cannot but be impressed with the proportion of hand work. At every single step except one the man is the important factor. It is the individual who judges the heat, controls the hammer and finally influences the form of the finished article. This dependence upon skill makes it impossible to produce close and accurate standards as far as size and contour are concerned, and since this is not absolutely essential but little attention is paid to it. One anvil may and does differ from another in magnitude, and since the work to be performed upon it is not of the accurate kind as understood in some mechanical operations there is no good reason why any attempt should be made to follow a pattern closely. In order to satisfy the whims and to meet the various views of blacksmiths it is desirable to have a varied assortment of different widths, lengths and proportions for the same weight of anvils. This permits every blacksmith to find an anvil that suits his ideas. As has been said, in only one step is the workmanship overshadowed by a machine-namely, in forming the base. This is an operation which has little to do with the final product, as it is simply a pedestal supporting the working face, and the method here pursued in making it is merely introduced as a labor saving and economical operation.

The Wrought Iron Billet

The wrought iron scrap for the base and upper part of the body is cut to proper size, bundled, heated in a furnace and welded into a compact mass under the steam hammer, as shown in Fig.1. The billets then pass to a second furnace, after which their paths diverge. The one which is to be incorporated into the upper part of the anvil is rough hammered into shape and is then ready to receive the steel face. This is a flat bar of steel of such composition as will take a certain temper, maintain its elasticity and resilience, and be so tough as, with the aid of the wrought iron backing, to resist fracture. The billet and steel are heated in the same furnace and welded with the steam hammer shown in Fig.2. From this engraving it will be noticed that the workman handling the billet places it upon a roughly curved die which forms the lower part of the upper part of the body. The blank then passes to another steam hammer, Fig.3. with which the point is drawn down, the tail of the anvil rough shaped, and the given dimensions approached as near as may be practicable. The final operation on the top is done by hand as illustrated in Fig.4. The sides are here brought truly parallel, the horn is tapered and properly curved on the under side and the top practically finished. During these operations pritchel and hardy holes have been formed on the top of the anvil.

Another billet passes from the furnace to the hydraulic press shown in Fig.5. What we may term the anvil of this press moves toward the rear, reference being had to the engraving, so that the billet can be dropped into the opening above the ram. The anvil is then moved back to a central position, water turned on and the ram is forced upward, completing the lower part of the anvil, with the exception of a slight burr due to excess of metal which has been squeezed out around the edge.

This finishes at one operation the bottom, which is next ready to be welded to the top. The small coke heated furnace shown in Fig.6 supports both top and bottom parts, the parts to be welded being placed next the fire. When the proper heat has been attained the base is placed upon the anvil of a steam hammer, the top centrally held upon it, as illustrated in Fig.7, when the two parts are perfectly incorporated.

The next step, and one of the most important, as far as the appearance of the finished anvil is concerned, is illustrated in Fig 8. From the welding hammer the anvil is taken to a cinder box, and by means of sledges the joint is thoroughly hammered down, the surfaces are sprinkled with water and smooth hammered to remove the scale, after which the anvil is placed face down in the furnace and when the steel portion has reached a dull red heat it is placed in the hardening tank, Fig.9, and streams of water thrown upon it. The anvil is left in the tank until cool. Before hardening the anvil has been roughly ground on the face, and after hardening it is taken to large grindstones, where it is ground upon the edges and the face trued up.

Testing

At first glance it would seem unnecessary to test each anvil. This would appear to be of little consequence, as the material is first intimately known, the various operations are carefully performed, and the product should seemingly meet the requirements in every instance; but this is not relied upon. Each heat of steel for the face is incorporated in one or more anvils, and upon their completion they are tested in every way possible with the hammer. They must resist indentation with a hard heavy hammer, must be free from all tendency to chip on the sides or flake on the top, and even under the most severe treatment must remain free from all cracks. Not only is this policy pursued when each new heat is commenced, but it is followed in the case of every anvil turned out. After grinding the body of the anvil is painted and the face varnished and covered with heavy burlap, which prevents the face from being damaged in shipping.

The Hay-Budden Company are now turning out about 100 tons of completed anvils per month, and their trade extends over the entire world. In this connection it may be of interest to note that during the past few years the importation of anvils into this country has steadily decreased. By far the largest supply of foreign anvils has been received from Great Britain, which country in 1893 sent here 1,567,746 pounds; in 1894, 704,764 pounds; in 1895, 1,085,054 pounds; in 1896, 859,580 pounds, and in 1897, 605,211 pounds. In 1898 the total importation amounted to 777,903 pounds, valued at $47,797.

Made of Best American Wrought Iron
Faced with Best American Cast Steel
Forged by Best American Workmen

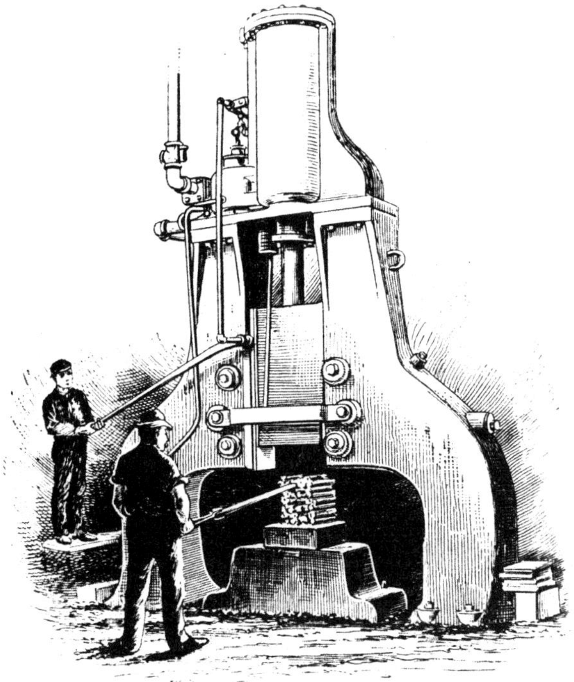

Fig. 1.—Hammering Wrought Iron Scrap.

A double-frame steam hammer at work. The workman in front of the hammer seems to be holding the faggot in place.

Fig 2.—Welding Steel Face Under the Hammer.

According to the text and drawing the body is roughed to shape and then the face plate is welded on. It sounds like they have the "cart before the horse" here. The body should be finished before the face plate is welded on.

Fig. 3.—Forming the Horn Under Steam Hammer.

The horn can be worked on after the face is welded on, but the heel cannot be worked on without ruining the steel face.

Fig. 4.—Finishing the Horn by Hand.

I can not imagine finishing the horn by hand on another smithing anvil. Concave rounded dies in the steam hammer would be more appropriate as preformed in the painting "The Last Anvil Makers."

Fig. 5.—Forming Bottom of Anvil in Hydraulic Press

A use of this hydraulic press would be a closed die operation. Dies would have to be changed for each base size.

Fig. 6.—Heating Top and Bottom Parts for Welding.

A large open forge fire.

Fig. 7.—Welding Top and Bottom Parts Under Steam Hammer.

You may notice that the single frame steam hammer being used in figures 2,3 &7, though crudely drawn is the same hammer as the one in use by "The Last Anvil Makers."

Fig. 8.—Finishing Anvil by Hand Hammering.

One man is holding a flatter or top swage while the other three men take turns striking as it is moved around.

Fig. 9.—Hardening Face of Anvil.

You will notice that water is being sprayed under pressure on the anvil face while the lower part is submerged in the tank of water.

Fig. 10.—The Anvil.

The completed anvil.

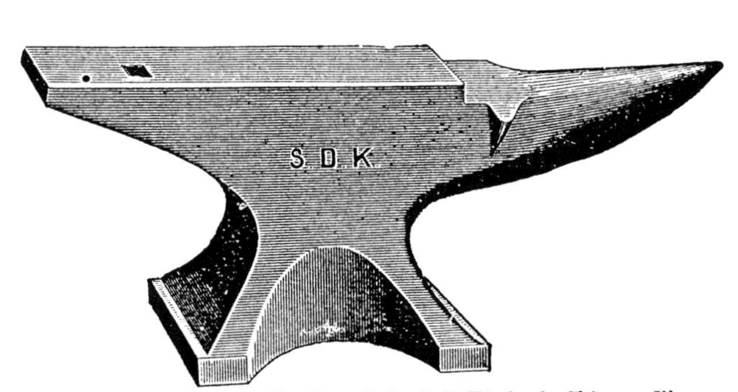

Farriers' Perfect Anvil. For sale by S. D. Kimbark, Chicago, Ill.

Early Hay-Budden. The Blacksmith and Wheelwright, November, 1894.

TRENTON

The Columbus Forge and Iron Company of Columbus, Ohio, produced the *Trenton* anvil between the years 1898 and 1952.[40] The name *Trenton* is unique among anvil makers. Why would a forge company in Columbus, Ohio, name their anvil after a city some distance away where a competitor made anvils? I thought the anvil was made in Trenton, New Jersey, upon seeing one for the first time, but soon found out that this was not the case. *Fisher & Norris*, the anvil makers in Trenton, New Jersey, since 1847, did not like the idea of a competitor using the name of their city and said so in several of their advertisements.

The American Blacksmith, December, 1910.

All the anvil makers I know have either used the name of the company owner(s) or named the anvils after characters or characteristics associated with blacksmithing, forging or strength. Such names as *Vulcan*, *Samson, Arm and Hammer, Hay-Budden* and *Peter Wright* come to mind.

Then why *Trenton*?

The last owners of the Columbus Forge and Iron Company did not know why the anvil was given the name *Trenton*, although they thought that the company's anvil sales agent, H. Boker and Company of New York City, had something to do with it. The name remained a mystery for a number of years and then I was given a photocopy of an old brochure from the Trenton Vise & Tool Works of Trenton, New Jersey, by Mr. John King (descendant of a company founder) dated 1888.

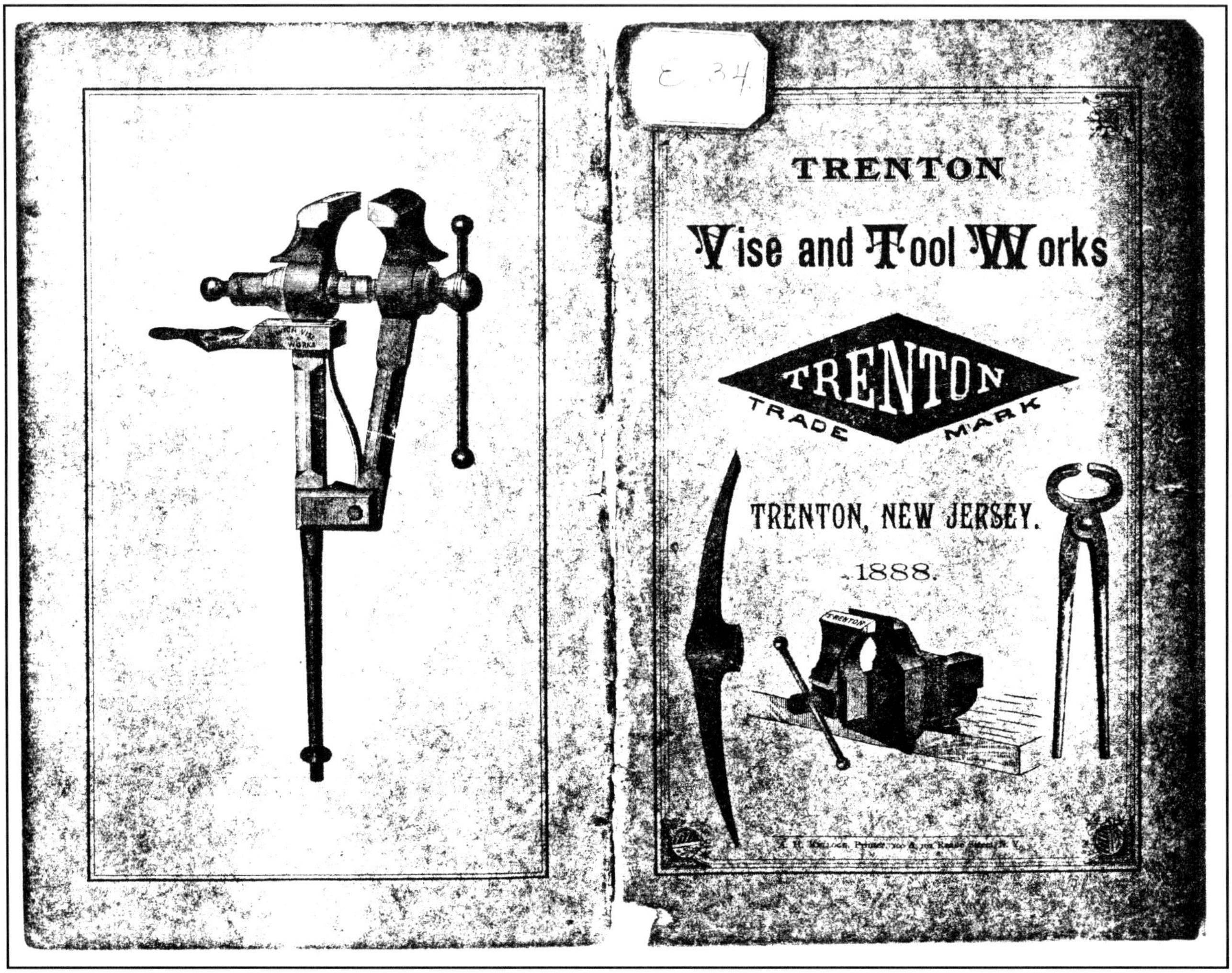

Front and back cover of the brochure.

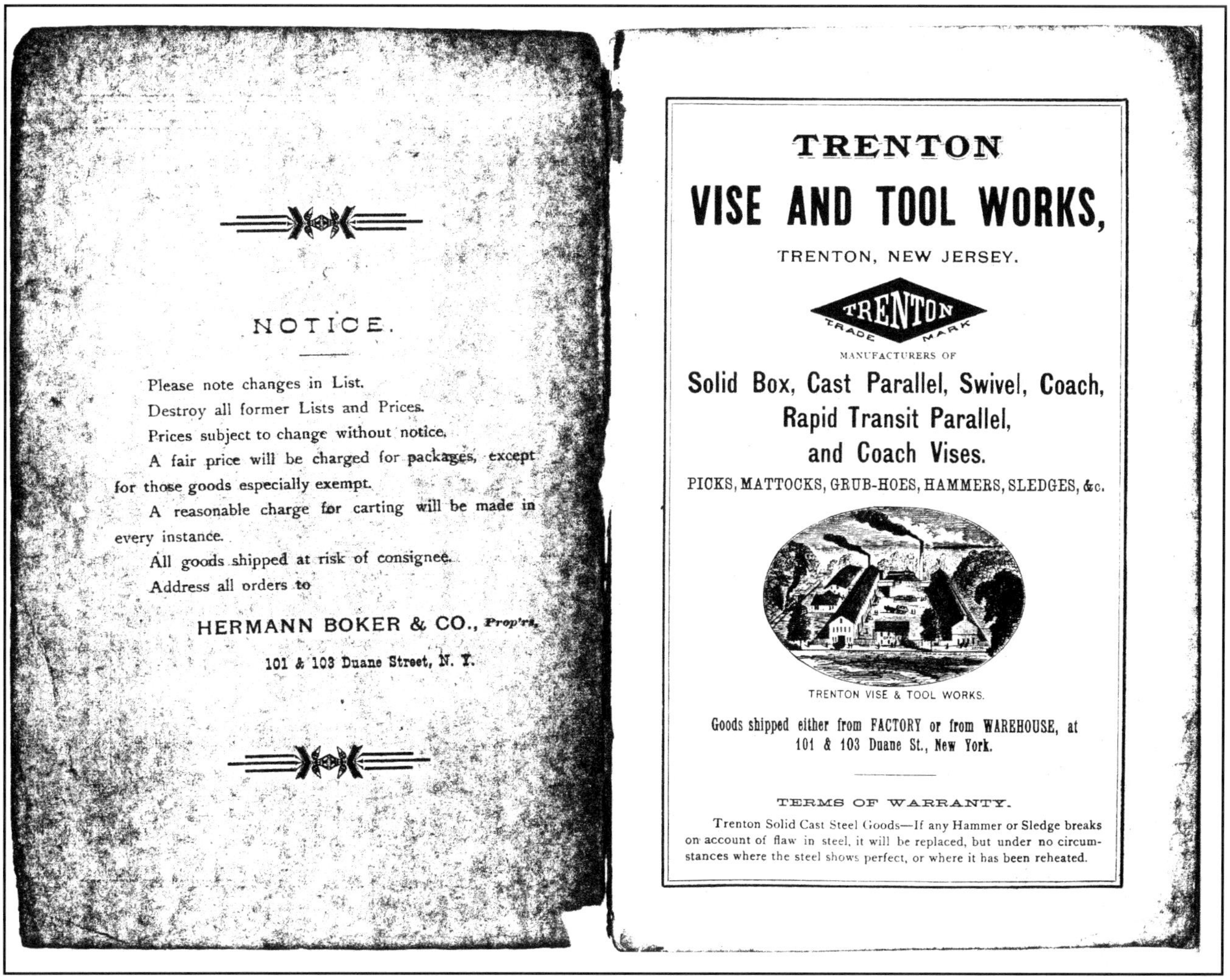

NOTICE.

Please note changes in List.
Destroy all former Lists and Prices.
Prices subject to change without notice.
A fair price will be charged for packages, except for those goods especially exempt.
A reasonable charge for carting will be made in every instance.
All goods shipped at risk of consignee.
Address all orders to

HERMANN BOKER & CO., Prop'rs,
101 & 103 Duane Street, N. Y.

TRENTON
VISE AND TOOL WORKS,
TRENTON, NEW JERSEY.

TRENTON TRADE MARK

MANUFACTURERS OF
Solid Box, Cast Parallel, Swivel, Coach, Rapid Transit Parallel, and Coach Vises.
PICKS, MATTOCKS, GRUB-HOES, HAMMERS, SLEDGES, &c.

TRENTON VISE & TOOL WORKS.

Goods shipped either from FACTORY or from WAREHOUSE, at 101 & 103 Duane St., New York.

TERMS OF WARRANTY.
Trenton Solid Cast Steel Goods—If any Hammer or Sledge breaks on account of flaw in steel, it will be replaced, but under no circumstances where the steel shows perfect, or where it has been reheated.

Inside front cover of brochure.

TRENTON

SWIVEL COACHMAKERS' VISE.

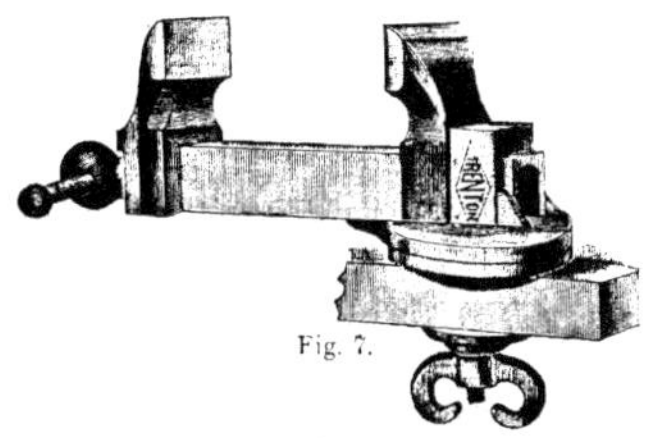

Fig. 7.

4 inch Jaw. Opens 9 inches $11.00

This Vise is interchangeable. (See Folio 6.)

OVAL SLIDE, OR FARMERS' VISES.

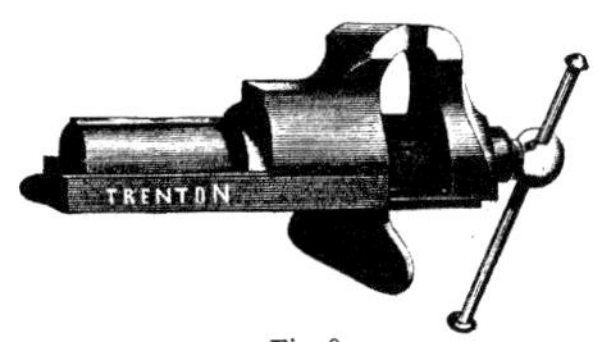

Fig. 9.

No.	2–0	0	1	2	3	4	
Jaws,	2	2½	3	3½	4	4½	inch.
List,	$2.50	2.70	3.60	4.65	6.70	9.65	each.

Discount, %

TRENTON

SOLID WROUGHT IRON STEEL-FACED ANVILS.

WARRANTED.

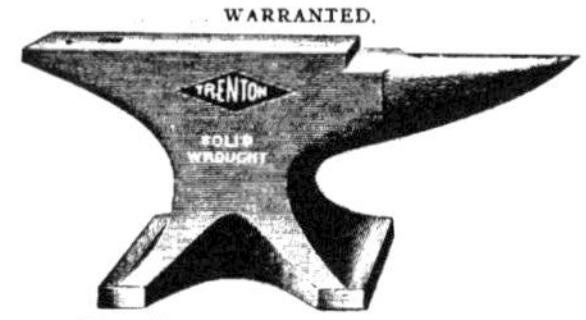

Peter Wright's Pattern.

85 lbs. and over,	. . . .	cents per lb.
70 to 84 lbs.	. . . . 1 " "	advance.
60 to 70 "	. . . . 1½ " "	"
50 to 60 "	. . . . 2 " "	"
40 to 50 "	. . . . 3 " "	"
30 to 40 "	. . . . 5 " "	"

HORSE SHOERS' ANVIL.

Price, per lb.

The Trenton Horse Shoers' Anvils are made in weights from 120 to 200 lbs., with face 3½ to 4 inches wide.

We also carry in stock Horse Shoers' Anvils without the clip horn, and make to order anvils of all descriptions. (See Horse Shoers' Vise, Folio 4.)

Pages 8 and 9 of brochure.

The trademark of this company was the same as that used ten years later by the Columbus Forge and Iron Company as the trademark on their anvils. The brochure says that the Trenton Vise & Tool Works made vises, picks, grub-hoes, hammers, sledges, etc. On the inside of the brochure are engravings of wrought anvils with the *Trenton* trademark on them. If the date 1888 was the earliest date for a *Trenton* anvil to appear in the advertisement, it might suggest that they were made by Hay-Budden Manufacturing Company under the *Trenton* trademark, as Hay-Budden had been making some anvils since 1886. However, I obtained an advertisement for the same *Trenton* anvil in the catalog of *Cook & Co.* of New York City, dated 1883. It will also be noticed that one of the anvils pictured in the brochure is an early farrier's pattern. S. D. Kimbark Company of Chicago, Illinois, who developed the farrier's pattern had these made in Germany since 1882.

If the Trenton Vise and Tool Works had made anvils it most certainly would have advertised the fact, but they did not because Hay-Budden Manufacturing Company began producing the first wrought anvils made in this country in their factory about 1890. The Trenton Vise and Tool Works was either partially or wholly owned by H. Boker & Company (Boker makes knives today). This leads me to believe that the anvils sold by the Trenton Vise and Tool Works were made in Germany and possibly England and sold under this company's trademark. H. Boker & Company had strong relationships with their homeland, Germany. In fact, Herman Boker, who founded H. Boker & Company in 1837 after establishing a successful enterprise in the United States, went back to Germany in retirement, leaving his sons to continue the business.[41] John Boker of H. Boker & Company (a great, great grandson) is still a broker in New York (1991). I have a small anvil (57 lbs.) with *H. Boker* stamped on it and the weight is in kilograms; undoubtedly a German-made anvil which resembles these early *Trentons.* H. Boker & Company may well have been S. D. Kimbark's importing agent and was also selling these imports under the *Trenton* trademark.

The Trenton Vise and Tool Works was still in business as late as 1897. The wholesale catalog of the Canton Hardware Company of Canton, Ohio, still carried their vises at that time, as did the Montgomery Ward and Sears Roebuck catalogs of the 1890s.

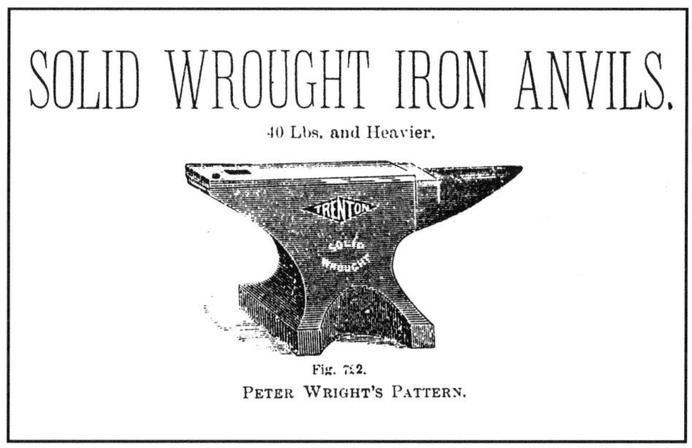

Cook and Co. catalog, 1883, New York, N.Y. This is the earliest ad I have for a Trenton anvil; 15 years before the Columbus Forge and Iron Co. started producing Trenton anvils. (From Mr. Hunter Pilkinton)

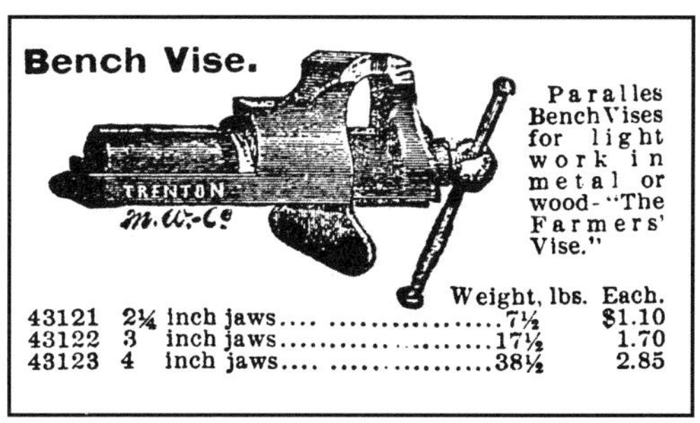

Montgomery Ward and Co. catalog, 1895. Notice "Trenton" on the side of the vise, the same as in the brochure.

Model Vises.

STATIONARY. SWIVEL.

Nos. 30 to 37.

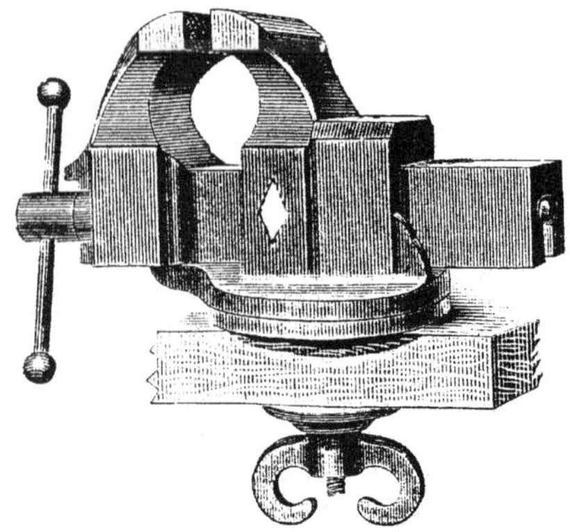

Nos. 40 to 45.

No. 30.	2	inch Jaws, opens	2¾	inches,	Stationary,	weight	13½	pounds	each,	$3 00
No. 31.	2½	" "	3¼	"	"	"	18½	"	"	4 00
No. 32.	3	" "	3½	"	"	"	26	"	"	5 00
No. 33.	3½	" "	4	"	"	"	31	"	"	5 75
No. 34.	4	" "	5	"	"	"	43	"	"	7 50
No. 35.	4½	" "	5	"	"	"	51	"	"	9 00
No. 36.	5	" "	5	"	"	"	67	"	"	13 00
No. 37.	6	" "	6	"	"	"	110½	"	"	20 00
No. 40.	2	" "	3	"	Swivel,	"	15½	"	"	3 50
No. 41.	2½	" "	3¼	"	"	"	20	"	"	4 50
No. 42.	3	" "	3½	"	"	"	29½	"	"	5 50
No. 43.	3½	" "	4	"	"	"	34½	"	"	7 00
No. 44.	4	" "	5		"	"	47	"	"	8 50
No. 45.	4½	" "	5	"	"	"	56½	"	"	10 50

The Canton Hardware Co. catalog, April 1, 1896, Canton, Ohio.

Bench Vise.

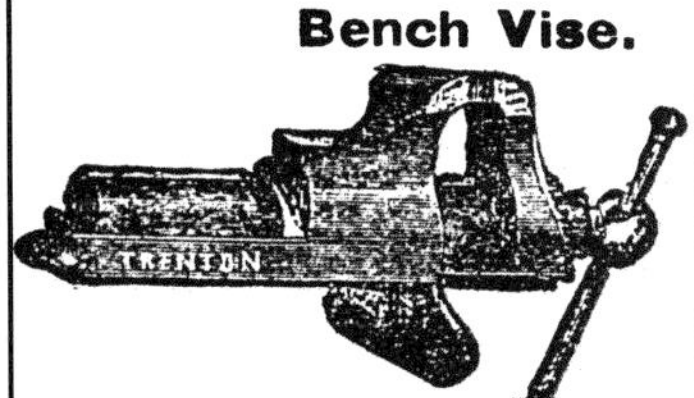

Parallel Bench Vises for light work in metal or wood. "The Farmer's Vise." No. 11683. 2¼ inch jaws; weight, 7½ lbs. Price, each ..**$1.00**

No. 11684. 3 inch jaws: weight, 17½ lbs. Price, each..**$1.70**

No. 11685. 4 inch jaws; weight, 38½ lbs. Price, each..**$2.85**

Sears Roebuck and Co. catalog, 1897.

Anvils.

Fig. H 428.

"Peter Wright" Anvils, 100 lbs. each and over	per lb.	....
"Wilkinson" " " "	"	....
"Eagle" " " "	"	....
"Trenton" " " "	"	....

Prices of Anvils weighing less than 100 lbs. will be quoted on application.

M. M. Buck Mfg. Co. catalog, 1900, St. Louis, Mo. The "Trenton" listed here was undoubtedly a German made anvil since it takes time to make up a catalog. The other anvils listed had been on the market for many years as had the German "Trenton."

126-pound German made "Trenton." (R.A.P. photo)

This suggests that the Trenton Vise and Tool Works must have gone out of business for some reason in the late 1890s, but H. Boker & Company retained the company's copyright to the *Trenton* trademark. It further suggests that when H. Boker & Company contracted to become the sales agent for the Columbus Forge and Iron Company, Boker must have suggested that the forge company use a trademark on their anvils that was already known rather than to try to break into the anvil market with a new name. Had the *Trenton* name not been adopted, the anvil would probably have been called the *Indian Chief*, the *David Buel*, or have used the company's forging trademark (FI) which appears on all of their forgings. Of all the products that the Columbus Forge and Iron Company made only the wrought anvil used the *Trenton* trademark. Their cast railroad anvils of 650 and 750 pounds were marked with *Indian Chief* as were the blacksmith or leg vises they made.

BUEL PATENT TRENTON

HIGH GRADE

SOLID WROUGHT ANVILS.

WARRANTED.

Superior shape.
All new material.
No scrap used.
Perfectly ground faces.
Straight holes.

Long pointed horns.
Long faces and tails.
Faces hard.
Edges tempered to prevent chipping.
Clear ring.

Watch this space each month for the appearance of a series of pictures that will be of interest.

COLUMBUS FORGE AND IRON CO, - - Columbus, Ohio.

The Blacksmith and Wheelwright, October, 1900. This was the second ad for the Columbus Forge and Iron Co. to appear in the journal.

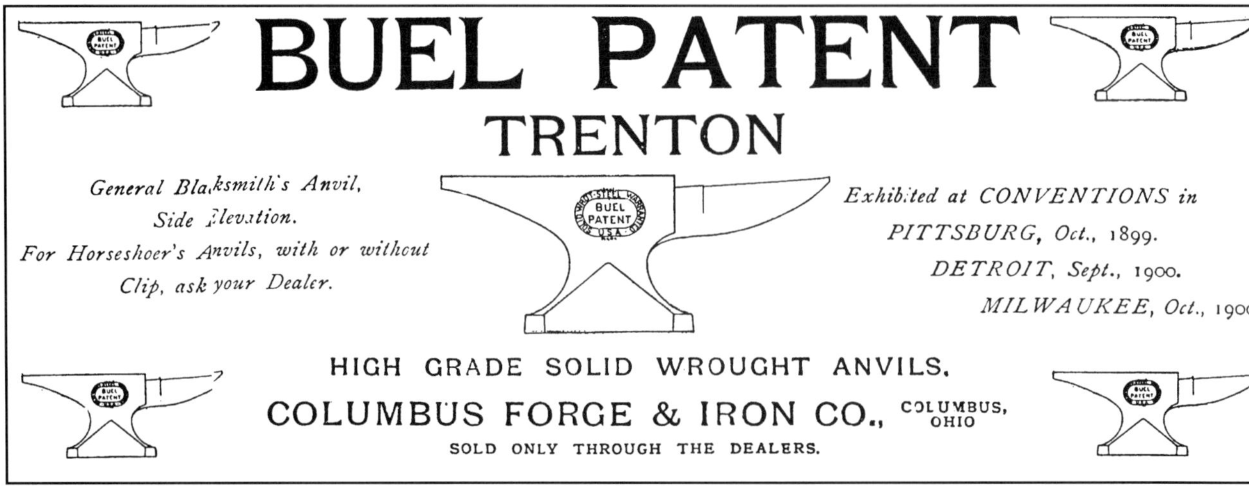

The Blacksmith and Wheelwright, November, 1900. This was the third ad to appear in the journal.

28-pound "Buel Patent" anvil made by the Columbus Forge and Iron Co. (Mr. Gill Fahrenwald photo)

Their miniature and craft anvils carried the (FI) trademark and their second advertisement of October, 1900, in *The Blacksmith and Wheelwright* journal called the anvil the *Buel Patent Trenton*. Their advertisement in the same journal of November, 1900, shows a sketch with *Buel Patent* on the side of the anvil instead of *Trenton*. One small anvil made about 1900 has *Buel Patent* in an oval as the trademark, the same as in the advertisement.

The first advertisement to show the *Trenton* trademark alone was in this same journal of March, 1901. After that time, the *Trenton* trademark remained the company's symbol of anvil excellence until the last anvil was made.

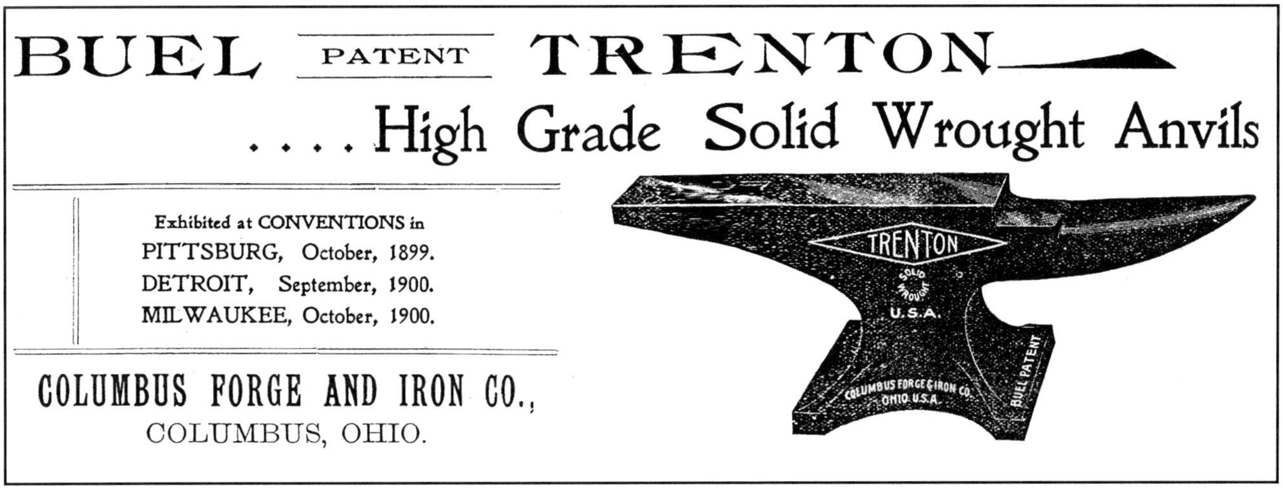

The Blacksmith and Wheelwright, March, 1901. This was the fifth ad to appear in the journal.

It was my good fortune to make contact with the Columbus Forge and Iron Company's last owner, Mr. David Tilton (grandson of one of the early founders), who still had many of the company records, including the record books of almost all the anvils that they made from 1920 on except for possibly a book of about 1941 to 1943. Mr. Tilton has allowed me to use these records pertaining to their anvils for this publication. With Mr. Tilton's assistance, I was also able to interview Mr. Karl Wright (a descendant of Peter Wright?), the last wrought anvil maker in this country. From these records the history of anvil-making by the Columbus Forge and Iron Company can be reconstructed with some certainty.

In the 1890s, Mr. David Buel, who owned a small forge shop (and had possibly worked for Hay-Budden Manufacturing Company or Peter Wright) which was known as the Columbus Forge and Iron Works, Columbus, Ohio, came up with an idea to produce a high-quality, low-priced wrought anvil. Instead of forging the anvil base as other anvil makers did he proposed to cast the base of a weldable material, then forge weld it to the forged body. This would constitute a considerable savings in time when manufacturing an anvil, thereby, lowering the cost. He applied for a patent on December 17, 1898, for this process, and it was granted on August 8, 1899.

Before applying for the patent Mr. Buel acquired a number of supporters and formed the Columbus Forge and Iron Company which was incorporated on January 25, 1898.[42] Undoubtedly, the company had been making anvils before applying for the patent because one advertisement says they exhibited this anvil in Pittsburgh, Pennsylvania, in October, 1899 shortly after the patent was granted.[43]

United States Patent Office [44]

David Buel, of Columbus, Ohio
Anvil

SPECIFICATION *forming part of Letters Patent No. 630,323, dated August 8, 1899.*
Application filed December 17, 1898. Serial No. 699,544 (No. model)

To all whom it may concern:

Be it known that I, DAVID BUEL, a citizen of the United States, residing at Columbus, in the county of Franklin and State of Ohio, have invented a certain new and useful Improvement in Anvils, of which the following is a specification.

My invention relates to improvements in anvils of the class having the trade designation of "solid wrought" or "solid welded." Anvils of this class are known as "high-grade" anvils, and are so designated because of their solidity, clear ring, resulting from the wrought-metal backing and a hard-steel face, the desirable long slender horn and heel made possible by the elasticity and strength of the wrought metal, and the strong rebound of the hammer, which is desirable and which is made possible by the employment in the anvil construction of "live" metal as distinguished from "dead" metal, the latter term being ordinarily applied to cast metal, the physical qualities of which prevent such elasticity and rebound.

Low grade anvils have heretofore been constructed of cast metal capped by a hard-steel face; but such anvils lack the desirable qualities of the high-grade article because the horn and heel in order to stand any amount of wear must be made short and thick and the anvil is without ring or life.

The production of high-grade anvils is attended by considerable expense, and the form and dimensions of the wrought-metal bases have been limited by the requirement for the employment of expensive machinery and the necessity for disproportionate labor in their production. In order, therefore, to produce a high-grade anvil at a greatly-reduced cost and embodying certain features of utility not found in any other high or low grade anvils at present known to the art, I have welded into a single structure a wrought-iron anvil-body capped by a crucible-steel or other comparatively hard-metal face and having a broad heavy cast base. By this combination of metals I am enabled to produce the desirable slender horn and heel, because the body from which they spring is wrought metal, and at the same time the expense attached to the formation of the base is greatly reduced and limitations of form and dimension of said base are removed. I therefore produce an anvil having all the desirable attributes of a high-grade anvil at a cost little exceeding that of the production of a low-grade anvil and having a broad heavy base of dead metal, which will be seated more firmly because of the lack of vibration under the blow of the hammer. These objects I accomplish in the manner illustrated in the accompanying drawing, in which the figure is a perspective view of my improved anvil.

Similar letters refer to similar parts.

***A** represents the base or bottom portion of an anvil, **B** the upper or body portion, and **C** the* *top face or working surface of said body, the separating-lines of these several portions of the anvil being indicated in the drawing by dotted lined **d** and **d¹**.*

*The base or bottom portion **A,** which may be the form indicated in the drawing or of any other suitable form or size, I produce of cast-steel--that is, of steel cast in a mold--or other metal capable of being welded.*

*The upper or body portion **B,** together with its extended horn **B¹**, I produce of wrought metal, either iron or steel--that is, of metal worked either by rolling, hammering, or pressing into the desired contour or outline. The upper working face **C,** which in the drawing I have shown separated from the body by the dotted line **d¹**, I produce of crucible-steel or steel that can be hardened. These three parts or portions of the anvil are united by welding one to the other, the welding-points being indicated or approximately indicated by said dotted line **d¹**. Although I have specified the production of the base by casting the same in a mold, it is obvious that said base might be cast in the approximate form desired and afterward worked into a somewhat-modified shape or form without departing from the spirit of my invention.*

From the construction described it will readily be seen that the production of an anvil with a cast-steel base admits of said base portion being produced in shapes and sizes which would be impracticable of production by forging, as in wrought-steel. It will also be seen that the mold or cast construction of the base admits of the production of a substantial and high-grade anvil at a much-reduced expense over that incurred in producing the same of wrought or worked metal.

Having now fully described my invention, what I claim, and desire to secure by Letters Patent, is--

As a new article of manufacture, a solid wrought anvil comprising a wrought-metal body portion having a slender horn and heel, a comparatively hard-metal face, and a broad, cast-metal base, the whole being welded to form a solid structure, substantially as specified.

DAVID BUEL

In presence of--

C.B. Shepherd,

P.S. Karsbuer

The patent and drawing courtesy of the Smithsonian Institution.

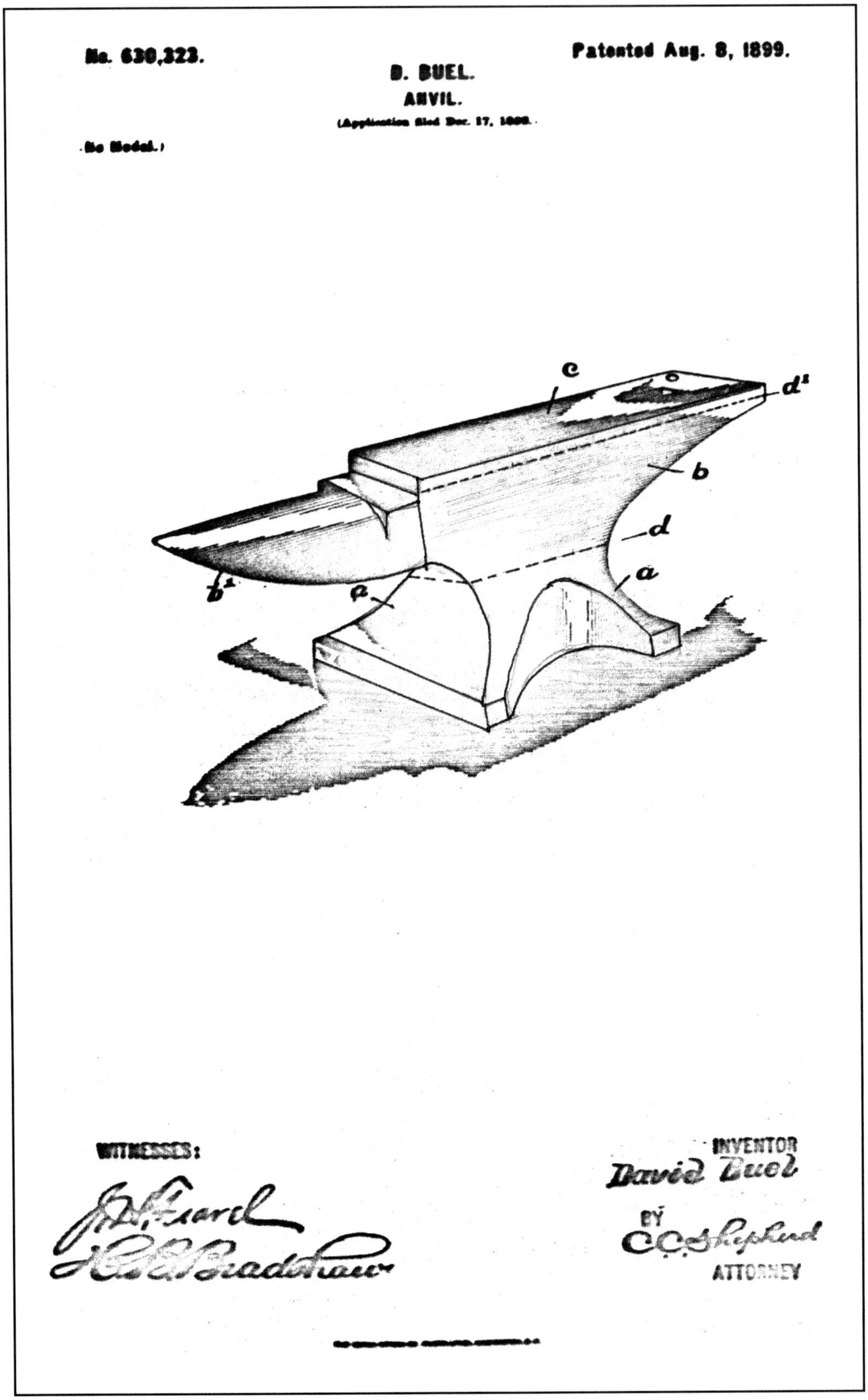
No. 630,323.
Patented Aug. 8, 1899.
D. BUEL.
ANVIL.
(No Model.)
c
d'
b
d
a
b'
a
WITNESSES:
INVENTOR
David Buel
BY
C. C. Shephard
ATTORNEY

Incidentally, the "Iron" in the company's name refers to wrought iron, not cast iron, because they were strictly a forging company. Any castings they used (which were many) both of cast iron or steel, as well as the anvil bases, were bought from foundries.

The body was made of one piece of low-carbon steel or wrought iron, whichever was less expensive. The steel plate was of Swedish tool steel when that was obtainable.

We do not know if the same methods and materials were adhered to at all times, but as with the other American anvil makers substitutions were made when their regular supply was unavailable or inadequate, as was the case during World Wars I and II. All the anvil makers used the least expensive material to produce their desired result. Anvil competition was very keen around 1900.

All the Buel patent specifies is that the base be made of a weldable material which in most cases would have been low-carbon steel. Wrought iron would have been used for the body when its price warranted it, as it forge welded to the base better than low-carbon steel.

The Buel patented process, as used by the Columbus Forge and Iron Company, brought the price of wrought anvils down to the lowest level in history, thereby reducing the price of all anvils. One old catalog (1895) said that for many years the price of wrought anvils had been ten to twelve cents per pound, but by 1903, American wrought anvils were selling for 7 1/2 cents per pound.[45] Sears Roebuck offered a U.S. made wrought anvil for 7 1/4 cents per pound.[46] I am certain that this Sears anvil was made by the Columbus Forge and Iron Company. Indeed, anvils were at their lowest price in history, but the price per pound has been rising ever since. The only wrought anvil available in this country now (1991) is a German import which sells for about four dollars a pound plus freight. I am sure that the other anvil makers at the turn of the 20th century were not overjoyed with the introduction of the Buel-patented *Trenton.*

At the turn of the century and shortly thereafter, blacksmithing and especially horseshoeing were reaching their peak, thereby creating a large market for anvils. All the major anvil makers seem to have prospered at that time and a number of new manufacturers of anvils came into the market, but the Columbus Forge and Iron Company, became one of the leaders.

The cast bases for the *Trenton* anvils were made in a range of standard sizes to fit anvils of a specific weight spread. For example, a number 23 base weighing 21 pounds was to be used on anvils of 45 to 50 pounds, and a number 41 base weighing 75 pounds was used on anvils ranging from 170 to 200 pounds. The largest standard cast base listed is a number 72 which was used on anvils of approximately 600 pounds. I have to assume that anvils over 600 pounds used a forged base. This is a reasonable assumption since there were only several hundred anvils weighing more than 600 pounds ever made by them so this would not justify the making of foundry patterns.

The company's use of these standard base sizes led me to an interesting observation, while I was recording *Trenton* anvils. Some of the bases I have recorded seem out of proportion to the body, being either too large or too small. The records list sixteen different base sizes, and I am sure that the company did not have an adequate supply of every base at all times. My observations suggest that at times, in order to expedite an order, a substitution of a larger or smaller base was supplied to a body that did not fall into the base's weight range. This shows up more on the earlier anvils than on the later ones, when there was a greater demand and orders had to be filled fast to compete.

During my recording of *Trenton* anvils, I have observed that more of them have broken at the waist than on any other make. These may be the forged low-carbon steel-bodied anvils that were welded to low-carbon cast-steel bases. I have recorded as many *Trentons* as *Hay-Buddens* made before 1920 and many more *Peter Wrights* than either of the others, but have found only one *Hay-Budden* anvil and no *Peter Wright* anvils broken at the waist. That is probably the price the Columbus Forge and Iron Company had to pay for using a low-cost base, but when they are arc-welded back together, they are as good as new.

Karl Wright (the last anvil maker) told me that at various times even when an anvil was finished and set aside to be shipped, a loud cracking noise could be heard and the steel plate would jump off of the body. An inventory taken in 1948 in the anvil shop of the anvils and materials associated with anvil making shows a number of scrapped anvils. What the problem was it does not say, but it most likely had to do with the shape not filling out in the closed dies or heat-treating problems, because by this time the anvils were plateless and the body was electric-arc-welded to the base.

I am sure the other anvil makers had similar problems. All in all, the *Trenton* has proved to be an excellent anvil and was the last wrought anvil to leave the market in this country.

The Columbus Forge and Iron Company produced somewhat over 225,000 anvils through the years, the majority being made before 1920. There is no way of telling how many *Trenton* anvils were imported by H. Boker & Company into this country; however, it must have been a considerable number because I have recorded a good many of them. As of April 16, 1920, the Columbus Forge and Iron Co. had produced 176,609 serial-numbered *Trenton* anvils, in twenty-two years. The Boker *Trentons* did not have serial numbers. The highest serial number I have listed is 225,551, so between 1920 and about 1952, the next thirty-two years, they produced 48,952 anvils. It is well to note that from the 1920s on this company had about the whole American wrought anvil market to itself. It is also a good indicator of how quickly blacksmithing disappeared.

After dropping their anvil line about 1952, the Columbus Forge and Iron Company continued as a forging job shop until June of 1982.[47] At that time the property was sold to the Capital City Products Company because of economics and the equipment was auctioned and sold off. It should be mentioned that Capital City Product*s* and the Columbus Forge and Iron Company started at approximately the same time and grew side by side. Capital City Products became one of the largest processors of eatable oils in the country and is still growing. Where some of the finest anvils in the world were once made is now (1991) a parking lot. The same is true of the Columbus Anvil and Forging Company (*Arm & Hammer* 1985). I was told by a person living in Brooklyn, New York that the same is true of the Hay-Budden works on North Henry Street (1994).

I photographed the site of the Columbus Forge and Iron Company in 1985 when most of the buildings were still standing. My pictures were in color and I had intended to get them in black and white so that they would reproduce better for this book, but I procrastinated for one reason or another. When in 1990 I felt I had better get these pictures before it was too late, I learned that I was too late. Is there a moral here?

Mr. Tilton, the last owner, gave me a tour of the old plant in 1985 when I took the photographs. While looking at this and that I was shown a small building joining the forge shop. The roof was caving in, letting in some light, and there in one corner were a number of long tongs. A piece of wire was wrapped around their handles to hold them together and from this wire hung an old faded shipping tag. Some writing was still legible and on close inspection, the words "Anvil Tongs" could be read.

In the same building or one next to it (I was so excited, I can't remember which), among a buildup of leaves, there were blacksmith vise boxes and screws (nuts and screws) scattered about the brick floor. The company had not made anvils since 1952 and vises since 1934.[48] Could it be that someone in the company thought that there was a possibility that blacksmithing might have a resurgence?

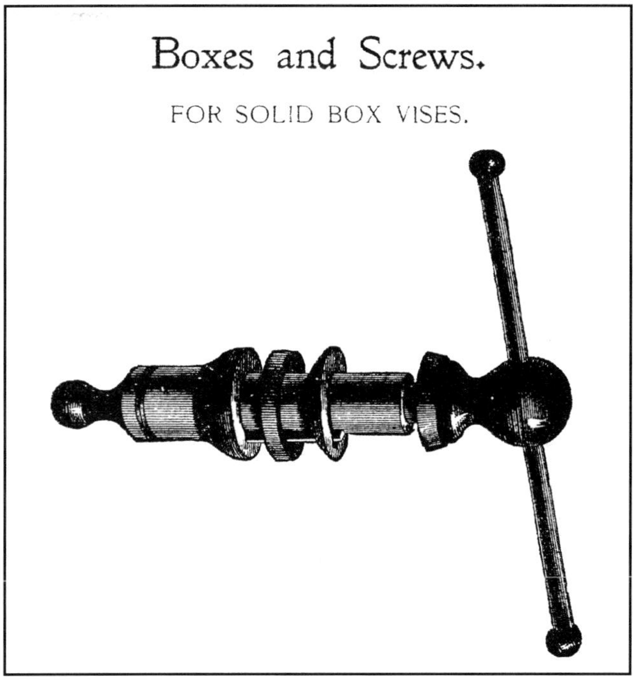

The type of nuts and screws found on the brick floor.

The company records show that in February of 1904 Mr. David Buel was given a six month's leave of absence with full pay. He reappears briefly in the records and by late 1905 appears no more. The records suggest that sometime between 1905 and 1907 Mr. David Buel, who held the largest amount of stock in the company, sold his shares of stock to Mr. David W. Singleton (Mr. Tilton's grandfather), whose family controlled the company until it was liquidated in 1982. I suggest that Mr. Singleton was the plant superintendent from the beginning and possibly also with the Columbus Forge and Iron Works. He is probably the person who had the trouble with Joe Long, or the story of Mr. Long would not have been preserved (see Columbus Anvil and Forging Company).

Another spinoff of the Columbus Forge and Iron Company was the Ohio Horseshoe Company, makers of pitching horseshoes. It was acquired by the company in 1921 and to my knowledge, is still operated by a great grandson of Mr. Singleton, Mr. John King. The Ohio Horseshoe Company made the first forged pitching horseshoes in the country.

The forge shop I saw and photographed in 1985 was built in 1922. At about this same time, the company started changing over from mostly open die forgings to closed die forging under the direction of Mr. Earl M. Tilton, Mr. Singleton's son-in-law, and a mechanical engineer. No one living in 1991 knew what the older forge shop looked like, but it is my guess that it was basically a wooden structure similar to a picture I have of the old Columbus Anvil and Forging Company forge shop (see Columbus Anvil and Forging Company).

The Columbus Forge and Iron Co. forge shop in 1985. (R.A.P. photo)

As the photo shows, the forge shop of 1922 was constructed of steel with a sheet-metal skin; "cook in the summer, freeze in the winter." Anyone who has worked in a large forge shop or foundry knows what I mean. Mr. David Tilton said that he could remember only one man who passed out because of the heat.

A few of the anvil makers' names have been preserved from the Columbus Forge and Iron Company. I have already mentioned Mr. Karl Wright, their last anvil maker. Mr. Wright was with the company from 1934 until his retirement in 1979.[49] His father, Mr. Arthur (Cy) Wright, started with the company in 1900 and finally retired in 1950.[50] If only I could have interviewed him. During his working years just about the whole span of wrought anvil making took place in America. There was the anvil maker, Tom Long, who probably started with the Columbus Forge and Iron Works and stayed on until 1899 or early 1900. I was told of an anvil maker by the name of Charles Zulty who was with the company from approximately 1900 until about the 1940s.[51] He was said to be one of the best hammer men the company ever had. He was a bit of a show-off and could bring the head of a steam hammer down and just touch the crystal of a pocket watch without breaking the crystal. Breaking the crystal and not harming the watch was considered quite a feat among steam hammer men. Another anvil maker, Mr. Joseph King (Mr. Earl Tilton's son-in-law), remembers was a Mr. Anderson. On a slip of paper among the records are the anvil makers' rate of pay for January 1, 1946. Five anvil makers are listed. They are Mr. Cy Wright, Mr. Karl Wright, Mr. Cheesman, Mr. "Doggie" Taylor and Mr. Jack Ruble. Mr. Joseph King said that the anvil maker would often put the first initial of his last name to the left of the weight on the front of the foot. Many of the *Trenton* anvils I have recorded have, Ws, Zs and Ts as well as other letters stamped to the left of the weight. These letters probably stand for some of the people mentioned above. One very large anvil (900 pounds) has "WZ" to the left of the weight. Could this mean that Wright and Zulty had combined to make this anvil? About 1916, the letter or letters to the left of the weight are gone, but show up again from time to time. I have not recorded enough *Trenton* anvils made after 1920 to be sure that the letter marking was no longer consistently stamped on at least some of these later anvils.

The following statistics should be interesting. Here is the cost of making a 175-pound anvil in 1946.[52] The cast base cost was 6 cents per pound, the steel body - $2.70, welding rod - 13 cents, cutting gas - 18 cents, labor - 35 cents per 100 pounds; for buffing - 30 cents per 100 pounds; for grinding - 50 cents per 100 pounds; for welding - ? and 10 cents to paint and cover. The hammer labor was 73 cents for blocking, 50 cents for forming the horn and the overhead was $8.00 per hour. The 175-pound anvil sold for $32.38 wholesale less a commission to the broker. At that time, they could produce about 1800 pounds per day of anvils above 100 pounds, and 1200 pounds per day of anvils under 100 pounds.

From 1898 until 1927, all their anvils were made in three pieces--forged body, cast base and tool steel face plate. The following is the procedure as Mr. Earl Tilton describes it in a letter.[53]

> *Old Method*
>
> *Heat suitable size billet (.20-.25c) to forging temperature. Block under open frame hammer and loose forming die. Weld tool steel plate to face of anvil using fluxing compound. Draw horn to proper shape. Heat lower part of top and top of base to welding heat together with heated plate. Place together with fluxing compound under hammer and strike, welding top and bottom, pounding weld line with air hammer. Heat treat face, quenching in brine solution. Finish and grind under surface grinder.*

This is a very brief description he wrote at the request of a man who understood forging practice. For those of you who are not familiar with forging practice I will translate the quotation. Those of you in the know may skip over this section, but if you do you could miss something of interest.

The *suitable billet* is one piece of low-carbon steel of a size for a particular-weight anvil. *Forging temperature* is about 2,000 degrees for this type of steel. *Block* consists of bringing the billet to the proper width, drawing out the heel to the length and thickness and roughly drawing out the horn (see *open frame hammer* under Glossary). The hammers were steam-operated. The *loose forming die* has the approximate shape of the heel and horn in it, but depends on the manipulation and the number of blows of the steam hammer head on the billet to arrive at a finished size. All this depends on the skill of the anvil maker. On a large anvil, there might be five people working together at the steam hammer. They were known as the "Hammer driver," "Heater," "Anvil maker" and "Helpers." The "Anvil maker" manipulated the billet which was held in a pair of large tongs with the help of one or two of the "Helpers." The "Heater" kept track of the material being heated in the furnace or forge and of the piece being forged. The "Hammer driver" operated the steam hammer controls under the direction of the "Anvil maker."

A small anvil, 200 pounds and under, would require two people--the "Hammer driver" and the "Anvil maker." In 1926 a team of five could produce 2500 pounds of anvil bodies weighing over 60 pounds each in a ten-hour day. If the anvil bodies averaged 125 pounds each (200-pound anvil), they put out 25 anvil bodies per day for one steam hammer.[54]

The *tool steel plate* of the proper proportions for the particular-weight anvil is heated to its welding temperature (a lower temperature than the body) as is the semi-finished body, using separate furnaces or forges. The face of the body and the underside of the plate would have a slight crown so that when the pieces are brought together with the flux on their surface, the center will touch first. Thus the flux and slag would be forced to the outer edges of the plate and body when struck by the steam hammer's top die, thereby producing a solid clean weld. This is a difficult weld and unfortunately the weld did not always take or hold. A loose plate might show up later when the anvil was put to use. I have seen a number of these, but don't recall any of them being *Trentons*. Here again, much depended on the skill of the anvil makers. After this, the horn and table are brought to the proper finished shape. At this time, the hardie and pritchel might be punched out.

The same procedure was involved in welding the base and body together except that in this case, a thin *heated plate* is used along with fluxing compound. The thin plate is probably made of a special wrought iron that welds easily. This plate is placed between the base and body as they are aligned under the steam hammer head. It must be remembered that the base was cast of any weldable material of unknown carbon content, and may have included other alloys that were not so weldable. The body was made of low-carbon steel in most cases, which does not forge weld as easily as wrought iron. The thin plate assured a good bond between body and base, as it adhered to both readily. Each company had their own special process for making these welds.

Pounding weld line was performed in the old days with sledge hammers and fullers. The *air hammer* which came later was hand-held, using a fast reciprocating motion like the action of a jack hammer. The anvil was held in a vise-like grip by the head of the steam hammer while this *pounding* was performed. The *line* was a mushrooming protrusion extending completely around the waist and had to be blended into the body of the anvil. This blended line can be seen in many cases on American-made wrought anvils, especially under the horn and heel. The line between the face plate and body is also readily seen on American-made anvils. This is because the face plate is forge welded on with one or two blows of the steam hammer, thereby not mushrooming the plate as would happen if the face plate was forge welded on with hand-held sledge hammers as the Europeans, including Peter Wright, did it. The Europeans had to dress the edges of the face plate after welding, whereas the American manufacturers did not. That is why the face plates on European anvils blend in so well with the body. The English tried to blend in both the face plate weld and the waist weld and did an excellent job. Any touching up of rough spots and any straightening took place at this time, as the base and body would be at white heat and easy to work.

Mr. Earl Tilton does not mention putting in the hardie and pritchel holes.

This photo was taken at the Columbus Anvil and Forging Co. The "hammer driver" is almost out of the photo to the far right. (Photo courtesy of Mr. J.E. Finneran Jr.)

As stated before they were either punched in at the time time the plate was put on or drilled and broached before heat-treatment.

Heat-treatment consisted of bringing the face plate to the proper temperature (1500 degrees Fahrenheit in most cases) in a forge and then plunging the anvil face down into a tank of swirling *brine solution.* Water could be used in the summer. Depending on how hard the plate became, there might be a drawing process after hardening. The records suggest a draw, although other manufacturers did not draw. Sometimes just the edges were drawn to prevent chipping. The record does not indicate whether each anvil was tested for hardness at this time or not. Look up *draw* in the Glossary if it is unfamiliar to you.

I am not sure what is meant by *finish*, because nothing can be done to the face after heat-treatment except to grind it smooth. If the face was really rough after welding, it would be machined flat before heat-treatment. The face was ground smooth by a huge grinding wheel made of Ohio sandstone that was about six inches wide and about six feet in diameter. Mr. David Tilton said he could still remember the strange sound that the wheel made as it passed over the anvil's face.

An old planer mill similar to the one used by the Columbus Forge and Iron Co. to grind the face of their anvils.

The horn was then buffed with a homemade rig that I suspect was similar to a flap sander. Some of the later horns appear to be finished with a hand or disk grinder. The anvil was weighed and marked, the trademark and serial number stamped in, then painted black and tagged for shipment. "Cover" is mentioned in the records and this must have had to do with protecting the face during shipment to the customer.

In 1927, the Columbus Forge and Iron Company started experimenting with a body and face of one piece of tool steel. Hay-Budden did this in 1908. While they were putting some of these new two-piece anvils on the market, the records show that many of the anvils were made by the older method, especially the larger ones. It was not long, however, before all of their anvils were made without face plates, with few exceptions

They repaired older anvils of their own make and that of other manufacturers, so they still had to plate these when necessary. The tool steel body was more expensive, but it reduced labor costs and there was no plate to come loose. I was told that Swedish steel was used until World War II for this plateless body and face.

It seems a bit strange that at about this same time, 1927-1928, Hay-Budden no longer made their plateless anvil and *Trentons* should become plateless. Maybe the Columbus Forge and Iron Company bought the patent rights or the patent had run out.

Here is how Mr. Earl Tilton describes the new process:

Final Procedure *- Solid Tops.*

> *Tops were made of billets about C.60, Mn. .70-.75 (we drop forged tops for 60 lbs., 100 lbs. and 125 lbs. and bases of proper weight). Other sizes we forged tops by old method and used cast steel bases. Bases were about .20-.25 carbon. The top of the base and bottom of the top was scarfed for about 1/2 inch all the way around, placed together, and electric welded. Tops were heat treated, giving about 100 hardness under scleroscope. Proportions were - tops about 65%, bottoms about 35% of weight.*

Translation: .60 carbon is considered a medium-carbon steel and can be heat-treated. .70-.75 manganese adds a work-hardening element to the steel so that the more it is hammered the harder it becomes, up to a point. The smaller anvil tops (60, 100 and 125 pounds) were made in closed dies, requiring less skill and giving a more uniform size and weight. The anvil record books show that they made hundreds of the same-weight anvils in succession whereas by the old method, the same-weight anvils varied as much as ten pounds either way. It appears that they also forged the bases for these smaller anvils using closed dies. The tops for the larger anvils (up to 1,000 pounds) were made in open dies. Here Mr. Tilton reveals that the cast bases were of low-carbon steel. *Scarfing* is to put a chamfer or angle on a piece of metal either by forging, grinding or cutting. The one-half inch scarf on both pieces would preclude welding to the center of the body and base as in the older process. Therefore, a deep line of weld, about one-inch wide and one-half inch deep around the periphery of the waist held the two pieces together. This also reduced the anvil's ring. *Electric welded* could mean two different electric processes at that time. One was similar to oxyacetylene welding except that a carbon electrode or electrodes were used to produce an arc flame that melted the metal to be welded while filler metal was introduced from a hand-held metal rod. This is the oldest meaning of electric welding. The other is to use a single electrode which produces an arc flame between itself and the metal to be joined, melting the metal to be joined and also providing the filler metal. In this method the electrode is consumed and is replaced with a new electrode until the job is finished. This last method is generally called "stick welding" and was just being perfected in the 1920s. I believe this last method was the one that is referred to.

I know some of these early electric welds at the waist did not hold. I recorded a *Trenton* anvil made in 1934 that had been rewelded by its owner. I asked the owner when it broke. He said he had used the anvil for about ten years and then one day while he was forging a piece on it, it just popped loose. After he rewelded it, it was as good as new.

The *scleroscope* was a device that dropped a steel ball on the surface to be tested for hardness. The harder the surface tested, the higher the ball would bounce or rebound up a type of ruler or scale. The reading was taken at the top of the bounce. The record books show that after August, 1935, they tested every anvil. They may have tested before 1935, but did not record it.

On all their anvils, they tried to keep two-thirds of the weight in the top part of the anvil. Mr. Joseph King told me that before an anvil was finished, the tip of the horn was brought up or down so that it was in the same plane and height as the face. This meant that if a piece of straight metal were laid flat on the face of the anvil, it would touch the tip of the horn. This would be of help to the smiths, but I doubt that the horns remained in this condition for long if the smiths used the horn very much. My recording of *Trentons* has confirmed as much. Outside of using no face plate and electric welding at the waist, the other finishing processes remained the same.

While the company was experimenting with the plateless top anvil (1927), they bought a Swedish solid-cast-steel anvil from Montgomery Ward and Company to compare with their *Trenton* anvils. In the fall of 1925, Montgomery Ward had stopped buying anvils from the Columbus Forge and Iron Company under Montgomery Ward's trademark, *Lakeside,* and had started importing them from Sweden. Wards must have been buying the Swedish anvils at a lower price. Montgomery Ward was selling the *Lakeside Trenton* in the 1922-23 Ward's catalog for $14.50 for a 100-pound anvil. In their catalog of 1925-26, $16.00 for the same 100-pound anvil. After a comparison for size and hardness, the data in the records suggest that the *Trenton* was a better anvil than the Swedish anvil. What ever came from this, the records do not reveal, but Montgomery Ward did not buy any more anvils from the Columbus Forge and Iron Company. Sears continued to buy anvils from them from time to time into the late 1930s. Swedish anvils took much of what was left of the American anvil market throughout the late 1920s and 1930s.

During and after World War II many *Trenton* anvils were sold to the United States military. Mr. John Boker said that the last ones he handled were to be used, "for shoeing Army mules."

I am told that about 1946, a number of freight carloads of *Trenton* anvils were shipped to Russia.[55] No wonder I can't find many of the later ones. Also, Mr. Joseph King told me that during World War II, the U.S. Army Air Force bought a number of 25-pound anvils. The Air Force said they were to be used on B24 Liberator bombers. What would an anvil be used for on an airplane? The records bear out the sale.

The Columbus Forge and Iron Company produced anvils in a wide range of weights. Their brochure advertised anvils of the standard blacksmith's pattern from 10 to 1,000 pounds. In a later brochure, the 10 to 1,000 pounds was crossed out and 25 to 600 pounds written above.

The largest *Trenton* anvil made after 1920 weighed 685 pounds, made in 1924. The next weighed 662 pounds and I have recorded this one. It was also made in 1924 and belongs to a man in New England.[56] The company could make a wrought anvil weighing up to 1,200 pounds. The 1910 photograph taken at the Ohio State Fair shows a 1,200-pound anvil in the center of their display. The face plate, I am told, was put on in one piece for these large anvils by a 8,000-pound double-frame steam hammer.

Display of forgings by the Columbus Forge and Iron Co. at the Ohio State Fair in 1910. All types of forgings beside anvils can be seen on the wall and around the anvils. The vises are to the left. There is a huge axle in the lower forefront. It appears that the company was making both open and closed die forgings.

(photo courtesy of Mr. David Tilton)

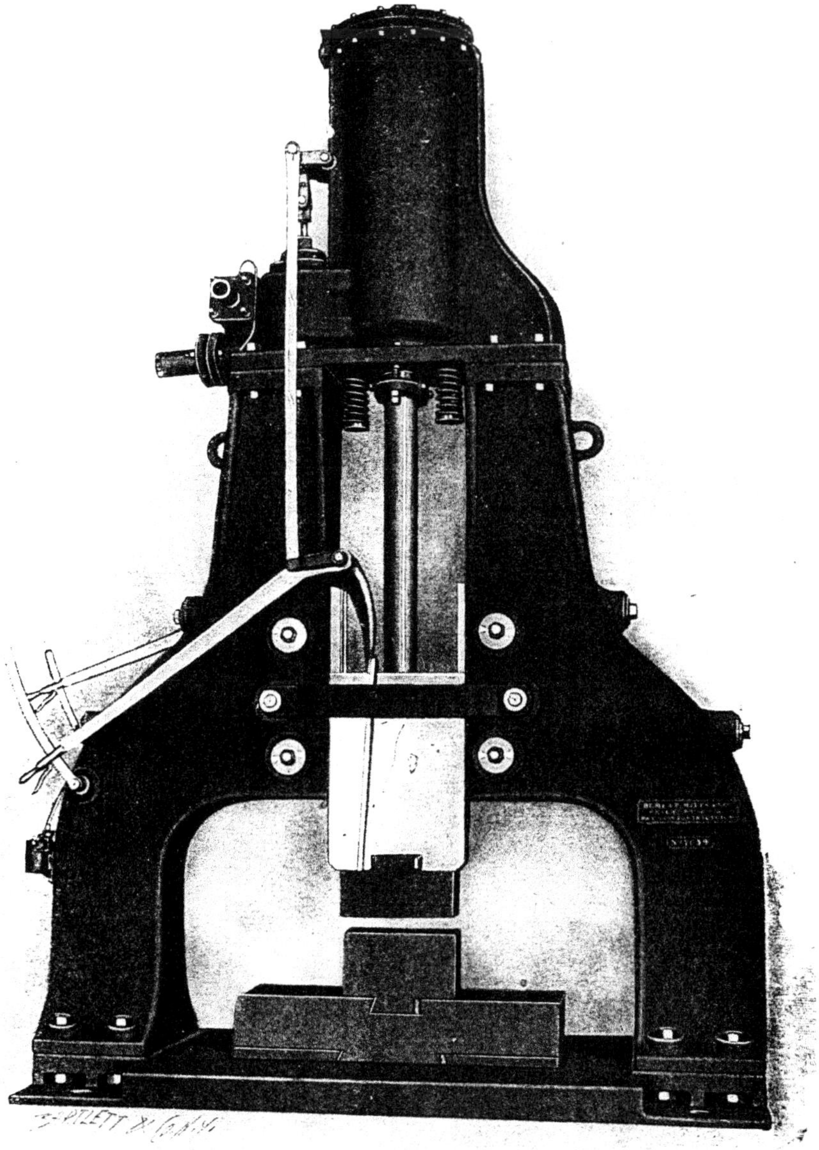

5,000-POUND BEMENT DOUBLE-FRAME STEAM HAMMER.
Niles-Bement-Pond Company

A double-frame steam hammer similar, but smaller than the one used by the Columbus Forge and Iron Co.

The 1,200-pound anvil is probably the largest wrought blacksmith's anvil ever made. The *Fisher* weighed 1,400 pounds, but is was cast. A friend of mine recorded a *Trenton* anvil in Washington state for me, and reported it to weigh 900 pounds. Its serial number indicated that it was made about 1900. Mr. Joseph King told me that about 1941 or 1942 the company received a phone call from a shipyard on the east coast. Someone wanted to know, "what is the largest anvil that you can make?" Mr. King said, "1,200 pounds." The person said, "I will take it." The anvil was made and shipped. Unfortunately, the record book for this time period is missing! The books I have ended with January, 1941, with anvil serial number 202445 and start again in November 1943 with serial number 210226. Missing are 7,781 anvils, one of which is the 1,200-pound anvil. At times the books indicated who bought the anvil, and it might have in this case. It most certainly was not scrapped in World War II and there is a good possibility that it still exists.

The Columbus Forge & Iron Co.

MANUFACTURERS OF

TRENTON SOLID WROUGHT ANVILS

COLUMBUS, OHIO, U. S. A.

H. BOKER & CO., Inc.

SOLE SALES AGENTS

101 DUANE ST.

NEW YORK CITY, U. S. A.

Front of the brochure. (Brochure courtesy of Mr. David Tilton.)

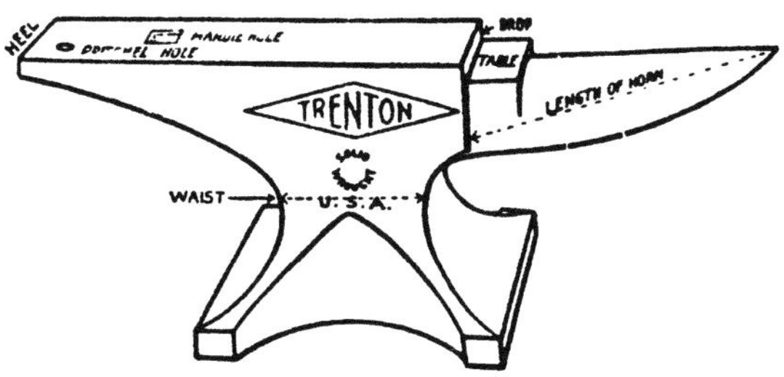

Blacksmith's Pattern

Principal Dimensions of Regular Blacksmith's Anvil

Face	Horn	Waist	Heel	Drop	Hardie Hole	Approximate Weight
3 3/8x14	9 3/4	5	13/16	13/16	7/8	100 lbs.
3 1/2x15 1/2	10 1/2	5 1/4	7/8	7/8	15/16	125 lbs.
4 x17 1/2	11 1/2	5 1/2	15/16	15/16	1	160 lbs.
4 1/4x19	12	6	1	1	1 1/16	200 lbs.
4 1/2x19 1/2	12 1/2	6 1/2	1 1/16	1 1/16	1 1/16	225 lbs.

Other sizes from ~~10~~ 25 lbs. to ~~1000~~ 600 lbs. in proportion.

Principal Dimensions of Standard Horseshoers' Anvil

Face	Horn	Waist	Heel	Drop	Hardie Hole	Approximate Weight
3 1/4x15	10	5	13/16	13/16	7/8	100 lbs.
3 1/2x16	11	5 1/4	7/8	7/8	15/16	125 lbs.
3 3/4x18 1/2	12	5 1/2	15/16	15/16	1	160 lbs.
4 x19 1/2	12 1/2	6	1	1	1 1/16	200 lbs.
4 1/4x21	13 1/2	6 1/2	1 1/16	1 1/16	1 1/16	225 lbs.

Back of the brochure.

TRENTON TRADE MARK

American Anvils

(RING LIKE A BELL)

TRENTON ANVILS Are forged, not from scrap, but from new solid billets of Special High grade Tool Steel.

These Anvils are endorsed by leading blacksmiths and horseshoers everywhere, who express a decided preference for them.

Features of special note are:

1. Long, perfectly shaped horn and long heel
2. Base with greatest possible area, or spread, insuring greater stability and absence of tipping when heavy work is done at the end of horn or heel.
3. Base which does not require strapping down and which stands firmly without shifting or rocking.
4. Special temper of edges to prevent chipping.
5. Faces level and true, and of correct proportion for practical work.
6. Hardie Holes and Pritchell Holes and other mechanical details of exact accuracy.
7. **Trenton Anvils** are unequalled for superior shape, graceful design, stability on the block and absence of superfluous weight in the base.

Trenton features.

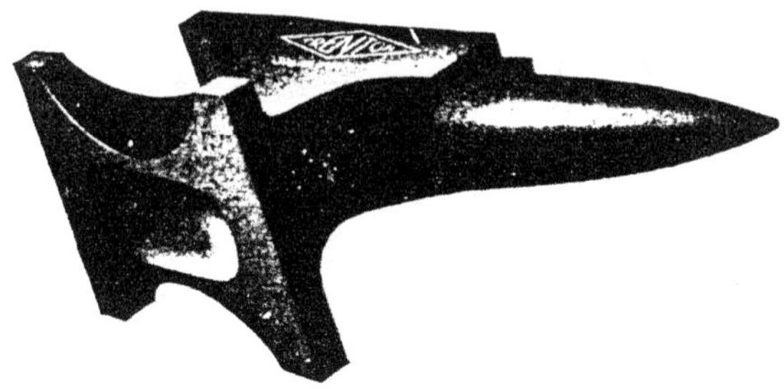

Bottom View, Showing Base Hollowed Out, Permitting Anvil to Stand Firm Without Rocking

General Blacksmith's Anvil

Top of General Blacksmith's Anvil

Smith's pattern anvil.

Farrier's Clip Horn Anvil

Farrier's Clip Horn Anvil
(Top View)

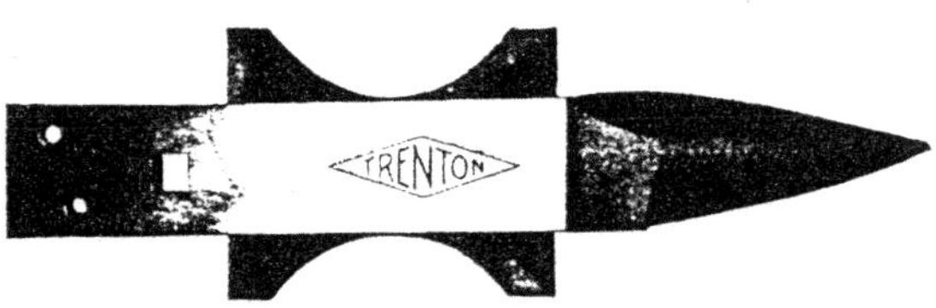

Farrier's Plain Horn Anvil
(Top View)

Farrier's Swelled Clip Horn Anvil

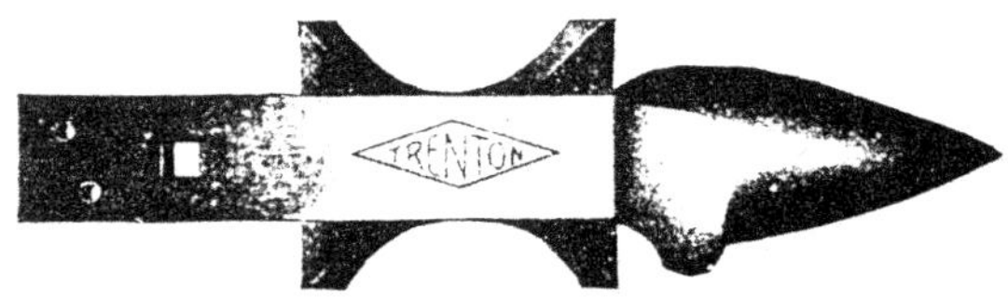
Farrier's Swelled Clip Horn Anvil
(Top View)

Farrier's Swelled Plain Horn Anvil

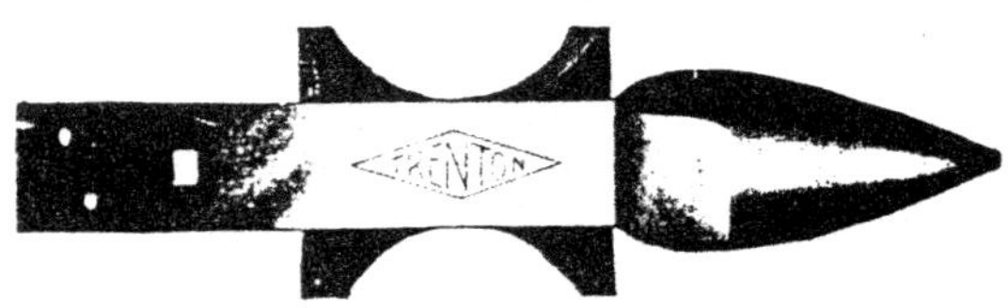
Farrier's Swelled Plain Horn Anvil
(Top View)

Farrier's pattern anvils.

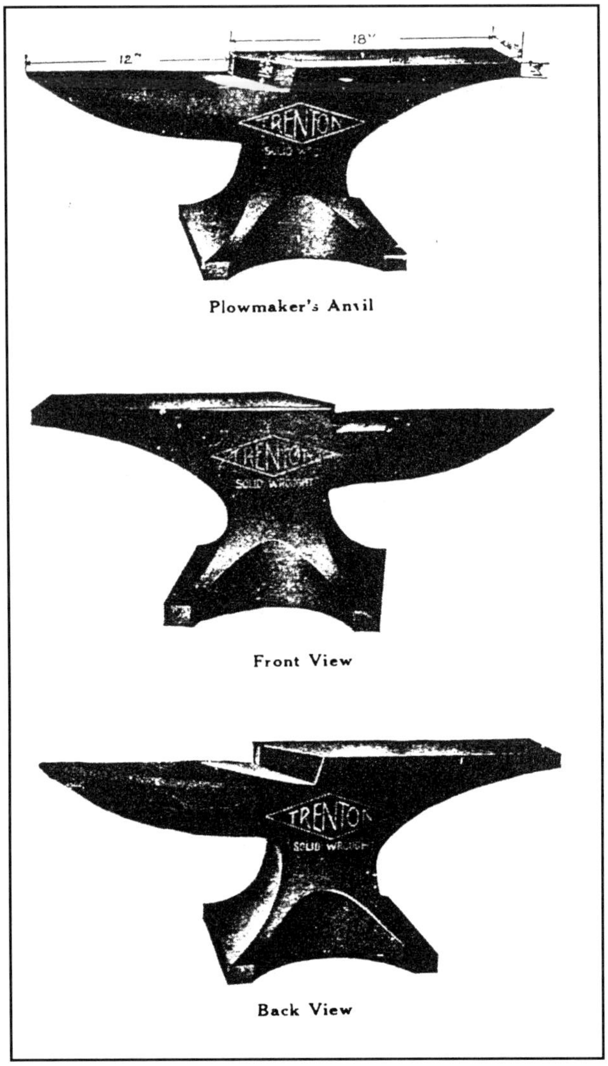

Plowmaker's anvil.

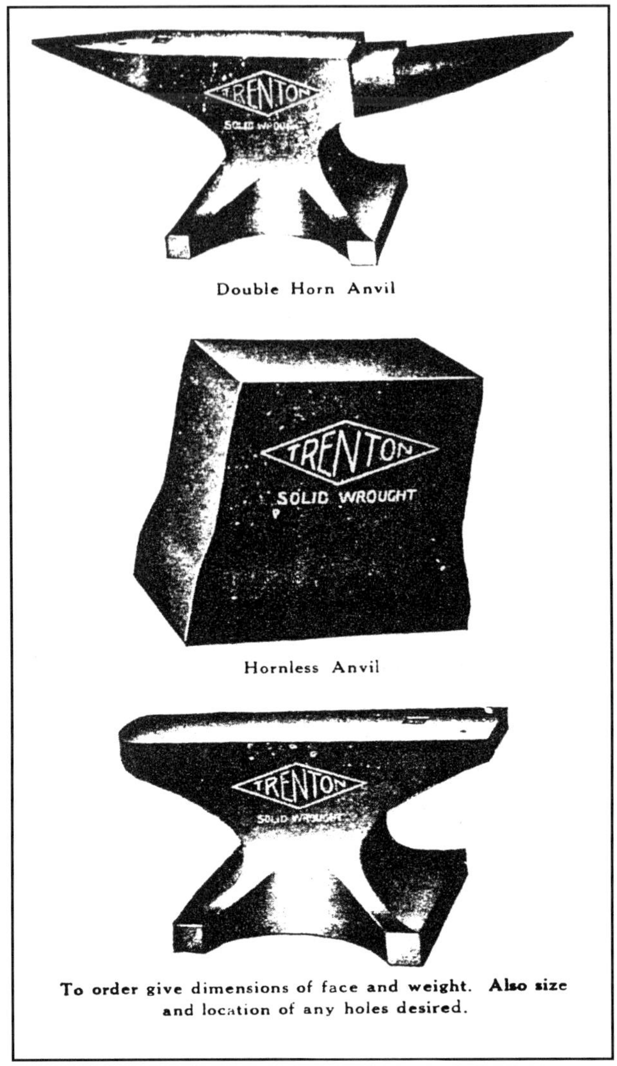

Various pattern anvils.

Speaking of scrapping anvils, I will make an observation at this point about anvils in general. I have read and heard it said that during the late 1930s and during World War II, many anvils were scrapped; in the 1930s, for economic reasons, and during the war because of patriotic fervor. I was told by several people that in a *Life* magazine of the early 1940s there is a photo of a large pyramid of anvils ready for the scrap yard. Whether this is true or not I cannot verify, but if it is true, then most of the anvils scrapped must have been of the cast variety such as *Fisher, Vulcan* and *Columbian*, because in looking for anvils to record, I have uncovered a tremendous number of wrought anvils considering the number made and shipped all over the world. Most of my research has only covered one small corner of the Midwest. Furthermore, of the anvils that I have recorded, most were made before World War I and would be the ones most likely to have been scrapped, but of course most anvils were made before World War I. When recording anvils I generally ask the owner where and when he acquired the anvil. In a considerable number of instances, the junk yard is mentioned and they were obtained during the 1930s, 1940s and 1950s. An anvil in decent condition has in most cases, I believe, brought a better price as an anvil for the junk yards than as scrap. Hopefully the 1,200-pound *Trenton* and 1,400-pound *Fisher* did not meet this fate.

When I was considering teaching a class in hand forging I was in need of a number of anvils and after having bought one new English anvil I realized the need to find some used ones because buying new anvils would be too costly. At that time, to me an anvil was an anvil was an anvil, and one was as good as another. It was mentioned in one book that used anvils were most likely to be found in junk yards along with other blacksmithing tools, so I visited the closest junk yard and asked about blacksmithing tools. The person I talked to said that they did not have any tools that he knew of, except that there was an anvil in the warehouse against a back wall. He showed me the anvil. It appeared to be strictly a junker. The horn with or without a table was missing and I could tell that the face was soft because the edges were mushroomed over rather than chipped, and it had a lot of little nicks in the face. However, the face was relatively flat, it appeared to weigh over 100 pounds and I was desperate. He asked $5.00 for it so I bought the anvil. I found that it weighed 120 pounds. As I learned more about anvils, I discovered that this junker was an old English anvil made before 1830 and possibly before 1800. The English weight marks I found on the side after cleaning it indicated that it originally weighed 137 pounds. We never did use it, but historically speaking, it has become a very valuable anvil to me. I can see exactly how it was made, how many pieces of iron it took to manufacture it and why it broke. I relate this incident to illustrate a point. Why wasn't the old junker scrapped long ago? It certainly was of no value except as a weight. A full-gauge piece of railroad track would have made a better anvil. The junk yard had it for years and knew they had it. I believe a high percentage of the wrought anvils still exist in this country. They are just not where they can be seen. Incidentally, I have never found another anvil in any scrap yards which I have visited. In 1995 I located a complete anvil in an antique shop just like the old English junker, and it has a table.

The "Old English" junker. Note the weld lines where the feet join the body. Also the added metal where the horn was. (R.A.P. photo)

Complete "Old English" anvil. (R.A.P. photo)

After 1928 the smaller anvil bodies of 200 pounds and under were drop forged in closed dies and not many anvils over 200 pounds were made. By about 1950, the Columbus Forge and Iron Company wanted to stop making anvils that could not be made in closed dies, but still wanted to supply orders for larger anvils. Mr. Earl Tilton asked the Columbus Anvil and Forging Company if they would make the large anvils for them and the Columbus Forge and Iron Company would make the smaller anvils for both companies. The Columbus Anvil and Forging Company used only single-frame open-die steam hammers. Mr. Finneran, owner of the Columbus Anvil and Forging Company said that they were too late, because they had dropped anvil making entirely and had scrapped the tooling just one week before.

Since the early days anvil making had not been the most profitable of enterprises. Although both *Trenton* and *Arm and Hammer* anvils continued to be made into the 1950s they were produced mainly because of tradition and to fill the small domestic need. As long as they were not showing a loss they had continued to make them. I was told by Mr. Joseph King that when the closed dies washed out for the several sizes of anvils they were then making (washed out dies misshape the item), they decided that it would cost more to replace the dies than could be recovered from anvil sales. Thus, they dropped anvils from their product line. The last American-made wrought anvils passed into history.

The company forged some small eight- to ten-pound craft anvils after that time and also forged some paper weight miniatures for advertising purposes. I have not run across any of these craft anvils in the field.

In 1972, Brodhead-Garrett Company of Cleveland, Ohio (a national supplier exclusively for schools), sent a letter to the Columbus Forge and Iron Company wanting to know about their anvils and if they could place an order with them for various sized anvils.[57] They were about 20 years too late.

Their last brochure stated that the Columbus Forge and Iron Company would make anvils of any size or shape to the customer's specifications. I have recorded *Trenton* blacksmith's, farrier's, plowmaker's and sawmaker's patterns, but have not located a hornless, or double-horned pattern anvil as listed in the brochure. I know that they made them because they are indicated in the record books. One entry reads "Special-nickel plated face-bronzed." It weighed 53 pounds. I have one *Trenton* anvil that is a cross between a smith's and a farrier's pattern. The buyer must have wanted the narrow face and large horn of the farrier's pattern, but wanted the table and only one pritchel of the smith's anvil. Some of the farrier's patterns have the hardie hole in the horn. Many variations may be found.

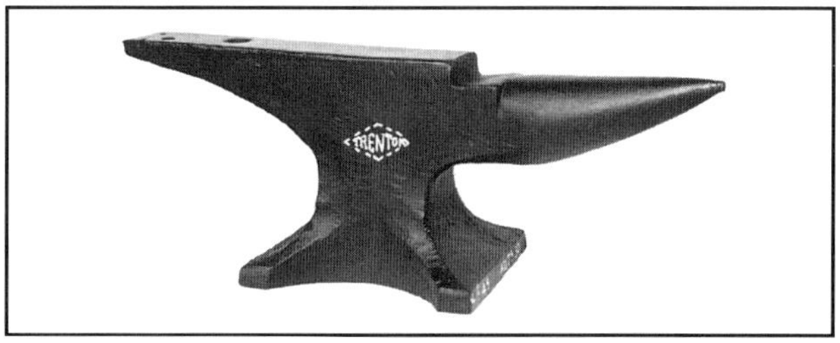

145-pound Trenton anvil. A cross between a farrier's and smith's pattern made ca. 1913. (R.A.P. photo)

I took a survey of all the anvils in the record books made between June, 1935, and January, 1941, and found that the most popular weight was 150 pounds, with 100 pounds being second. The majority of the anvil weights were in increments of 40, 50, 70, 80, 100, 125, 150, 175, 200, 250 and 300 pounds. There was almost one of each weight in between the major ones. If this was displayed on a line graph, it would look like the stock market does at times. They made 45 anvils over 300 pounds in that time period, the largest being 500 pounds, and three of 20 pounds. I noticed after a number of entries, the word "Stolen." These were always 100 pounds or less. If I ever record one of these,

COLUMBUS FORGE AND IRON COMPANY

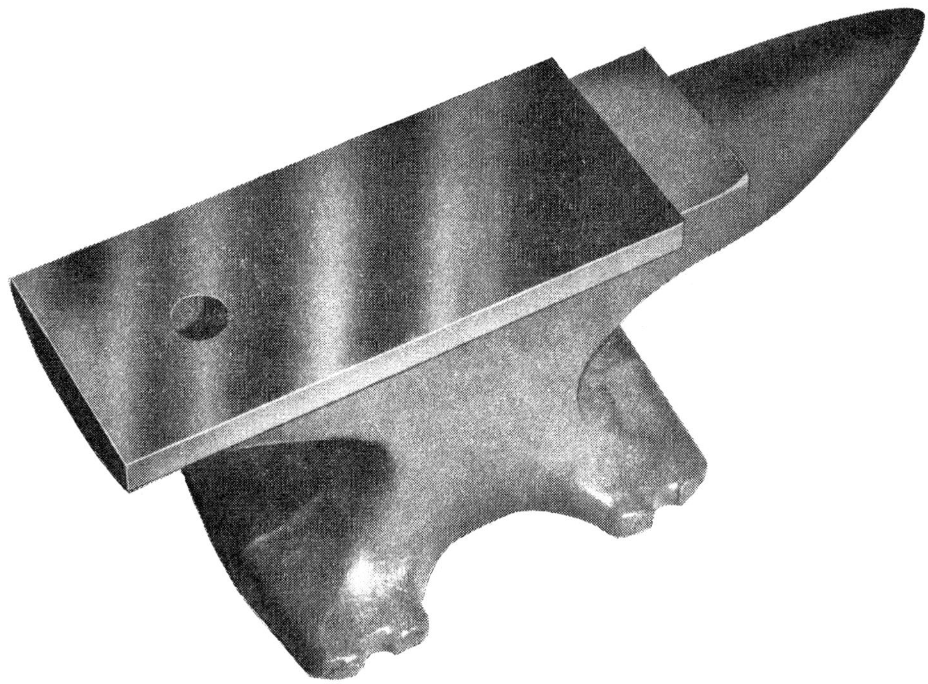

The Spreading Chestnut Tree Is Gone,
The Smithy Anvil Rings But Little More,
Yet Handed Down The Skill Goes On And On,
Through Father, Son and Time, Now As Before.

Author Unknown

The Anvil — Symbol of quality, strength, and the proud heritage that is ours at Columbus Forge and Iron is available only in the home workshop size. Featuring a milled, polished and lacquered face, the anvil above weighs 6½ pounds, is 8″ overall and 3½″ high. Bright red enamel finish and good quality steel make it a useful as well as magnificent item. Drop forged, of course.

Since 1898

FI

THE COLUMBUS FORGE & IRON CO.

COMMERCIAL **DROP** FORGINGS

544 W. 1st Avenue | AX 9-3588 | Columbus 8, Ohio

The forge craft anvil approximately 6 1/2 pounds, 1959. (Photo courtesy of Mr. David Tilton) it will tell me something about its first owner. Another entry had "Freak" behind it.

The Columbus Forge and Iron Company made anvils under a number of different names for various hardware companies and catalog stores. Sometimes the *Trenton* trademark can still be seen beneath the hardware company's trademark. I have recorded the following brand names on Columbus Forge and Iron Company anvils, but in most cases have not discovered the buyer:

Acme - S/N 66690, (1907); Sears & Roebuck trademark used on their anvils for many years.

Ajax - S/N 49094, (1904); I was told that the anvil was a second of the Columbus Forge and Iron Company, but did not affect its usefulness. However, the records show many of these made one after the other, so it must have been some hardware company's trademark also.

A T C, U S A - S/N 124449, (1914); unknown.

Black Prince - S/N 192770 (1926) A trademark of Shapleight Hardware Company, St.Louis, Missouri. The Columbus Forge and Iron Company made over 300 of these anvils in the 1920s and30s.

Buel Patent - S/N A1713, (1900), S/N 10718, (1899). An early Columbus Forge and Iron Company trademark for some anvils.

CFI - S/N 314?4, (1902), S/N 113355, (1912); The Columbus Forge and Iron Company's forging trademark.

Cole - S/N 69039, (1907); unknown.

Derby - S/N A24712 (1901), S/N 15931, (1900); Farrier's anvils.

Indian Chief - These were bridge or railroad anvils made of cast iron or cast steel. They were made for the C&S Railroad in two weights, 650 and 750 pounds.

Indian Chief - S/N A81940, (1908); A Columbus Forge and Iron Company trademark.

Lakeside - S/N183427, (1923); A Montgomery Ward trademark.

Phoenix - S/N A47254, (1904); unknown.

Reliable - S/N A61275, (1906); unknown.

Roberts Anvil - S/N 13726, (1900); unknown

Village Forge - Unknown. They made hundreds of these in the 1920s and 1930s.

Vulcan - S/N A8072, (1899); unknown. I have recorded two of these and they were both made about 1899

. *???? S D E* - S/N 134678, (1915); unknown.

There are probably other brands made before 1920, but these are the ones that I have been informed of and discovered so far.

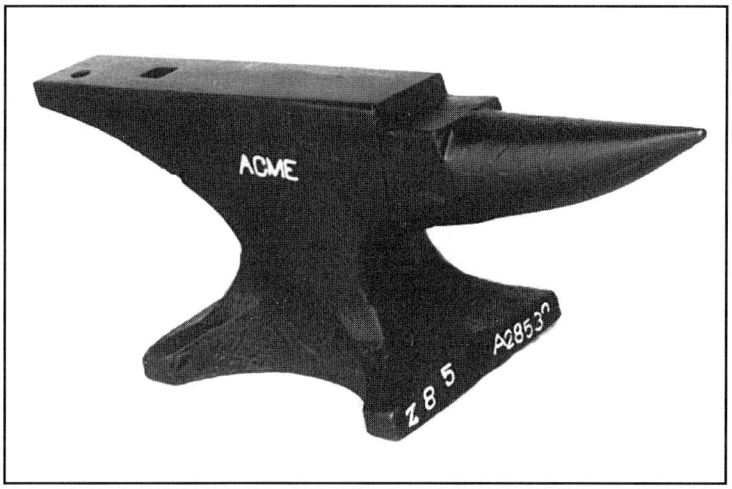

85-pound Acme anvil made ca. 1910. (R.A.P. photo)

213 -pound Ajax anvil made ca. 1904. Part of the Trenton triangle can still be seen. (R.A.P. photo)

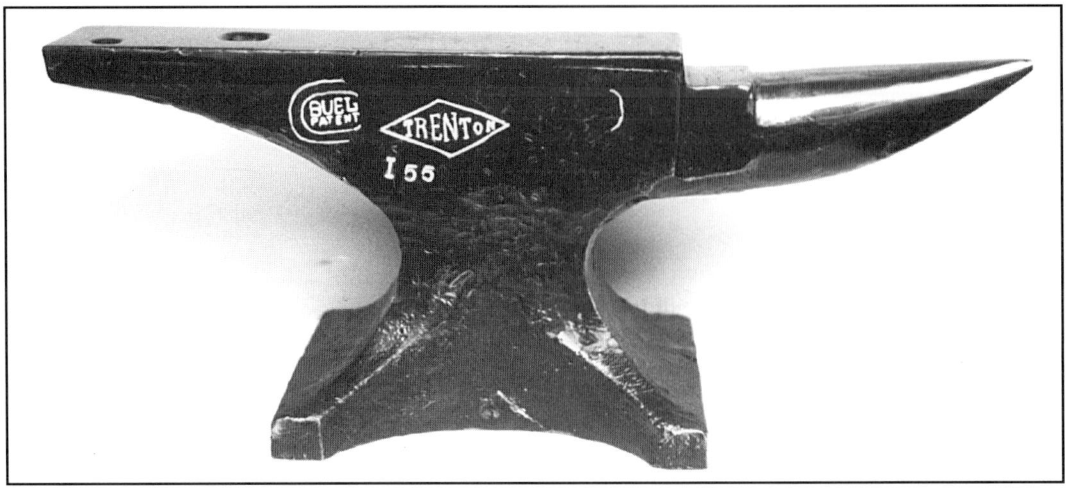

155-pound Buel patent Trenton anvil. (R.A.P. photo)

Pattern for the "Indian Chief No. 70 C & S Railroad anvil." (R.A.P. photo)

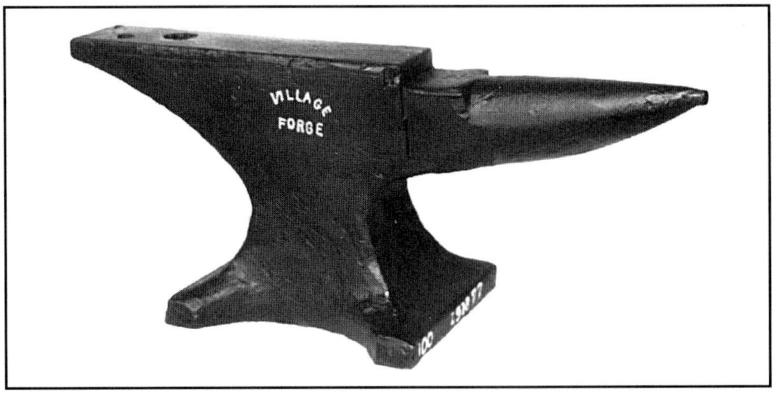

100 -pound Village Forge anvil made January 12, 1926. (R.A.P. photo)

164-pound German Trenton. This anvil has "German" stamped on the foot under the horn. (Mr. Steve Jacobson photo)

A Columbus Forge and Iron Company anvil is usually not hard to identify whether a trademark is on the side or not. The *Trenton* trademark on the side does not necessarily mean that the anvil was made by the Columbus Forge and Iron Company. As I explained before, *Trenton* anvils were imported for fifteen to twenty years before 1898 under the *Trenton* trademark.

Here are some of the characteristics of the imported *Trenton* anvils that distinguish them from those made in this country. They do not have a serial number; they do not have *U.S.A.* on them below the trademark; the weight is always on the side below the trademark; the word *Patent* may be below the trademark, and they are always flat under the base. They may also have *Solid wrought* in a circle like *Peter Wright's* on the side below the trademark. Like the *Peter Wright,* some of them have a step on top of the front and back foot and the base is always forged. I suggest that the anvils with the step and that have *Patent* under the trademark were possibly made by Peter Wright or Henry Wright. The ones without the step or *Patent* and have no handling hole in the front of the foot were probably made in Germany.

The earliest Columbus Forge and Iron Company *Trenton* looks very much like the foreign makes except that they have a serial number and *U.S.A.* below the trademark. Some of the early American *Trentons* have the step on the feet, the four handling holes, the forged base that is flat beneath, the weight on the side, and the *Solid Wrought* in a circle on the side. I suggest that the Columbus Forge and Iron Company used the *Peter Wright*-style *Trenton* as their model when they first started producing anvils. Some of these early American *Trentons* may have been made by the Columbus Forge and Iron Works before 1898.

Very soon after 1898, the Columbus Forge and Iron Company's *Trentons* became distinctive. First the base was cast steel and a parting line of the mold can usually be detected running vertically under the horn and heel from the waist to the center of the top of the foot. This line cannot be seen on the front of the foot under the horn because this area was ground smooth in order to put the weight and serial number on. The line can often be seen on the front of the foot under the heel, however.

These bases are hollowed out beneath, which looks like a large hour glass. The hollow or depression can be almost an inch deep in the center, but usually not. The hour glass shape was used on the cast bases until about 1910. At first the weight was on the side, but probably within a year, it was moved to the left on the front of the foot under the horn and after this only the *Trenton* trademark and sometimes *Solid Wrought* remained on the side. The serial number is always to the right on the front of the foot. The weight may be prefixed with a letter or letters, which was explained earlier. The serial number may be prefixed with an ***A***, but not necessarily. As far as I can determine, the ***A*** stands for "anvil." I was told that it stood for *Ajax,* meaning a second, but too many of the earlier *Trentons* that I have recorded have it and I could find nothing wrong with them. The *A* disappears after 1912 or 1913.

Under the base of a 1899 Columbus Forge and Iron Co. anvil. (R.A.P.photo)

Only the Columbus Forge and Iron Company put the weight and serial number on the front of the foot in that order. Sometimes this is the only way a *Trenton* can be recognized or distinguished from an *Arm and Hammer* anvil when all other markings are missing. Some of the very large *Trentons* have forged bases and are flat underneath, as the 900-pound one made in 1900.

As an example, let us assume that you are trying to identify an anvil you think might be a *Trenton,* but are not sure who made it. If the anvil is in good condition (body and base) and the trademark and serial number are on it, it was made by the Columbus Forge and Iron Company. The weight could be on the side under *U.S.A.* or under *Solid Wrought,* which is in a circle. If there is no serial number and no *U.S.A.,* it is an import. The older *Trentons* with the serial number I assign to the year 1898 and the highest serial number I have recorded for this group is 6589. The first one I have recorded for the year 1899 is serial number A6698. It has a cast base with the hour glass-like depression beneath. It has four handling holes and only the *Trenton* trademark on the side.

The company's anvils remained like the last one until about 1904. At that time the hole in the front of the foot was deleted. Between the years 1904 and 1907, there are at times one, two or three handling holes depending on the size of the anvil. After 1907, there is only one handling hole and it is in the waist under the horn. By 1909, all handling holes are gone except on very large anvils which may or may not have one. Starting in 1907, the hour glass depression starts to disappear, replaced by a depressed oval which I call the *Trenton* oval. Some Swedish anvils have it as well, as do *Arm and Hammer* anvils. After 1910, only the oval appears under the base and remains so until the last anvil was made.

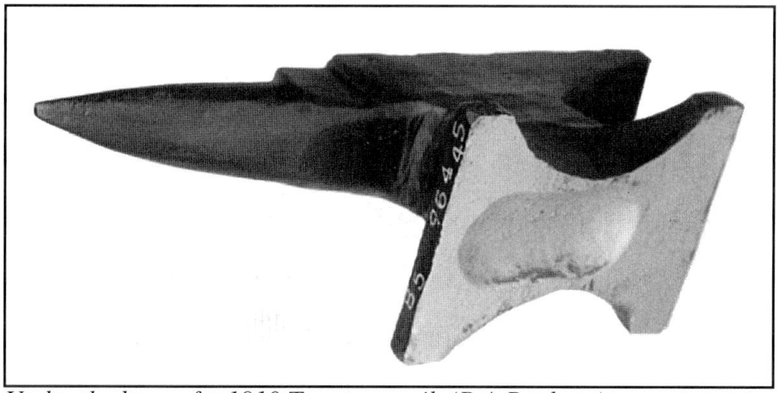

Under the base of a 1910 Trenton anvil. (R.A.P. photo)

As explained before, about 1927 the face plate began to be replaced by a solid-forged tool-steel top and electric welded at the waist. The electric weld bead can be seen, especially under the horn and heel where it was not ground down even with the body surface. At this time, the trademark was changed slightly, too.

The diamond shape around the name was changed from solid lines into broken lines, like so many dashes. The letters of the *Trenton* name were also curved rather than being straight as they were before.

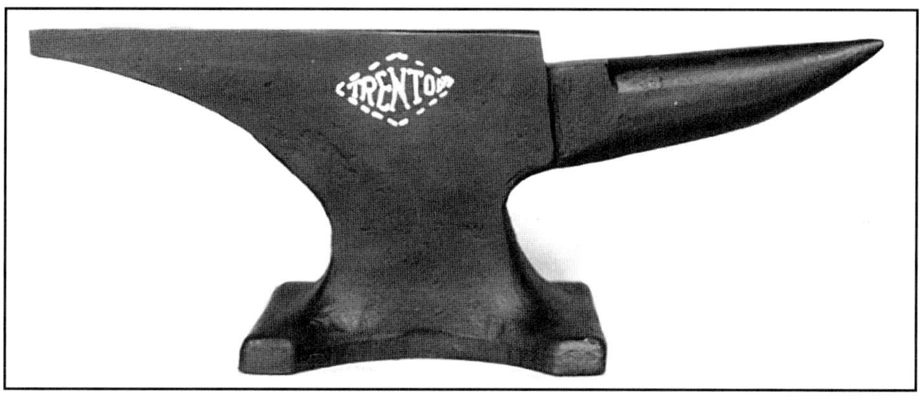

105-pound Trenton anvil made July 6, 1927. (R.A.P. photo)

By the late 1930s and later, the ***N*** in the center of the *Trenton* trademark, which is the highest letter in the diamond, was made to look like an ***X***. A number of times people have told me that they have discovered a new anvil named *Trexton*. I do not have an explanation for changing the ***N*** to ***X***. Mr. Tilton said that the **N** in the stamp probably washed out, and the stamp maker put the ***X*** in by mistake.

The "Trexton" trademark.

By at least the late 1920s, the cast feet of the base no longer had sharp edges on top, but were rounded off as seen on the 105-pound *Trenton.*

There should be enough information here to identify a *Trenton* anvil, but if not, you are welcome to contact me as long as I live or unless I contract Alzheimer's.

I have a blacksmith friend who acquired a 400-pound anvil from a large steel fabricating company that went out of business in Detroit, Michigan, in the 1980s. The people of the company told him that they had made the anvil and as there was no name on it, he accepted their word. He told me about the anvil and I told him that I had my doubts about their having made it. To make an anvil that large is difficult, and it would have been cheaper to buy one rather than to make just one. Some time later I had a chance to see the anvil. It was nicely made and had a face plate. It looked too professionally made to me to be a homemade job, although I am sure this company had the equipment to do it. The first thing I did was to tip it up and look under the base. There was the *Trenton* oval. I wire-brushed the front of the foot and numbers began to appear. The anvil turned out to be a *Trenton* made in 1925 and listed in the anvil record books I have.

The best proportioned anvil I have ever seen in my judgment is an 88-pound *Roberts Anvil* made in 1899 by the Columbus Forge and Iron Company.

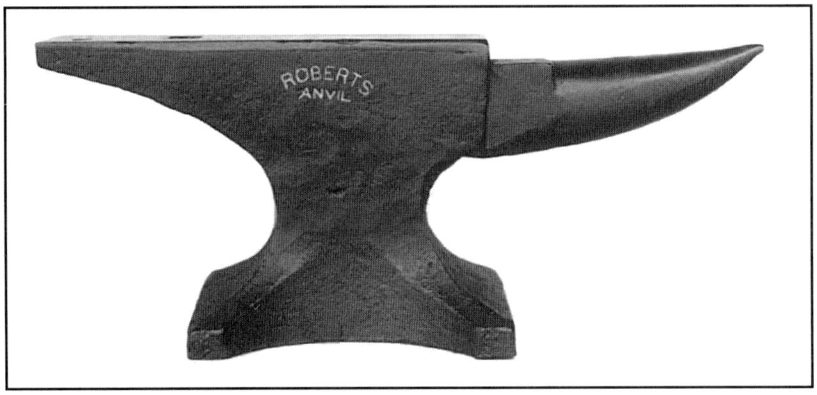

Roberts anvil. (R.A.P. photo)

In the 1940s, a symphony in Columbus, Ohio, used *Trenton* anvils wired to an amplifier for a concert in which the *"Anvil Chorus"* was played. I am sure that was a unique experience.

The following is a breakdown by serial numbers of the year in which *Trenton* anvils were made by the Columbus Forge and Iron Company The calculations before 1920 may vary by one or two years.

Serial Number	Date	Serial Number	Date
0-4000	1898	179713-183238	1922
4001-12000	1899	183239-186528	1923
12001-20000	1900	186529-188290	1924
20001-26000	1901	188291-191738	1925
26001-34000	1902	191739-194975	1926
34001-42000	1903	194976-195355	1927
42001-50000	1904	195356-196467	1928
50001-58000	1905	196468-197444	1929
58001-66000	1906	197445-198190	1930
66001-74000	1907	198191-200199	1931-1935
74001-82000	1908	200200-201638	1936-1938
82001-90000	1909	201639-202429	1939-1940
90001-98000	1910	202430-210375	1941-1943
98001-106000	1911	210376-216185	1943-April 1945
106001-114000	1912	216186-221625	April 1945-Feb.
114001-122000	1913		or Mar. 1947
122001-130000	1914	221626-222501	Feb. or Mar.1947-
130001-138000	1915		March 1948
138001-146000	1916	222502-223202	March 1948-Dec. 1948
146001-158000	1917	223203-223867	Dec. 1948-Oct. 1949
158001-170000	1918	223868-224057	Oct. 1949-July 1950
170001-176000	1919	224058-224465	July 1950-July 1951
176001-178436	1920	224466-224838	July 1951-March 1952
178437-179712	1921	224839-225551	March 1952-June 1953 ?

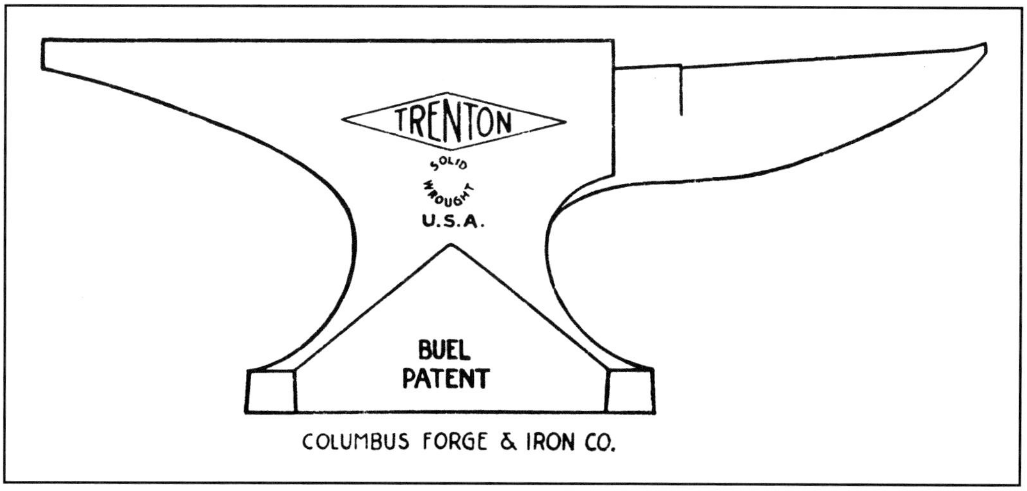

The following ads are a good cross-section of the ads used by the Columbus Forge and Iron Company.

The 1st Columbus Forge and Iron Co. ad to appear in The Blacksmith and Wheelwright, September, 1900.

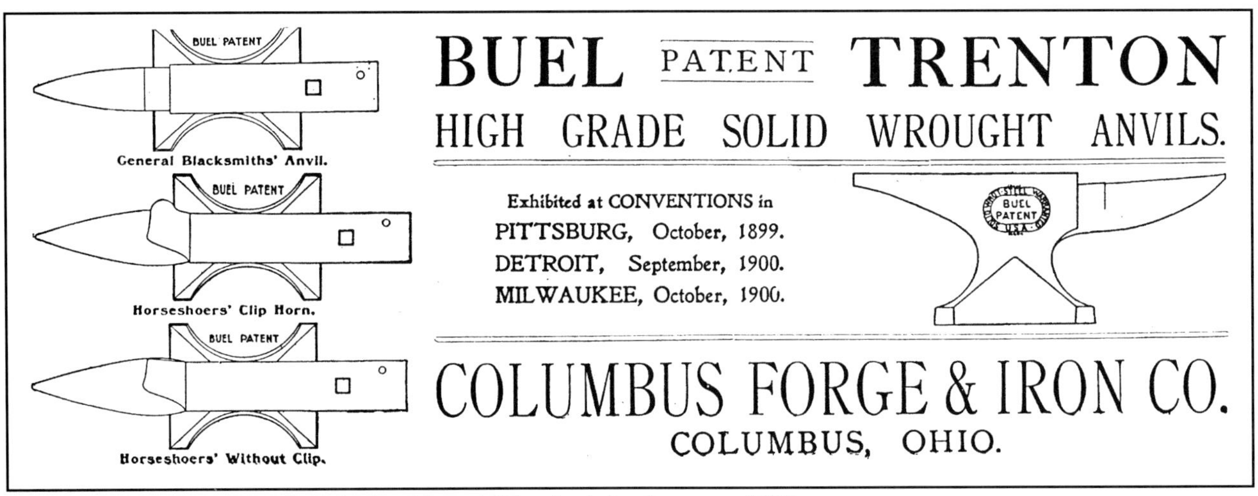

The 4th ad to appear in The Blacksmith and Wheelwright, January, 1901.

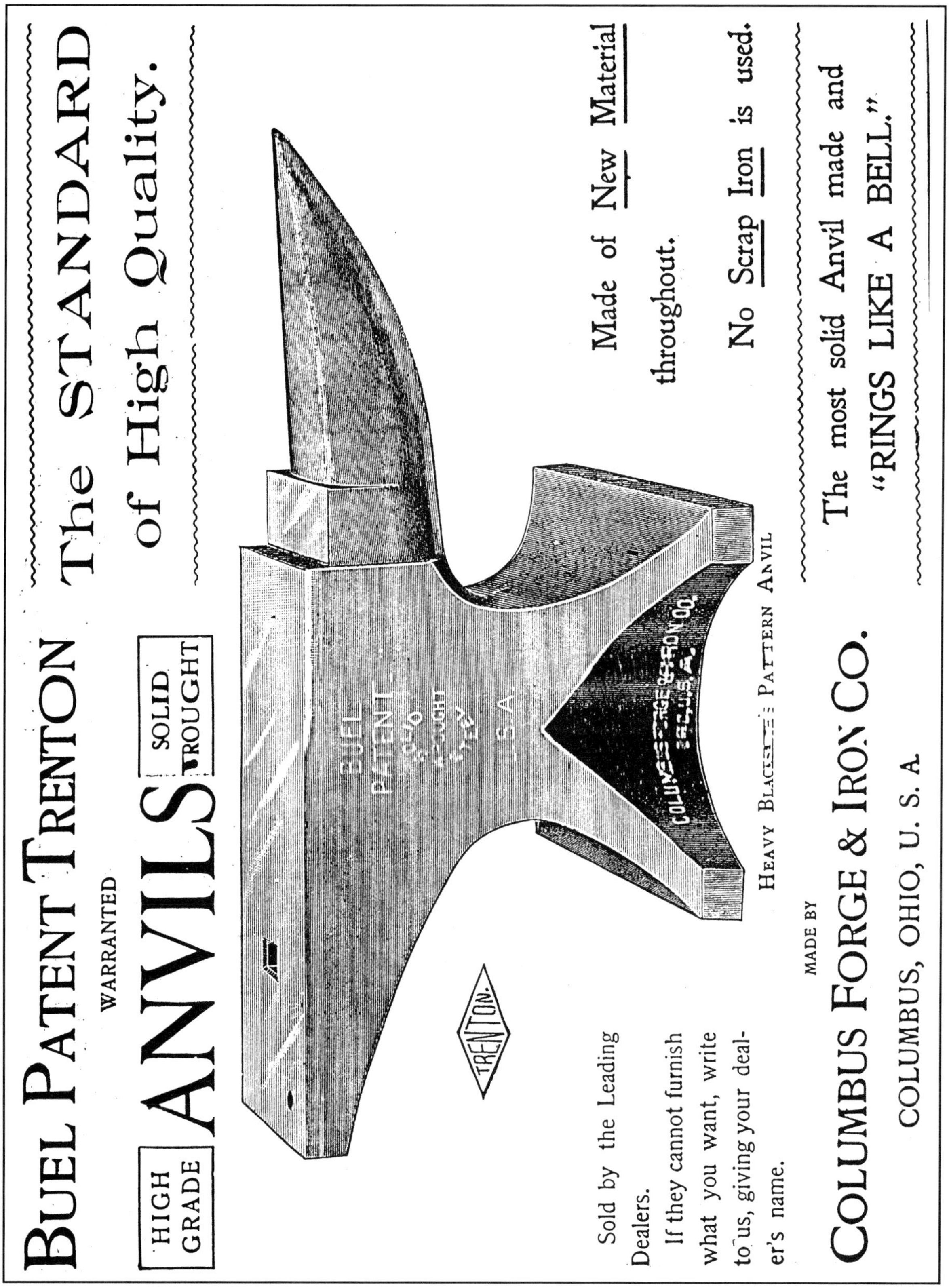

The 7th ad to appear in The Blacksmith and Wheelwright, August, 1901.

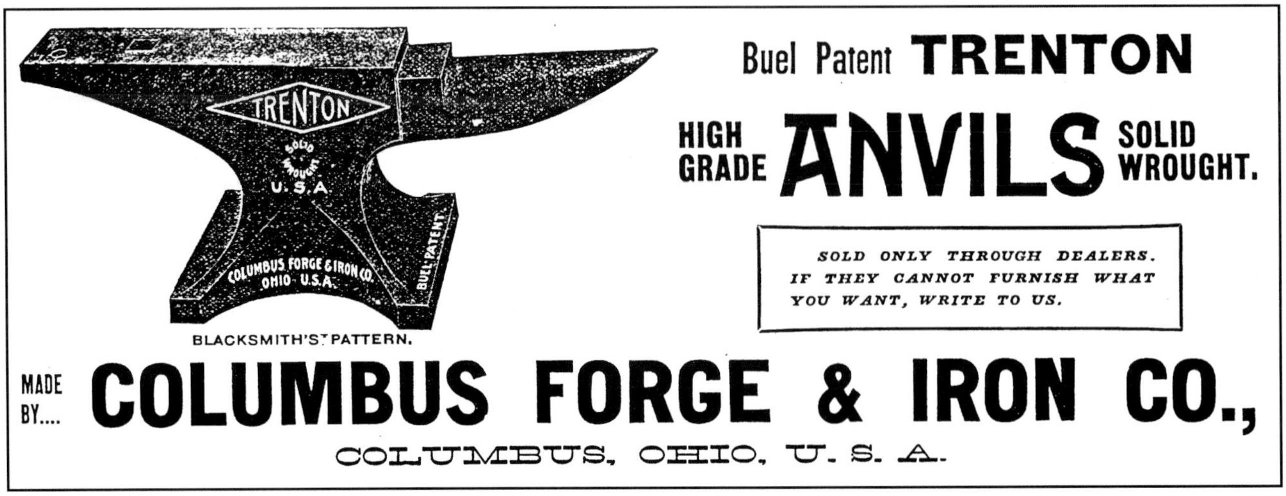

The 6th ad to appear in The Blacksmith and Wheelwright, April, 1901.

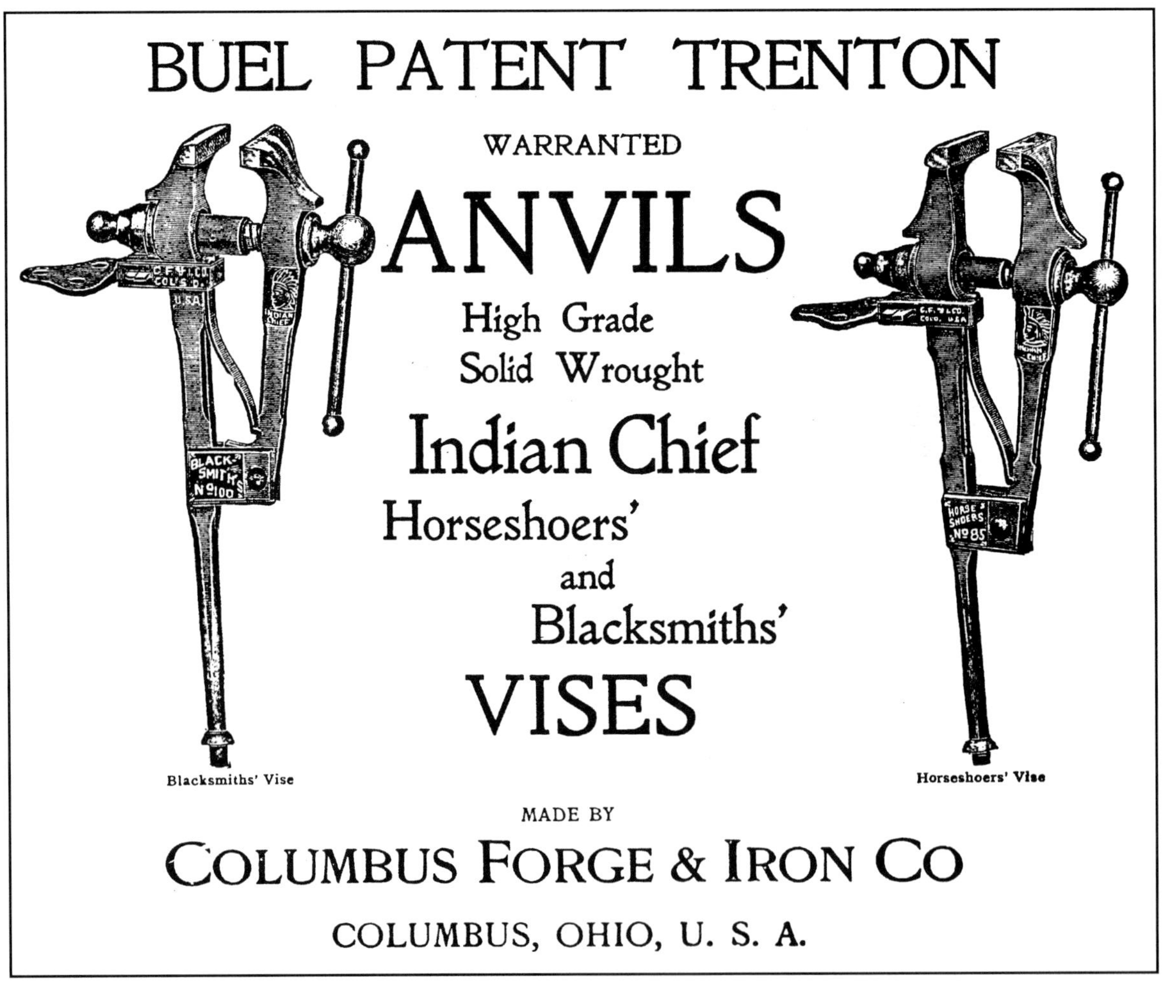

The Indian Chief vises, The Horse Shoer's Journal. January, 1904.

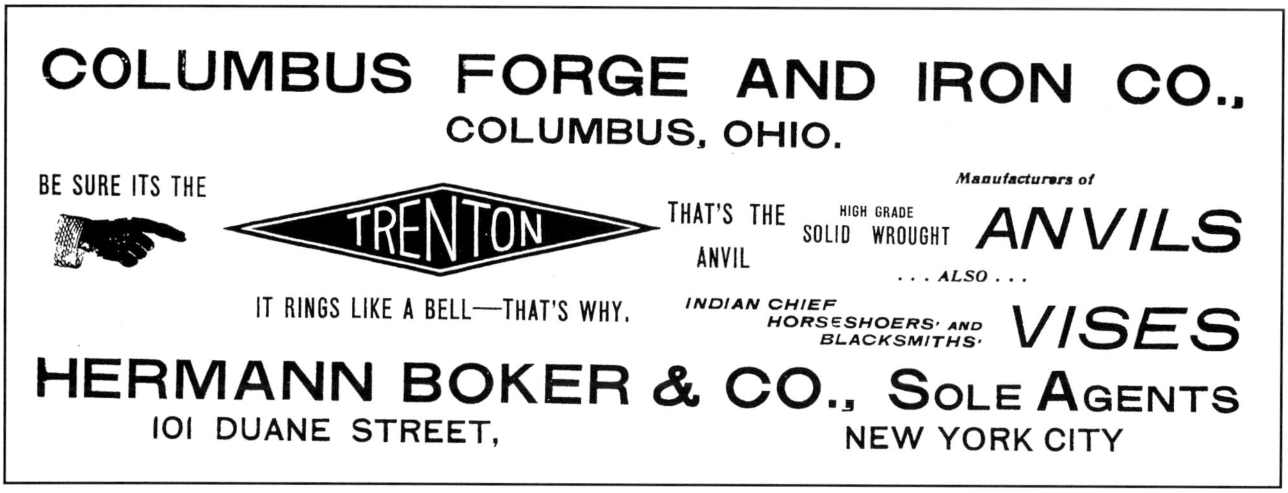

The Horse Shoer's Journal, September, 1905. Notice that Herman Boker & Co. is the "Sole Agent."

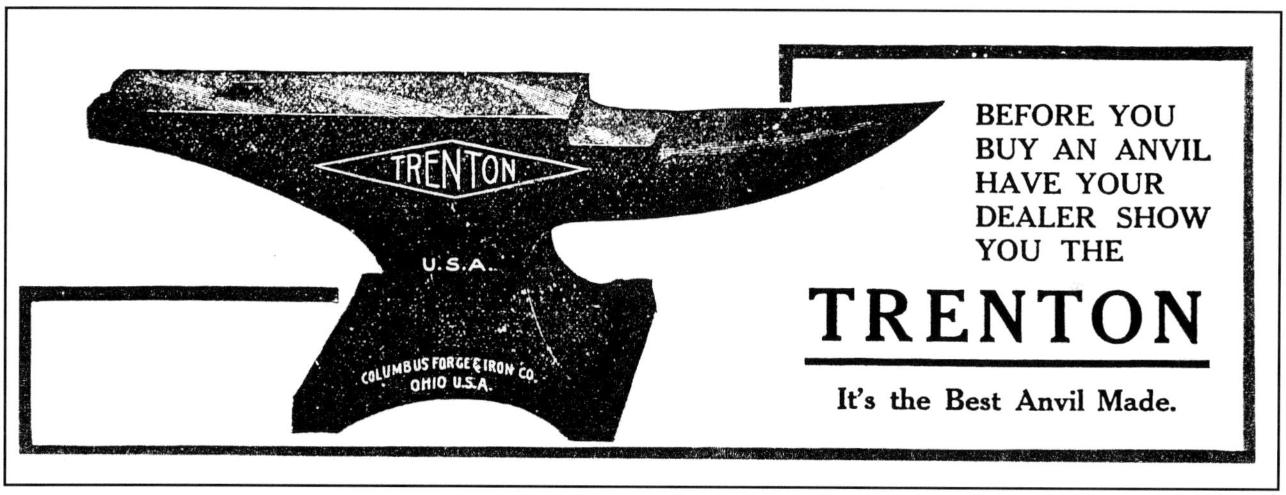

The American Blacksmith, June, 1907.

The American Blacksmith, September, 1908.

The American Blacksmith, March, 1909.

The American Blacksmith, June, 1911.

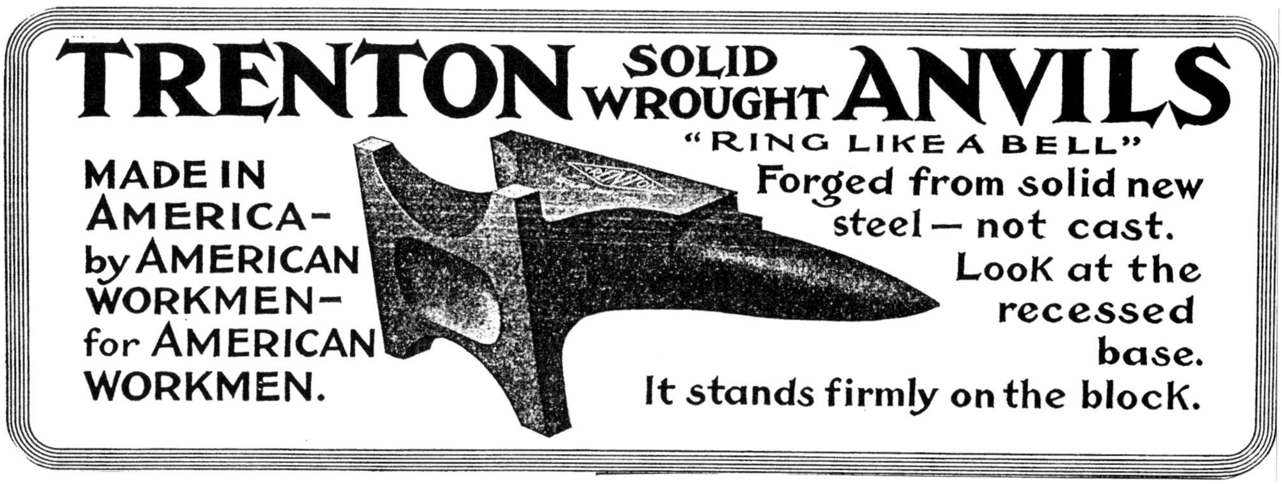

The American Blacksmith, May, 1912.

TRENTON ANVILS.

Body Forged in Two Pieces and Welded at the Waist, Face One Piece of Special Tool Steel, Tempered to Prevent Chipping.

Nos.	50	60	70	
Dimensions of face	11×3	11½×3⅛	12×3¼	in.
Size of hardie hole	13/16	13/16	13/16	in.
Size of pritchel hole	9/16	9/16	9/16	in.
W't about	50	60	70	lbs.
Nos.	**80**	**90**	**100**	
Dimensions of face	12½×3⅜	13×3⅜	14×3⅜	in.
Size of handle hole	⅞	⅞	⅞	in.
Size of pritchel hole	9/16	9/16	9/16	in.
W't about	80	90	100	lbs.
Nos.	**125**	**150**	**200**	
Dimensions of face	15×3½	16½×3¾	4¼×19	in.
Size of hardie hole	15/16	1	1⅛	in.
Size of pritchel hole	9/16	9/16	⅝	in.
W't about	125	150	200	lbs.
Per pound			**AB $0.18**	

GUARANTY—Should this anvil break at the waist or its steel face become loose within a year from date of purchase, it will be replaced without charge. All claims will be referred to the manufacturers. These anvils are tempered in a manner to suit the majority of smiths. No claim for defective temper will be allowed.

Hilbard, Spencer, Bartlett & CO. catalog, Febuary, 1913. Notice the price.

The American Blacksmith, March, 1914.

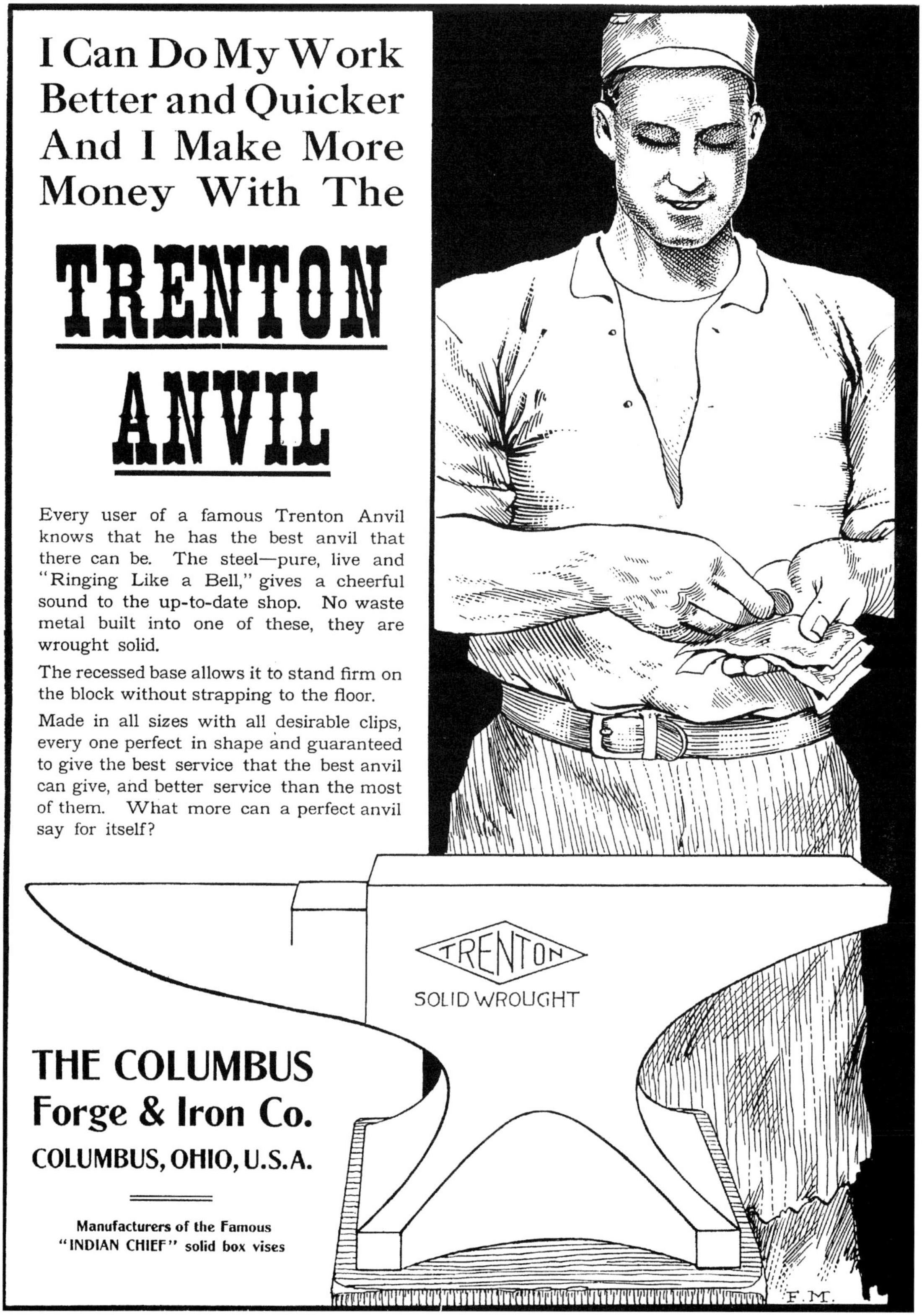

The Blacksmith and Wheelwright, April, 1918. This ad was originally 16 X 11 inches and took up a whole page of the journal.

ANVILS IN AMERICA

The following ads from The American Blacksmith between December, 1918 and April, 1921 use only three different graphics, but different captions for each graphic. The three graphics depicted are from May, 1919 (No. 1), March, 1920, (No. 2) and May, 1920 (No. 3). I will mark the captions with the appropriate number.

The American Blacksmith, May, 1919. (No.1)

An Irishman---

stood one day, watching a workman breaking large stones. As a final blow struck a huge rock and broke it, he asked the worker, "Say now, wuz it th' furst blow that cracked that rock, or wuz it the last wan?"

The workman looked up.

"Sure" he said, "it wasn't the furst wan and it wasn't the last wan, nor yit all those in between. It was all of thim put together".

And So It Is With Your Anvil

It isn't just the big heavy blows, that wear out an anvil, nor the little ones, but the constant, hammer, hammer, hammer that wears out the poorly built anvil.

TRENTON ANVILS

are built solidly and honestly clear thru to the core. They are made of solid steel and ring like a bell.

If you haven't a Trenton, Ask Your Dealer—

He Knows.

The Columbus Forge & Iron Co.

COLUMBUS, OHIO

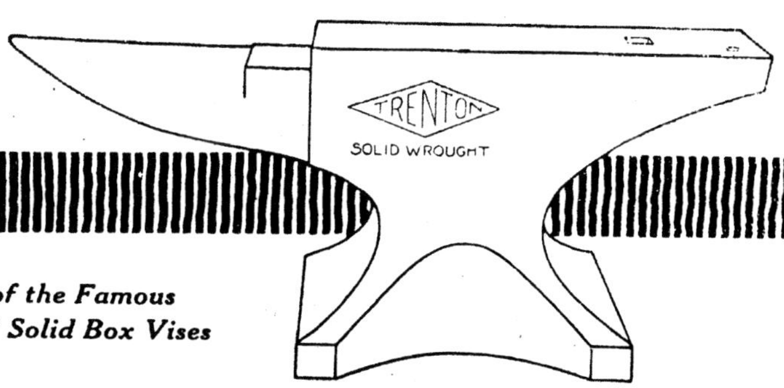

Manufacturers of the Famous "INDIAN CHIEF" Solid Box Vises

The American Blacksmith, March, 1920. (No. 2)

What Does It Profit a Man--

if business be good, yet his shop equipment be poor and his gains are quickly turned into loses, through the continual replacement of worn-out cheap tools?

When you by a **Trenton** anvil, you buy a product of honest American workmanship. And what is better, they are as honestly priced, as they are honestly made.

They're pure steel, solid clear through, and they 'ring like a bell'.

Ask Your Dealer, He Knows

The Columbus Forge & Iron Company

Columbus, Ohio.

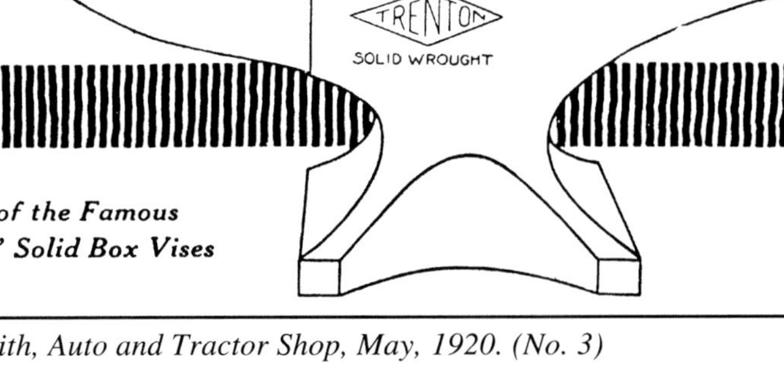

Manufacturers of the Famous
"INDIAN CHIEF" Solid Box Vises

The American Blacksmith, Auto and Tractor Shop, May, 1920. (No. 3)

DECEMBER, 1918 (#1)

Liberty and Democracy founded on Truth, Honor and Brotherhood may live until the sands of the desert grow, until the waters of the seas dry up, until your **Trenton Anvil Crumbles to Dust**.

A solid wrought Trenton will give you pounds of satisfaction for every ounce of its weight. The day you sign your order for a Trenton Anvil you sign a bond of freedom from anvil troubles and perplexities.

JANUARY, 1919 (#1)

Never in History

has humanity faced a more joyous season and New Year than this-

1918-1919

May it find you with a full measure of health and wealth and with a contentment of heart as true as the joyous ring of a **Trenton Anvil.**

Contentment in one's work can come through the use of only good tools. The Trenton Anvil is a good tool that will free you from anvil troubles and perplexities. A Trenton Anvil on the block means contentment in the heart.

FEBRUARY, 1919 (#1)

Not an ordinary anvil-

A Trenton Anvil is not an ordinary anvil by any means. It is not cast from scrap metals but **wrought** from pure, live steel --- metal with **vitality and life.**

That's why Trentons are so universally appreciated - why they are recognized, by smiths who know, as the best there's to be had in anvils.

The anvil that "rings like a bell!"

MARCH, 1919 (#1)

The men behind the guns

are the important factor in warfare.

But the anvil behind the man is the important factor in good smithing. Trenton Anvils have proved themselves time and time again, through long years of service, worthy of the best craftsman in the land.

Built of steel that's purest, they are full o' life and pep, with a spring and vitality that keeps 'em ringing like a bell.

APRIL, 1919 (#3)

Keeping Pace With the Time

---a good shop

---good tools

---a good anvil

---a TRENTON!

Look at that old anvil o' yours. "Pretty well battered," you'll say, "but still on the job." Maybe so but you can't do your **best** work on an old, battered, worn out anvil any more than you can with tools that are run down at the heel.

Better see your jobber right away. And be sure you ask him about the TRENTON---

the one that "rings like a bell." There's only **one** TRENTON, and if you want the very best there is to be had in an anvil, she's the only one for you.

JUNE, 1919 (#3)

Business Comes To The Well Equipped Shop

Your shop equipment should consist of good tools and a good anvil---

A TRENTON.

How's the anvil your [you're, author's note] using - is it pretty well battered? You cannot expect to do good work on an old battered worn out anvil any more than you could with dull and worn out tools.

You need the BEST today - You cannot afford to use anything else - than a TRENTON. Call upon your jobber and ask him about the TRENTON. There is only ONE TRENTON and she rings like a bell.

JULY, 1919 (#1)

Your Old Anvil

Would not have to be replaced now, had you purchased a **Trenton** in the first place.
This time play safe ---**Buy a Trenton.**

Service is wrought right into **Trenton Anvils----**
They are built of the purest steel and have a ring of satisfaction. You cannot afford to use inferior tools. Modern shop methods demand first class tools---
The **Trenton** is the best for any purpose.
Your jobber knowing the quality of the **Trenton Anvil** has them on hand.

Ask him about it.

AUGUST, 1919 (#2)

They're Bound to Come---

to the repair shop turning out dependable repairs.

You can do dependable work --- ONLY with good tools and a good anvil --- A TRENTON.

There is no better Anvil made for Repairman or Smithy --- than a TRENTON ---
She is Steel through and through and rings like a Bell.

SEPTEMBER, 1919 (#1)

Treat 'em Rough

The rougher you treat a "TRENTON ANVIL" the louder will be the ring of satisfaction.

You simply cannot wear them out. "TRENTON ANVILS" stood the test. They are built for Service.

Call on your jobber and ask him about the "TRENTON" --- the one that "rings like a bell."

If you want the best there is to be had in an anvil --- The "TRENTON" is the one for you.

NOVEMBER, 1919 (#1)

STRIKES!!!

SPEAKING OF STRIKES --- The strikes of the mighty smithy delivered upon a TRENTON ANVIL do not affect it -- But each strike HELPS his business because BETTER work can be turned out. Is yours a TRENTON?

TRENTON outlasts others --- She is steel through and through and rings like a bell. Your jobber KNOWS the TRENTON --- ask him.

DECEMBER, 1919 (#1)

We Wish You a Merry Christmas
and
A Happy and Prosperous New Year

We also wish that before the new year starts to wane you will be doing your work on a "TRENTON."

Your next Christmas will then be a merrier one -- ask your jobber he knows.

JANUARY, 1920 (#2)

How Are You Fixed For 1920--------

Are you ready for the repair work to be done during the year.----

If you have good tools and --- a **"TRENTON"** you are all set -- and able to turn out dependable repairs---
TRENTON ANVILS are made of pure steel and are solid thru, and thru' and ring like a bell. -- Don't worry through another year with your old chipped anvil --- Get a **TRENTON---**

Ask Your Dealer He Knows

FEBRUARY, 1920 (#2)

Built for the Finest Shop in the World

The American Shop

As Americans, we demand two things of our tools, **Dependability** and long efficient life.

You can turn out dependable work on a **TRENTON.** She's steel thru and thru and rings like a bell.

If you haven't a **TRENTON** on the block--- ask your Jobber about the TRENTON------

He Knows.

APRIL, 1920 (#1)

SOME ANVILS are MADE-------

TRENTON ANVILS are **built.** There's a heap of difference. They are built of the purest steel and have a ring of satisfaction.

And remember, there's only one **TRENTON.** If you want the very best there is to be had in an anvil she's the only one for you.

The Anvil that "Rings like a Bell."

Ask your Jobber Today --- He Knows

JUNE, 1920 (#1)

AN ANVIL----
OR ONLY A SHELL?

Notice the little word on the anvil pictured above, "TRENTON." Therein lies the one big reason why you should choose This Anvil. It's steel clear to the core.

The Anvil that "Rings like a Bell!"
ASK YOUR JOBBER TODAY --HE KNOWS

AUGUST, 1920 (#3)

THERE ARE ANVILS-------
AND ANVILS------

But the anvil for your work is a TRENTON --- the one reliable anvil that is built of purest steel and is solid thru and thru.

There is only one TRENTON and she rings like a bell.

Ask Your Jobber Today ---He Knows.

SEPTEMBER, 1920 (#3)

THE ANVIL YOU PURCHASE.

Should meet the requirements necessary to do good work --- a lasting surface.

There are no soft spots on ---
THE TRENTON --- She is steel thru and thru and rings like a bell.

If you haven't a TRENTON Ask Your Jobber.
HE KNOWS.

OCTOBER, 1920 (#3)

ANY WORK----

that requires the use of an anvil --- deserves the use of a good anvil --- Here is where the **TRENTON** rings in.
She rings like a bell because she's solid thru and thru.

Ask Your Jobber Today --- He Knows.

NOVEMBER, 1920 (#3)

It is Not Only Our Aim to Build Anvils----
But to Build Only the BEST.

When you purchase your next anvil --- BETTER SAY---
I want to buy a **TRENTON.**
You're sure, then of receiving full value for your money. TRENTONS are steel thru and thru and ring like a bell.

If you haven't a TRENTON Ask Your Jobber.
HE KNOWS.

DECEMBER, 1920 (#3)

A GOOD IDEA ----

Let your anvil be a TRENTON.

She's steel thru and thru and
Rings like a bell.

Next time you go to your jobber ask him about the TRENTON.

HE KNOWS

JANUARY,1921 (#3)

The Knack of Good Smithing

Is in the Daily Use of a "TRENTON"

She rings like a bell because she's solid thru and thru.

Ask Your Jobber Today --- He Knows

FEBRUARY, 1921 (#3)

IF YOU WANT -------

an anvil with a smooth surface
an anvil with no soft spots
an anvil that always gives satisfaction

YOU WANT A "TRENTON"

She's steel thru an' thru an' rings like a Bell

Ask Your Jobber Today --- He Knows

MARCH, 1921 (#3)

Spring time will soon be here, with it will come
an abundance of repair work ---
The smooth surface of a TRENTON is what
you want to do the forging on.

She's steel thru an' thru an' rings like a bell.

Ask Your Jobber To day --- He Knows

APRIL, 1921 (#3)

How do you do---
without a TRENTON
The smooth surface of a TRENTON ANVIL is just as adaptable to rough forging as it is to fine forging.
Both are done best on a TRENTON.

She's steel thru an' thru an' rings like a bell ---

Ask Your Jobber Today --- He knows

Starting with July, 1921 the ads change in size.

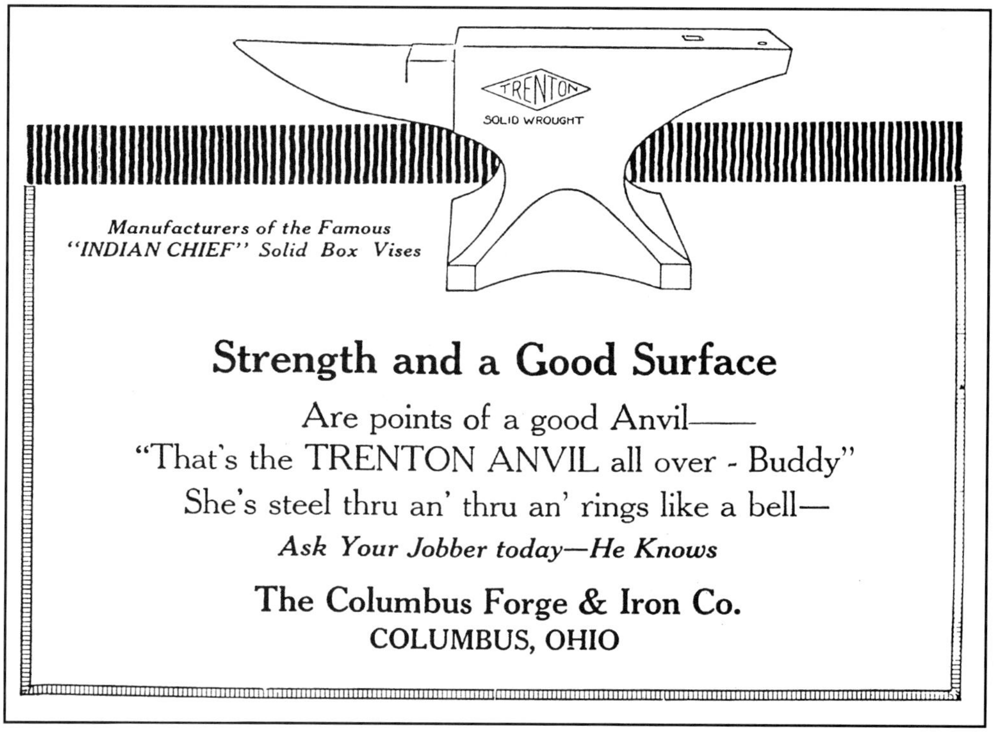

The American Blacksmith, Auto and Tractor Shop, July, 1921.

TRENTON

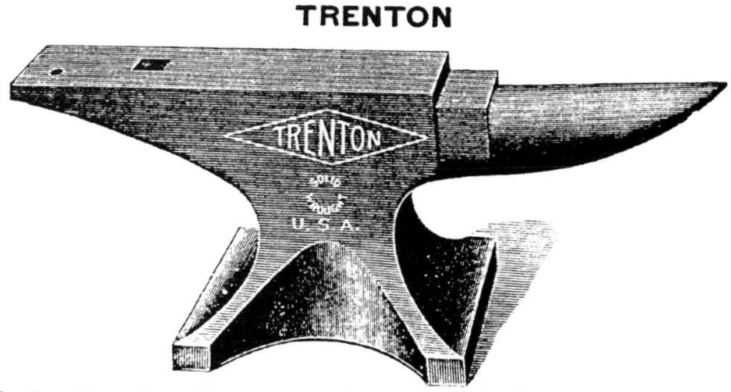

Solid forged wrought iron body with tool steel face and horn; made of highest grade tough steel, fully warranted.

Approx. wt. each, lbs.	40	50	60	70	80	100	125	140	150	175	200	250	275	300
Approx. size of face, ins.	2¾x9	3x9½	3¼x10	3⅜x11	3½x12	3⅝x13½	3¾x15	4x16½	4x16½	4¼x17	4⅛x18	4¾x20	4¾x20	5x20
Approx. length over all, ins.	15	16	17	18½	20½	22½	25	27	29	28	29½	32½	34	35
Approx. height, inches	8	8	8½	8½	9	10	10⅜	11	11¼	11¼	11½	12	12½	13
Size of Hardie hole, inches.	¾	¾	¾	¾	¾	¾	⅞	⅞	⅞	1	1⅛	1⅛	1⅛	1⅛
Size of Pritchel hole, ins.	7/16	7/16	7/16	7/16	7/16	7/16	7/16	9/16	9/16	9/16	11/16	11/16	11/16	11/16
Pound	$0.38	.37	.36	.35	.35	.35	.35	.35	.35	.35	.35	.35	.35	.35

California Hardware Co. catalog, 1921, Los Angles, Ca. Notice the price per pound.

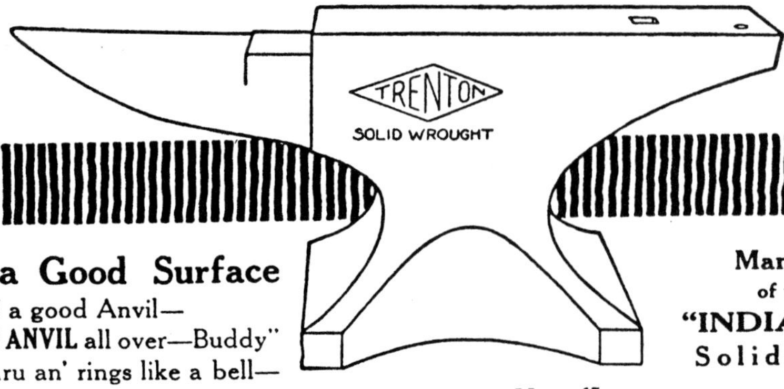

Strength and a Good Surface
Are points of a good Anvil—
"That's the **TRENTON ANVIL** all over—Buddy"
She's steel thru an' thru an' rings like a bell—

Manufacturers
of the Famous
"INDIAN CHIEF"
Solid Box Vises

Ask Your Jobber today—He Knows

THE COLUMBUS FORGE & IRON CO. **COLUMBUS, OHIO**

The American Blacksmith, Auto and Tractor Shop, January, 1922.

"Trying to do good work on a battered, worn out anvil is like tryin' to bring up a seal-skin wife on a calico salary. It's mighty hard an' difficult."

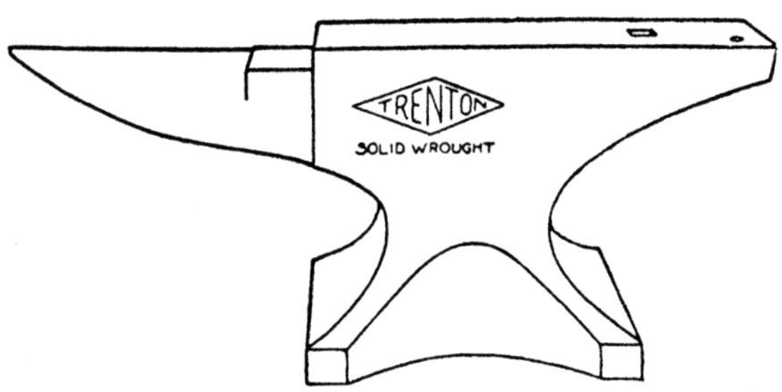

We suppose that Trenton Anvils will wear out —but it takes longer than usual for a Trenton to become "battered and worn". It's because solid, honest quality is built into every one of 'em.

The American Blacksmith, Auto and Tractor Shop, April, 1922.

Seven Points of Trenton Superiority.

1—Rings like a Bell.
2—Perfectly shaped Horn.
3—Long, graceful heel.
4—Large, firm base.
5—True, level face.
6—Special Temper prevents chipped edges.
7—Solidly forged of new steel.

If your jobber won't supply a Trenton, make a fuss about it to us.

We also make the famous Indian Chief Solid Box Vises

Columbus Forge and Iron Company

Columbus, Ohio.

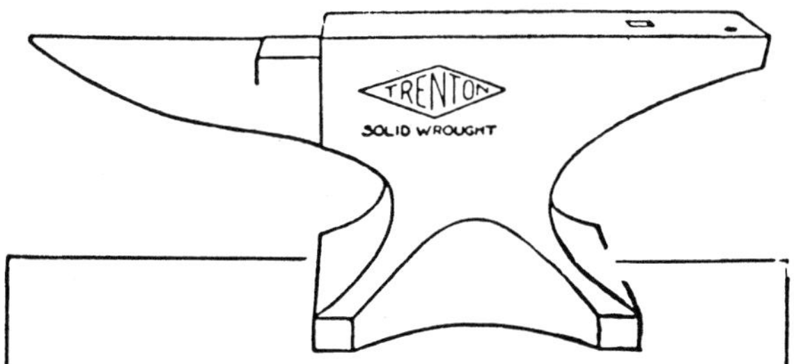

Strength and Good Surface

are points of a good anvil—

"Thats the TRENTON ANVIL all over—Buddy"

She's Steel thru an' thru—
An' Rings like a Bell—

Manufactures of the famous

"Indian Chief"

Solid Box Vises

Ask your jobber today—He knows.

The Columbus Forge & Iron Co.

Columbus, Ohio.

The American Blacksmith, and Motor Shop, December, 1924.

Trenton Anvils

Forged from two pieces of new, solid ingot steel, welded at the waist, and faced with one solid piece of special, high grade tool steel. Has a long, perfectly shaped horn and long heel. Special tempered edges to prevent chipping.

Faces level and true. Hardie and pritchell holes and other mechanical details of accuracy.

Size Pounds	Extra Over Base Price per Lb.	Size Pounds	Extra Over Base Price per Lb.	Size Pounds	Extra Over Base Price per Lb.
80 to 425	Base	60 to 69	$.01	20 to 29	$.08
427 " 625	$.00½	50 " 59	.02	10 " 19	.15
626 " 800	.01	40 " 49	.03	....	...
70 " 79	.00½	30 " 39	.05	....	...

W.J. Holliday and Co. catalog, 1927, Indianapolis, In.

RESERVED FOR THE

TRENTON ANVIL

"IT RINGS LIKE A BELL"

THE

Columbus Forge & Iron Co.

COLUMBUS, OHIO

H. BOKER & CO., INC., Sole Sale Agents
101 Duane Street - - - - New York City

The Blacksmith and Wheelwright, October, 1925. From Mr. Keith Sommer. The American Blacksmith journal was discontinued in the 1920s and The Blacksmith and Wheelwright, in the 1930s.

Blacksmiths Anvils

Trenton—Steel Face

Forged from solid billets of new Special High Grade tool steel. Long shaped horn and heel. Hollowed base permits anvil to stand firm without rocking when heavy work is done at end of horn or heel. Edges are specially tempered to prevent chipping. Furnished with Hardie or Pritchell holes, accurately shaped.

Wt., Each. Lbs.	25	50	75
Size, Face. In.	2¼x9½	3x11	3¼x12
Hardie Hole, In.	⅝	¾	¾
Each	54.00	54.00	54.00

Wt., Each. Lbs.	100	125
Size Face. In.	3¾x14	3¾x14
Hardie Hole, In.	1	1
Each	54.00	67.50

Note—Write for complete information on sizes not listed. We can furnish anvile from 10 to 1000-lbs.

Vonnegut Hardware Co. catalog No. 53, 1953, Indianapolis, In. Notice the weight of anvils that can be supplied. They were selling for .54 cents per pound. H. Boker was still the sales agent.

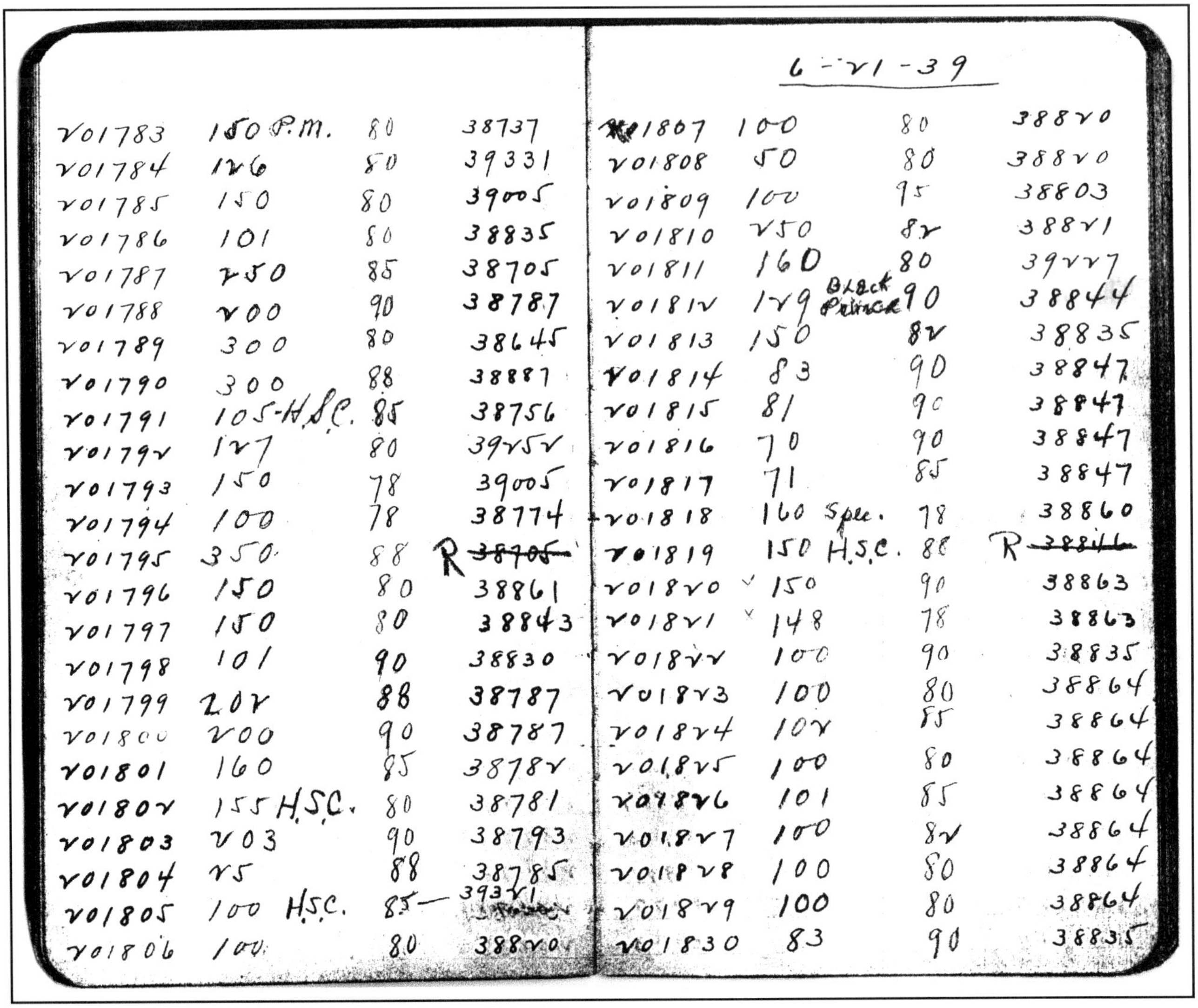

Serial	Weight	Hardness	Order
201783	150 P.M.	80	38737
201784	126	80	39331
201785	150	80	39005
201786	101	80	38835
201787	250	85	38705
201788	200	90	38787
201789	300	80	38645
201790	300	88	38887
201791	105-H.S.C.	85	38756
201792	127	80	39252
201793	150	78	39005
201794	100	78	38774
201795	350	88	R ~~38705~~
201796	150	80	38861
201797	150	80	38843
201798	101	90	38830
201799	202	88	38787
201800	200	90	38787
201801	160	85	38782
201802	155 H.S.C.	80	38781
201803	203	90	38793
201804	25	88	38785
201805	100 H.S.C.	85	39321
201806	100	80	38820

6-21-39

Serial	Weight	Hardness	Order
201807	100	80	38820
201808	50	80	38820
201809	100	95	38803
201810	250	82	38821
201811	160	80	39227
201812	129 Black Prince	90	38844
201813	150	82	38835
201814	83	90	38847
201815	81	90	38847
201816	70	90	38847
201817	71	85	38847
201818	160 Spec.	78	38860
201819	150 H.S.C.	88	R ~~38846~~
201820	150	90	38863
201821	148	78	38863
201822	100	90	38835
201823	100	80	38864
201824	102	85	38864
201825	100	80	38864
201826	101	85	38864
201827	100	82	38864
201828	100	80	38864
201829	100	80	38864
201830	83	90	38835

The above two reproduced pages are from the Columbus Forge and Iron Co. anvil book No. 28, from June 5, 1935 to January 7, 1941.

The left column on each page is the anvil serial number (stamped on the front of the foot to the right). The next column is the anvil's weight. The third column is the sclerorscope or hardness number. They tested each anvil for hardness. The column to the right is the company's order number. "H.S.C." stands for horseshoer's clip anvil. *Black Prince was* made for the Shapleigh Hardware Co. of St. Louis, Mo. The Columbus Forge and Iron Co. made several hundred anvils with the *Black Prince* trademark. "Spec." was the designation used for a special anvil made to someone's specifications. "6-21-39" shown on the right side of the page is the date these anvils on this page were either shipped or when they were logged in the book. The page to the left and the one before it, were dated, "5-8-39." "PM" stands for plowmake's anvil. I'm not sure what the "R" stands for, possibly a returned anvil to be repaired.

"Buffalo" FORGES

Ready to Make or Mend

The forge in the machine shop is frequently regarded as a necessary evil; smoky, dirty, creating unpleasant and unwholesome gases, it is usually housed as far as possible from the shop proper.

Not so here—as you can see by the photograph the Buffalo Forge in the round-house of the Hocking Valley R. R. (Columbus, Ohio) holds the center of the stage. You will note too that the air is clear and the wall unstained by smoke or gas.

The Buffalo Blower System which draws off all smoke and fumes makes this a "clean" forge, well deserving a good place in a clean, well lighted, modern shop.

Let us send literature on Buffalo Forges and ventilating systems.

Machinery journal, March , 1922. This is a Buffalo Forge ad which shows an Indian Chief railroad or bridge anvil in use in a railroad shop.

The smaller 650-pound Indian Chief railroad anvil featured in a Buffalo Forge Co. ad. (Photo courtesy of Mr. David Tilton)

A photo taken at an Industrial Show at Lockbourne Air Force base. (Now Rickenbacker Air Base), Columbus, Oh., June, 1950. Notice the two anvils. (Photo courtesy of Mr. David Tilton)

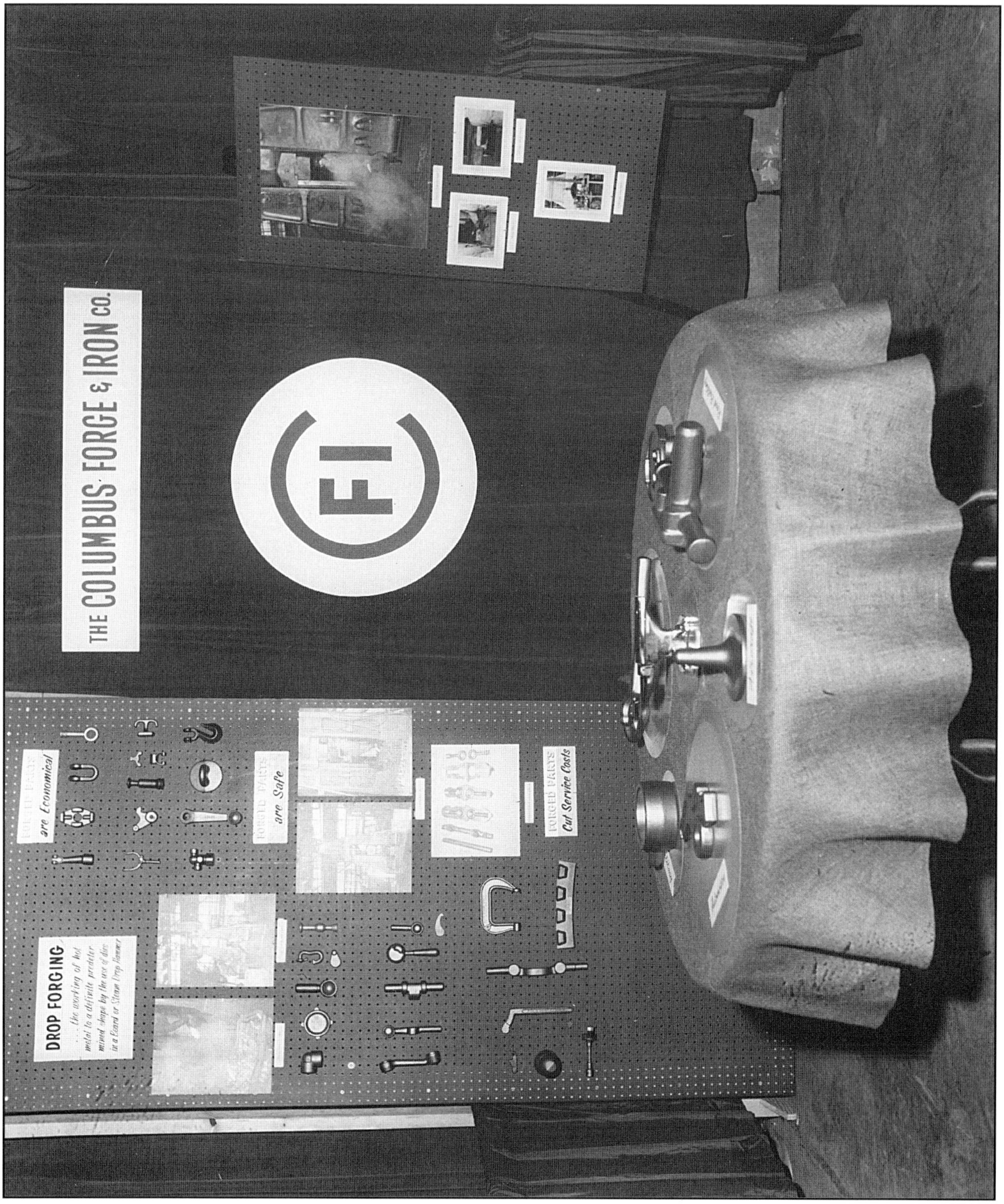

A display at the Ohio State Fair in 1962. The craft anvil is in the center of the table. The mounted photos show various steam hammer operations. Compare this photo with the one taken at the Ohio State Fair in 1910.
(Photo courtesy of Mr. David Tilton)

The miniature advertising anvil for the Columbus Forge and Iron Co. It is a steel drop forging. Anvil from Mr. David Tilton. (Photo by Mr. Gill Fahrenwald)

The company's logo on the reverse side of the miniature anvil. (Photo by Mr. Gill Fahrenwald)

Companies Make Their Wares By Forge Here

By ERNEST LEOGRANDE

The strains of the anvil chorus have faded away in the past few years, but in Columbus they still ring out loud and true.

Two companies here still carry on with the making of forged anvils, that is, those worked on individually and not cast in molds. The two companies are the Columbus Forge & Iron Co., 544 W. First-av, and the Columbus Anvil & Forging Co., 117 W. Frankfort-st.

Of the two, the former is the older, having been started in 1898, two years before the Columbus Anvil & Forging. They and a British forging company are probably the only firms in the world still putting out forged anvils.

THE MASTER anvil maker at the Columbus Forge & Iron is Karl Wright of 2191 Morse-rd. Karl has worked on the iron blocks with the distinctive "horns" since 1933 when he began by welding. The anvils are made in two pieces, top and bottom, and welded together.

Karl worked beside his father, A. J. (Si) Wright, who started work for the company 52 years ago at its very beginning. Si retired two years ago at the age of 75 and left Karl, who is now 36, to carry on.

There is a market for the anvils throughout the world. A recent shipment went to Manila, half-way around the world.

EARL M. TILTON of 2390 Arlington-av, is president of the company and his son, David E. Tilton of 2341 Edgington-rd, is his assistant.

While strolling through the Mexican section of Los Angeles on a recent visit there, the father and son saw an anvil they thought they recognized. Sure enough, it was one of theirs.

An old Mexican workman was using it to shape ornaments for the tourist trade. He told the Tiltons it had been given him by his father.

Mr. Tilton Sr. saw the face of the anvil was scarred and offered to reface it for him free, even though the guarantee had long since ran out.

"My father would turn over in his grave!" the workman said. "And besides, my hammer is used to the bumps. I couldn't work the same without them."

THE TILTONS count anvil-making as 10 per cent of their business, but a steady 10 per cent. The other 90 consists of forged iron and steel pieces.

The scale of sizes for the anvils ranges ordinarily from 25 to 300 pounds. They have made some as heavy as 1000 pounds, though, and

Continued on Page 6B, Col. 1.

FATHER AND SON, president and assistant, are Earl M. Tilton and David E. Tilton. They are officers of the Columbus Forge & Iron Co., an old-time Columbus industry. It still produces wrought anvils in addition to machine and drop forgngs.

AN OLD MASTER is Karl Wright, 2191 Morse-rd. Karl's father also made anvils at the Columbus Forge & Iron Co. Karl soon may be the only anvil maker in the country.

First part of a newspaper article from Columbus, Ohio.

Strains of Anvil Chorus Still Ring Out in City

Continued From Page 5B.

once, during the depression, there was a shipment that weighed only six and a half pounds.

These last had to be drop-forged and were fashioned for recreational use.

During World War II the company made thousands of anvils for Army engineers. Today, as before the war, they ship mainly to commercial buyers.

Blacksmithing also was done in the shop as late as 1937, but it now has been discontinued.

MR. TILTON is not quite sure of the original ownership of the company, but in 1902 his father-in-law, D. W. Singleton, took over. In 1917 he himself stepped into the job of managing the firm.

In the office shared by the two Tiltons, there is an oil painting, done in 1883, showing an old-time blacksmith operating a water-driven forging hammer.

"Most people wouldn't be interested in that picture," Mr. Tilton Sr. said, "but it appeals to me. Everything in it is entirely accurate."

As the number of anvil makers dwindles, perhaps someone in later years will look at a Columbus Forge & Iron wrought anvil and speak as approvingly of its workmanship.

The second part of a newspaper article from Columbus, Ohio. ca. 1951. There are a few errors in the write-up as can be found in most newspaper articles. The part about the Mexican workman is referred to in Appendix E. (Article courtesy of Mr. Mike McGee)

CHAPTER 6
Miscellaneous American Anvils

Scythe Anvils and Hammers

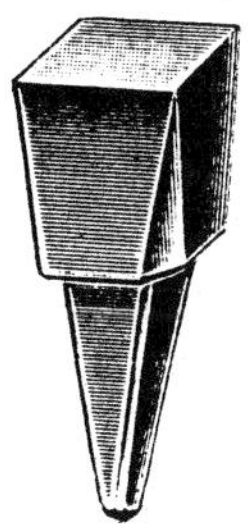

For Austrian Scythes

Medium Weight

Scythe Anvil and Hammer
. . . . per Set **1.50**

Cast Bench Anvil.

Fig. 553.

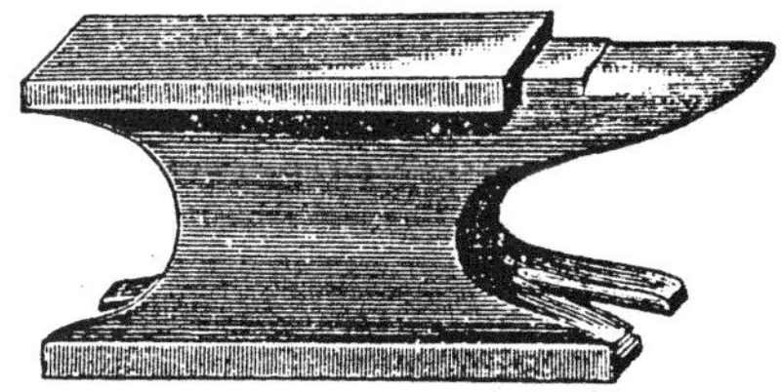

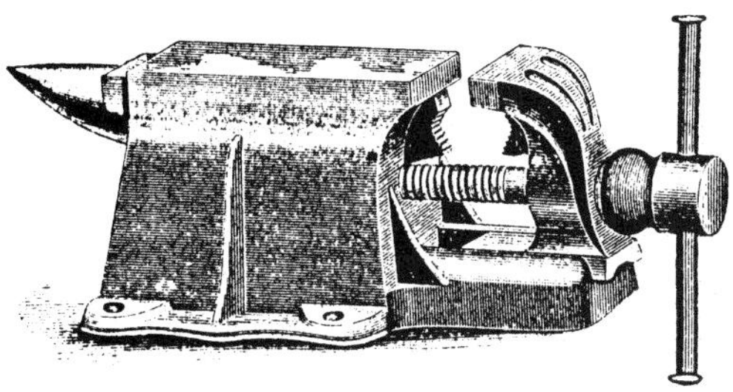

Anvil and Vise Combined.

Fig. 552.

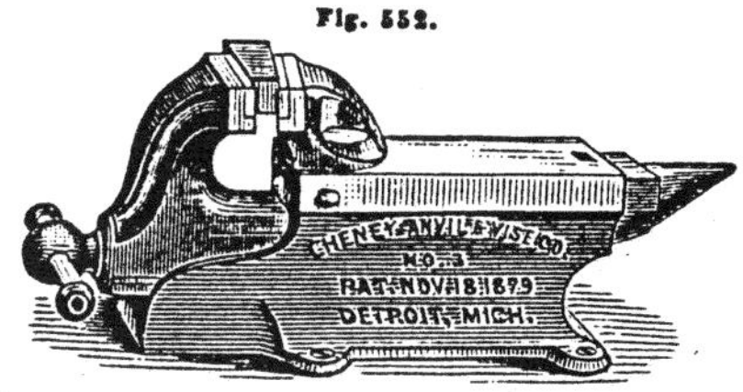

Cheney's Anvil and Vise Combined,

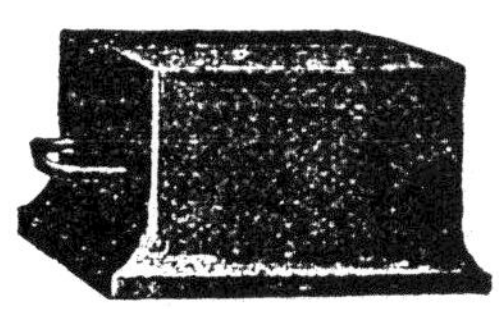

SAW ANVILS

Freight, Fourth Class

Hardened Iron Face; Used by Sawyers to lay Saw on, while hammering

Nos.	P335H	P335H4
Face, In.	10x6	10x6
Height, In.	5	6
Weight Each, Lbs.	90	110
Per Pound	$0.60	.60

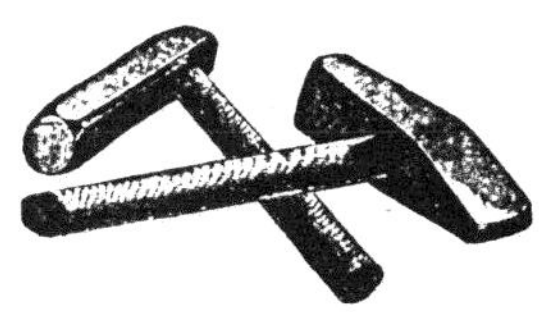

SAW MAKERS HAMMERS

Solid Steel Face, Hickory Handle

ROUND FACE

	Each
No. 335L—Assorted 2½ to 3 Lbs.	$8.30
No. 335M—Assorted 3½ to 4 Lbs.	11.60
No. 335N—Assorted 4½ to 6 Lbs.	16.00

SQUARE FACE

	Each
No. 335P—Assorted 2½ to 3 Lbs.	$8.30
No. 335R—Assorted 3½ to 4 Lbs.	11.60
No. 335T—Assorted 4½ to 6 Lbs.	16.00

Open Stock

Unknown

I have looked at any number of all-cast-steel anvils on which there were no trademarks or other indications as to who made them. Many of them resemble major brands in the smith's and the farrier's pattern. My best guess is that some people or steel foundries made up the patterns and then either had them cast or cast them, possibly in the manner that the *Clark* anvils were cast. Many of them have nicely finished, hardened faces, and as far as I can tell are as good as any cast-steel anvils made today. It is just a shame that the makers did not mark them. Several people have sent me photos and dimensions of unmarked anvils, hoping that I could identify the maker or tell them something of its history. However there is very little to be said.

If you own or locate one of these anvils and it is of cast steel, you can be sure that it was made in the 20th century and most likely after about 1930 and before the 1960s. There were few anvil makers left by the 1930s.

Miscellaneous Anvils

Some of the following anvil names may be just names, and not an anvil make at all or a misrepresentation of an actual anvil brand. I will indicate with each entry what I know at this writing about these names.

American-Ross: I found the name of this anvil in the book *Foxfire 5*, page 174.[1] "So we went down there, and he said, ' I'd love to swap you my big anvil for your little one.' So we made an even trade. He took the little and I took the big one. It's sitting down there now. It's *American Ross,* they don't never wear out." This is as much as I know about this anvil.

Baker: This anvil name was given to me by a machinery salesman. It might have been *Boker* (see H. Boker).

Bay State: This anvil appears in advertisements in several hardware catalogs that I have seen, but I have yet to locate an anvil with this name. The ads say it is cast iron.

BAY STATE*
FARMERS' CAST.
Contains a large percentage of Steel, which makes them Hard and Tough; Body and Horn Japanned Finish; Face is True and Well Polished; Square Hardy Hole near the Heel.

Nos.	B7	B10
Dimensions of Face, inches	3 1/8 x 10 3/4	3 3/8 x 12 1/2
Height over all, inches	8 1/3	9 1/4
Length over all, inches	18	20
Size of Hardy Hole, inches	7/8	7/8
Weight each, lbs.	70	100
Per lb.	$0.14	.14

Packed Loose.

Simmons Hardware Co. catalog, 1930, St. Louis, Mo. (From Mr. Dave Poppke)

Boston: This wrought anvil was made for a hardware company in Boston, Massachusetts. *Boston* is the only word that is clear. What I can make out is as follows: *H & ??? Boston Mass 179.* It has four handling holes, but does not look like any American or English brand. It is or was in the machine shop boiler room at Greenfield Village, Dearborn, Michigan. It may be a German make or made by the American Wrought Anvil Company. I have also received a data sheet on another anvil with the following trademark:

Butts & Ordway, Boston Ma.
A,
161

Butts and Ordway was a hardware company in Boston. Could these two anvils be made for the same company?

Bradley: I was told by someone (I do not remember who) that the Bradley Company of Syracuse, New York, makers of the Bradley power helve hammers, made a cast-iron anvil at one time. I have found no conformation of this yet.

Columbia: The H. Channon Company of Chicago, Illinois, advertised an anvil by this name in their 1908 catalog. The ad says it is a wrought anvil. It was probably made by Hay-Budden.

COLUMBIA ANVILS

Solid Wrought. Crucible Steel Face. Warranted.

Our Columbia Anvils are made of all wrought iron doubly refined and consequently are **Solid Forgings.**

The Crucible Steel on face, which is **the very best** that can be bought for the purpose, is welded to the wrought iron. They are the best anvils on the market, none excepted.

General Blacksmiths' Pattern

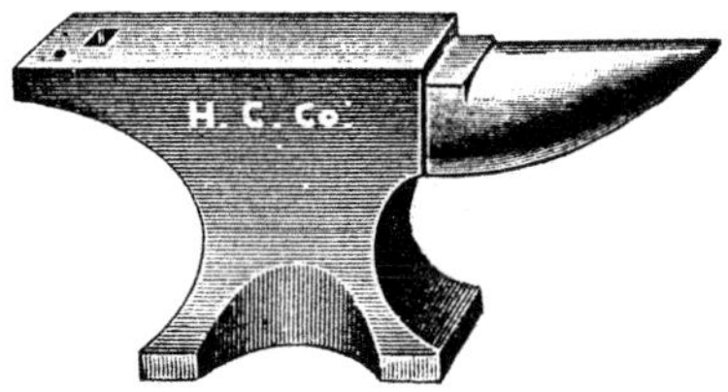

COLUMBIA FARRIERS' CLIP HORN ANVILS

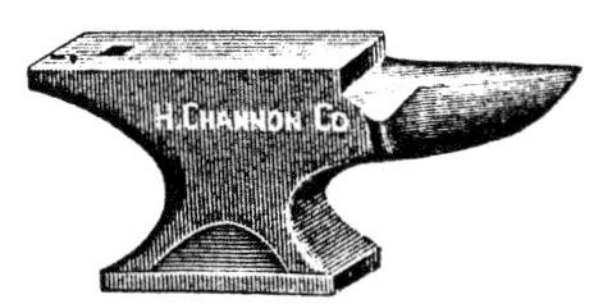

Made from 80 to 400 lbs. Generally with 3½ **and** 3¾ face. Drop on step from ⅞ to 1 inch. Add 1 cent per lb. to Blacksmith Pattern prices.

H. Channon Co. catalog, #33, 1908, Chicago, Il.

Cranberry C. M. McClung & Company, Knoxville, Tennessee, advertises this anvil in their 1903-04 catalog. It is my guess that it was made by Hay-Budden, although they advertise *Hay-Budden* anvils in the same catalog.

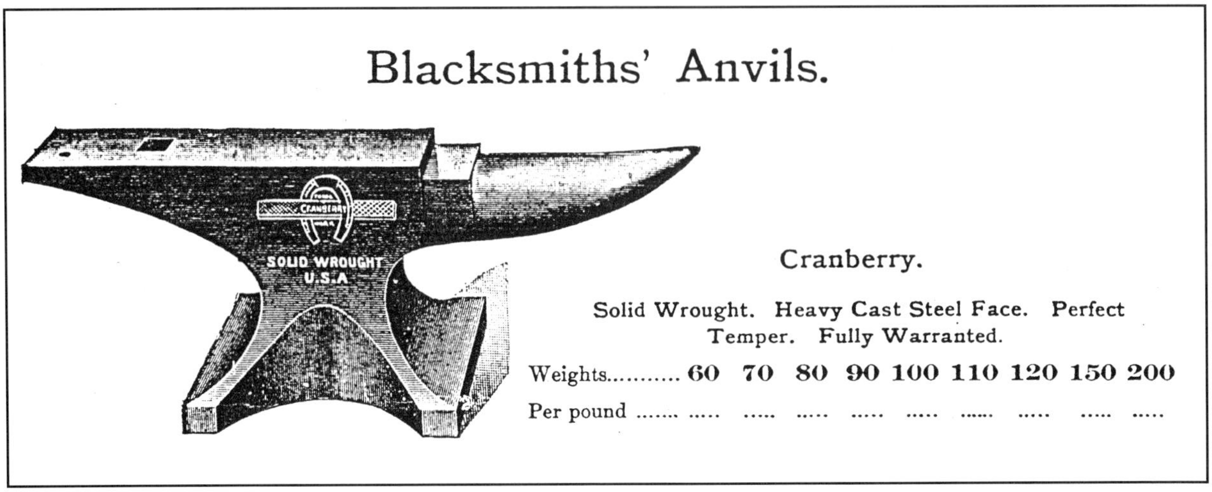

C.M. McClung and Co. catalog, ca. 1903-1904, Knoxville, Tn. (From Mr. John Catchings)

Cray Brothers: Cray Brothers was a large hardware company in Cleveland, Ohio. They advertise an anvil in their 1901 and 1926 catalogs under their own trademark. Most likely the anvil was made by the Columbus Forge & Iron Company of Columbus, Ohio.

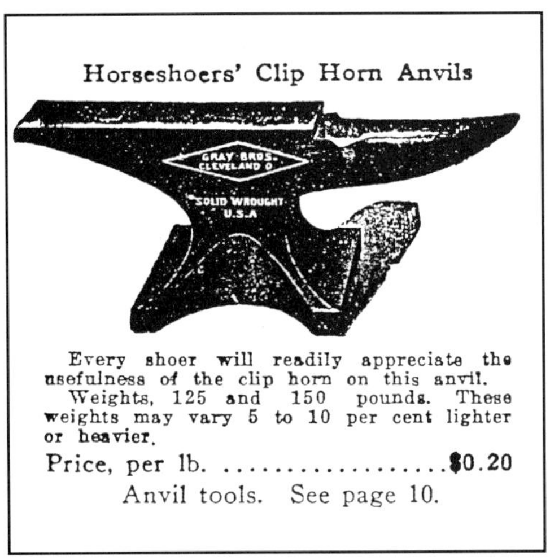

Cray Brothers catalog, 1926, Cleveland, Oh. (From Mr. John Catchings)

Colonial: This was the trademark of the Schwabacher Hardware Company, Seattle, Washington. They advertised a solid-steel anvil under this name. I recorded a *Colonial* anvil made by the American Wrought Anvil Company, Brooklyn, New York. The *Colonial* was stamped just above the *American.* However, the *American* is a wrought anvil and not cast steel. The catalog was dated 1920, long after the *American* anvils were off of the market. The anvil advertised was probably of Swedish make[2] (see chapter 5).

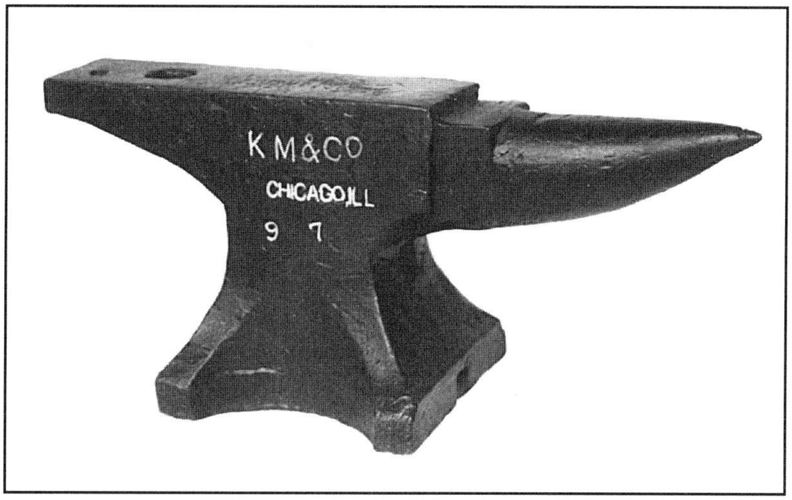

97-pound KM & Co. (R.A.P. photo)

K M & Company: I have recorded two of these anvils; one weighed 97 pounds and the other one 147 pounds. They have *K M & Co.,Chicago, Il.* on the side and then the weight. The smaller one is a smith's pattern, the larger one a farrier's I am almost sure that this anvil was made by the American Wrought Anvil Company of Brooklyn, New York.

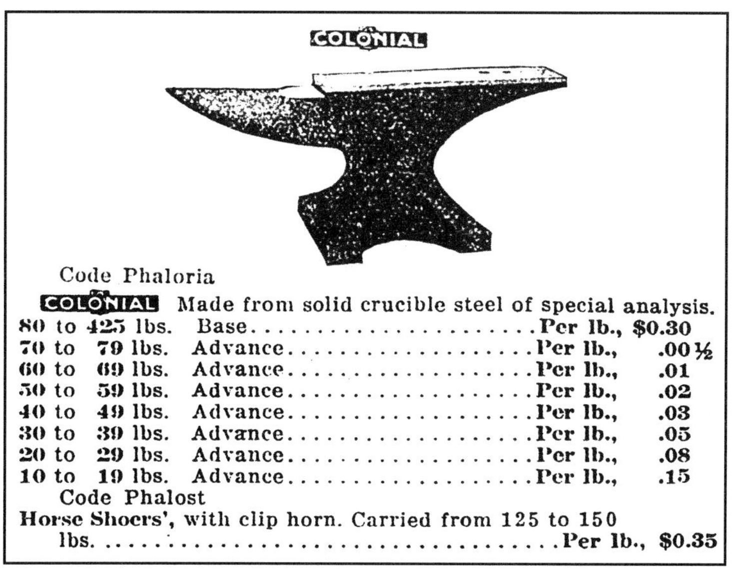

COLONIAL

Code Phaloria

COLONIAL Made from solid crucible steel of special analysis.

80 to 425 lbs.	Base	Per lb.,	$0.30
70 to 79 lbs.	Advance	Per lb.,	.00½
60 to 69 lbs.	Advance	Per lb.,	.01
50 to 59 lbs.	Advance	Per lb.,	.02
40 to 49 lbs.	Advance	Per lb.,	.03
30 to 39 lbs.	Advance	Per lb.,	.05
20 to 29 lbs.	Advance	Per lb.,	.08
10 to 19 lbs.	Advance	Per lb.,	.15

Code Phalost

Horse Shoers', with clip horn. Carried from 125 to 150 lbs. .... Per lb., $0.35

Schwabacher Hardware Co. catalog, 1920, Seattle, Wa. (From Mr. Gill Fahrenwald)

Nelson, Big 4: (dated 1892). This anvil was advertised in an auction paper from Iowa in 1995. I did not attend the auction.

Oak Leaf: *Oak Leaf* is one of the trademarks of W. Enders Hardware Company. The wrought anvil was advertised by the Simmons (Keen Kutter) Hardware Company, St. Louis, Missouri, in their catalog of 1913, 1921-1922 and 1930. Shapleigh Hardware Company of St. Louis, Missouri, also carried *Oak Leaf* anvils.

BLACKSMITHS' ANVILS.

OAK LEAF*.

Solid Wrought Body and Horn, Japanned Finish, with Hardened and Tempered Flat Steel Face, Welded on and Polished; Square Hardy Hole near the Heel.

Nos.	E100
Dimensions of Face, inches	3⅝x13⅓
Height over all, inches	10
Length over all, inches	22½
Size of Hardy Hole, inches	¾
Weight each about, lbs.	96 to 106
Per lb.	$0.40

Packed Loose; Face and Horn Burlapped.

Simmons Hardware Co. catalog, 1930, St. Louis, Mo. (From Mr. Dave Poppke)

Macarco Anvils

"THE CHOICE OF THE CRITICAL"

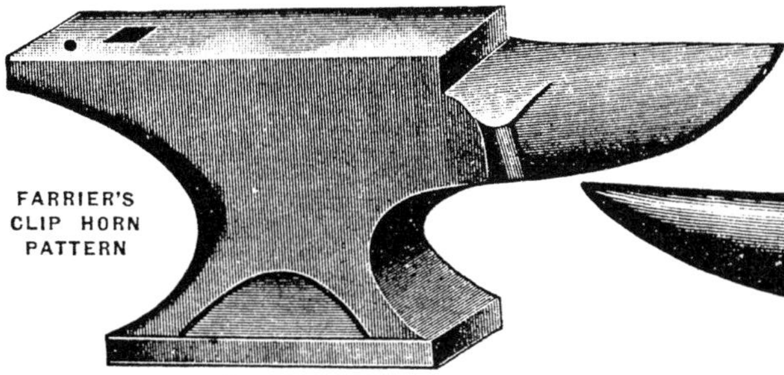

FARRIER'S CLIP HORN PATTERN

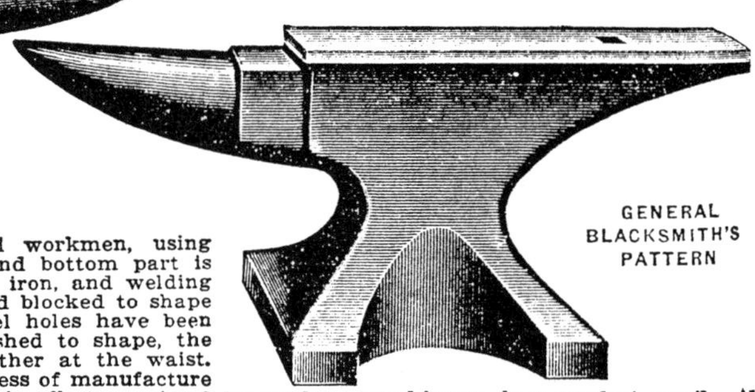

GENERAL BLACKSMITH'S PATTERN

The Macarco Anvils are made by skilled workmen, using nothing but the best materials. The base and bottom part is built up by piling up carefully selected scrap iron, and welding into a solid mass. Then it is put in a die and blocked to shape while still hot. After the hardy and pritchel holes have been punched, the heel squared and the horn finished to shape, the two pieces, base and upper, are welded together at the waist.

The Macarco Anvils are handled during process of manufacture by experienced workmen and the result of their efforts, assisted by modern machinery, is a perfect anvil. An anvil with horn and heel in perfect proportion, large base that does not require stamping, tempered edges to prevent chipping, level and true tool steel face of correct size, also hardy and pritchel holes and other mechanical details of correct accuracy.

Size No.	Nominal Weight, Pounds	Variation in Weight, Pounds	Size Face, Ins.	Purpose for Which Recommended	Size Hardy Hole	NET PER LB. Far'rs' Pattern	NET PER LB. B'smiths' Pattern
0	75	70 to 80	3½x12½	Light Repair Work	⅞	$0.22	$0.22
1	100	95 to 105	3½x13½	Farmers' Size	⅞	.22	.22
2	125	120 to 130	3¾x15	General Blacksmiths' and Horseshoers' Contractors' Sizes for Ordinary Work	1	.22	.22
3	150	145 to 160	4 x16		1 1/16	.22	.22
4	200	195 to 210	4¼x17½		1⅛	.22	.22
5	250	240 to 260	4¾x19	Machine and Heavy Forge Work	1⅛	.22	.22
6	300	290 to 310	5 x20		1¼	.22	.22
7	350	340 to 360	5½x22		1¼	.22	.22
8	400	390 to 415	5½x24		1¼	.22	.22
9	500	475 to 525	6 x26		1⅜	.22	.22

Mc-Master-Carr Supply Co. catalog, 1937, Chicago, Il.

Macarco: The McMaster-Carr Supply Company, Chicago, Illinois, listed a wrought anvil with this trademark in their 1937 catalog. I found the ad interesting. The first part of the ad was indicative of an anvil made about 1900, whereas the last part suggests a later-model anvil. Until I can locate one of these anvils with *Macarco* stamped on it, I will not be sure who made it, although by 1937 there were very few wrought anvil makers.

Smith & O'Leary, Chicago: As I mentioned in another section, when I first discovered this anvil I was confused. The anvil was of the *Old English* style, yet it was stamped with an American trademark and the weight was in actual pounds. The trademark was in the form of a circle, **O.** The weight was 141pounds. The anvil looked like a *Mousehole;* therefore I thought that possibly Mousehole Forge had made the anvil for the Chicago concern. Then I found the following advertisement in the May and June 1884 issues of *The Blacksmith and Wheelwright* journal:

> *Have you any broken anvils? If so send them to Smith & O'Leary, Steam Forge Works, 132 & 134 W. Lake St., Chicago, who will make them as good as new. Dressed, Ground and Tempered, 4cts. or new Steel Faced, Ground and Tempered, 6.5 cts. Mfrs. of Vises, Boxes, Screws and Champion Pinch-bars. Send for catalogue.*

The "4" and "6.5" cents refer to price per pound of anvil, I am sure. A new English anvil cost from ten to twelve cents per pound at that time. In an article in *The Blacksmith and Wheelwright* journal for January, 1898, titled: "Repairing Anvils," the author mentions Arthur J. O'Leary of Chicago. Their prices were as follows: "...dressing, tempering, and grinding, 4 cents per pound; 'Steeling' [new face plate], 6.5 cents per pound, and 'new horn or heel', 7.5 cents per pound." The anvil that I recorded was made by Mousehole Forge and repaired by O'Leary. I also have one of Smith & O'Leary's leg or blacksmith vises (see chapter 3, Smith & O'Leary for photo).

Snyder & Oaks: First I found the patent for this anvil, then Mr. Gill Fahrenwald located a person who owned one. He sent me six color slides of the anvil. From looking at the patent, I doubted that this anvil was ever produced, but I was wrong. Regardless of how many were ever made, at least one is in existence. From the slides the anvil appears to be cast steel and the real anvil looks exactly like the patent shows.

The patent was granted in 1901 and the anvil is cast steel, which makes *Snyder & Oaks,* along with the *Columbian*, the first cast-steel anvils made in this country.

UNITED STATES PATENT OFFICE

NELSON HOWARD SNYDER, OF NEW JERSEY.

ANVIL

SPECIFICATION forming part of Letters Patent No. 667, 593, dated February 5, 1901.

Application filed July 30, 1900. Serial No. 25,308. (No model.)

To all whom it may concern:

Be it known that I, Nelson Howard Snyder, a citizen of the United States, and a resident of Newton, in the county of Sussex and State of New Jersey, have invented a new and Improved Anvil, of which the following is a full, clear, and exact description.

The purpose of the invention is to provide an anvil especially adapted for horseshoers, or blacksmiths and so made that all horseshoe construction and the ordinary work of a blacksmith may be successfully done without necessitating the operator to be constantly moving to or from the ends of the anvil, and, further, to so construct the anvil that extended bearing-surfaces are provided flush with the upper face of the body of the anvil, which extended bearing-surfaces are especially useful when working upon axles and when forming the toe and side clips of horseshoes.

The invention consists in the novel construction and combination of the several parts, as will be hereinafter fully set forth, and pointed out in the claims.

Reference is to be had to the accompanying drawings, forming a part of this specification, in which similar characters of reference indicate corresponding parts in both figures.

Figure 1 is a perspective view of the improved anvil, and Fig. 2 is a plan view of the same.

The bed A of the anvil is preferably made longer than ordinarily and the horn 10 is given an outward curve or stands at an inclination to the body of the anvil instead of parallel therewith, as is customary, and this inclination is given to the horn 10 in order that the horn shall not interfere with the use of tongs or other implements or with the metal being treated when said metal is being operated upon on a wing 11, which wing extends at a right angle from the front side of the anvil. The upper face of the wing 11 is flush with the bed of said anvil, as is shown in Fig. 1, and most of the work incident to horseshoeing and blacksmithing may be performed upon this wing 11, is provided with an opening 12 for the "hardy" usually employed for cutting iron and with an opening 13, adapted to receive a punching-tool, which opening may also be utilized for threading bolts. These two openings 12 and 13 are duplicated at the right-hand end of the bed of the anvil, where they are shown designated as 12^a and 13^a , so that the majority of the work that formerly had to be performed at the end of an anvil under construction may be executed at the center and wing 11. The side surfaces of the wing are straight, as well as the front side surface of the body of the anvil, so that where the sides of the wing connect with the front side of the body of the anvil a perfect right angle is obtained, and the portions of the anvil may be utilized in the same manner as a two-armed square. The wing 11 is also adapted for use in making toe-calks for horseshoes, as the calks may be formed over the edges of the wing instead of at the right-hand end of the anvil, as has been the practice heretofore.

At the left of the wing 11 and between said wing and the horn 10 a series of grooves is formed in the bed of the anvil. These grooves, three of which are shown and designated as 14, 15, and 16, are graduated either in length or width, or in both, and these grooves 14, 15, and 16 are angular or **V**-shaped in cross-section and extend to and through the front side wall of the anvil. These grooves 14, 15, and 16 are adapted to receive steel calks when said calks have been temporarily attached to a hot shoe and enable the calks to be welded to the shoe without interfering with the sharpness or original shape of the calks.

At the rear of the grooves 14, 15, and 16 and at the left of the wing 11 a rear extension 17 is formed from the bed A of the anvil, the upper face of this extension being flush with the upper face of the anvil-bed A and the upper face of the wing 11, and this projection 17 is perfectly located quite close to the horn 10. The projection 17, which is practically a small table, is utilized for forming the toe and side clips on horseshoes, for which purpose the side edges of the bed have been heretofore employed. At the right-hand side of the wing 11 another groove 18 is formed transversely in the bed A, and this latter groove also extends out to the front side edge of the anvil. The groove 18 is semicircular or rounded in cross-section or is made more or less oval in cross-section and rounded at its inner end. This groove 18 is employed to receive a punch adapted for turning a shoe, and therefore the operator does not have to carry the shoe to the end of the anvil to turn it, as has been the practice. The groove 18 is also available for forming the body portions of bolts and the like.

The wing 11 in addition to its other functions is particularly serviceable when an axle is to be treated, since it enables the operator to provide a long bearing for the axle, and the operator may manipulate the axle to much greater advantage than when placed across the narrow trans verse surface of an anvil-bed.

Having thus described my invention, I claim as new and desire to secure by Letters Patent -

1. An anvil, having a wing extending from one side surface at a right angle to the said surface, the upper face of the wing being flush with the bed or upper face of the body of the anvil, and a projection extending from the opposite face of the bed or upper surface of the body of the anvil flush therewith and at one side of said wing.

2. An anvil having a wing extending from one side surface at a right angle to the said surface, the upper face of the wing being flush with the bed or upper face of the body of the anvil, a second projection extending from the opposite face of the bed or upper surface of the body of the anvil flush therewith and at one side of the said wing, and a horn which is curved in a direction away from the said wing, as specified.

3. An anvil, a wing extending at a right angle from one side face of the anvil, the upper surface of the wing being flush with the bed or upper surface of the body of the anvil, the said anvil being also provided with a projection from the side opposite that at which the wing is located and to the left of the said wing, a horn which is curved in a direction away from said wing, the bed of the anvil being provided at the left-hand side of the wing with a series of transverse grooves adapted to receive metal calks, to facilitate welding said calks to a shoe, the bed of the anvil being further provided at the right-hand side of the wing with a transverse groove, transversely semicircular or oval, all of the grooves extending out to the side of the anvil at which the wing is located, for the purpose described.

In testimony whereof I have signed my name in the presence of two subscribing witnesses.

NELSON HOWARD SNYDER.

Witnesses:
Levi II. Morris,
Veranus M. Rundle.

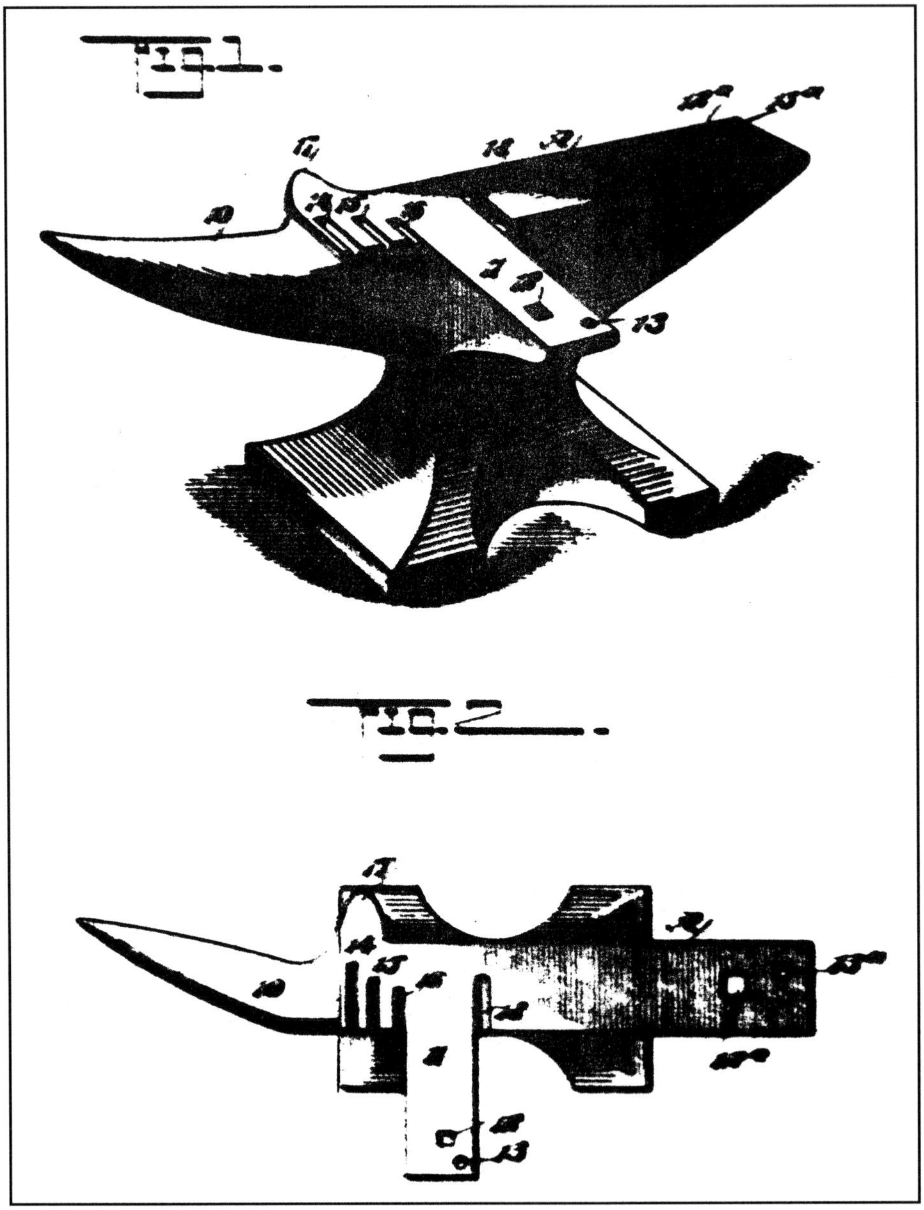

Patent and drawing courtesy of the Smithsonian Institution.

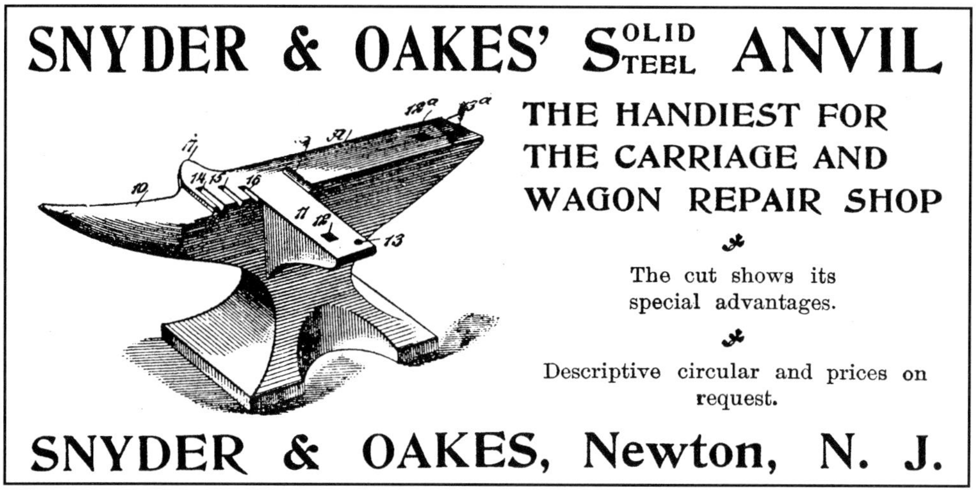

The Blacksmithsmith and Wheelwright, December, 1903. (From Mr. Ken Wilson)

Special and Unusual Anvils

There were many special and unusual anvils made for various purposes. It seems that every trade had a special anvil. This list, I suspect, only scratches the surface.

Many of the following are so-called anvils that I have seen, photographed or have seen advertisements for. Most of the anvils and stakes are designed for a particular trade or function. I am including the stakes because they were called anvils, were used by traditional craftsmen, and are all connected with the working of metal in one way or another.

Anvil-Vise: Strictly speaking, these are more vise than anvil. However, the older makes provide a larger anvil profile than the vise does. I have had people send me photos of the anvil part of these with the moveable vise part missing. They wish to have the anvil identified. I have also seen these in flea markets with the vise part missing and the remaining part labeled as an anvil.

The anvil part will make a fairly good craft anvil for very light work, but you would not want to do any serious forging on one. All the anvil - vises that I have seen are hollowed out in the body, and are made of cast iron. A piece of railroad track dressed up a little would be far superior to any of these so-called anvils.

ANVIL AND VISE COMBINED

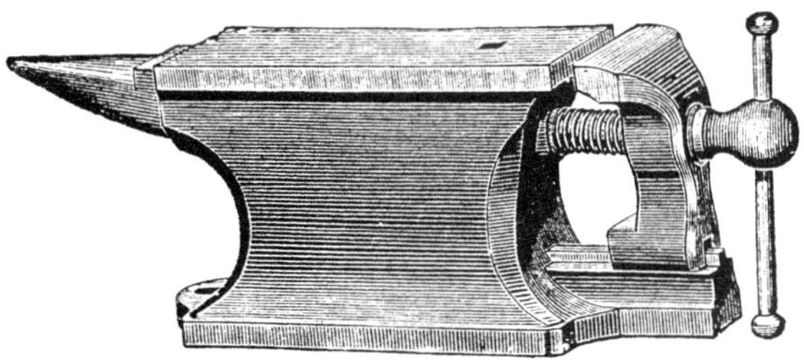

The face of the Anvil is chilled, rendering it of sufficient strength and hardness to withstand all ordinary usage; Wrought Iron Screw, with Brass Wire Spring to throw the Jaw out.

No.	Face, inches.	Jaw Vise, inches.	Weight, lbs.	Each.
1	10½×4	4	40	$5.25
2	8 ×3¼	3¼	25	4.25
3	6 ×2¾	2¾	14	3.50

Chilled Anvil, 4×2½ inch Face; Vise with 2 inch Jaw; Wrought Iron Screw to Vise. A Cold Chisel and Punch are sent with each Vise, and also two Coach Screws to fasten it to the bench. Weight, 16 pounds.

No. 16 each, $3.00

Two in a box.

ANVIL, VISE, AND DRILL

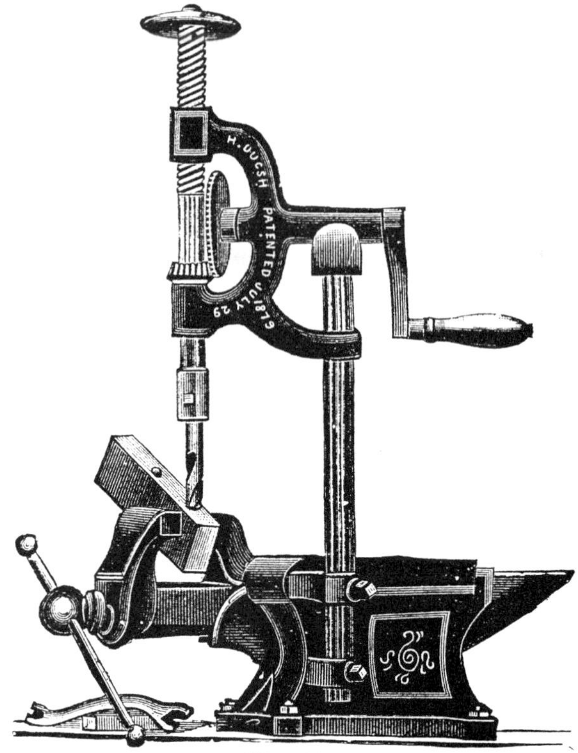

No. 80

The Anvil has a Steel Face 4×8 inches; Vise Jaws, 3½ inches wide, and Steel Faced.

The Steel Drill Press has an Adjustable Chuck holding Drills up to ¼ inch. With Chuck removed the Spindle will hold Drills with ½ inch Shanks.

			Each.
Anvil, Vise and Drill, weight,	80 lbs.,		$18.00
" and Vise,	"	60 lbs.,	10.00
Drill Press,	"	20 lbs.,	8.00
Offset Vise and Drill,	"	61 lbs.,	15.00
" " only,	"	41 lbs.,	7.00

Russell & Erwin catalog, 1899, New Britain, Ct. (From Mr. John Catchings)

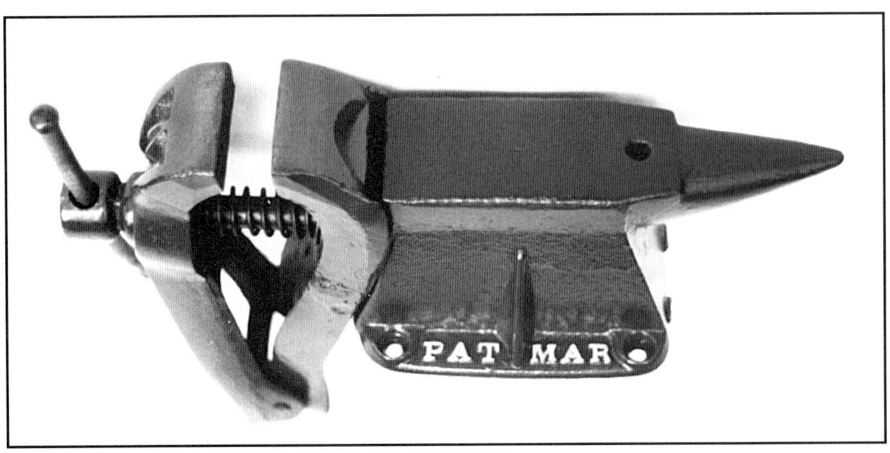

Anvil vise combination. (R.A.P. photo)

Vise part removed from the combination. (R.A.P. photo)

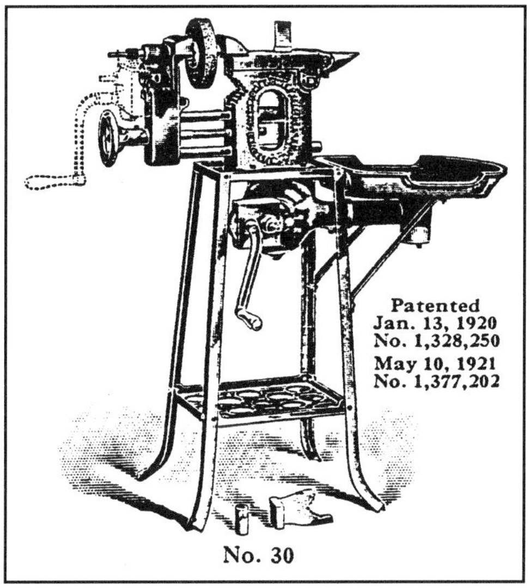

The Champion Blower & Forge Co. catalog, 1935. Lancaster, Pa.

VISES AND ANVILS.

COMBINATION.

CHENEY.

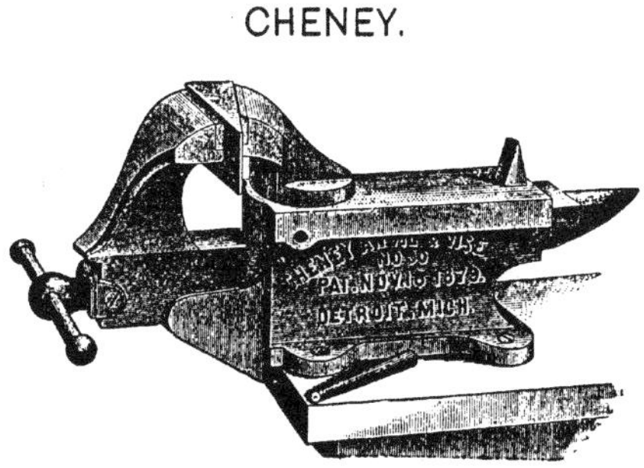

Nos.	10	20	30	40
Face of anvil, inches	4½x2	6½x3	8x3½	8½x4
Width of jaws, inches	2½	3½	4	4½
Jaws open, inches	3	4	5	6
Weight, crated, pounds	18	31	43	57
Each	$3.50	4.50	5.50	6.50

FARMER.

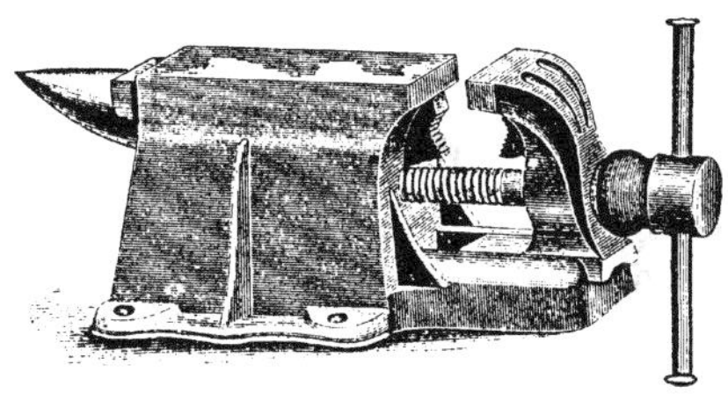

No.	1
Face of anvil, inches	7x
Width of jaws, inches	3
Jaws open, inches	3¼
Weight, pounds	33
Each	$4.5

Farwell-Ozmun-Kirk & Co. catalog, January, 1901.

Axe Maker's Anvil: I have no idea what one of these anvils would look like, although they are mentioned in a number of publications.[3] They probably look like a hornless anvil. Before the mid-19th century, I am sure that the smith used his blacksmith's anvil to make axes.

Bench Anvils: These small anvils weigh from about 5 to 25 pounds and look like smith's anvils but are stubbier. Most of them are made of cast iron, but almost all of them are solid throughout. They are also called "craft anvils" because of the fact that they are usually used for light craft work, such as riveting, straightening or bending small rods or for cracking nuts.

The Columbian Vise Company still makes a bench anvil weighing about 10 pounds and have done so since the early 1900s. Most of the bench anvils that you see today are either made in China or Taiwan. Many foundries made them in the past, as did some of the anvil makers. One of the more interesting ones is the *Adamant.* I have recorded one of these. From their ad you will notice that it appears to have been made from a piece of railroad rail. However, the one I saw appeared to be a single forging. Montgomery Ward carried them in their 1900 catalog, as did many hardware companies as early as 1881 and as late as 1915 (see Craft Anvils).

HILL'S SOLID STEEL ANVILS

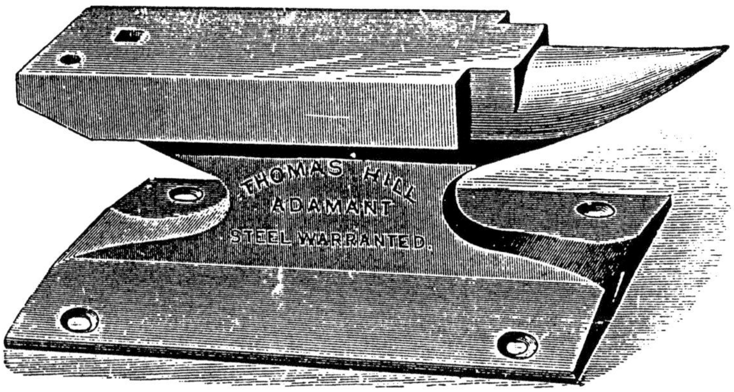

FORGED FROM SOLID STEEL WITH A TEMPER BEST ADAPTED TO AN ANVIL.

Face, sides and horn polished; the base japanned black.
Length of horn two inches in addition to length stated below.

Nos.	1	2	3
Length	5 in.	6 in.	7 in.
Width	1½ in.	2 in.	2 in.
Weight	4 lbs.	5½ lbs.	9 lbs.
Price	$2.10	$2.70	$3.10

Miller Falls Co. catalog, 1915. (From Mr. Hunter Pilkinton)

William Dixon Co. catalog, 1933, Newark, N.J.

VILLAGE BLACKSMITH

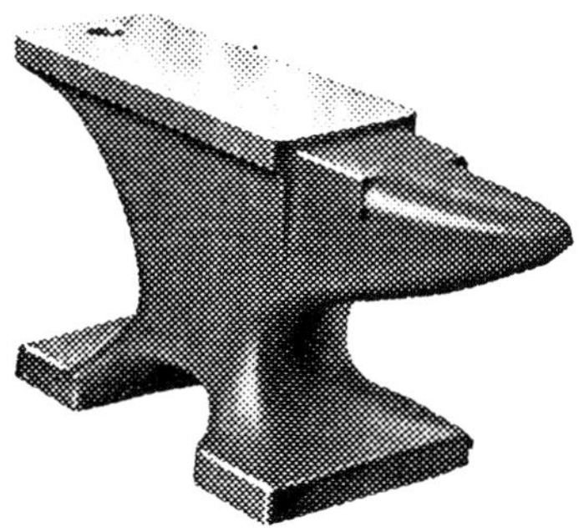

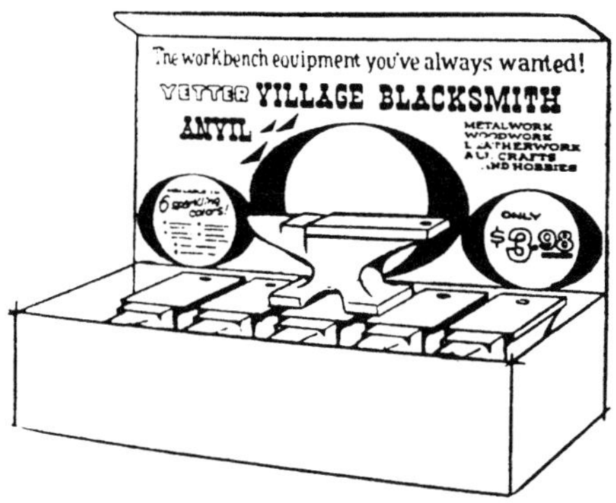

Farm and home size anvil for use in metalwork, woodwork and leatherwork.

Sturdy steel construction with baked enamel finish.

Size 8x3 7/8x3 inches.

Display consists of one each of following colors: Gold, red, blue, green, gray and orange.

T71-56 .... Display $49 50
(Display of 6 will not be broken)
One in shp ctn; wt each 48 lbs.

Belknap Hardware & Mfg. Co. catalog, 1961, Louisville, Ky.

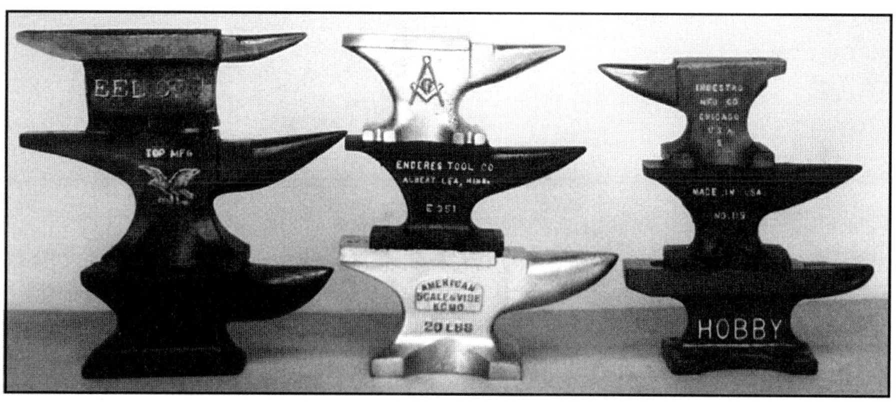

Nine bench anvils weighing 5- to 20-pounds. (R.A.P. photo)

Bridge or Railroad Anvils: These so-called anvils have nothing to do with bridges to my knowledge. They take their name from the fact that there is a large hollow space going through the body from one side to the other, and the face of the anvil "bridges" this hollow space. They were mainly in railroad forges and repair shops as a type of anvil bending fixture. The hollow allowed certain bends to be made on railroad equipment that were difficult to perform on a large smithing anvil.

Most of them were made of cast iron in local foundries for particular railroads. I know that both the Columbus Forge and Iron Company and the Columbus Anvil and Forging Company made them. The Columbus Anvil and Forging Company called theirs a "Mule." I have recorded several cast-steel ones made by unknown foundries. They generally weigh from 400 to 1000 pounds, but most of them are in the 700-pound range.

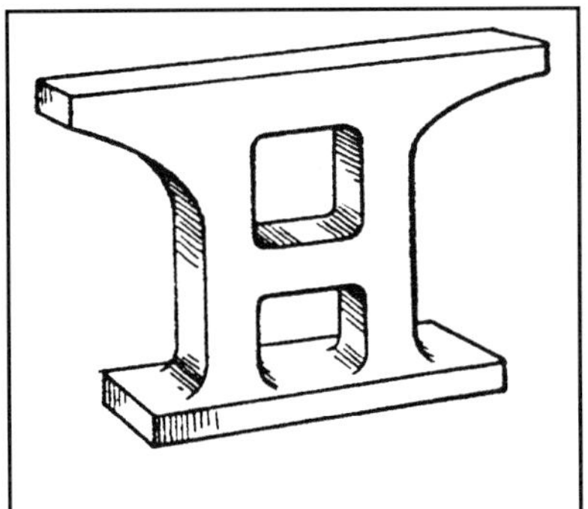

Fig. 7

Another type of anvil fig. 7, is sometimes seen in shops where much bending is done. Loops and such work can be handled to better advantage on an anvil of this shape, and jigs and forms may also be fastened to it with greater ease than to the common shape. (Practical Shop Work, 1906)

945-pound cast-iron railroad anvil. From the Monterey, Tenn. railroad yard. (R.A.P. photo)

Bridge anvil at the Museum of Appalachia. (R.A.P. photo)

Other anvils that can be called bridge anvils are the ones used in the oil fields. I know of two companies that made them; one was the Oil Well Supply Company, Pittsburg, Pennsylvania, and another company in Ohio. They were built lower to the ground than the railroad anvils. The hollow through the anvil is arched at the top, giving a good impression of a bridge. They are made either of malleable iron or semi-steel because they are not "dead" when struck, and the anvil that I have has seen much use and is still in good condition.

300-pound bridge anvil used in the oil fields. (R.A.P. photo)

Cooper's Anvil: The photo shows that these are more stake than anvil, although the faces on some are shaped like a smith's anvil. The cooper's anvil is used to bend, punch and rivet the metal hoops that hold a wooden barrel together. They require a wooden block as a base for support, and are easily moved from place to place. They generally weigh about 50 pounds without the wooden block. The one that I have has a tapered horn, and three sizes of holes through the face: 1/4, 1/2 and 3/4 inches. It also has two dimples in the center of the face, 7/16 and 1/2 inches wide, for riveting. An old Hay-Budden brochure lists them as being one of their products.[4]

Cooper's anvil. (R.A.P. photo)

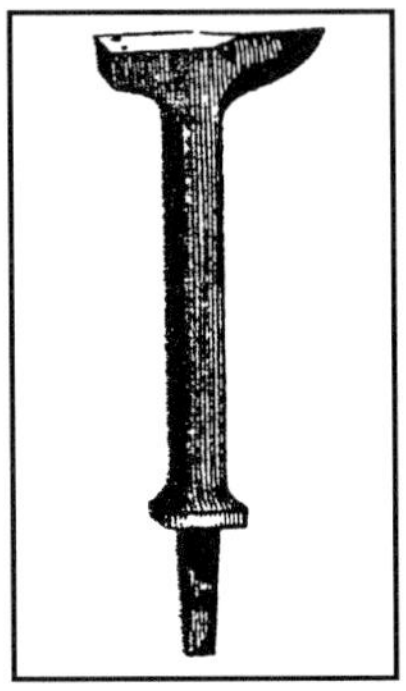

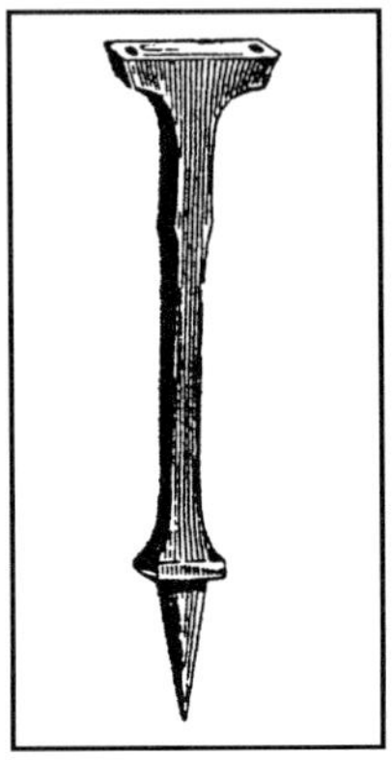

Two types of cooper's anvils.

Chainmaker's Anvil: Some of these were special anvils and some were modified smith's anvils. Fisher and Norris made most of the chainmaker's anvils found in America, whereas the English often just modified a smith's anvil for medium chains, 3/8 to 1-inch. For chains under 3/8 inches in diameter, a special stake was generally used. For chains from one inch in diameter and on up, several fixtures and jigs were used to complete the work on the chain.

I have recorded or seen five chainmaker's anvils (1995). Three were made by Fisher and Norris, one large and unnamed, and one 150-pound English anvil that was modified for chain-making purposes.

It is not my purpose to describe how chain was made years ago, but I will have to do so in part in order to describe how the anvil was used. For the medium-size chain (which is most common) the anvil will have a hole going through the body from one side to the other. The larger chainmaker's anvils have two holes. The holes are generally square all the way through, but if the hole is round, part of it on one side of the body will be squared up for at least one-half inch into the hole. The holes are square in order to keep the fixture that is mounted in the hole from turning.

In making chain, first the proper length for the link is heated and bent on the anvil's horn or special jig. The ends are brought to a welding heat, lapped, and welded on the anvil or special jig. The fixture that went through the body of the anvil is now used to finish and smooth up the weld joint. The part of the fixture that went through the body of the anvil has a fixed die on one end of it for the particular size chain. The other end is hinged with an arm that can be brought over the anvil and in line with the fixed arm. The moveable arm has a die on its end that just matches the fixed die. As soon as the weld is made the link that is welded is placed in the lower die, the arm is brought over, and the top die comes precisely on the weld joint. The back of the top die is then struck with a hammer, thereby dressing the weld joint to match the rest of the chain. This fixture can be changed to suit the size of the chain to be made.

If you discover an anvil with a square or partly square hole about an inch wide going through the body, it was most likely used for making chain as well as for general blacksmithing work.

Only one of the chainmaker's anvils that I have recorded still had part of the fixture attached to it. This fixture was wedged in place and could be changed easily.

If you have further interest in chain making, see *Chains and Chainmaking,* by Charles Fogg, Shire Publications Ltd, Album 69, 1981. Also see Fisher and Norris.

"Old English " made into a chainmaker's anvil. From Mr. Mike McGee's collection. (R.A.P. photo)

110-pound chainmaker's anvil. The smallest one Fisher made. (R.A.P. photo)

215-pound chainmaker's anvil made by Fisher. Photo taken at Mr. Bill Gichner's. (R.A.P. photo)

Cobbler's Anvil: This bench-size anvil was made for a shoemaker. The four stakes marked "A," "B," "C" and "D" slip into the face and the screw on the vise locks them in place.

Cobbler's anvil-vise. (R.A.P. photo)

Cone Anvil or Cone Mandrel: These are a cone, just as the name implies. They are generally two to four feet high. Most are hollow cast iron, but a few are solid, and a few are in two sections. The tapered cone allows for many different-size rings to be made depending on the cone's diameter at the top and bottom. Some have lugs at the base for attaching them to the floor; most do not. One manufacturer patented a slot coming down on one side of the cone so that it would be easier to hold a hot round object with tongs while it is being hammered perfectly round.

Blacksmiths' Cones or Mandrels

These are cast smooth and round, and are well adapted for forming rings, bands and similar work.

No.	Height Inches	Diameter at Base Inches	Diameter at Top Inches	Weight Each Pounds
1	36	8	1	55
1½	40	10	1	90
2	48	12	1	115
3	52	14	1	165
4	54	16	1¾	200
5	50	24	5	500

Price....................per Pound ....

Dental Anvil: The photo shows that this anvil has a sheet metal body with a round iron plate inserted into the top. They were made by a dental company in Buffalo, New York, most likely to be used for making dentures and dental plates. They have a handle on each side of the body so that the dentist, or whoever, could move it wherever necessary. The body was filled with sand to give it stability and when empty it is quite light.

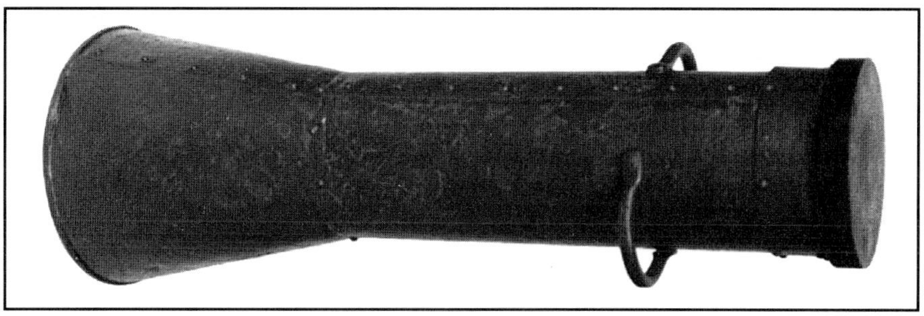

Dental anvil shown sideways. (R.A.P. photo)

Gun Anvil: Technically this is a swage block, but when the round or multi-sided swages were used for welding and shaping gun barrels in days gone by, they were referred to as "Gun Anvils."

"Gunmaker's anvil" at the Museum of Appalachia. (R.A.P. photo)

Cast Iron Swage Blocks

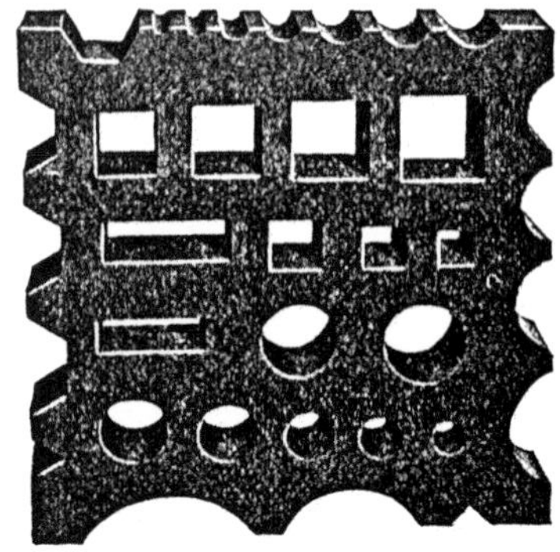

Made of strong iron, cast smooth and accurate to pattern.

No.	1½	2½	3½
Size........Inches	4x12½x12½	4x15x15	4¼x16x16
Weight....Pounds	100	148	190
Price...per Pound	.....	.....	.....

Swage block sometimes used as a gunmaker's anvil.

Harnessmaker's Anvil: This is more or less a bench anvil, specially designed to do the metal work when putting leather harness together and also for repairing the harness. The body does not taper from the face to the bottom, so it can be turned in any position in order to use the holes, dimples, and slots for punching, riveting, etc.

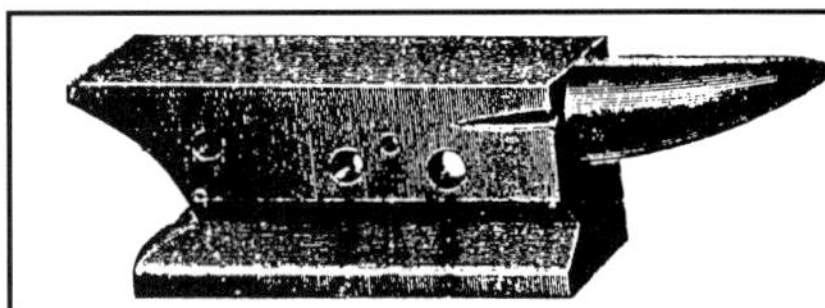

Harness Makers' Anvil

Harness maker's anvil. (Courtesy of Alvin Sellens.)

Jeweler's Anvil: The most common type looks like a miniature *Continental* pattern anvil. Some have a relatively heavy base and are free-standing. Others are made to slide into the jeweler's table, and locks. A few have a spike for driving them into some wooden base. They generally weigh less than a pound. Some plain ones are just a small stake and look like a scythe anvil, except that the face is highly polished.

Seven jewler's anvils. Size relative to the soft drink can. (R.A.P. photo)

Knee Anvil: As the name implies, this small anvil is cupped in the base so that it fits on the thigh part of a person's leg while they are sitting down. You can relax in your favorite easy chair and do light craft work, or crack nuts, etc.

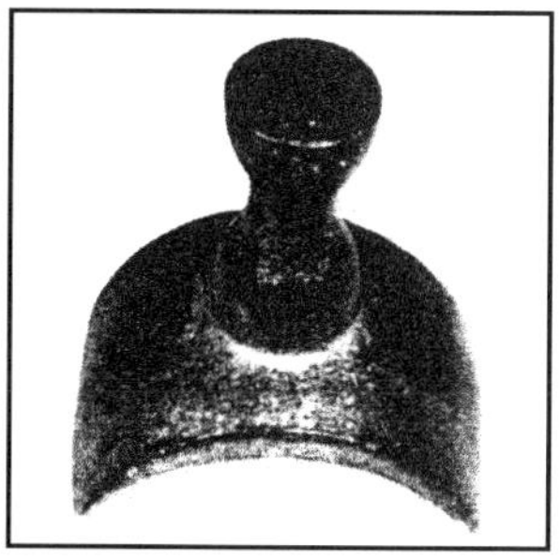

Knee anvil.

Knifemaker's or Cutler's Anvil: I have had photos of a number of these sent to me. They were used where a large number of different pieces of cutlery were made. Only part of the anvil is used directly; most of it has tapered slots in the face to take any number of different dies and jigs. They are heavy and free-standing and the ones that I have seen are very similar in shape, indicating that they were made by one or more of the major anvil makers, mostly in England. There is a good photo on pages 71 and 78 of *The Cutting Edge, an Exhibition of Sheffield Tools*, which shows them being used.[5]

Approximately 500-pound cutler's anvil.
(Mr. Jonathan Nathan photo)

Cutler's anvil. (M.D. Witzler photo)

Lap Anvil: This is similar to the knee anvil except that it rests on both legs and usually has a square tapered hole in the center to accommodate any number of stakes. It allows you to sit in a chair and do the same work that would be done on a bench or small anvil. Some have a flat raised surface in the center.

The lap anvil photo is part of a portable army shoe repair kit. (R.A.P. photo)

Nailmaker's Anvil: Professional nailmakers used headers in addition to this small stake-like anvil. It was fixed in place to facilitate more rapid nail-making of the most common nail sizes. Nails would be headed with the hand-held header over a hole in the anvil. The flat of the anvil is used for drawing out the nail and a hardie in the large hole to cut the nail to length.

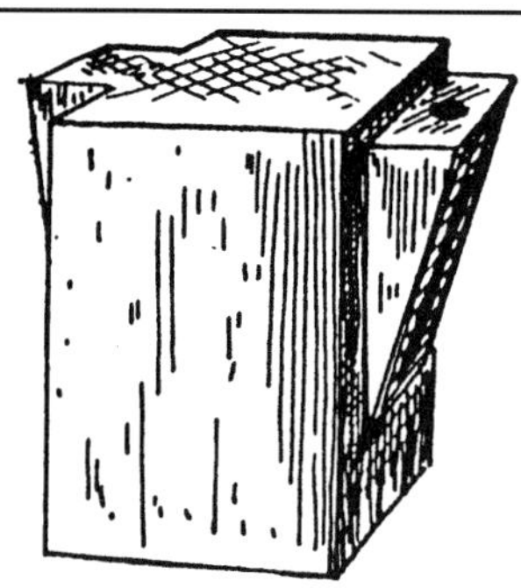

FIG. 58
Nail heading anvil, Roman 1 : 2

Taken from Book of Tools. (From Mr. John Catchings)

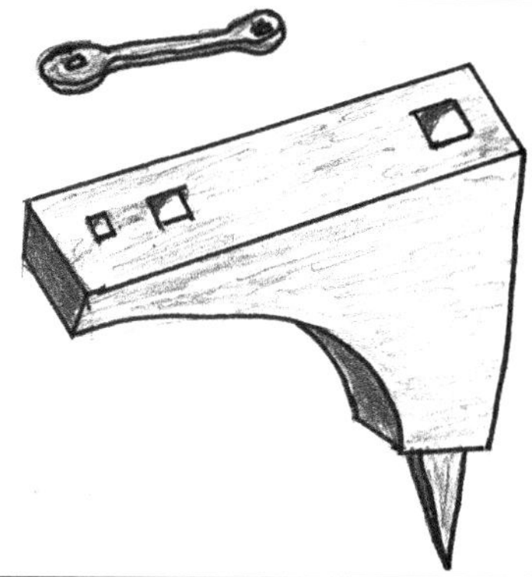

Nailer's anvil and header.

Plowmaker's Anvil: There are two types of this anvil. One is small, has a rounded head and is held by hand, much as a dolly is used in sheet metal work. Used with a hammer, it was carried into the field and used to dress the tip of the plow or point, cold. It weighs three or four pounds.

PLOW.

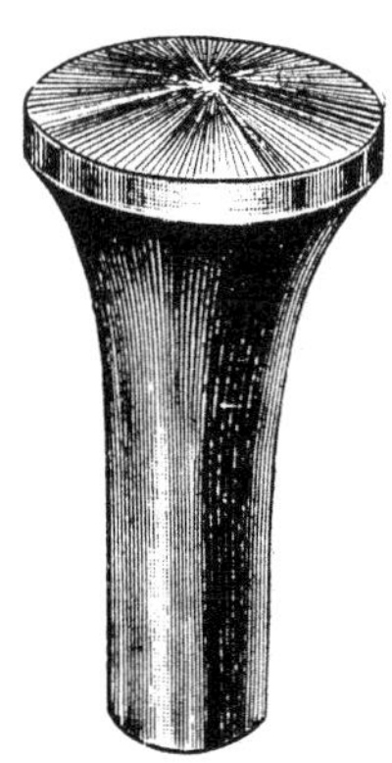

	Per Dozen.
Solid steel, weight, 4 pounds each	$10.00

Farwell-Ozmun-Kirk & Co. catalog, 1901.

The other type is essentially a smith's anvil with the face cut back at a sharp angle of about 45°, to better shape a plow shear when hot . All of the major American anvil makers made these anvils, but I doubt that they sold very many. The ones made by the Columbus Forge and Iron Company (*Trenton*) are essentially farrier's anvils; that is, there is no shoulder where the horn meets the body. They have two pritchel holes, but no clip horn. For a photo of a *Hay-Budden* plowmaker's anvil see *Hay-Budden* in chapter 5.

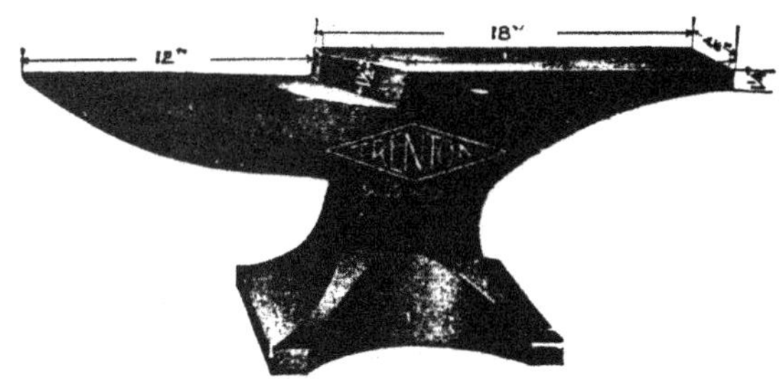

Plowmaker's Anvil

Front View

From the Columbus Forge & Iron Co. brochure

Mr. O. P. Wilcox patented a special plowmaker's anvil, but I have yet to find one or the patent.

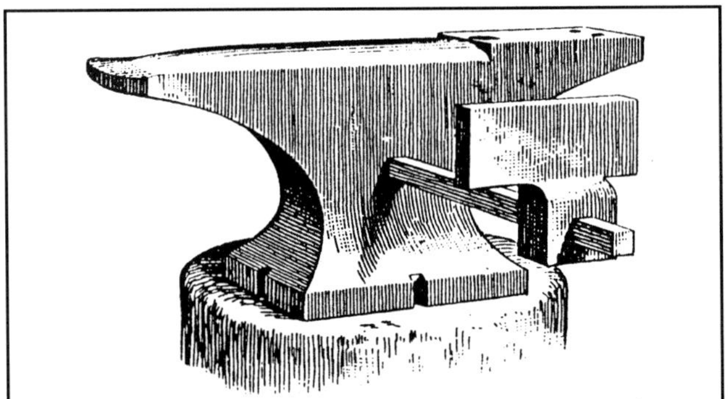

A Plowmakers' Anvil. Patented by O. P. Wilcox, February 28, 1901.

Railroad Track: Believe it or not, one company advertised short pieces of railroad track for bench anvils. The 1901 Farwell - Ozmun - Kirk Company catalog carried an advertisement for pieces of rail weighing ten pounds each at one dollar per dozen. They also sold regular anvils. Full gauge track is seven inches high. A foot or so of track when cut properly with a cutting torch makes an excellent beginner's anvil. The face of the rails naturally work-hardened as trains passed over them, and can be hardened by heating and quenching.

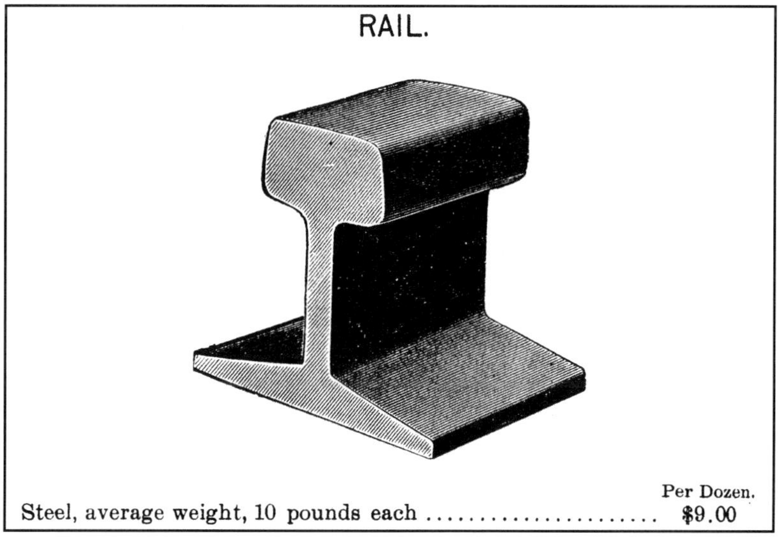

A piece of railroad track.

Railroad Track Anvil: This is a special anvil that slides onto a standard piece of railroad track. It was probably used extensively in the American West where a forge shop was not close and for quick repairs in the field. It weighs 92 pounds, is 9 inches long, 7 inches wide and 9 inches high. It has a 5/8-inch pritchel hole and a step down on the end.

Railroad track anvil.
(Mr. James Moon photo)

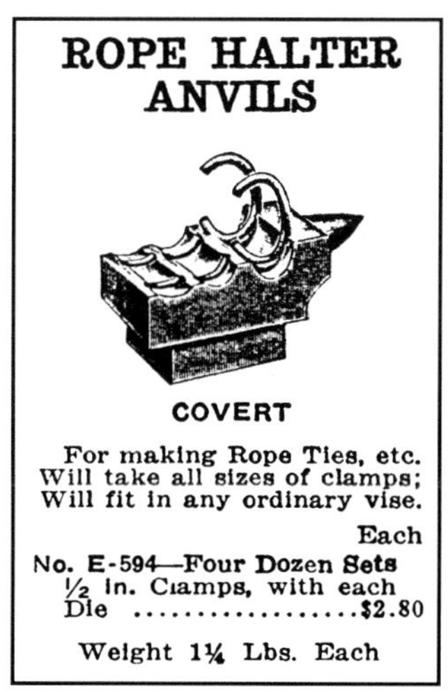

Marshall-Wells Co. catalog, 1924, Seattle, Wa. (From Mr. Gill Fahrenwald)

Rope Halter Anvil: This is usually a type of die or bending tool. You might say it is a vise tool, as it is clamped in a vise for use. It is used for bending pieces of wire around two or more pieces of rope to hold them together. It could also be used for making a light chain.

Sawmaker's Anvil: These anvils are nothing more than square or rectangular blocks of iron or steel with a hardened flat face. The older ones made by the major anvil makers weighed from 60 to 400 pounds. In 1985, the Hanchett Company was making them up to 1600 pounds (see Appendix J).

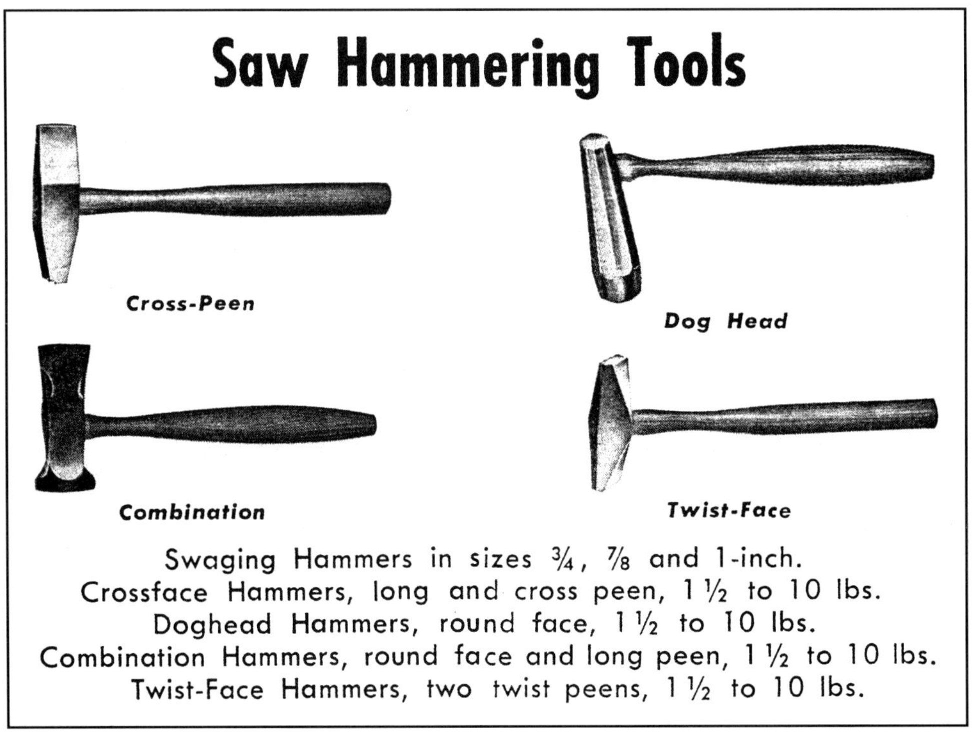

Part 1 of Hanchett Mfg. Co. ad.

Band, Gang and Circular Saw Anvils and Leveling Blocks

Leveling Block
Furnished with trunnion for ease in turning over on special order.

Round Anvil

Fracture of saw anvil showing one face chilled and hardened to a depth of about one-third the thickness of the block.

Square Anvil

Our Saw Anvils are a solid cast one-piece block made from a proper mixture to insure a dense iron block with a very hard face, chilled deep into the anvil body. This one-piece construction with deep chilled face results in a superior hammer action and skilled saw men demand this type of anvil. The faces are ground either flat for general band saw service, or crowned for general circular saw service. Be sure to specify when ordering whether for band or circular saws, and whether required flat faced or crown faced.

Standard leveling blocks are made from fine grained cast iron and are surfaced on both sides. They are intended for use in leveling band saws and are not intended for hammering. They are used in combination with the flat faced saw anvil so that hammering when required may be done on the anvil.

Hard-faced leveling blocks are really anvils with the dimensions of a leveling block. Hard-faced leveling blocks, therefore, must be furnished in thicknesses of 5″. They may be furnished in all the face dimensions of the leveling blocks, but must be 5″ in thickness, and the weights will increase accordingly. Hard-faced leveling blocks are carefully ground to precision limits on large surface grinding machines. They are used when it is desired to hammer band saws while the saw is on the leveling block.

HANCHETT
MANUFACTURING
BIG RAPIDS, MICHIGAN 49307
A/C 616—796-7678 TELEX 22-6458

Division Of
MWA
COMPANY

Part 2 of ad for the Hanchett Mfg. Co. of Big Rapids, Mi., As of 1997 this company is still in business.

60-pound Fisher sawmaker's anvil dated 1920. (R.A.P. photo)

ANVILS IN AMERICA

Scythe Anvil: These are small stake-like anvils with mostly square faces. They were used (primarily in Europe) to sharpen scythes and sickles by hammering the blade edge. They were carried into the field along with a hammer. To use them the anvil was driven into any convenient piece of stable wood. It is said that a scythe or sickle sharpened in this manner would retain its sharpness two or three times longer than one sharpened with a stone or file. The hammering not only drew out and sharpened the blade, but also work-hardened it, whereas the stone or file only put a sharp edge on the blade.

Four different scythe anvils. (R.A.P. photo)

Stake Anvils, Beak-Irons, Bick Irons or Bickerns: The main difference between these and smithing anvils is that they do not have a base. They generally have a long leg with a spike on the end which is driven into a heavy wood support or held in a special iron holder. Most of them are of the *Continental* pattern; that is, they usually have two horns, one flat and tapered, the other round and tapered. They can weigh from a few pounds up to about 300 pounds. They are used in the same way as the smith's anvil in the larger sizes. The smaller sizes are generally used for bending and light work.

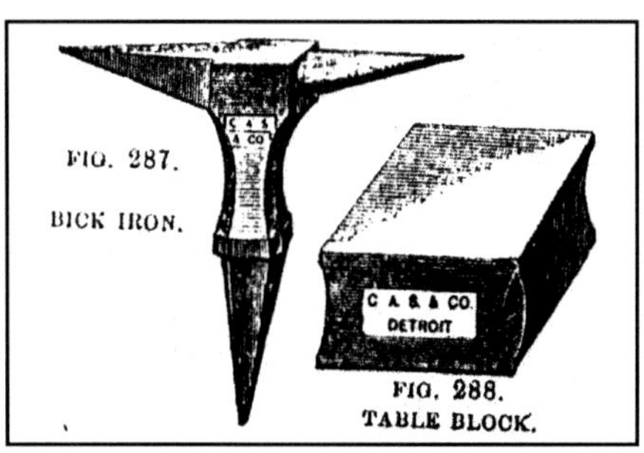

Charles A. Strelinger & Co. catalog, 1895, Detroit, Mi.

Silversmith's Anvil or Stakes: These are of various shapes and sizes. They have round, square, rectangular, convex, and concave faces which are usually highly polished. Most of them fit some type of stake holder. These same stakes are used by most whitesmiths, copper and brass smiths, pewterers, etc.

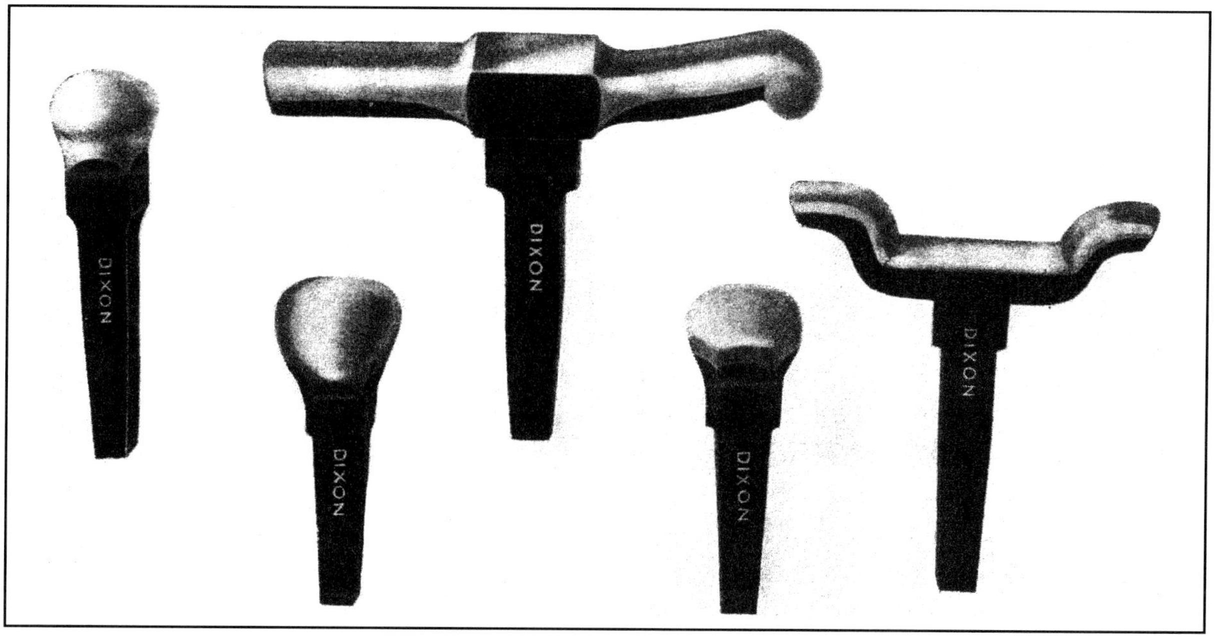

William Dixon Inc. catalog, 1933, Newark, NJ.

Swage Anvil: These are swage blocks with a smooth flat surface on one of the rectangular sides and a horn on the same side. The ones that I have seen appear to be very old and were made of cast iron.

Cast-iron swage anvil with a horn. At Colonial Williamsburg, Va. (R.A.P. photo)

98-pound swage anvil. Taken at Mr. Bill Gichner's. (R.A.P. photo)

Tinner's Anvil or Stake: This is one of the tinner's tools that resembles a stake anvil. They weigh about 50 pounds and could be used for light blacksmithing work as well as for sheet metal work. They are either steel or steel-faced wrought iron.

50-pound tinner's beak horn stake. (R.A.P. photo)

Miscellaneous

Approximately 1,000-pound anvil at Old Economy, Ambridge, Pa. (R.A.P. photo)

This is a one-of-a-kind anvil found at the Old Economy Museum, Ambridge, Pennsylvania. When I first saw a photo of it I thought it must have been a German anvil brought over by the Harmonist Society in 1803.[6] In 1994 I had a chance to visit Old Economy and examine this anvil. The body turned out to be cast steel and the two steel horns must have been placed in the mold when the anvil was cast. It did not make sense to me. Anvils were not made of cast steel, as far as I know, in 1800. I was at a loss as to what to think until I met the director of the museum and asked him about the anvil. It seems that the anvil was donated to Old Economy Museum about 1960 by one of the steel mills in the area. Apparently one of the mills made it for some special purpose. The face has seen a lot of use and abuse. Just to look at this photo you would think that it was ancient, especially knowing where it is located. It weighs about 1000 pounds. Incidently, it was possible to make an anvil of cast steel around 1800, but most unlikely. Crucible cast steel was being used after 1750 in Europe, but was very expensive and no one in their right mind would go to the expense to use so much steel in an anvil. Cast steel was not even used for face plates on wrought-iron anvils. I did find out that the actual anvil used in the blacksmith shop at Old Economy was a *Moueshole.*

133-pound anvil when put together. Thought to be a ship's anvil. (R.A.P. photo)

The body is wrought iron with a 1" thick face plate. It weighs 63 pounds. (R.A.P. photo)

The base is cast steel and weighs 70 pounds. The fact that the base is cast steel indicates that the anvil was made sometime in the late 19th or early 20th century. If it is a ship's anvil it was probably used on a sailing vessel. Owned by Mr. Daniel Huges. (R.A.P. photo)

CHAPTER 7

Anvil Odds, Ends And Trivia

(Courtesy of Mr. Douglas Freund)

ANVILS IN AMERICA

This section is concerned with information relative to anvils in general. As the title suggests, most of the material is just odds, ends and trivia about anvils, that I have discovered myself or have found out from others. Some of it, especially in Anvil Stories, may be an exaggeration of the truth. I take no blame for the information given to me by others without proof and will give credit to the source, if known. I am looking forward to adding much more to this chapter in a future edition.

MARKS ON ANVILS AND THEIR MEANING

I have often wondered why so many of the anvils I have recorded have the tips of their horns flattened. I could not imagine any operation on an anvil that would require hammering in the tip of the horn. Then one day, while working at the forge and talking to someone, I backed into the tip of the horn. This horn was not blunt, and it hit me right in the thigh. I now believe I have the answer to the mystery of the blunt horn tips. After the smith has stabbed himself a few times he takes his hammer and destroys the anvil tip that is trying to destroy him.

To determine whether the smith who owned an old anvil shoed horses, look for an indentation on the heel of the anvil, opposite the pritchel hole. After a smith has made a great many horseshoes on his anvil this spot is worn in the heel. It is surprising how many old anvils are worn this way . Some are shallow and some are deep, depending on how many shoes the smith made.

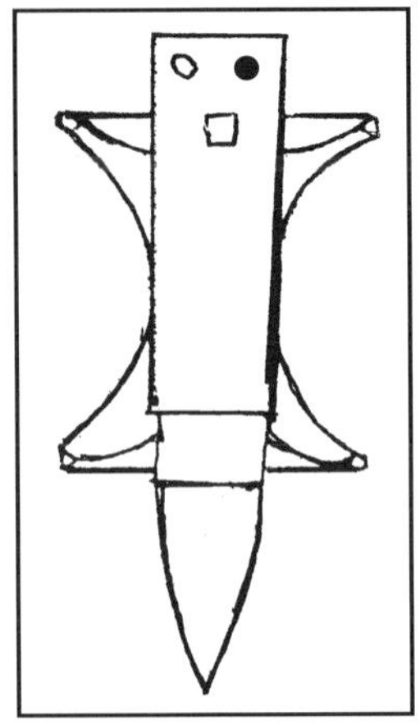

The spot opposite the pritchel hole.

Sharp cut-like marks on the feet and sides of an anvil tell you that the smith was making and heat-treating chisels, or just heat-treating them. He was using the anvil to test the temper of the tool. I have seen anvils on which the feet were almost cut off by this practice.

If there are a lot of sharp punch marks on the body of the anvil, you know that the smith was testing center-punches for temper.

If the trademark is completely missing on an anvil made in the last hundred years, there is a good possibility that the smith did a lot of metal bending.

Chipped edges on the anvil mean that the face was either too hard or that the operating smith was not the person you would have taken your work to. There is a good chance that most of the edge chipping took place after the anvil had passed through a few hands and was no longer used by the perpetrator to make a living.

If an anvil breaks at the waist and is welded back together, the weld beads are usually apparent on the sides of the anvil. However, if there is just a wide weld bead under the horn and heel at the waist, you can be almost certain that the top and base were put together that way at the factory. All of the side arc-weld was ground off flush with the body, but it was too much trouble for the manufacturer to remove the beads from under the horn and heel.

A lot of wear around the hardie and pritchel holes indicates much hole punching or bending was performed using these holes. Much use of loose anvil tools will also wear the edges of the hardie hole.

Old anvils have a way of telling the story of their use. It is too bad that they cannot talk; what a story most of them could tell.

STRIKING THE ANVIL

If you have ever watched blacksmiths working at their anvils, very often you will notice that some of them, and I might say most of them, after striking the piece being worked on four or five blows will hit the anvil a slight blow, making the anvil sing (if it is a good wrought or steel anvil).

There are a number of reasons or supposed reasons for striking the anvil after every few blows. These "reasons" differ with each smith.

Probably the oldest reason was to ward off evil. When the devil or his demons heard the ringing of the anvil they did not want to get too close to the blacksmith because they were afraid of what he might do to them. It was also good for keeping away witches. In fact every Saturday night before closing time, the smith struck the anvil three times for good measure to keep evil away from the shop until Monday morning. After a time it became tradition and if the smith did not strike the anvil after a certain number of working blows, he was not thought well of.[1]

Here is a reason I heard from an Amish man.[2] He told me that when soldiers crossed a bridge, they did not walk in step, because if they walked in step the rhythm of the marchers could cause the bridge to collapse. The same was true with anvils. Should the smith hammer away on the work piece and create a rhythm, he could crack the anvil. Therefore, striking the anvil every fourth or fifth blow breaks up the rhythm.

One author thought it gave the smith a slight pause to think before striking the next series of blows. Another person told me that striking the anvil every so often gave the smith a break or momentary rest and allowed him to work much longer without fatigue than if he struck every blow on the work. A third person stated that striking the anvil every fourth or fifth blow helped maintain a rhythm, therefore making the work go easier and faster.

Here is what Holmstrom has to say about striking the anvil.[3]

"Some smiths have the foolish habit of beating on the anvil empty with the hammer, they will strike a few blows on the iron, then a couple of blind beats on the anvil, and so on. This habit has been imported from Europe, free of duty, and that must be the reason why so many blacksmiths enjoy the luxury." In other words he says it is a waste of time and energy and just a bad habit. What do you think?

When there are a number of people in the forge while I am demonstrating I often strike a noisy anvil next to the one I am using to get some of the people's attention, especially children when they start wandering about the shop. A good ringing anvil will hurt your ears and make you sit up and take notice.

As a hobby smith, striking the anvil every fourth or fifth blow is the traditional thing to do, as "those in the know" expect it, and you can use it to pass on some of the reasons smiths in the past did it. A demonstrating smith should be a good story teller.

If you know of any other reasons for striking the anvil off of the work every few blows, feel free to pass them along.

TRUTH IN ADVERTISING ANVILS

Each anvil maker advertised his anvil as the best on the market for the price, even the low-grade anvil makers. *Vulcan,* a low-grade anvil, advertised their anvil body as being made from the best charcoal cast iron. The fact that the body was cast iron is very true, but the use of charcoal to make it is more than doubtful. Even well into the 20th century their ads continued to make this claim. You might ask, "why would they advertise using charcoal?" Using charcoal to make iron produces a better product then coke. Coke contains many impurities that charcoal does not contain, but charcoal is much more expensive.

VULCAN

Body made from best Lake Superior charcoal iron; the face is one solid piece of high grade tool steel, thoroughly welded to the body, and then accurately ground, tempered and polished. The horn is covered with, and its extremity made entirely of tough, untempered steel.

Nos.	Weight Lbs.	Size of Face, Inches	Size of Hardy Hole, in.	Each
00	5	1½x 4	½	$13.40
1	15	2¼x 5¼	½	20.00
3	30	2¾x 7¼	½	22.60
5	50	3 x 8⅝	⅝	26.60
7	70	3¼x11	¾	33.50
10	100	3½x12½	¾	41.50
13	130	3¾x13½	1	48.00
15	150	4 x14½	1	57.00
17½	175	4½x15⅛	1	67.00
20	200	4½x16	1⅛	75.00
25	250	4½x17½	1⅛	94.00

The Bostwick-Braun catalog, 1955, Toledo, Oh,

A *Trenton* advertisement said,"You have tried the rest now try the best." All the other makers said essentially the same thing, but the best ad, depending on how you look at it, is one that Sears Roebuck ran in their catalog from about 1902 to 1909.

American Wrought Anvils.

HAVING SECURED THE EXCLUSIVE SALE of one of the best wrought anvils in the world, we have agreed to take the entire output, and in order to do so, we have reduced the price to where it will enable us to sell more anvils than any firm in the United States.

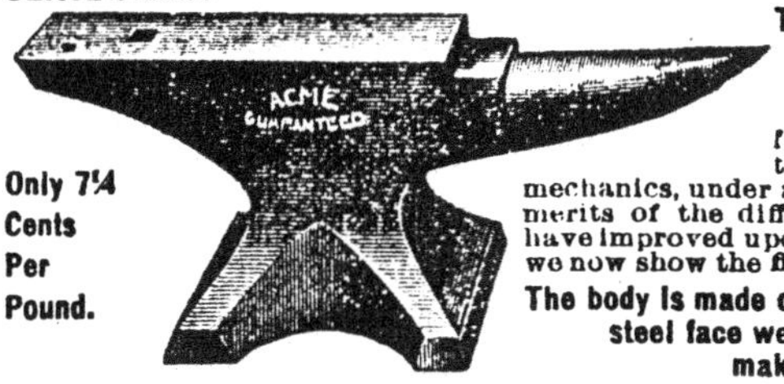

Only 7¼ Cents Per Pound.

THE TIME HAS PASSED WHEN WE MUST GO TO ENGLAND FOR OUR BEST ANVILS.

Having sold both the American and English anvils for a number of years, shipping them to every part of the United States, having been used by all kinds of mechanics, under all circumstances, we are in a position to know the merits of the different makes. In our American Wrought Anvil we have improved upon all foreign makes in shape, style and finish, until we now show the finest high grade anvil produced by anyone.

The body is made of the best quality wrought iron, with an extra heavy steel face welded to the body by an electric process, which makes it practically a solid steel anvil.

THE FACE is tempered by a special process, known only to the makers of this anvil, which warrants it against flaws, chipping, or being too soft. This, by the way, is the most difficult obstacle to overcome in anvil making. Hundreds of anvils, made of good material, are made useless by poor tempering, being either too hard or too soft.

THE FACE IS TRUED AND SHAPED BY A SPECIAL MACHINE so that every anvil is mechanically perfect; no hollow or uneven places. The horn is extra long with the Peter Wright shape, pattern and curve, nicely rounded. The hole for the hardie is made true and straight, so you experience no trouble by hardie or flatter sticking or not setting level. While these points seem small, the defects of nine-tenths of all anvils, both foreign and American make, now being sold, rest in the fact that these points are not looked after. We cannot always furnish even weights; please allow five pounds either way.

NOTICE—Never leave a hot piece of iron for any length of time on the face of an anvil, as this will surely draw the temper and thus make it soft. A great many anvils are ruined in this way.

No. 24R988 Weight, 80 pounds and up, per pound........ 7¼c
Weight, 70 pounds to 80 pounds, per pound........ 8½c
Weight, 60 pounds to 70 pounds, per pound........ 9½c
Weight, 50 pounds to 59 pounds, per pound........ 10½c

Sears Roebuck Co. catalog, 1902, Chicago, Il.

In the first place most of the *Acme* anvils I have recorded were made by the Columbus Forge and Iron Company (*Trenton)* and a few by Hay-Budden Manufacturing Company. I know for a fact that both of these companies, from at least 1900 on, always sold anvils under their own trademarks while making them for others too. It is possible that Sears might have taken the total output of the American Wrought Anvil Company, but to this day I have yet to discover one. Also many of the *Acmes* made for Sears Roebuck were manufactured by Columbus Forge and Iron Company and Hay-Budden Manufacturing Company between 1902 and 1909.

The ad gets better. It says that the face plate is welded on by an electric process. I wondered about this after reading the ad for the first time. If electricity was used it must have been the electric resistance process. I live near one of the largest resistance welding manufacturers in the United States, and it so happens that I know one of the engineers who works there. I called him up and asked what it would take in the way of electricity to resistance weld, say, a 4-inch by 16-inch by 1/2-inch thick steel plate to a 4 x 16-inch steel or wrought-iron surface, such as an anvil. He hardly hesitated before answering. He said it would probably take the total output of electricity from Cook Nuclear Power Plant to do the job. Cook produces megawatts of electricity.

How could Sears make such a claim when it was virtually impossible? My guess is that electricity was just coming into common use around the turn of the century and if electricity was used to do some type of work, it was better than any other way. Electricity was a seller.

There are other examples of anvil makers or their agents putting out false or misleading ads, but Sears is the most blatant I have found. It shows that an honest tool like the anvil can be misrepresented like so many other items in our society.

ANVILS IN AMERICA

An advertisement from the H. Channon Company catalog of 1925 goes the other way for their *Bullock* anvils. Every *Bullock* that I have recorded was made by the Columbus Anvil and Forging Company, manufacturers of the *Arm and Hammer* anvil. They are far superior than the advertisement represents them. *Arm and Hammer* anvils were not made from scrap, but from the best solid pieces of wrought iron obtainable. Hand tools were not used to finish an *Arm and Hammer* anvil. The description given was taken from a write-up for the *Peter Wright* anvil.

Bullock Warranted Anvils

Wrought Base, Solid Crucible Tool Steel Face

General Blacksmiths' Pattern

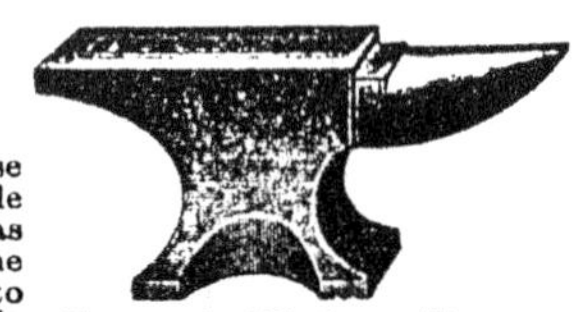

Farrier's Cliphorn Pattern

Our Bullock Anvil is strictly an American product, and because of this is sold on its merits alone and not upon tradition. While the principal features of manufacture are essentially the same as the Peter Wright, it varies in one important particular, i. e., the face ***is made in one piece,*** whereas the old English method was to weld the face on ***in sections about six inches in length,*** a fair sized anvil requiring several sections to complete the face.

With the single exception of the face, the Bullock anvil is made in exactly the same manner as the famous old Mousehole and Peter Wright, but with newer and better equipment.

The process of manufacturing the Bullock anvil is as follows: The base or bottom part is formed by piling up scrap iron, carefully selected, of course, and welding into a solid mass. While still hot it is placed in a die and blocked to shape, the power used being steam. The whole of the upper part including the horn and tail, is also forged from selected scrap. Next the tool steel face is welded on, the hardy and pritchel holes punched the tail or heel squared and the horn finished to shape. The two pieces, base and upper part are then welded together at the waist. The anvil is then thrown on a finishing block and finished so far as forging goes, with hand tools.

The Bullock is an anvil with horn and heel in perfect proportion, large base that does not require strapping, tempered edges to prevent chipping, level and true face of correct size, also hardy and pritchel holes and other mechanical details of correct accuracy. In addition to this it has the proper resilience between anvil and hammer and a real musical ring. Its proportions overall are correct, its design is graceful, and above all it is stable on the block. Everything considered it is

The Ideal Anvil for Particular Workmen

Size No.	Nominal Weight, Pounds	Variation in Weight, Pounds	Size of Face, Inches	Purpose	Hardy Hole	Farrier's Clip Horn Extra per Lb.	General Blacksmiths' Extra per Lb.
160	75	70 to 80	3 1/2 x 12 1/2	Light repair work.............	7/8	$0.01	$0.00 1/2
161	100	95 to 105	3 1/2 x 13 1/2	Farmers' size.................	7/8	.01	Base
162	125	120 to 130	3 3/4 x 15	General blacksmiths' and horse-	1	.01	Base
163	150	145 to 160	4 x 16	shoers', contractors' sizes for	1 1/16	.01	Base
164	200	195 to 210	4 1/4 x 17 1/2	ordinary work.............	1 1/8	.01	Base
165	250	240 to 260	4 3/4 x 19		1 1/8	.01	Base
166	300	290 to 310	5 x 20		1 1/4	.01	Base
167	350	340 to 360	5 1/2 x 22	Machine and heavy forge work.	1 1/4	.01	Base
168	400	390 to 415	5 1/2 x 24		1 1/4	.01	Base
169	500	475 to 525	6 x 26		1 3/8	.01	$0.00 1/2

H. Channon Co. catalog, 1925, Chicago, Il, (From Mr. Ken Wilson)

ANVIL MYTHS

The myths that have surrounded the blacksmith in years gone by are many. Here was a man who had control of the four known elements of the world. Earth, from which he received his iron; fire, which separated the iron from the soil and by which he worked the iron; air, which he used to make the fire hot enough to separate the iron from the earth and to forge it and water, used to cool or temper the iron to make weapons of war and tools.

The striking of the anvil, as mentioned earlier, helped to ward off evil. It was also thought that the anvil had curative powers.[4] The first step was to place the sick child on the smith's anvil. Then the blacksmith proceeded to strike the horn of the anvil a number of sharp blows, thereby driving out the sickness. Another cure was to place the sick child in the smith's cooling tub or slack tub, then put the child on the anvil, while three blacksmiths with the same name passed pieces of hot iron over the child. I wonder how these "cures" helped out.

My daughter sent me this clipping from a California newspaper.[5]

> *In early Russia, the village blacksmith was a man of high calling. Something in the nature of a witch doctor. It was to him the locals turned to perform 'white magic' against evil spirits. And they took solemn oaths with their hands on his anvil in matters of marriage and other agreements.*

Maybe marriages should be performed over an anvil today; they might last longer. There is more about anvil marriages in Anvil Stories. I am sure there are other myths about anvils, but these are all I have collected so far.

ANVIL STORIES

I suppose that if all of the stories that center on anvils were told, they would fill several large volumes. Thankfully I have not collected that many stories yet. Some are long, some are just one or two lines, but I think most of them will be of interest to the reader.

Gretna Green

I first came across this name on a miniature anvil. All that was inscribed on one side of the anvil were the words "Gretna Green" and on the other side it said, "The Famous Anvil." I asked the dealer who owned it what he knew about the name. He said it was a salesman's sample of an anvil manufacturer. My anvil research had not progressed very far at this time, but I had my doubts about what the man said. I asked several anvil collectors I had met if they had ever heard of Gretna Green. One man said he thought it was some place in England. At the university where I taught there were several professors from England. I asked one of them if he had ever heard of Gretna Green. He laughed and said "sure!" Gretna Green is a village just across the border from England in Scotland. Often couples who eloped went there from England for a quick marriage in olden times. Now it is a tourist attraction. He said the village blacksmith shop was one of the first buildings across the border and the couples would be married by the blacksmith over his anvil. He said that the miniature anvils are available there, as souvenirs.

Later on I found another miniature with the following inscription on the side. " The Blacksmith's Anvil, Gretna Green, over which marriages were performed."

Gretna Green miniature anvil. (R.A.P. photo)

In more recent years I have received material from Mr. John Catchings and Mr.Gill Fahrenwald which tell the story or history of Gretna Green and the famous anvil. In the chapter on the *Mousehole* anvil I related that in some information from England it said that the "famous anvil" was a *Mousehole.*[6] From the material sent to me giving the history of the place it seems that the blacksmith and anvil story is just that, a story concocted around the turn of the century to put a little romance and credibility into the marriages performed in Gretna Green. I am still concerned about that anvil and hope to one day find out if it really is a *Mousehole.*[7]

Thieves and the Big Anvil

This story was related to me third-hand. It came about because I was talking to a couple blacksmith friends about some of the difficulties encountered in handling and moving very heavy anvils. I do not even remember who told this story, but the essentials of the story are still in my memory.

A store owner kept a large anvil of about 500 pounds in front of his business to attract attention. A fellow who was interested in blacksmithing happened to be driving by one day and noticed the anvil. He stopped to have a look at it, as most smiths would do. He had a little time on his hands so he proceeded to find the owner of the business and ask about the history or background of this anvil. In the course of the conversation the smith asked the owner if the anvil could be purchased. The owner said he did not want to sell it, although it was just an attention-getter as far as he was concerned.

The smith gave the owner his phone number and said if he ever decided to sell the anvil, to please call him. With that he went on his way. Every once in a while when the smith was in the neighborhood he would drive by the business just to look at this large anvil.

Some time went by, maybe a month or two, when the smith received a phone call from the owner of the anvil. The owner asked if the smith was still interested in buying the anvil. Of course he said "yes" and a price was agreed upon. However, the smith was sort of surprised that the owner would change his mind so soon after telling him he had no intention of ever selling the anvil. He asked the owner why he had changed his mind. The owner said that two attempts had been made to steal the anvil, andhe was afraid the third attempt might be successful. The second attempt had taken place only a few days or so before, convincing him that he better sell it and get some money before the anvil was gone, leaving him with nothing.

I do not remember if I was told what took place in the first attempt to steal the anvil, but I think the anvil had been moved somewhat. Whoever tried to pick it up did not realize how heavy it was and was frightened away before he or they could get help to lift it. The second attempt was the last straw.

The second attempt started with a phone call to the owner from the police late one evening. The owner lived some distance from his business. The police said a neighbor who lived close to the business had called them saying that a car was making a lot of noise spinning its tires on the asphalt pavement, but was not moving. This was taking place near the owner's business, and that he had better come to the scene, because his anvil was involved.

When the police came on the scene they found two men trying to remove a large anvil from the trunk of a Ford Pinto, while the driver kept trying to move the car, but continued to burn rubber. The police arrested the men and called the owner of the business.

Here is what the owner found had happened. The three men had come late that night in the Ford Pinto hatchback with the intention of taking the anvil and putting it in the trunk of the car. If you will recall the small size of a Pinto and the size of its trunk you would say that these three must have been very stupid or drunk. If the anvil did fit in the trunk they would not have been able to close the lid, but I suppose that did not worry them. Anyway, the three men managed to pick the anvil up and bring it to the open trunk. But as they were trying to lift it in, it slipped from their hands, the horn going into the trunk first. The horn went through the bottom of the trunk (I guess the spare tire was missing) and into the asphalt of the road. They tried to pull the anvil out, but the anvil's position in the trunk made it difficult for three people to get a good grip. You can imagine trying to reach over into the back of a Pinto to remove 500 pounds of iron, especially when it is in the shape of an anvil. Having failed to loosen the anvil, they jumped into the Pinto and attempted to drive away with the anvil's heel sticking out of the back. They did not realize that the point of the anvil's horn was several inches into the asphalt.

The driver put the car in gear and released the clutch, but the car would not move forward and the engine stalled. He tried backing up but that did not help, so he revved up the motor in an attempt to force the car to move, but all he managed to do was to burn rubber tires. This was when the neighbor heard the noise, looked out of his window, observed the scene, and called the police.

Next the two men who were not driving got out of the car and attempted to lift the rear end, but as soon as the car moved a little forward the two had to let go and the anvil point dug into the pavement again. I don't know how far they managed to move the Pinto in this manner, but before they had burned all of the rubber off of the tires and roused the whole neighborhood the police arrived.

It is too bad someone with a video camera was not there to catch the action before the police arrived. When I heard the story I laughed until I cried, just imagining what those three thieves were going through in an attempt to escape in that Pinto. It would make a good story for the "Three Stooges."

Cartoon drawn by Nathan Greene.

The Bridge Anvil

This next story happened to me shortly after I began researching and collecting anvils.

If someone had video taped this scene it might have been just as funny as the previous story. But it was not funny to me, as I am sure the previous scene was not funny for the three thieves.

A friend of mine noticed railroad memorabilia for sale in a newspaper devoted to antiques. One of the items listed was a railroad anvil (generally referred to as a bridge anvil). He thought he would do me a favor and he showed me the advertisement. The ad only said that it was a large railroad anvil, gave the price, and phone number. My curiosity got the better of me so I called the number, and a lady answered the phone. I asked her for particulars on the anvil, as it was not sold. She said she thought it weighed around 500 or 600 pounds and had a large square hole going through the center of the body. From her description it did not sound like a smith's anvil, so I asked her to send me a photo of the anvil, as I might be interested in buying it. When the photo arrived I was looking at an anvil the shape of which I had never seen before; in fact, I have never seen another like it since. The lady wrote me that the anvil came out of the railroad repair shop in Monterey, Tennessee, and had been there for many years. The shop was to be demolished and everything had been removed except this anvil. The man in charge said to those standing around " Anyone who can pick the anvil up can have it." The lady's husband was there and he said that he picked the anvil up and put it in the back of his pickup. The lady wrote that on seeing the anvil she figured he was just making up the part about him picking the anvil up himself. I know he was. This had taken place some years before and her husband was now deceased. Maybe the anvil was the cause of his demise.

I phoned her and we dickered on the price because I would have to drive about 300 miles to get to her home. We agreed on a price and she gave me the directions. It was in the middle of December, so depending on the weather, I would call her just before we were ready to leave.

I had a van at the time, so my wife and I took out the rear seat and loaded the van with a one-ton chain hoist, several heavy boards, some chain and several lengths of one-inch pipe. I was going to roll the anvil on the pipe up the boards into the back of the van, using the hoist attached to the middle seat to do the pulling. I figured that I could slide or maneuver a 500 or 600-pound anvil across the floor to the bottom of the boards that were angled into the van.

We, my wife and I, left before noon on a typically cloudy winter day and headed south. About half way to our destination we experienced generator trouble. The battery was overcharging so I almost turned around to try to get home. But first I tapped the voltage regulator with a hammer, figuring the circuit breaker points were sticking. It did the trick and we went on. We arrived at our destination while it was still light, but no one was home. There was a little snow on the ground and a cold drizzle of rain had started.

About fifty or sixty feet from the house was a sharp drop leading to a valley. Next to the drop-off was a low-roofed building with its door facing the drop-off and within about seven or eight feet of the drop-off.

Shortly after dark an elderly lady (over 70) drove up to the house and introduced herself as the person I had been communicating with concerning the anvil. She said the anvil was in the low building. I asked if there was some other doorway into the building besides the one by the cliff. She said there was not. The ground was slippery and it was down a slope to the side of the door which I would have to back up to at a right angle. Fortunately there was a slight cover over the door so when I gingerly backed the van down the slope the back end came just under the cover, keeping off the rain. My left rear wheel was within eighteen inches of the cliff on icy snow. The shed had a rough concrete floor and must have been fifty by forty feet in size with just a few lights. On the farthest side from the door reposed this large anvil.

I went over to it and attempted to move it, but my best effort failed even to budge it. The base of the anvil is 18 x 28 inches, which created a lot of friction. I tried to lift one end of the anvil, but was unsuccessful. I have moved 700-pound barrels of oil around and had managed to move a 450-pound smith's anvil by alternately shifting each end. Therefore, I could not understand why I could not move this 500 or 600-pound railroad anvil. I finally resorted to the one-ton chain hoist. The lady found a ladder, so I attached the hoist to a ceiling joist and proceeded to try to lift the anvil without pulling the roof down. I saw that the joist would not support the whole weight of the anvil so I lifted one end and put the pipes under the base of the anvil, which was flat. By moving the hoist from joist to joist to relieve some of the weight, we managed to slowly roll the anvil on the pipes to the door over the rough floor. The elderly lady and my wife helped push the anvil and change the pipes. Without their help I would never have been able to move the anvil by myself.

At the door we had another problem. As previously mentioned the door was at right angles to the rear door of the van, so the boards could not be extended into the building. I had to sort of drop the anvil out of the door down to the ground at the lower end of the boards yet keep the anvil off the ground so that I would be able to pull the anvil up the boards on the pipes and into the van. To solve the problem the lady found a couple of boards which we used to let the anvil down out of the building, always keeping one or two pipes under its base. The chain hoist was attached to a rafter and the anvil so that it would not get away from us. When the anvil stopped at the bottom of the boards I removed the hoist and attached it to the rear-most seat of the van and to the anvil. By this means the anvil shifted around and was pulled up the boards on the pipes and into the van. This whole process took about three hours and my wife and I had not eaten since we left home. It was about 10 o'clock that night when we managed to drive away and I gave a sigh of relief. I was not sure that the van would get up the icy slope from the shed and I feared that the rear end of the van might slip off the edge of the cliff in the process. But with the weight of the anvil in back, the van came right up without slipping.

We got something to eat and spent the night at a local motel, then proceed home the next day through a snowstorm. I had originally planned that we could pick up the anvil about 6 or 7 o'clock that evening and go home the same night.

At the university where I taught we had a two-ton forklift, so I removed the anvil from the van with the forklift and placed it on a one-ton capacity scale that I had at the metals laboratory. I set the scale weights at 500 pounds, but the scale arm did not move. I put another 100-pound scale weight on, still no movement, then another 100, but still the arm did not move. I thought there might be something wrong with the scale, therefore I removed the anvil and checked the scale with known weights. There was nothing wrong with the scale; in fact, this scale is accurate within one-half of a pound. I put the anvil back on the scale and continued to add weights. The anvil finally weighed out at 945 pounds. No wonder I could not slide it around!

If I had had any idea beforehand of how much the anvil weighed and the trouble it took to get it I doubt very much if I would have bothered with it. The anvil now resides in front of our house, and my wife has said that she is going to use it for my tombstone.

945-pound bidge anvil. The soft drink can indicates the size. (R.A.P. photo)

The Mountain Anvil

A colleague I knew at the university had an anvil. After starting my anvil research I asked him where he had found the anvil. He related this story.

He was a logger in northern California some years ago and while bulldozing a logging road up on the side of a mountain he heard something clang like metal on the front of his bulldozer's blade. He stopped the machine and went around to the front of the blade to see what he had hit. After dragging some of the dirt away, he discovered that the metal object was an anvil of about 135 pounds. He cleaned it up and kept it, using it in the aircraft shop where he was now teaching.

He let me borrow the anvil to see if I could determine who made it and how old it was. I took all the data from the anvil I could think of and filed it. After my research had continued for over a year and I had gathered much more data on all makers of anvils I pulled this data sheet out of the file on anvils. The anvil turned out to be a *Hay-Budden* made about 1898.

What would a 135-pound anvil be doing in the woods high up on a mountain side? My colleague said there was no old road where he found the anvil, but there was one higher up on the mountain. Maybe it fell off of a wagon using the higher road and rolled to the spot and the driver of the wagon did not know where it fell off and could not find it. Your guess is as good as mine.

The "mountain anvil" owned by Mr. Stan Sorenburger. (R.A.P. photo)

Pete Paulson's Anvil

Here is a story as written by Peter Fels and sent to me by him in 1994. The large anvil he is speaking of weighs about 250 pounds.

Even though he hadn't used it in years, as his junkyard came to dominate his business, he'd never consider selling me his big old anvil and I understood why. But I kept an eye on it and lusted after it with its fine surfaces and massive size. I'd borrowed a hundred pounder for some years, then lucked into a Peter Wright from off a farm (which I still have) but always kept an eye out for that old anvil and when Paulson abruptly sold out and retired I went to the guy who'd bought the place who waved an arm at the empty yard and told me he'd scrapped out everything to pay off the loan (there would have been some really wonderful stuff) and I was morose for days. But the anvil reappeared along with an odd little square anvil that had been nearby the big one in a SLO junque shoppe. They wouldn't sell either and worse yet they abused them both ... nothing I could do.

Eventually, with the planning dept. ragging them and their lease nearly up I waved my little life savings at them and they sold them both to me. The small little cube anvil somehow survived almost unmarked. I still don't know what special purpose it was made for but I plan to polish it, in time, for fancy work. The big one is my main anvil and I lay hammers on it most every day. It suffered under its intermediate owners. The corner was gouged with a cutting torch, and face and horn scarred. When I'm better with a hammer, or braver perhaps, I'll try to repair its scars... soon now. I bought the 27 dollar a pound rod for the job a while back but the combination of self-confidence to pre-heat, prep and lay those fancy sticks and the frustration of those funky surfaces that mar my products, hasn't quite overcome my awe at the old anvil's venerable prescience.

The Dangerous Anvil

The following story was sent to me by Mr. John Lemoine of Whitinsville, Massachusetts. I will paraphrase Mr. Lemoine's letter.

It seems that a fellow by the name of Dave, the main character of this story, who lived in Oxford, Massachusetts, met Mr. Lemoine in an antique shop in Oxford in 1964. The shop's owner, Jack Williamson, introduced Mr. Lemoine to Dave. This antique shop has long since gone out of business. This story came about from that meeting.

Dave was a collector of clocks, watches and pocket watches. He also had a great interest in anything antique including blacksmithing. Dave had a next door neighbor (John never learned his name) who had retired after having been an industrial blacksmith. He had worked in one of the shops in Worcester, Massachusetts, for many years. Dave told this neighbor that if he would make him a few forged items from time to time he would buy the tools and help him set up a small forge shop in the neighbor's garage. The old smith had never owned his own tools and these tools would be his. The old smith agreed and Dave bought the necessary items for the shop.

For several years the old smith made things for Dave and for himself and others, I suppose. However, Dave being the kind of giving person that he was and who liked everything just so, decided that the anvil he had bought for the shop was not good enough. He proceeded to look for a better one to give to the old smith.

The old smith went to Florida every winter and during this winter Dave located and bought a nice 240-pound late model Peter Wright anvil. There had been a heavy snow fall the day the anvil was to be delivered. The fellow who was to deliver the anvil did not think he could get into Dave's driveway so he dropped the anvil off at the road by the end of the driveway. When Dave came home and cleared the driveway he found the anvil. Dave was a slightly built man and had suffered while a prisoner of war during World War II. He was not supposed to lift or move heavy objects, so he thought he would leave the anvil where it was until spring and the neighbor returned and then have it moved to the shop.

When spring came and it was about time for the old smith to return home, he noticed a truck pulled up to the neighbor's garage and the old smith's son loading the blacksmith equipment and other items into the truck. The son waved to Dave, finished loading the truck and drove away. Dave went over to his neighbor's house and was told by those who were there that the old smith had died that winter while in Florida. The family had sold all of the tools in the garage that spring. Dave told Mr. Lemoine that he had no regrets as he had given the tools to his neighbor.

"Dave now had his own anvil," Mr. Lemoine said. Dave told Mr. Lemoine that he had fond memories of hanging out in the neighbor's garage shop watching him work at the forge and anvil. He also had a number of nice souvenirs made by the old blacksmith.

The large anvil lay at the end of the driveway for about two years and then one day things began to happen. Dave's wife was backing their new station wagon out of the drive. She was not used to the big vehicle as she usually drove a smaller Ford Falcon which was in a body shop being repaired. She ran a little off of the drive as she reached the highway and BANG!, drove the exhaust pipe into the anvil. The whole exhaust system had to be replaced. Wouldn't you know it within about another week she did the same thing again. The station wagon, as Mr. Lemoine says "was their third car, the good Sunday, ride-around, trip car."

After these mishaps Dave, with probably some help, moved the anvil about ten feet away from the driveway. However the following winter Dave's nephew using his Jeep and snow plow, was pushing snow away from the drive to make room for more snow from future snow storms when, BANG! he ran into the anvil. The anvil which was lying on its side was burried under the snow and frozen into the ground. Dave paid for the damage to the Jeep and felt that he had to get rid of the anvil before he had more expenses to pay out.

It was shortly after this last episode while in the antique shop that Dave met Mr. John Lemoine. When he found out that John was interested in anvils, a deal was made. John said that Dave's wife was overjoyed when he traded Dave a Waltham #18 size-River Side- Pocket watch for the anvil.

John said he eventually traded the Peter Wright anvil, two vises, a post drill and a small forge for a 416-pound *Hay-Budden* anvil which he still uses.

MUSIC

I have noticed that on a number of occasions, orchestras have used anvils as part of the percussion section.

Mr. Tilton of the Columbus Forge and Iron Company told me that in the 1940s the Columbus orchestra borrowed a *Trenton* anvil from them, wired it to an amplifier, and used it in playing of *The Anvil Chorus* during a concert. Some day I'm going to have someone line up enough of my anvils so that they can be used to play a tune. It would be something like a bell choir.

In the Broadway musical *The Music Man* an anvil salesman followed the leading character around trying to expose him as a fraud. He sold *Gibraltar* anvils and at the end of the show the bag he was carrying his sample anvil in slipped out of his hand, falling on his foot.

TRIVIA

Here is a piece that appeared in the Winter 1992-93 *The Anvil's Ring* journal, titled "Anvil Town."

> *I recently spent a month in Switzerland about 90 miles southeast of Zurich in a village called Appenzell (The Abbots Cell). In the village, old anvils are used as yard and business decorations. Within a two block area there must be nearly 100 anvils. All antiques, some must be 200 years old and I saw 15 to 20 different styles. Very few, if any, are the U.S. style. Many are obviously made for special work such as armor, swords, battle axes, etc. All are somewhat beat up; however, many are quite usable. Appenzell is a great place to study one of the largest concentrations of anvils in the world.*[8]

The National Agricultural Center, Bonner Springs, Kansas, claims to have the largest collection of anvils in the world, more than 300 anvils, from 2 ounces to 520 pounds.[9]

Back in 1985 I was in a hardware store, talking to several people about anvils, when a man who said he was a school teacher entered the conversation. He said that while he had been in "Supply" in Vietnam they were expecting a ship load of corrugated sheet iron, but instead a ship loaded with anvils arrived. If this is true I wonder who made the anvils and what happened to them.

When I was in the military in December, 1950, in the Philippines, our troops in Korea were retreating as the Chinese entered the war. We read in the *Stars and Stripes* that the troops were short of ammunition and could not halt the Chinese. What should come in but a ship load of Christmas trees. I know because the Christmas trees came to us and there were enough for almost everyone on the base. We were so mad; our guys were dying up there for lack of ammo and we get a load of Christmas trees. None of the fellows in my outfit would put the trees up. So even though it seems incredible, when the government is involved, strange things can happen. It makes the school teacher's story believable.

I read somewhere (I do not remember where) that a lot of anvils were delivered to an island in the Pacific and the people there having no other use for them were using them for seats.

Mr. Tilton told me that right after World War II the Columbus Forge and Iron Company shipped a couple of boxcar loads of *Trenton* anvils to Russia. No wonder it is hard to locate the late model *Trenton* anvils. I wonder what the Russians used them for?

I was told by a historian that during the late-19th century sailing ships coming to the United States from England often used anvils as ballast when the anvils were part of their cargo.

I heard the following story from Mr. John Lemoine.

During the American Civil War (1861-1865) when Union cavalry rode into a Southern town, one of the first things they did was find the blacksmith shop or shops . If the smith had not escaped and hidden his anvil the Union troops would take a sledge hammer and break the horn and heel off of the smith's anvil if they could. If successful it would help prevent the smith from shoeing Confederate horses, thereby rendering the Southern Cavalry less effective.

I visited the Museum of Appalachia in Tennessee in the late 1980s and talked to some of the employees there about the anvils the museum had. One person said that they had an anvil off of which the Union forces had broken the heel during the Civil War, so the story went. I inspected this anvil and sure enough it was an old English make made between about 1830 and 1850, which would fit the Civil War time period. This sure adds credence to the story.

Many years ago, while dredging out New York harbor, workmen brought up a *Mousehole* anvil. This information came to me by way of England.[10] I wonder if someone had been attached to that anvil when it was dropped in the harbor?

There is a church on Chicago's North Side that has a 250-pound, double-horned *Sheffield* anvil on the altar in the church. The church was started, built and pastored by a Welsh blacksmith around the turn of the century.[11]

Mr. Lester Six sent me two photos of St. Peter's A. M. E. Church in Weirton, West Virginia. In three of the windows of the church are images of an anvil with a cross going through each of them.

The cross in an anvil in a West Virginia church window. (Mr. Lester Six's photo)

Stone anvils are sometimes found in cemeteries, used as grave stones. I have seen a number of photos of the anvil headstones, especially from England. It leaves no doubt as to what the fellow did for a living.

ANVIL LOGOS, MOTIFS And CARTOONS

I was surprised to find out that a lot of people who were born in the last half of the 20th century do not know what an anvil is. If they do know they can only associate it with horseshoes. However, those who watched much T. V. in their youth generally recognize anvils from the T. V. series *Little House on the Prairie* or from the cartoon when the Coyote tries to drop an *Acme* anvil on the Roadrunner. Those who do not know what an anvil is certainly have not paid much attention to the world around them because the anvil motif is extensively used, or maybe I just think so because I am always looking for it.

An antique dealer, Mr. Howard Busse, who lives not far from me, has named his shop Anvil Antiques. As I was driving through a city in northern Indiana, I noticed a sign in front of a shop; it said Blacksmith Distributing. I had to stop and find out what kind of anvil or anvils they had. It turned out to be an upholstery shop and they did not know a thing about anvils. I asked why they used that name then. They said it just sounded good to them.

A concern in California has the registered name Anvil Cases, Inc. They manufacture heavy duty trunks and cases to ship valuable items in, such as a museum might use when transporting objects to be displayed, etc.

A turn-of-the-century post card advertising Waverly bicycles sold for $12,500 in the1980s or early '90s . It shows a half-dressed girl sitting on a bicycle and resting one elbow on an anvil.[12]

Many post cards, Christmas cards and Valentine cards from the turn-of-the-century use the anvil as an outstanding or central part of the card's picture. These are collectibles today.

Most recently a number of satirical cartoons have used the anvil as subject matter.

One shows a parachutist falling through space having just pulled the rip cord, and an anvil pops out of the pack that is attached to his back. On the anvil it says: "Stress."[13]

Another cartoon published in 1992 shows George Bush's leg chained to a large anvil as he is about to start a race. It shows Clinton and Gore already off and running. On the anvil it says, "The Economy."[14]

Still another shows a paratrooper ready to jump out of an airplane. In his lap is a large anvil. Under the drawing it says: "Although he was nervous about his first parachute jump, Roland knew nothing would go wrong as long as he had his lucky anvil."[15]

Another from the *Cincinnati Post* dated 1992 shows a police lineup with Joe Camel smoking a cigarette, the kid from *The Simpsons* with his hands in his pockets, Wile E. Coyote holding an *Acme* [author's emphsis] anvil and Yosemite Sam from Looney Tunes cartoon, with a firecracker in his hand, all standing behind the glass. In front, a policeman is saying to a kid: "Which one influenced your bad habits?"[16]

My favorite is from Larson's "The Far Side." It shows the fat kid swinging under a tree. Anvils are hanging from the branches of the tree and one has already fallen and is embedded in the ground close to the kid. The woman is standing aside with her hands on her hips saying; "All right, Billy, you just go right ahead!... I've warned you enough times about playing under the anvil tree!"[17]

I really like this one, too. It is dated 1995, and is by Mark Parise and titled "Off The Mark." An inventor comes into the patent office and puts an anvil on the administrator's desk, saying: "It's an invention I call the 'Anvil'. Not only is it a perfect tool on which to hammer and shape metal, but it'll also come in handy if, say, a coyote needs something to drop on a roadrunner."[18]

A clothing company in High Point, North Carolina, calls their concern, "Anvil Brand Factories." They are or were the manufacturers of work cloths. Their trademark is an anvil.[19] I do not know if they are still in business.

A 1950 ad from Mr. John Catchings.

Other examples of the anvil motifs are as follows:

- A clothing manufacturer in Mullins, South Carolina goes by the name " Anvil Knitwear."
- A motel in Jackson Hole, Wyoming is named the "Anvil Motel." (courtsey Jan McGowan, editor)

A number of organizations use the anvil motif on the name of their journals.

- An American farrier's magazine goes by the name *Anvil.*[20]
- The Artist-Blacksmith's Association of North America calls their journal *The Anvil's Ring.*[21]
- The Bostwick-Braun Company of Toledo, Ohio, has always used the anvil as their trademark on the cover of their catalog.

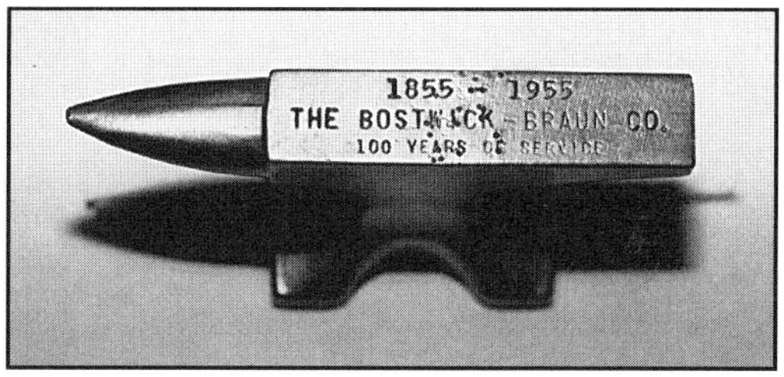

Bostwick-Braun miniature anvil. (R.A.P. photo)

MISCELLANEOUS

"Caught in an anvil shower" and "Roasting marshmallows over an open anvil" are both Christmas cards designed by and sent to me from Mr. Gill Fahrenwald.[22]

"Caught in an anvil shower." From Christmas card furnished by Mr. Gill Fahrenwald.

"Roasting marshmellows over an open anvil." Christmas card furnished by Mr. Gill Fahrenwald.

A poem taken from *The Blacksmith and Wheelwright* journal of July, 1891, by Edwin Ralph Collins, in *Texas Siftings* is entitled:

The Anvil

Many hours have I stood watching
By the village blacksmith's anvil,
Where, upon the glowing metal,
Dealt he blow on blow like rain,
Forging bolt or bar or horseshoe,
Turning shear or pruning sickle,
Or a link for some great chain.

How sparks at each concussion
Flew in bright and burning showers,
As the hammer, merrily ringing,
Fell, and rose to fall again;
While beneath it all the anvil,
Sturdy, strong and firmly seated,
Never seemed to heed the strain.

Many men have I seen standing
Standing, strong and firmly anchored,
While upon them, like the anvil,
Fell the blows of fate amain;
Turning they not from searching sorrow,
Nor the very king of terrors,
Proudly steadfast, true in manhood,
Showed not where the blow had lain.[23]

Here is another, better- known poem by an unknown author, probably from the 19th century.

The Anvil - God's Word

Last eve I passed beside a blacksmith door
And heard the anvil ring the vesper chime
Then looking in, I saw upon the floor
Old hammers, worn with beating years of time.

"How many anvils have you had," said I
"To wear and batter all these hammers so?"
"Just one," said he, and then, with twinkling eye
"The anvil wears the hammer out, you know."

And so, thought I, the anvil of God's Word,
For ages skeptic blows have beat upon;
Yet, though the noise of falling blows was heard,
The anvil is unharmed, the hammers gone.[24]

Ad sent in by Phillip Blundell.

Appendix

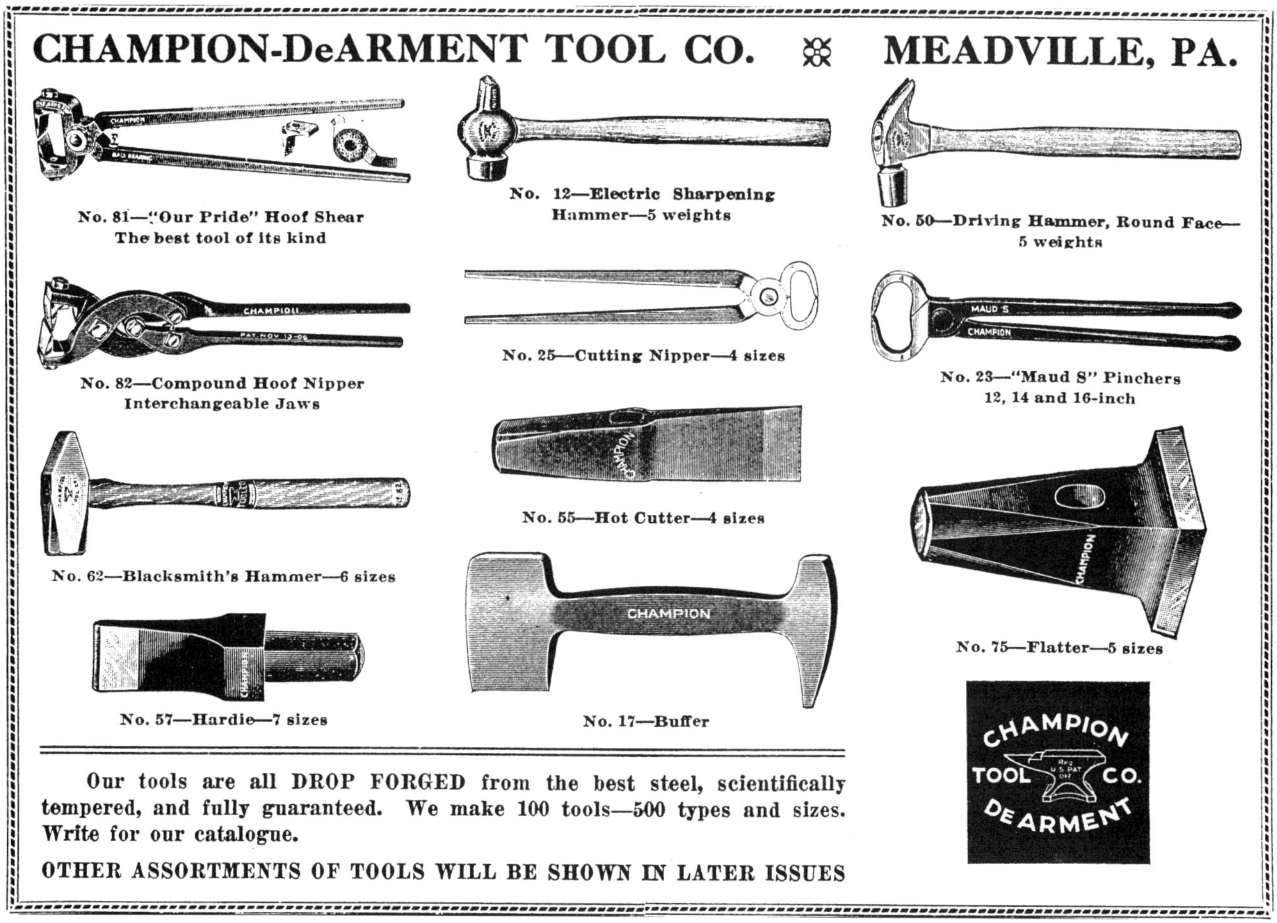
CHAMPION-DeARMENT TOOL CO.
MEADVILLE, PA.
No. 81—"Our Pride" Hoof Shear
The best tool of its kind
No. 12—Electric Sharpening
Hammer—5 weights
No. 50—Driving Hammer, Round Face—
5 weights
No. 82—Compound Hoof Nipper
Interchangeable Jaws
No. 25—Cutting Nipper—4 sizes
No. 23—"Maud S" Pinchers
12, 14 and 16-inch
No. 55—Hot Cutter—4 sizes
No. 62—Blacksmith's Hammer—6 sizes
No. 75—Flatter—5 sizes
No. 57—Hardie—7 sizes
No. 17—Buffer
CHAMPION
TOOL CO.
DEARMENT
Our tools are all DROP FORGED from the best steel, scientifically tempered, and fully guaranteed. We make 100 tools—500 types and sizes. Write for our catalogue.
OTHER ASSORTMENTS OF TOOLS WILL BE SHOWN IN LATER ISSUES

YOU WILL NEVER FAIL
To Make a Strong, Clean Weld If You
Use Perfection Welding Compound.
We invite you to give our Compound a thorough test, and will ship any amount to any address for that purpose. If it does not prove just as represented we pay all expenses.
TRADE MARK
P W C Co
Perfection Welding Compound will do the trick quickly and neatly. It makes a stronger weld than any other compound manufactured, as proven by the tests it has undergone. Send for free sample.
PRICE CONSISTENT WITH QUALITY.
PERFECTION WELDING COMPOUND CO., SCRANTON, PA.

Appendix A

Quick Identification of Major Anvil Groups and Makes

The following characteristics of anvils should help in quickly identifying them as to the approximate time period of manufacture, material, country of origin in most cases, and in many cases, the maker.

If you can find a name on the anvil, look it up in the index of this book and then you will know as much as I know. On the other hand, if all you can make out is a word such as "Sheffield" or "N. Y." for example, you have an English anvil in the first instance and a *Hay-Budden* in the second. Any number of anvils have defied me as to their maker, because so few markings are left, but they are easily identified as to material, time period of manufacture, and in most cases country of origin.

How to identify an anvil with no discernable markings or trademark:

When was it made?

Does it fall into the category of *Colonial, Old English,* or *Modern* in style? Look these categories up under Dating of Anvils.

Is it a *Modern* style? This tells us that it was made after 1852. If it is *Old English*, and has a hardie and a pritchel it was made after 1830. If it is *Old English* without a pritchel it was made between about 1800 and 1830. If it is *Colonial*, that is without a pritchel or table, it was made before 1800. Most of the *Old English* style anvils with a pritchel and hardie excepting *Mousehole* were made between about 1830 and 1850. All *Colonial* and *Old English* style anvils have at least three handling holes; one under the base, and one under the horn and heel in the waist (see Dating of Anvils).

What is it made of?

Next check for the type of metal. With the anvil on a wooden surface tap the horn or heel with a hammer. If it gives a clunking sound (no ring) you have a cast-iron or steel-faced cast-iron anvil. If it rings to some extent, you most likely have a wrought-iron, cast-steel, or in a few cases a ductile-iron anvil. Most *Colonial* and *Old English* will be of wrought iron with a steel face. On a cast-steel anvil you can usually find the mold parting lines under the horn and heel. Most cast-steel anvils were made after 1900. Your chances of buying a ductile iron anvil used would be very unlikely as they are of recent production and of the farrier's pattern. They have been made since about 1960. Also cast-steel and cast-iron anvils rarely have any handling holes. The exception is the *Fisher.*

What is its country of origin?

All English and German anvils from *Colonial* to *Modern* style are flat under the base. Almost all American- made wrought anvils except the very early ones (before 1900) are partially hollowed out under the base, with either a kind of hourglass shape or an oval. Almost all *Colonial* and *Old English* style anvils are of English make. The early *Fishers* and *Vulcans,* plus some "no-name" cast-iron anvils, are of the *Old English* style, too. Virtually all American-made wrought anvils have a serial number on the front of the foot under the horn. The one exception is the *American,* but it is quite rare. All Swedish anvils that I have recorded are cast-steel and in most cases without handling holes and can be flat or slightly depressed under the base. The German-make anvils made for American hardware companies are the hardest to detect. They usually look like American-made anvils, and can easily be

confused with the *American* and very early *Hay- Buddens* and *Trentons.*
Of course the earliest *Trentons* were German-made.

Who manufactured the anvil?

Here are some distinguishing characteristics of the better-known makes.

Mousehole **(English)** *Old English* style, flat under the base, three handling holes on models before about 1900. After 1900 there are four handling holes, with a serial number on the front of the foot under the horn. The underside of the horn is usually sharp where it attaches to the body. It was made in Sheffield, England, from about 1600 to 1933.

Peter Wright **(English)** Before 1852 they looked like any other *Old English* style anvil. After 1852 they always have four handling holes and are the *Modern* style. The top of the feet have a flat running from side to side under the horn and heel. Only a few of the very early American anvils have these flats and are rare.

Wilkinson **(English)** Early ones are of the *Old English* style. Toward the end of the 19th century they were made in the *Modern* style. The later ones have four handling holes, but no flats on the feet. Without some markings they are very hard to discern. Crossed cannons are the most outstanding marking, if indeed they are cannons.

Henry Wright **(English)** Looks just like a *Peter Wright.* Without the name *Henry Wright* you cannot tell the difference, but they are much rarer.

American **(American)** Has four handling holes, but if the word *American* is not stamped on the side it can be confused with other makes.

Arm and Hammer **(American)** The first models had four handling holes (1900), but by about 1910 there were no handling holes. Without markings it can be confused with the *Trenton.* Has a cast-steel base in most cases and relatively thin waist. The most outstanding characteristic is the roughness under the heel.

Hay-Budden **(American)** The shape was modified three times. Look for a number from 0-9 stamped on the waist under the horn; only *Hay-Budden* did this. The models after 1908 are distinctive in their horn and body shape, and do not have a face plate.

Trenton **(German and American)** Up to 1898 they were made in Germany. The German-made ones do not have a serial number on the front of the foot. The ones made shortly after 1898 have "U. S. A." as part of the trademark. They have a cast-steel base, except some early American and all German ones. Both *Arm and Hammer* and *Trenton* (after 1900) have long round horns. Except for the very first ones both makes have depressions under the base.

Columbian **(American)** Cast steel. No markings except an inverted triangle with *C* in it and a letter of the alphabet on the other side. I have never recorded a *Columbian* that did not have the trademark.

All the *Old English* style anvils made in England look almost exactly alike with the exception of *Mousehole.* I have recorded almost thirty different trademarks of the *Old English* style. Without fairly good markings it is virtually impossible to know the maker. Only the material, method of manufacture, country of origin and time period can be determined.

For closer identification study the drawings, trademarks, and photos of the various makes.

APPENDIX B

Buying a New or Previously Owned Anvil

There are several reasons for buying an anvil: to do blacksmithing in a shop or at home, as a collector, just to have one for nostalgia's sake, and last, as an investment.

What do you plan on using the anvil for? If it is to be used for serious smithing I suggest the buyer check out the new anvils first. In most cases they are more expensive than a used anvil, but of late (1996) I have seen some high-quality used anvils selling at blacksmith suppliers for as much or more than the new ones. The only new forged anvil that you can buy is the German-made *Peddinghaus* and it is of the *Continental* pattern, running in weight up to 274 pounds. Most of the other brands are of cast steel and most of the American makes are the farrier's pattern. A smith's pattern English anvil can be had up to about 500 pounds. As of 1995 almost all of these are in the high three-dollar and four-dollar per pound range, plus shipping.

When buying a new anvil be sure you know what weight you want. Don't get one too light. For general smithing I would recommend one of at least 150 to 200 pounds. I use a 300-pound anvil for the little demonstrating that I do. If you travel with your anvil, an 80- to 100-pounder will be what you want.

Once you have settled on a weight and the brand you intend to buy, find out what kind of guarantee the dealer or maker gives. If you pay four dollars a pound for an anvil and find out it has a soft face with no return privilege, you are going to be very upset. I saw some new anvils for sale at a blacksmith meet in 1995, and I checked one for hardness and found it soft.

Unless you subscribe to a farrier's magazine or know someone connected with blacksmithing, you will have a hard time locating a new American-made anvil. The industrial dealers generally handle the German *Peddinghaus* and the one English brand, if they handle any at all. Check under Modern American Anvils (Chapter 4) for the American anvil makers' addresses if you are interested in buying a new American anvil.

Buying a previously-owned or antique anvil is another matter. Regardless of who you buy it from, there are no guarantees.

Investing in a used anvil or anvils is fine if you buy it right and if you enjoy owning it or them, but I would not buy an anvil for speculation purposes, as you will probably be disappointed. The only people that will ever be interested in paying more for a used anvil than a new one would be collectors, not users. Anvils are not the greatest things to collect, as they are usually large and heavy, and have a limited appeal to most people except those of us with the "anvil disease."

Collecting anvils is not as expensive as collecting cars, art, etc., but for the average person it could get a little expensive if you collect very many of them. Why someone would collect anvils just to be collecting I do not know unless they are part of a tool collection. But if you do collect them I am sure this book will be of help to you in defining your reason for doing so. The few people I know who have a collection of anvils collect other tools as well. For example, the large collection at the National Agricultural Center is part of a larger collection put together by a blacksmith over many years and donated to the Center as a permanent display.[1] There is nothing I can tell you to look for or look out for if you are putting together a collection of anvils except to take care of them in the condition in which you find them. I will keep emphasizing the fact that most of them are antiques and should be treated as such.

If you want your anvils to look good and you do not intend to use them, wire-brush the loose rust off, fill any flaws with auto body putty, sand the putty even with the original surface, and spray the anvil with flat black or wrought-iron black paint. It will not only make the anvil look great, but will preserve it in its present condition without harming it in any way. I have seen very old anvils on which someone ground down the faces in order to take out the sway, and they have gone right through the plate in spots to the wrought-iron body, making the anvils useless for work. The anvils are now just museum pieces at best. They were hoping to sell them as users. It would have been better to have sold them as a museum pieces in the first place; at least they would have been in the condition in which they were last used.

If you want to buy a used anvil to put into use, here are a number of things to look for before you buy.

First, is the face of the anvil in good shape? If you plan to do flat work and not much bending over the edges, a few rough, chipped or rounded edges will not hurt. Leave them as they are. Always leave the anvils with badly chipped or cracked faces alone. Do not buy them!

Second, check for a loose face plate if the anvil has a face plate. To do this tap the face plate all over lightly with a hammer. If the face is loose you will get a totally different sound where the loose spot is.

Third, watch out for anvils that are painted on and around the face. The paint can hide all kinds of flaws, even body putty.

Fourth, check the face plate for hardness. You can use a ball-peen hammer or a good-sized chisel to do this in the field. If you use the chisel, hold the chisel in your hand sharp end down four or five inches above the anvil's face, usually somewhere over the heel. Open your hand and let the chisel free fall on the face. If the anvil is hard it will not make a mark, but may knock off a little rust if the face is rusty. If the anvil is soft the chisel point will leave a sharp line in the face. That is the reason you want to test the face on the heel. The owner may not like you putting a mark on the working part of his worthless anvil. It is better to check the hardness of an anvil you are thinking about buying when the owner is not present, if he knows the face is soft. If he knows the face is hard he will not mind your testing it.

Using about a 16 ounce ball-peen hammer you do essentially the same thing. Give the anvil a sharp rap with the ball end of the hammer. If the anvil is soft it will leave a nice round dimple in the face. If the face is hard the hammer will just rebound or maybe knock off a little rust.

Fifth, if the anvil has a face plate be sure to check its thickness. Anvils from 100 to 200 pounds should have at least 3/8- to 1/2-inch of face plate. If it is any thinner there may be spots where the body metal is almost coming through the face plate; therefore, 1/2- to 3/4-inch thick would be preferred. A 3/8-inch face plate is often found on anvils weighing under 100 pounds. I think the makers finished them to that thickness.

Sixth, a rusty face can also hide a number of small dents and flaws. If the face is heavily rusted you had better watch out. Although the face may be hard and in great shape, some of the pits under the heavy rust can be quite deep, and they are usually deepest where most of the work is performed.

Seventh, look at the wear around the hardie and pritchel holes. If you plan on using them much they should be in fairly good shape. You might try an anvil tool in the hardie hole to see if it fits snugly without sticking. You do not want to repair either hole because welding can soften the face in that area.

The face is the most important part of the anvil, so make sure it suits you and take good care of it.

To check for any other flaws or hollow spots within the body of a wrought anvil, put it on the ground or on wood and rap the horn or heel lightly. If the ringing stops about as quick as it starts you will know there is a hollow spot or flaw within the body of the anvil. A wrought anvil with no flaws will have a clean ring that will continue for a few seconds. Later anvils such as the *Trenton* and *Arm and Hammer* lost some of their ring because the body is arc-welded to the base for about an inch all around, leaving a hollow in the center. The fact that some anvils have a slight flaw or hollow in the body will generally not affect the anvil's performance, but I prefer an anvil without flaws. If a wrought anvil is dead when struck, be careful about buying it, as eventually it may break at the waist, or if it was made by the built-up method, the horn or heel may eventually fall off.

If the price is right you may not mind a flawed anvil; just be sure the face plate is tight and the face is hard.

You will notice I have said nothing about the horn, table or feet. The reason is that they are made of wrought iron or mild steel on a wrought anvil and can be repaired easily with an arc welder and a disk grinder or sander, almost regardless of their condition (see Anvil Repair, Cleaning and Care).

If you are buying a cast-steel anvil be sure the horn, table, and body are in reasonably good shape as well as the face, because the whole anvil is tool steel with a hardened face. Rewelding on the horn, etc., require a special rod.

A steel-faced cast-iron anvil such as *Fisher and Norris, Vulcan, Southern Crescent,* etc. can be as useful as a wrought anvil. Be sure that they are in good shape with a hard face, because you do not want to have to repair them.

Good Hunting!

Appendix C

Setting Up the Smith's Anvil for Use

Almost every book that gives instructions on blacksmithing or about the blacksmith's anvil has something to say about how to place the anvil for use. However, some of the readers of this book may not have access to such material so I will essentially repeat the instructions here, along with my own twist.

First of all consider the position of the anvil. If you plan on working alone and without a helper, you should place the anvil within one step of the forge fire, giving the minimum time between fire and anvil. When using a helper for heavy work it may be desirable to have the anvil a little further away so both smith and helper can move freely around the anvil.

Right-handed smiths have traditionally worked on the anvil with the horn to their left. If you use the horn much you will find that this is the best way to position the anvil.

It would be vice versa for left-handed smiths.

The standard recommended height of the anvil is to have the knuckles of your hammer hand resting lightly on the face of the anvil with your arm hanging loosely by your side. This allows the hammer to come down flat-faced on the work you are doing, minimizing the fatigue on your arm when you are using the anvil regularly. However, for intermittent use it will be easier on your back if the anvil is a little higher. On the other hand, if much sledging at the anvil is to take place, have the anvil a little lower than the knuckle height.

The traditional mounting block for the anvil is an elm log slightly larger in diameter than the anvil's base, set three or four feet in the ground. Most any hardwood log would do however, such as oak, maple, hickory, etc., but not a soft wood such as cottonwood. I read somewhere that elm was recommended because it was not good for anything else, as it is difficult to split for firewood, and little used for any other purpose.

If the anvil is used as a bending tool, as it often is, you will want the anvil support anchored firmly to the ground or floor, regardless of what material it is made of. In this case you also want the anvil anchored to its support block. When a wrought anvil is bound down tight to the block most of the ring is lost, which is good if you use the anvil much. Anvils under one hundred pounds, regardless of use will need to be bound down to keep them from moving around. If you do not use the anvil for bending all you need is to drive four large nails (preferably handmade) into the support at each corner of the anvil base radius. This will keep any anvil from shifting and allows the anvil to be picked up easily if you want to move it. It also allows the anvil to give a full ring if that is what you want. Besides the four-nail method, there are all kinds of ways to fasten an anvil to its support. Four large staples driven into the block over each of the protruding feet works well and will still allow the anvil to ring if they are not driven down tight. There are all kinds of bolting and strapping methods; in fact I have seen anvils fastened in more ways than I care to recount. Later-model *Fisher* anvils had two lugs on the anvil base for fastening the anvil down. Some of the other cast anvils have a little notch in the middle of the rear foot to help in securing the anvil. Whatever method you choose is fine so long as it serves your purpose. Now you are ready to work!

In addition to wood supports, which I like best, there were at least three different factory-made cast-iron supports. I have a 350-pound Swedish anvil of 1980s vintage fitting perfectly on a factory-made cast-iron base made about 1900. There are angle iron supports, built-up board supports, concrete supports, etc. All of them are all right if they suit you.

Here is part of an article taken from *The Blacksmith and Wheelwright* journal of May, 1912, entitled "Some Anvil Blocks," by James F. Hobart, M. E. I found it interesting for that time period.

> *Chopping blocks and anvil blocks of decent proportions are becoming very scarce articles, nowadays, [I wonder where he lived?] but we won't worry about the former for as there is little wood to chop, not much block is necessary. But when we find anvils mounted upon contraptions made from plank and worse yet, from scantling spiked and nailed together, then we feel desirous of three feet of butt-log from George Washington's cherry tree.*
>
> *Recently, the writer saw an anvil which had been nicely mounted upon a heavy casting in place of the traditional block. So well did the casting fit the anvil, that it was asked if the base was designed for the purpose, but judge the surprise of the writer when informed that the anvil block was the base of a discarded cream separator from the milk depot!*

He goes on to describe how the anvil was attached to the separator base in three more paragraphs. Then he spends a full page and a half of the journal describing how to use an old flour barrel as a mold to make a concrete anvil base. This was one of the most stupid methods for making an anvil base that I have ever encountered. I would print the whole article, but I believe it would be a waste of the reader's time.

I bought an anvil that was attached to a cream separator base. I also bought an anvil that was tack welded to a large piece of pipe with a truck wheel to support the whole thing. I told the person that I bought it from that I wanted the anvil only, but he said if I wanted the anvil I would have to take the whole contraption. I had no way of cutting the anvil loose at the time so I took it all. The pipe and wheel are somewhere in my junk outside. It seems that most anything will make an anvil support.

Appendix D

Anvil Care, Cleaning and Repair

This is a tough subject for me, especially the repair part. I now consider all anvils made before 1960, and all wrought anvils, as antiques. I would like to see them all left in the condition in which they were found. But, I know that this is not going to happen, so I will try to be realistic.

When I say an anvil should be left as it is found, I mean it is better not to grind the face plate or the edges of the face plate, to surface-grind the plate, or to weld on the anvil. As far as cleaning up the anvils, I am all for that. I think cleaning enhances their antique value and tends to preserve the anvil as found. It is best to use a power wire brush to accomplish this, but it can be done with a hand-held wire brush, although the task will take longer. Wire brushing removes all the loose scale rust, allows you to locate any markings, and puts a nice dark, almost black patina on the anvil if you do not hit the base metal. If the anvil is then oiled or sprayed with a clear lacquer or some clear anti-rust preservative you have a good user, or a splendid-looking piece to show as part of your tool collection. If the anvil is left outside for show it is best to keep a coat of paint on it.

You may recall the anvil that the Smithsonian would not let me wire brush where the trademark was, which I mentioned in a previous chapter. This is an extreme example of leaving the piece as found, and although I might disagree with that, museums have their own set of standards, which is probably a good policy.

Should the anvil be in usable condition, I think it ought to be used, so long as it is not abused. I sometimes demonstrate on an anvil that is at least 250 years old. You are not likely to wear out a good anvil if it is not misused. In fact a good anvil will last many lifetimes even with continued but proper use.

There are a number of activities that can ruin anvils beside using them as an arc welding or oxyacetylene table. Striking the edges too hard, using a hammer or sledge that is too large for the anvil's size, and working on a very cold anvil can cause damage. One writer suggested that on anvils over one hundred pounds use only one pound of hammer for every ten pounds of anvil.[2] Therefore a forty-pound sledge can be used on a four hundred- pound anvil. I would not use a ten-pound sledge on a one hundred-pound anvil, however. Remember, anvils can be broken, especially when they are very cold.

I was told some years ago by a reliable person that one of the reasons the Germans lost on the Eastern front against the Russians during World War II was because the steel used in their vehicles was high in sulfur, making the steel very brittle when extremely cold. Therefore, many of the German tanks and trucks broke down as pieces of the steel snapped while they were being operated during the Russian winter.

I am sure some of the old blacksmith shops fell well below 0^0 F. inside at night during winter before the smith lit his fire in the morning. Starting off with some heavy anvil work the smith might easily have broken the horn or heel off of his anvil, especially the heel. I have seen numbers of anvils with missing horns or heels, and they were not, I am sure, always broken on purpose. If a building is 0^0 F. inside when the stove and forge fire are lit, the building might warm up quickly to 50 or 60 degrees, but the anvil will still be at 0 degrees or below. If you keep your "user" anvil in a cold place and plan on using it, warm it up first or use it lightly until it does warm up. One smith told me that he heats up old "sad irons" and places them on the anvil's face for a while in cold weather before using the anvil. Remember an anvil is also a heat sink and will draw the heat from the piece you are working on when the anvil is cold. I bought a set of "sad irons" afterward.

I have a 95-pound *Hay-Budden* anvil that I bought from a salvage yard. It is in perfect condition, except the heel is broken off an inch or so ahead of the hardie hole toward the horn. It is a clean break and there is not a hammer mark on the face. How could this have happened? I suggest that a smith was bragging on the quality of his anvil and said that it would not break like one of the older built-up ones would. (This anvil was made about 1899). It was probably winter and work had just started for the day. Someone else or the smith, in order to show how good the anvil was, took a sledge and gave a sharp blow ahead of the hardie, expecting the hammer to rebound, but instead the heel snapped off. The break, as I said, is clean and there are no flaws in the metal where the break occurred. I do not intend to repair it.

If you do not want to cry, do not work on your anvil when it is real cold!

Broken 95-pound Hay-Budden anvil. Complete with "Sniffy" the cat. (R.A.P. photo)

If you use the anvil often, you need only wipe an oily cloth over the face when you finish for the day. Wipe it off when you start again, and you will not have to worry about rust; that is, if the anvil is under cover. Even a skim of rust puts pits in the face of an anvil. Should the anvil be left for a long period of time without use in an unheated building, it is best to either put a coat of grease all over it or oil it and put a cloth or covering over it. Otherwise the anvil is going to sweat and rust when the warm moist days come in the spring, and also when quick warm-ups happen in the winter. If you make knife blades or plan on doing so, you will want to be extra careful with the polished face of your anvil. I am sure knife makers need no reminding.

Take care of your anvil, remember it is an antique, and you will be able to pass it on to be used for many generations to come.

For those of you who own or bought an anvil that needs some repairing before it can be used, I reluctantly give the following suggestions.

If an anvil is forged wrought iron, the body is easily repaired by arc welding, with the exception of the hardened tool-steel face plate. Any good rod, especially E. 6013 or E. 7014, are excellent for building up the horn, table or a piece missing on the foot.

All wrought anvils with a face plate have a body (includes the horn, heel and base) made of either wrought iron or mild steel which welds easily with these rods. I have added as much as two inches to the end of an anvil's horn on which the tip was broken off, and after grinding it down it looked as good as new. The metal I welded onto the horn was better than the original metal and will hold up longer. A special rod will be necessary for plateless and cast-steel anvils.

To do this welding properly you should be a pretty good welder. The areas to be welded should be cleaned with a power hand grinder or sander to get the rust off, etc. Pre-heat or warm the anvil a little, but keep some wet crumpled newspaper on the face while you weld. The water in the paper acts as a heat sink and will get warm, but will not allow the face to get hot enough to remove the temper. If you are not a welder, take the anvil to a good welding shop or to someone you know who is a good welder, but be sure you supervise the job.

If the anvil does not have a face plate, such as the *Columbians, Hay-Buddens* after 1908, and *Trentons* after 1927, it is best to use an all-purpose maintenance rod for repairing everything but the face. These rods cost considerably more than the rods mentioned, but if you do not know what kind of steel you are welding, the cost of the rod is minimal compared to the damage that might occur, as the whole body of these anvils is made of tool steel, with a hardened face. The tool steel these manufacturers used was not always the same. Almost all anvils made today are solid cast steel of some kind.

For anvils with a cast-iron body and a steel face plate such as *Fisher, Vulcan, Southern Crescent, Star* and *Badger,* you best forget about repairing them! They can be repaired, but it is difficult and generally not worth the time and effort.

Repairing the face plate of a wrought anvil or the face of a solid tool-steel anvil is a different proposition. This plate or face is always high-carbon steel in the older anvils, and either a medium- or high-carbon alloy steel in modern anvils. The face may be repairable, depending on the amount of damage. In former times there were shops that could forge-weld a new face plate on worn anvils and also temper them, but to my knowledge these concerns no longer exist. If a whole new plate is required it is best to buy another anvil, new or used. I know of one company today that can put face plates on smaller anvils (under 200 pounds), but the plate is not forge-welded on, but rather brazed on, and this can be a chancy proposition at best. I know of one person who had a new face plate made for his anvil by building up the face with pass after pass of special rod and then surface grinding the face. If he had to pay for this job he could have bought a new anvil. As of now, I suggest that should a new face plate be required, it would be best to forget it.

Many of the old anvils only need the edges of the face built up to make them serviceable again. I want to remind the reader again, you are dealing with an antique, and if it is at all possible that you can use the anvil as it is, leave it alone! If the anvil's face has a sway or belly in it, as many of the old wrought anvils do, live with it and use it to advantage rather than trying to grind the face flat again. To weld up these edges economically I recommend a work-hardening rod such as "Stulz Manganese XL" or E. 11018. I have used "XL" with good results, although it is not the easiest rod to weld with. I was told by an expert welder that E. 11018 welds easily and does a good job, but I have yet to try it. Ask your welding dealer for a rod that work-hardens, and has high impact resistance if you cannot obtain the ones mentioned. There are rods that are hard as welded, but I have never used any of these and I do not know how they would stand up to the impact of a hammer when the weld is on the edge of the anvil. I suspect the weld would either crack or break off.

The following procedure has worked well for me when building up edges of the anvil face. Of course this was before I began to realize the significance of anvils as antiques.

1. Clean the surface to be welded down to bright metal; no rust, oil, paint, etc. If the surface or edge is rounded and smooth, rough it up with a disc grinder. This will also clean the surface. Do not take off any more metal than you have to.

2. Preheat the whole anvil to 130^0 -180^0F., but not over 400 ^{0} F, or too hot to touch. I have accomplished this simply by putting two cement blocks parallel a short distance apart, then placing the anvil on top of the blocks and putting another block across one end of the first two. This makes a sort of oven. I then build a little fire under the anvil. It does not take long to heat up the anvil to any desired temperature, then remove the anvil from the fire.

3. Use some type of heat sink down the center of the face. Wet paper or cardboard can be used, well mushed up or shredded and kept wet.

4. Using the lowest heat possible on the welder for the size and type of rod, weld about two inches, then peen all over the weld smartly with the rounded end of about a 16 oz. ball-peen hammer. The welding process will maintain the heat in the anvil.

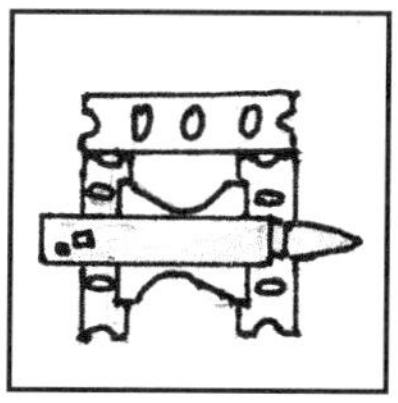

Method for heating an anvil

5. Keep building up the weld bead as in number 4 until reaching the desired results.

6. When you have finished welding and while the area being welded is still hot, try to square up and smooth the weld beads as much as possible by using the flat part of the hammer. This will shape and harden the metal and will also reduce the amount of grinding necessary.

7. Using a disc grinder, dress the weld beads up somewhat.

8. Hammer the weld bead some more so that when the grinding is finished the surface will be relatively work-hardened and smooth.

9. Finish the grinding or sanding with very light strokes and lightly grind the entire face of the anvil so that the welded edges blend in with the rest of the face.

10. I finish grind with a disc as follows: From side to side and then from front to rear or horn to heel and back. Then at 45^0 from side to side very lightly, then from horn to heel and back at 45^0 in the other direction as in the drawing. If you do this right it leaves a nice smooth pattern and an even surface.

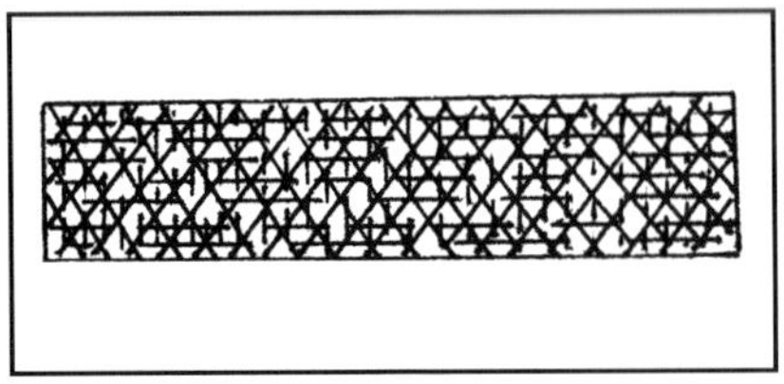

Grinding pattern

11. If you want a very smooth surface on the face, after crisscross-cross grinding, block sand the surface with silicon carbide paper or aluminum oxide cloth, finishing with 400 grit or until you get the surface finish you want. I have brought a repaired anvil face to a mirror finish in this manner.

12. If the edges dent or chip, they are easily touched up by running a short bead without preheating, and treat as before.

The original face will probably be a little soft just behind the weld for from 1/32- to 1/8-inch, but from checking on anvils I have repaired for other people over the years this does not seem to be a problem, and the welded edges are holding up as good as on a new anvil. I have no idea how well these anvils would stand up under industrial or continuous use, however.

I usually do not recommend welding up large pits in the face. If light grinding will not remove most of them, you had better leave the anvil alone. If the anvil has more than a 1/8-inch sway in the face it is best left that way. If you have to have an anvil with good edges, no pits, and a flat face, get one that is in that condition in the first place.

Just remember, you are dealing with an antique!!

Should you have an anvil that breaks at the waist, grind a 45^0 chamfer on both broken pieces about 1/2-inch in, all the way around. It should leave a gap that will be from 3/4- to 1-inch wide when the body and base are placed together again. Using a deep penetrating rod, tack weld four corners while the body and base are held in compression. Then proceed to fill the joint running bead after bead around the circumference until the vee joint is filled to just above the original surface of the body. Be sure to keep the face from overheating. Grinding flush or not grinding is optional.

APPENDIX E

The Ringing of the Anvil

This appendix begins with a quote from *Twentieth Century Toolsmith and Steelworker:*[3]

> *Don't have the anvil merely sitting on a block that is continually jumping up and down with every blow from the hammer, but have it well bound down to the block. But some will say, 'That stops it from ringing' or 'I can't work on an anvil that does not ring!' Reader, this is all nonsence, what has the ringing of the anvil to do with the work, it may be all right for the class that wants to make a lot of noise to let others know they are working, but it is of no use when it comes to doing the work with ease, both for the blacksmith and the helper...*

Here is another interesting quote from *Dictionary of American Hand Tools:*[4]

> *Many smiths were proud of the rhythmic ring of their anvils and considered the sound as a sort of professional trademark. Some makers of cast anvils apparently tried to downgrade this idea by advertising that their anvils did not reverberate and that therefore more energy was transmitted to the workpiece. Without joining the argument over which was best, most of us who have pounded iron would agree that dull hammer thuds issuing from the blacksmith shop leaves something to be desired.*

Whenever I go to a traditional festival of some type at which I think a blacksmith may be working, I stop and listen. I follow my ears to the ring of the anvil and am soon in conversation with the smith. Indeed the ringing anvil has a place in tradition, but there are also other aspects of the anvil's ring; some are of value and others require caution.

When wrought anvils were manufactured, the final test was to tap the horn with a hammer to determine whether there were any flaws in the metal or welds. A clear ring that continued for a few seconds indicated a first class anvil ready for the market. On the other hand if a clear ring was not heard it would be the inspector's judgement as to whether the anvil was to be shipped, reworked, or scrapped. Unfortunately some that should have been scrapped reached the market. After years of practice the inspector could tell exactly where the defect was in an anvil by tapping in various places. This was more difficult with the old built-up anvils because of all the weld joints. But on the two-piece anvil (so-called) there was only the joint where the face plate was attached to the body and where the body was welded to the base. Therefore, detecting defects in these was much easier.

It may sound strange, but no two wrought anvils ring exactly the same, regardless of whether they are made from the same metal and are the same weight. Of course, as has been mentioned, no two wrought anvils are exactly alike. Maybe that is why the smith enjoyed the ringing of his anvil; it was distinctive and made him feel singular.

Mr. Earl Tilton, president of the Columbus Forge and Iron Company and his son, Mr. David Tilton (the last president), were visiting California in the late 1940s.

The following quote was from a Columbus newspaper concerning their trip.

> *While strolling through the Mexican section of Los Angeles on a recent visit there, the father and son saw an anvil they thought they recognized. Sure enough it was one of theirs [a Trenton].*
>
> *An old Mexican workman was using it to shape ornaments for the tourist trade. He told the Tiltons it had been given him by his father.*
>
> *Mr. Tilton Sr. saw the face of the anvil was scarred and offered to reface it for him free, even though the guarantee had long since ran out. 'My father would turn over in his grave!' the workman said. 'And besides my hammer is used to the bumps. I couldn't work the same without them.*[5]

When the article from which the excerpt was taken was given to me in 1995, I called Mr. David Tilton and asked him about it and the encounter with the Mexican workman.[6] Mr. Tilton said he remembered the occasion well and that the man was young, not old. He said also that the main thing the man was worried about was whether refacing the anvil would make a change in the anvil's ring. The workman asked Mr. Tilton if the anvil would ring the same when repaired and Mr. Tilton said it would not. That is when the workman said, "My father would turn over in his grave!"

On the downside of the "anvil ring" is what it does to the smith's hearing as well as anyone else working in the shop with him. There is an article in *The Anvil's Ring,* entitled," "Beware the Anvil's Ring," by Stephen A. Mackizie.[7] With help Mr. Mackizie tested two different farrier's anvils for decibels while working on a horseshoe. Both anvils registered above 100 decibels on the A scale, especially on the horn and heel. Ninety is the accepted safe level for continuous noise for an eight-hour day and 115 for one-quarter of an hour, according to the article. I assume that the anvils were not bound down, for they were moved from inside the building to the tail gate of a pickup truck for further testing.

As the article suggests, if you use your anvil a great deal, check it to see if your hearing could be affected.

There are a number of ways to lower the sound level of an anvil without going to extremes. Putting a magnet under the heel will help, or wrapping a piece of rubber innertube around the heel is another. Binding the anvil down tight will effectively kill the ring, too. Placing a rod bent in a U in the pritchel hole will work too. You could use ear plugs, but they are inconvenient However, if you are demonstrating and want the anvil to "sing" to attract customers while you work, the ear plugs are a good idea. Your hearing is nothing to trifle with!

APPENDIX F

Anvil Patents [8]

The following patents concern anvils or improvements that are not mentioned in the text in conjunction with any of the anvil makes or trademarks. Rather than print the entire text of the patents I will only reproduce the drawings and summarize or comment on the text. All of the text for the patents will come first and be numbered, followed by all of the patents with corresponding numbers.

The patents are in order by date of application, from the earliest to the latest. They are reproduced by courtesy of the Smithsonian Institution.

Patent Index

1. "Machine for Making Anvils," Patent # 3,418, January 31, 1844.
2. "Improvements in Anvils," Patent # 60,745, January 1, 1867.
3. "Improvements in the Manufacture of Anvils," Patent # 95,577, October 5, 1869.
4. "Anvil," Patent # 327,566, October 6, 1885.
5. "Anvil," Patent # 358,756, March 1, 1887.
6. "Anvil," Patent # 386,755, July 24, 1888.
7. "Anvil," Patent # 543,397, July 23, 1895.
8. "Anvil," Patent # 826,499, July 17, 1906.
9. "Anvil," Patent # 846,548, March 12, 1907.
10. "Anvil," Patent # 880,335, February 25, 1908.
11. "Anvil," Patent # 921,787, May 18, 1909.
12. "Anvil," Patent # 953,305, March 29, 1910.

{See Patent 1}

United States Patent Office
John Taylor, of Shade Gap, Pennsylvania
Machine for Making Anvils
Specification of Letters Patent No. 3,418, dated January 31, 1844.

This patent proposes to make an anvil out of one piece of wrought iron in closed dies by the use of a drop hammer. I doubt that the process would have worked at that time, but the patent illustrates an attempt to lower the cost of producing an anvil at an early date.

{See Patent 2}

United States Patent Office
Lancelot Kirkup, Brooklyn, New York, Assignor To Himself,
Albert Palmer, Joel Parker, and W. M. Hudson.
Letters Patent No. 60,745, dated January 1, 1867.
Improvements In Anvils.
The Schedule referred to in these Letters Patent and Making part of the same.

I am including the entire text with this particular patent.

You will notice that this anvil would be plateless with a hardened face and the base of any suitable material. This process actually did not become practical until the Columbus Forge and Iron Company's patent of 1898 and Hay-Budden Manufacturing Company's patent of 1908.

To All Whom It May Concern:

Be it known that I Lancelot Kirkup of Brooklyn, Kings County, New York, have invented, made and applied to use, certain new and useful improvements in the Construction of Anvils; and I do declare the following to be a full, clear, and correct description of the same, reference being to the accompanying drawings and making part of this specification, and to the letters of reference marked thereon, in which - Figure 1 is a side elevation of my improved anvil. Figure 2, a top view of the same.

In the drawing, like parts are indicated by the same letters of reference.

The nature of my invention consists in constructing an anvil as hereinafter fully shown. The object of my invention is to produce an anvil more durable than the anvil now in use, and one more readily and economically manufactured.

To enable those skilled in the arts to make and use my invention, I will describe the same.

A** shows the upper section of an anvil, and **B** the lower section of the same; which sections **A** and **B** may be firmly united by the bolts or keys **C C,** or in any convenient manner. The upper section, **A** of the anvil is made of cast steel, in one piece, or of steel without any welding, instead of welding the ears and other parts separately onto the core, as in the ordinary mode of manufacture. The upper surface of this section is made large in area, but smaller in depth, making the same less weighty and more portable, and, when placed upon the lower section or block **B,** affording the same area of top as an anvil made in the ordinary way, but weighing twice as much. The lower section block, **B**, may be made of cast iron or any suitable metal having the ends of its upper surface inclined slightly to receive the upper section or cast-steel portion, **A**, fitting snugly upon the same. **C C** show bolts or keys passing through the section **B** and the section **A**, to retain the same in position upon the section **B.

The great advantage arising from the construction of an anvil as described, is that the upper section upon which the work is done is or may be made of solid steel, while the lower portion may be made of such cheap material as to make the perfect anvil no more expensive than these ordinarily used, but very much more durable.

Having thus described my invention; what I claim as new, and desire to secure by Letter Patent is - The anvil constructed substantially as shown for the purposes specified.

Lancelot Kirkup

Witnesses:

A. Sidney Deane,
Jos. Peven

(See Patent 3}

United States Patent Office
David Foster, of Sheffield, England.
Letters Patent No. 95,577, dated October 5, 1869, patented in England, June 4, 1868.
Improvement in The Manufacture of Anvils, and The top and Bottom Parts of Hammers etc.
The Schedule referred to in these Letters Patent and making part of the same.

In this patent a mixture of melted steel is poured into the mold first and then the melted cast iron follows. This is similar to the Fisher & Norris process of welding cast iron to steel except in the case of Fisher the steel is only heated before being placed in the mold, not melted. I have no way of knowing whether this English process was ever applied. If it did work it would still not be as valuable as the Fisher process.

{See Patent 4}

United States Patent Office
John Mackert, of Springfield, Ohio.
Anvil
Specification forming part of Letters Patent No. 327,566, dated October 6, 1885.
Application filed December 15, 1884, Serial No. 150,358. (No Model).

I do not see how a patent could be obtained for this anvil. It is essentially a *Continental* pattern anvil, nothing more, nothing less. Mr. Mackert patented an anvil that had been in use for years before 1884 on the European Continent. This pattern anvil is still made today (1996).

{See Patent 5}

United States Patent Office
Enoch Statham and Henry Price, of Newport, Kentucky:
Anvil
Specification forming part of Letters Patent No. 358,756, dated March 1, 1887.
Application filed September 4, 1886. Serial No. 212,757. (No Model).

The patent seems to indicate that this is a large two-piece anvil for use in a railroad shop or where large bars are worked on and bent. The anvil proper "A,B,C" fits into base "E". The patent says the anvil either sets on a floor or the ground. I suppose that this anvil could have been made in small quantities.

(See Patent 6}

United States Patent Office
Joseph O'Brian, of Boston, Massachusetts.
Anvil
Specification forming part of Letters Patent No. 386,755, dated July 24, 1888.
Application filed April 28, 1888. Serial No. 272,207. (No Model).

I certainly do not think this modification deserving of a patent. It would be interesting to know if anyone ever replaced part of a face plate in this manner.

{See Patent 7}

United States Patent Office
Malree W. McInturff, of Prosperity, Missouri.
Anvil
Specification forming part of Letters Patent No. 543,397, dated July 23, 1895.
Application filed April 1, 1895. Serial No. 543,933. (No Model).

---My invention relates to improvements in anvils, and particularly to an anvil intended for use in the blacksmith shops of mining companies.

The patent does not say what kind of metal would be used in making this specialty anvil. If it was ever made, cast steel would have been the preferred metal. A forged anvil of this design and features would have been very expensive.

{See Patent 8}

United States Patent Office
Josephus H. Green, of Marietta, Ohio, Assignor To
Joseph W. Green, of Lancaster, Ohio.
Anvil
No. 826,499[patent] *Specification of Letters Patent* *Patented July 17,1906*
Application filed May 10, 1904. Serial No.207,310.

---My invention relates to an improvement in anvils; and the object is to provide an anvil which has a much larger base to rest upon than the ordinary anvil, so that it will retain its position while accomplishing its functions.

A further object is to provide an anvil particularly adapted for oil-well tools when they require sharpening or shaping, it being my purpose to provide an anvil designed for three different sizes of bits; and to this end my invention consists of a triangular body portion the angles of which are fashioned in different sizes for as many size tools, so that facility is attained in the manipulation and fashioning of the tools adapted to be operated upon on this anvil.---

A specialty anvil strictly for use in the oil fields.

{See Patent 9}

United States Patent Office
Laurence F. Clar, of Willits, California
Anvil
No. 846,548 [patent] *Specification of Letters Patent* *Patented March 12, 1907*
Application filed January 5, 1905. Serial No. 239,707

---My invention relates to anvils, and more particularly to anvils having a removable face-plate, and has for its objects new and improved means for securing such face-plate to the anvil, and the further object of providing a face-plate which is composite in its nature, whereby the working face of the plate can be made of hard resisting material, whereas the body thereof can be of comparatively soft and easily-workable material.---

By the time this patent was issued there were excellent cast-steel anvils on the market and plateless wrought anvils were about to come on the market. An anvil with a replaceable face plate would have little chance of being marketed.

{See Patent 10}

United States Patent Office
Hubert Edward Sheward, of Upper Norwood, England.
Anvil
No. 880,335[patent] *Specification of Letters Patent,* *Patented February 25, 1908*
Application filed March 18, 1907. Serial no. 363,099.

This is a patent to reinforce a cast-iron anvil by having a hollow core filled with metal chips and cement to help strengthen the steel-reinforced rods. This also gives extra weight to the anvil. If these anvils were manufactured in England they were never exported to the United States.

{See Patent 11}

United States Patent Office
Warren E. Andrews, of Wormleysburg, Pennsylvania, Assignor of one-half to Albert H. Shaffer of Harrisburg, Pennsylvania.
Anvil
No. 921,787[patent] *Specification of Letters Patent* *Patented May 18,1909*
Application filed May 28, 1908. Serial No. 435,602

---It is an object of the invention to provide a device of this character employing a novel manner and method of securing the face plate and horn to the body of the anvil.

It is also an object of the invention to provide a novel device of this character which will possess advantages in points of simplicity, efficiency and durability, proving at the same time comparatively inexpensive to manufacture.

You can see that this anvil would be mainly a machining operation and would have been quite expensive to produce compared to anvils being manufactured at that time. I am sure that the patent is as far as this anvil went.

{See Patent 12}

United States Patent Office
George Washington Summers, of Anniston, Alabama.
Anvil

No.953,305[patent] *Specification of Letters Patent* *Patented March 29, 1910*
Application filed January 22, 1909. Serial No. 473,658

---This invention relates to anvils and more particularly to revolving anvils, the object being to provide an anvil which is so constructed that it can be easily and quickly turned and locked in its adjusted position.---

I wonder who would buy this anvil?

This is possibly the last anvil patent ever issued. The Columbus Forge and Iron Company and the Columbus Anvil and Forging Company did not patent the electric process for welding the body to the base. This was the last innovation to be used in manufacturing wrought anvils.

The following patent drawings courtesy of the Smithsonian Institution are in numerical order matching the previous numbers and dates in the text.

Patent 1

J. TAYLOR.

Machine for Making Anvils.

No. 3,418. **Patented Jan'y 31, 1844.**

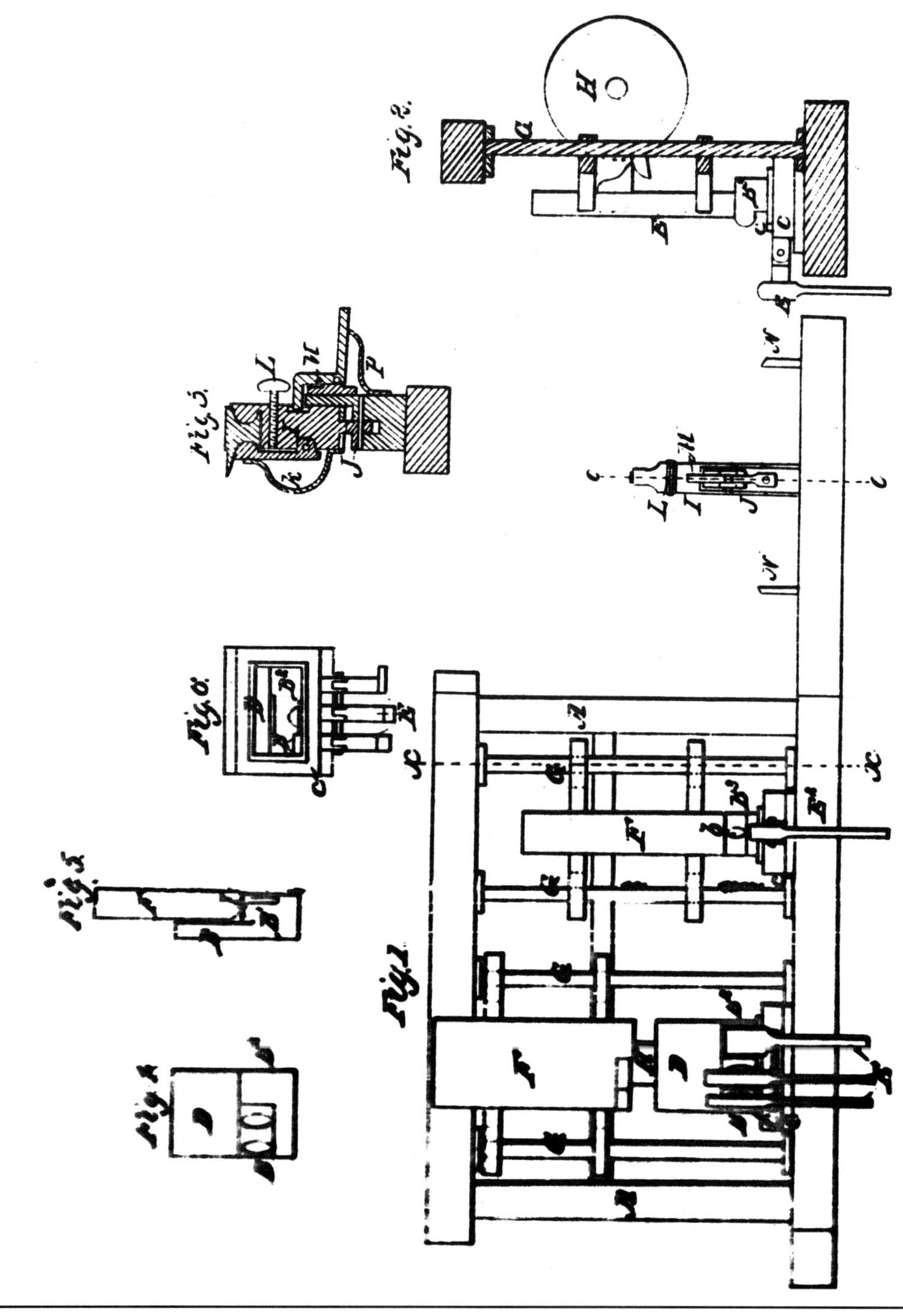

Patent 2

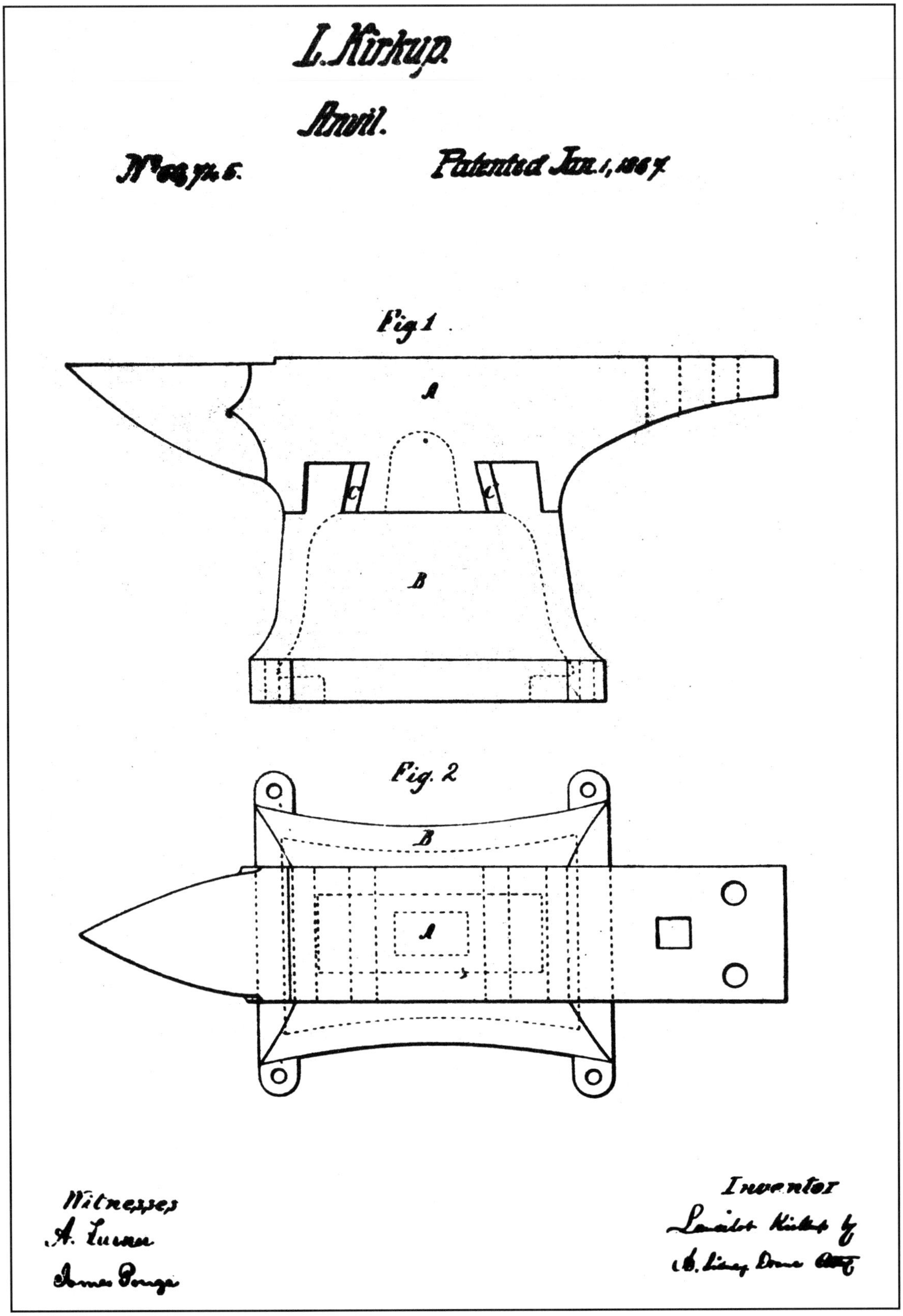
L. Kirkup.
Anvil.
Patented Jan. 1, 1867.
Fig 1.
A
C
C
B
Fig. 2
B
A
Witnesses
Inventor

Patent 3

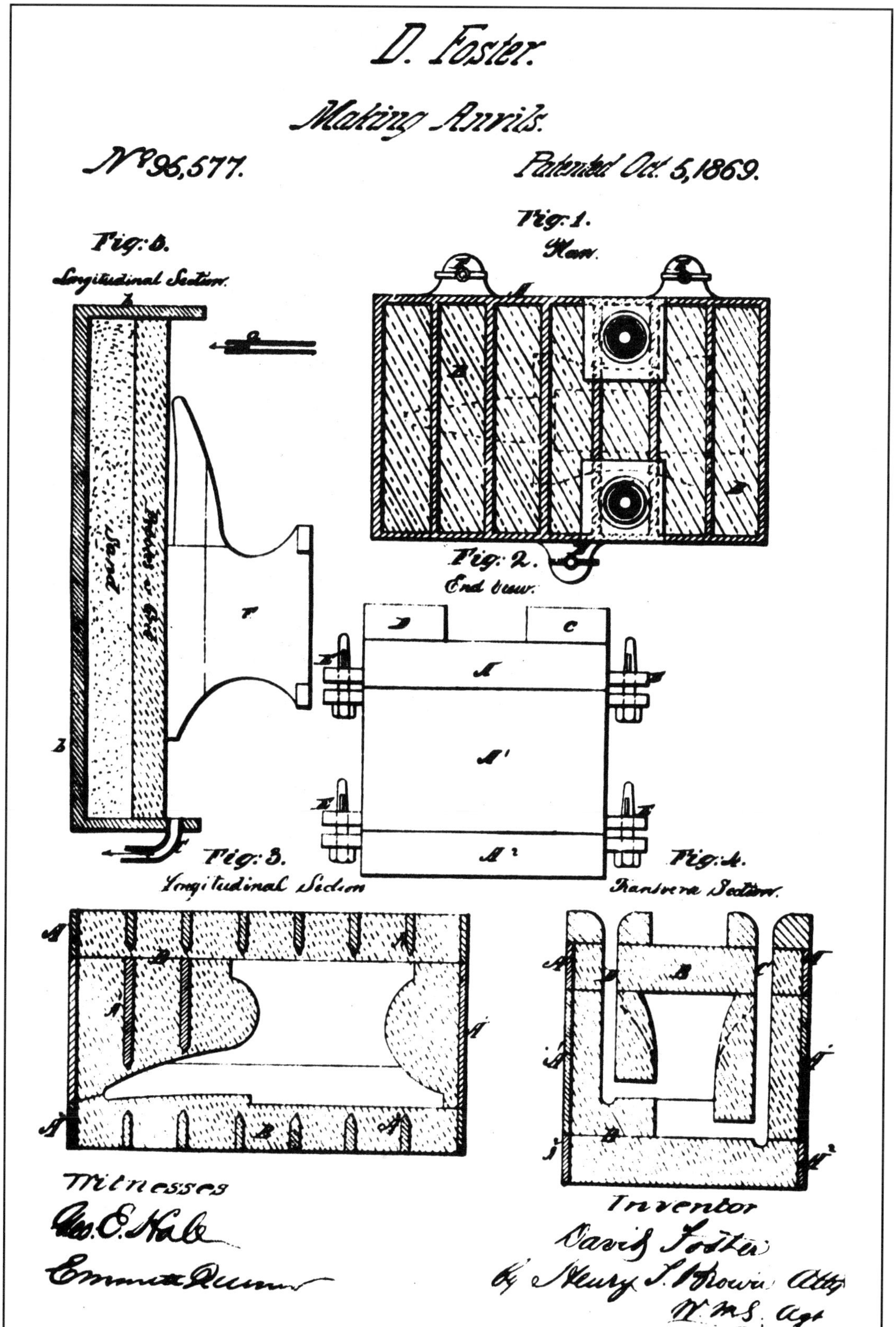
D. Foster.
Making Anvils.
No. 95,577.
Patented Oct. 5, 1869.
Fig: 1.
Plan.
Longitudinal Section.
Fig: 2.
End view.
Fig: 3.
Longitudinal Section
Fig: 4.
Transverse Section.
Witnesses
Inventor
David Foster

Patent 4

(No Model.)

J. MACKERT.

ANVIL.

No. 327,566. Patented Oct. 6, 1885.

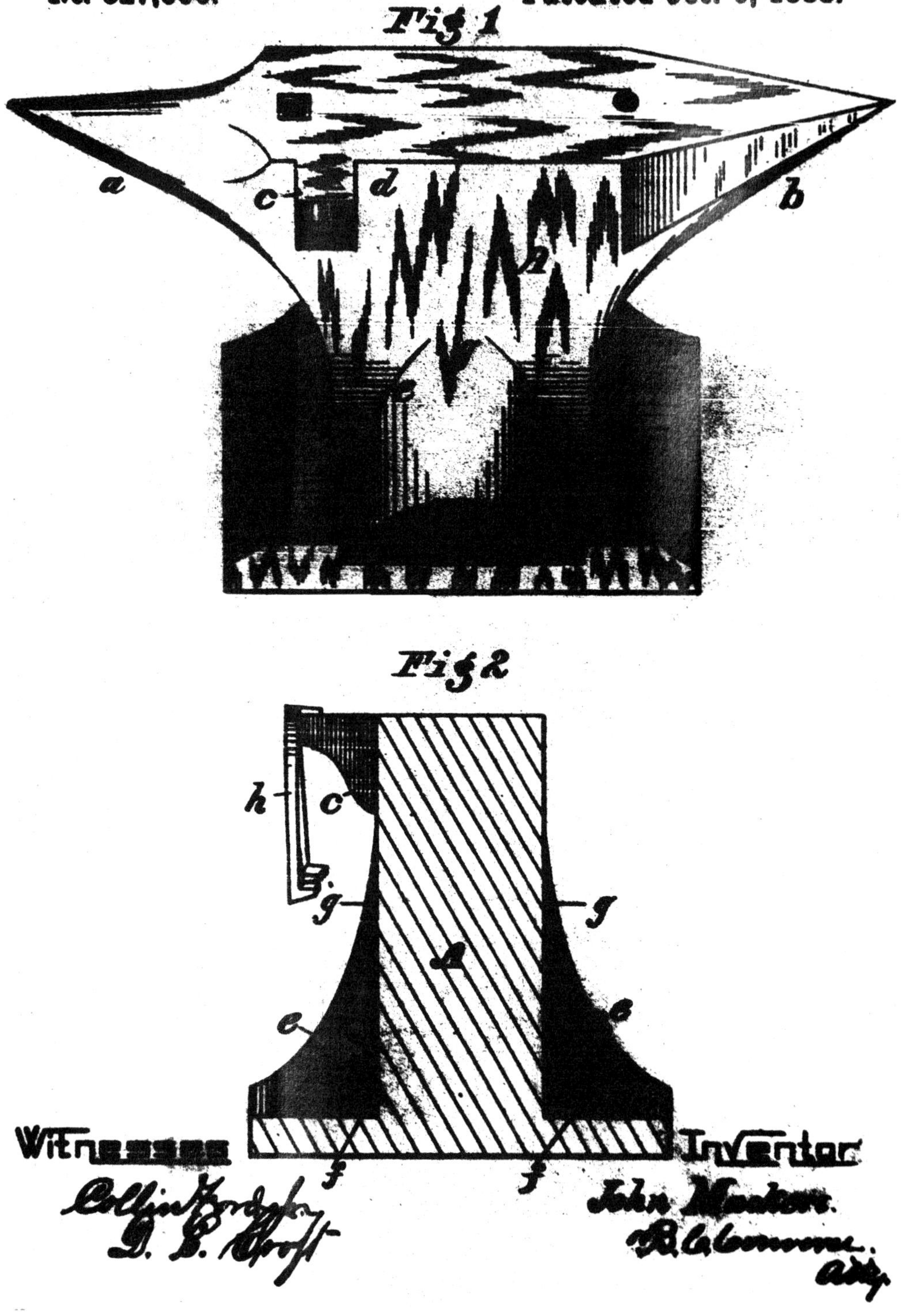

Patent 5

(No Model.)

E. STATHAM & H. PRICE.

ANVIL.

No. 358,756. **Patented Mar. 1, 1887.**

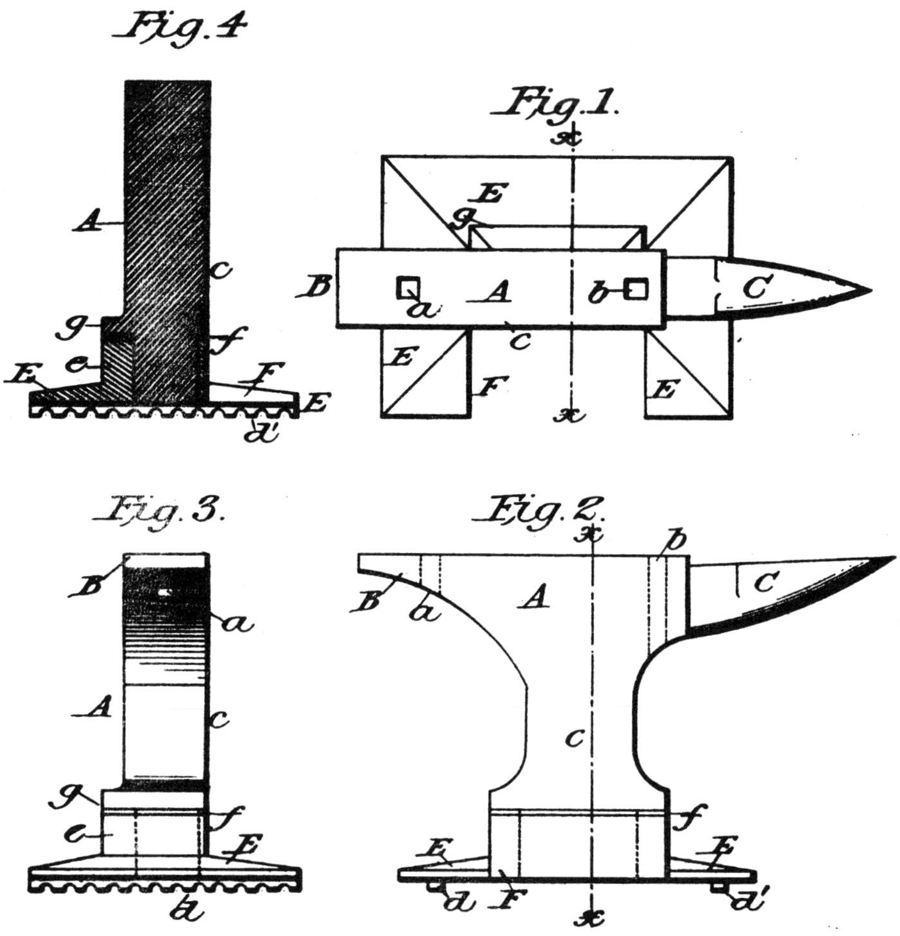

Witnesses:
Edw. J. Dempsey.
John Adams

Inventors:
Enoch Statham & Henry Price
by John E. Jones
Attorney

Patent 6

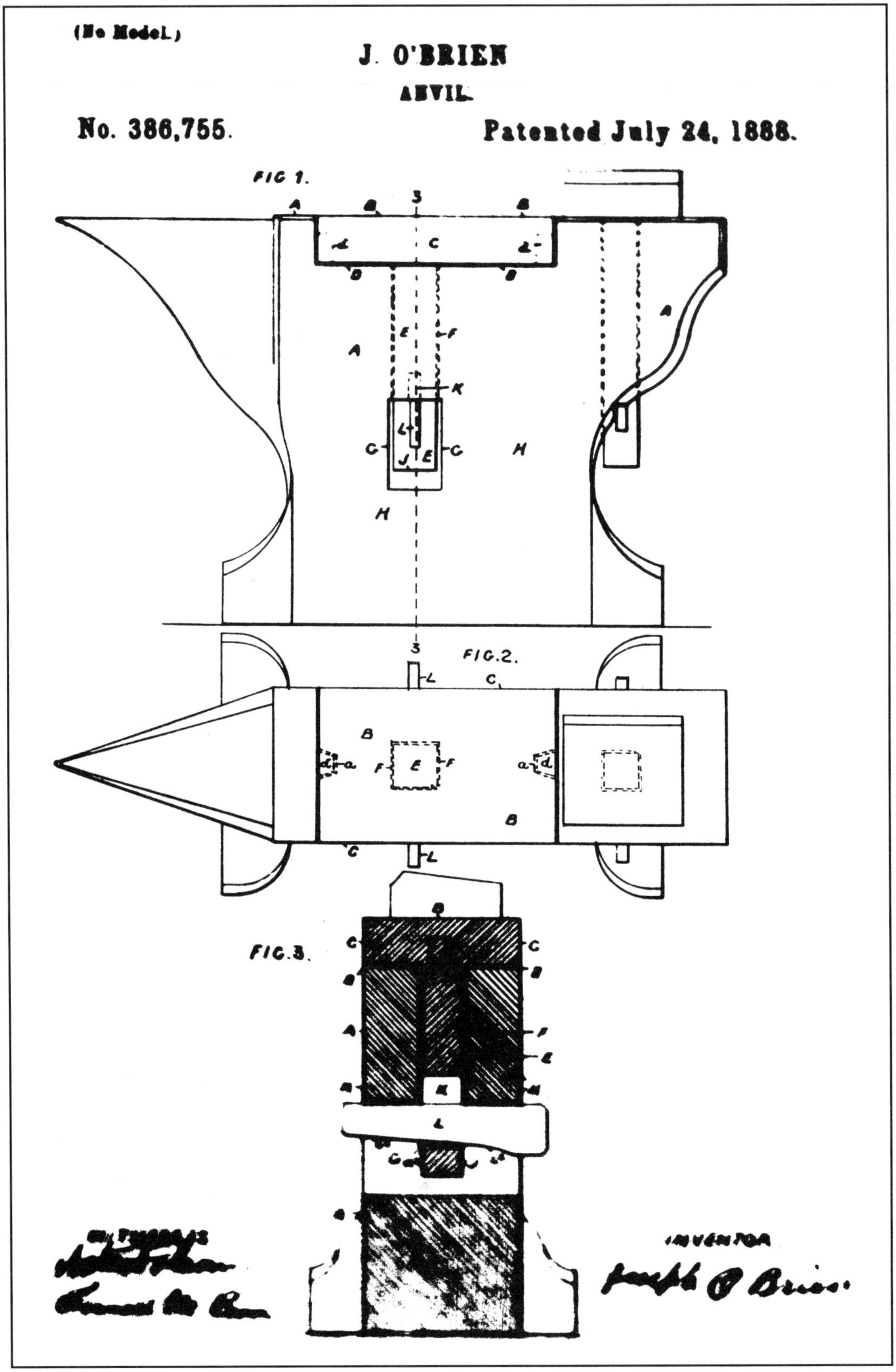
(No Model.)
J. O'BRIEN
ANVIL.
No. 386,755.
Patented July 24, 1888.
FIG 1.
FIG.2.
FIG.3.
WITNESSES
INVENTOR
Joseph O'Brien

Patent 7

(No Model.)

M. W. McINTURFF.

ANVIL.

No. 543,397. **Patented July 23, 1895.**

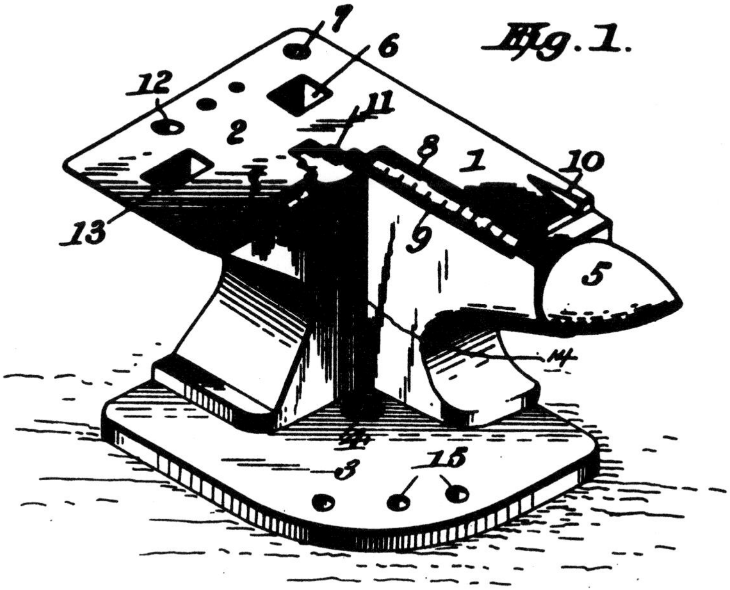

Fig. 1.

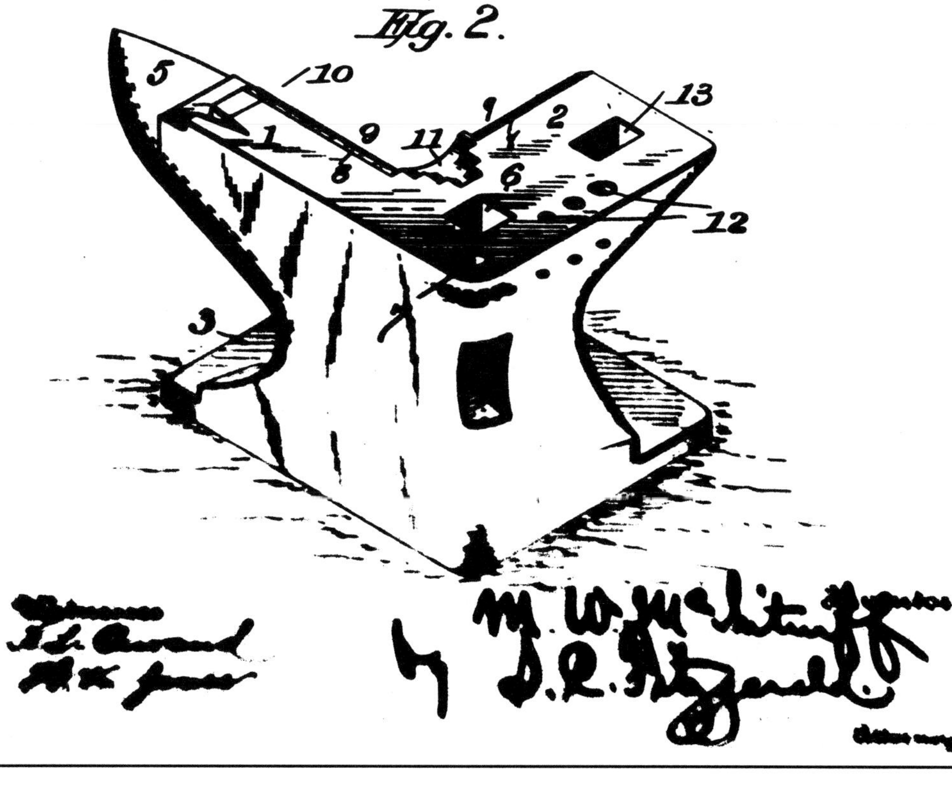

Fig. 2.

Patent 8

No. 826,499. PATENTED JULY 17, 1906.

J. H. GREEN.

ANVIL.

APPLICATION FILED MAY 10, 1906.

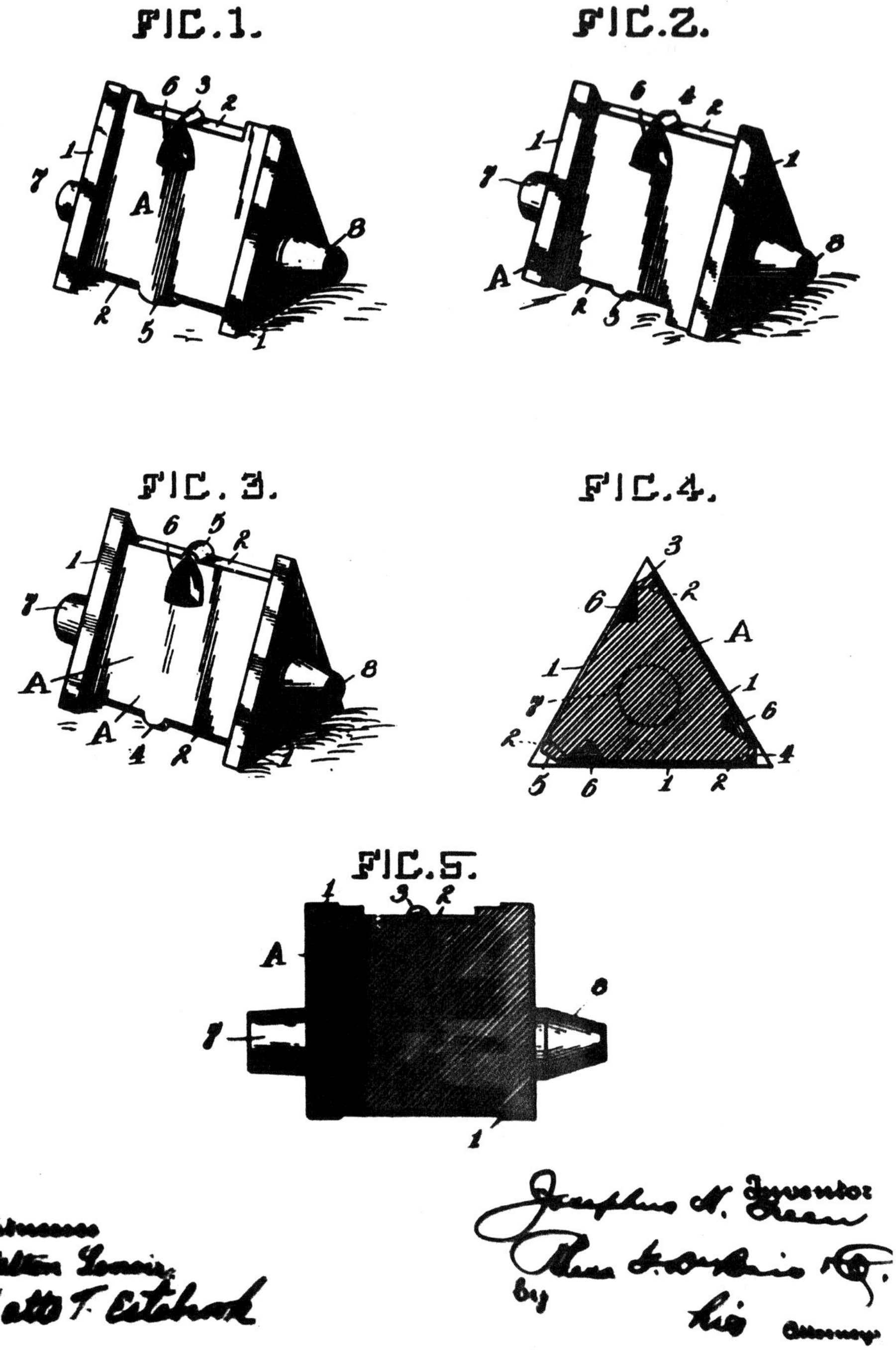

Patent 9

No. 846,548. **PATENTED MAR. 12, 1907.**

L. F. CLAR.

ANVIL.

APPLICATION FILED JAN. 5, 1905.

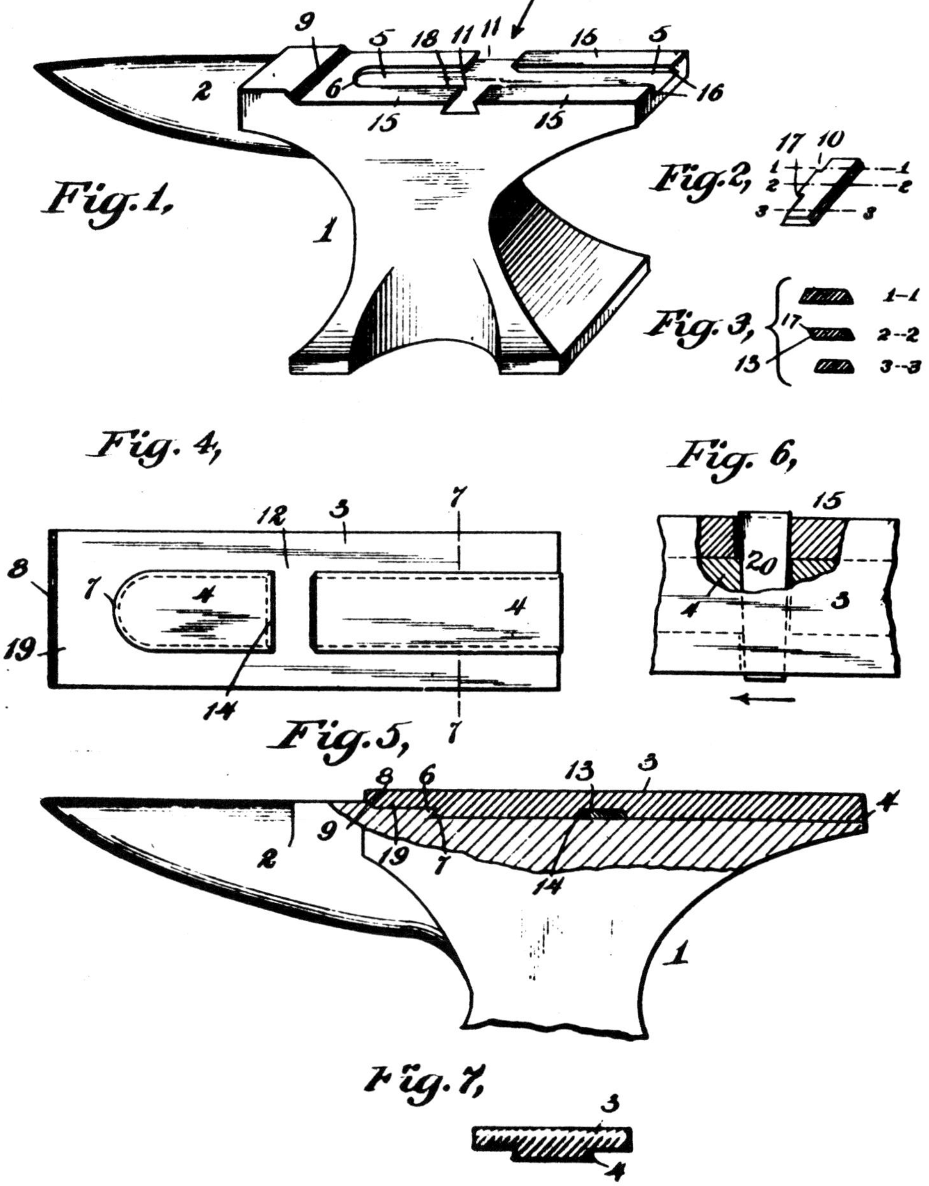

WITNESSES:

INVENTOR

BY

his ATTORNEYS

Patent 10

No. 880,335. PATENTED FEB. 25, 1908.

H. E. SHEWARD.

ANVIL.

APPLICATION FILED MAR. 18, 1907.

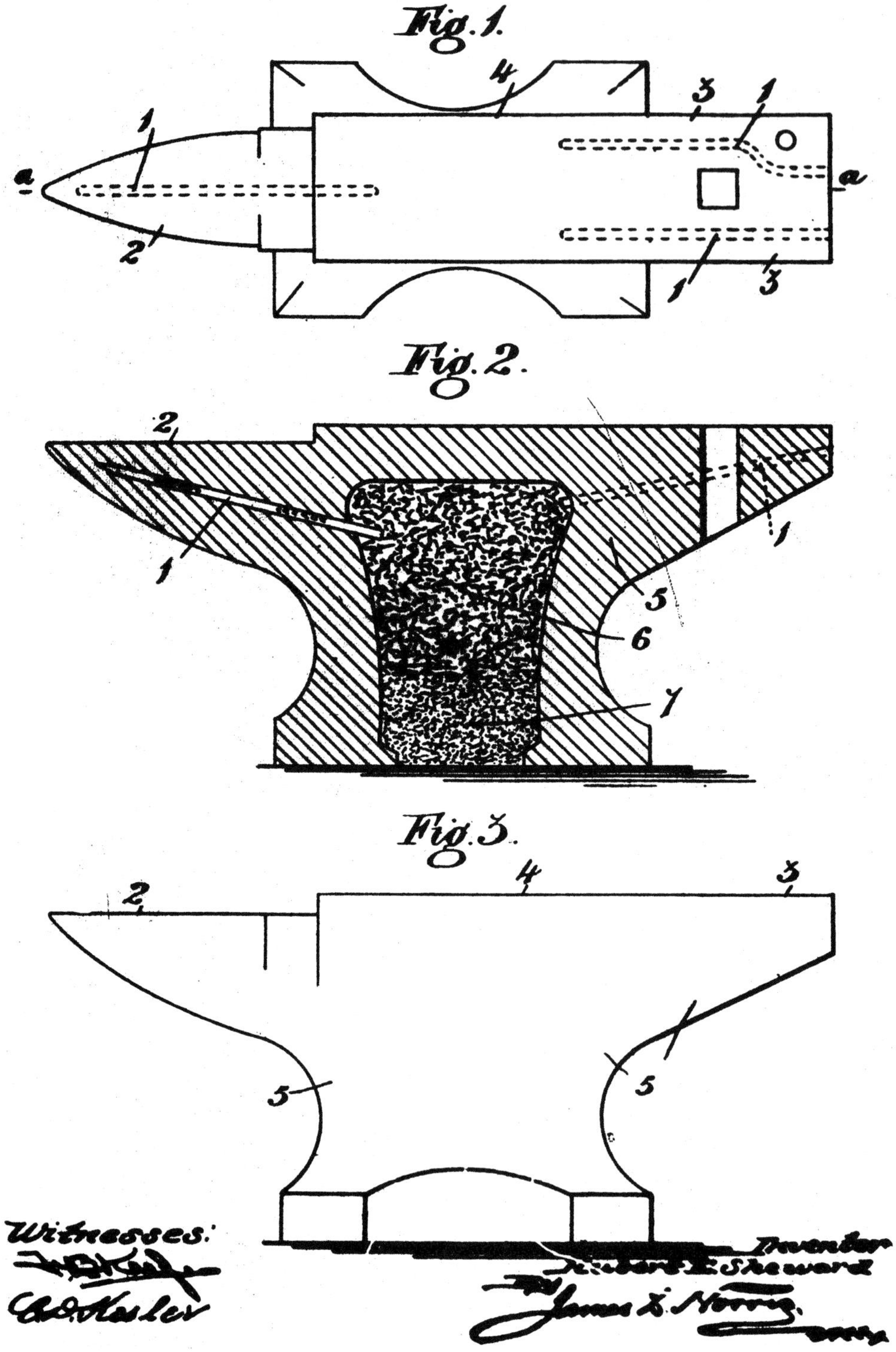

Patent 11

W. E. ANDREWS.

ANVIL.

APPLICATION FILED MAY 28, 1908.

921,787. Patented May 18, 1909.

Fig. 1.

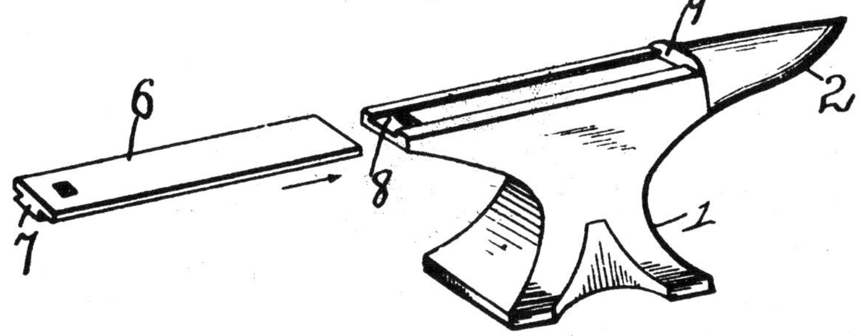

Fig. 2.

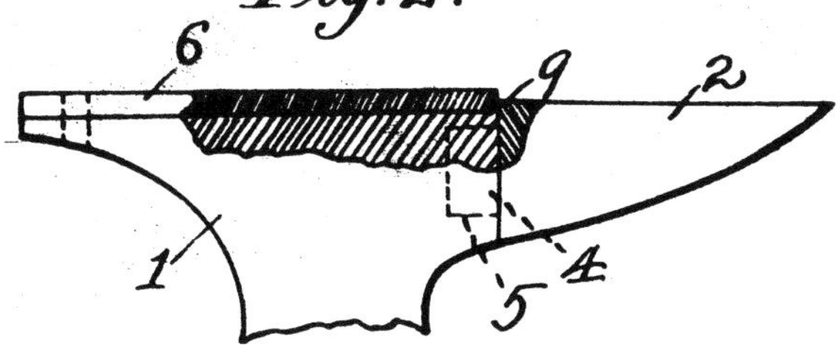

Fig. 3.

Witnesses

L. C. Oakley.

L. A. Sands.

Warren E. Andrews, Inventor

By Frank S. Appleman

his Attorney

Patent 12

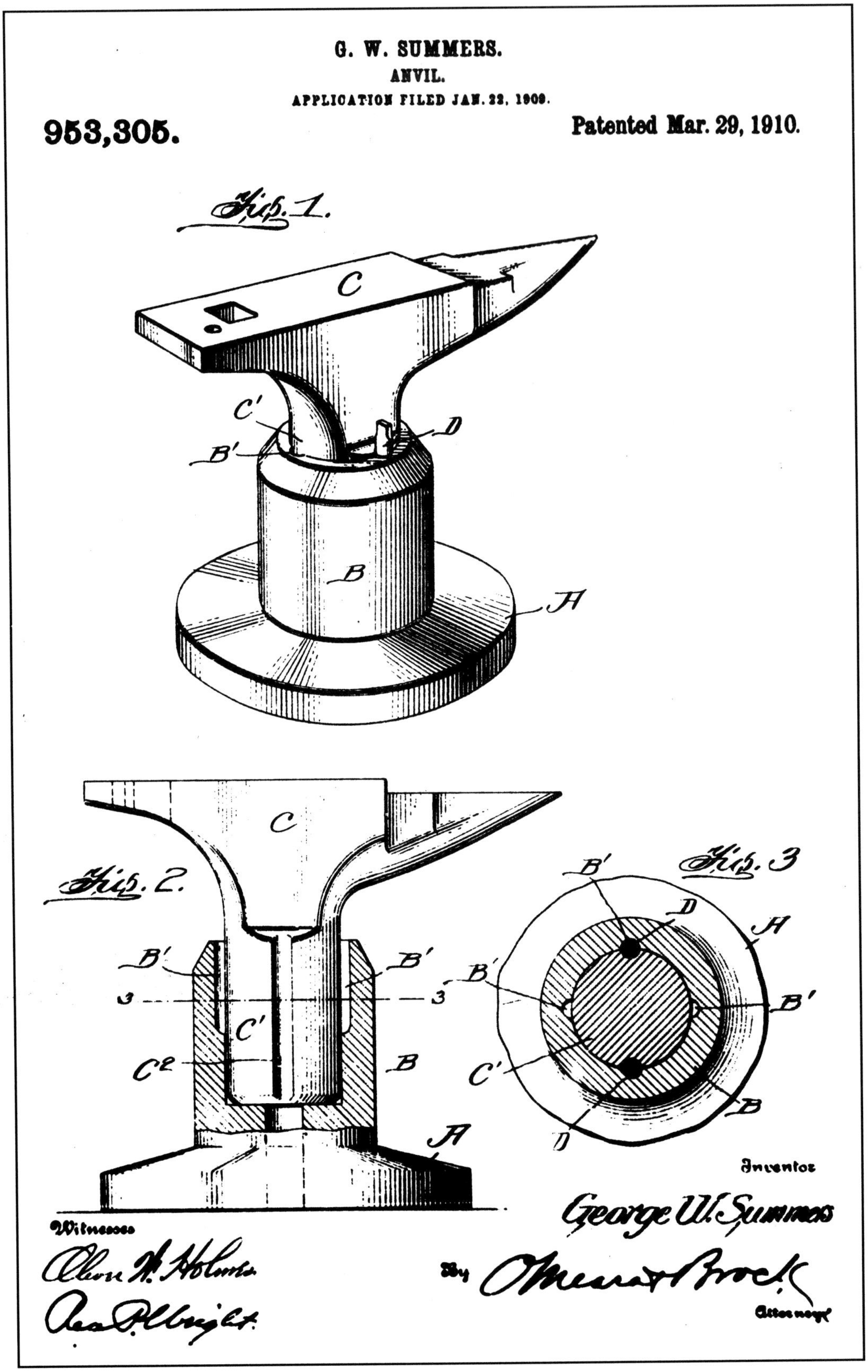
G. W. SUMMERS.
ANVIL.
APPLICATION FILED JAN. 22, 1909.
953,305.
Patented Mar. 29, 1910.
Fig. 1.
C
C'
B'
D
B
A
Fig. 2.
C
B'
B'
3
3
C'
C2
B
A
Fig. 3
B'
D
A
B'
B'
C'
B
D
Inventor
George W. Summers
Witnesses
By
Attorneys

APPENDIX G

Shooting or Blowing the Anvil

I first discovered shooting or blowing the anvil when reading through Alex Bealer's book, *The Art of Blacksmithing.*[9] I quote the piece here.

> *In the old days the anvil was used on holidays such as the fourth of July, or Christmas Day in the South, for a purpose far removed from blacksmithing. Men still alive today [1969] can remember as boys the excitement of 'blowing' the anvil in the street in front of the blacksmith shop on a gala day. Two anvils were needed for this operation. One was laid on the street face down, and the cavity in its base was filled with black powder. Another anvil was then placed right side up on the base of the first so that the cavities of the two coincided. A fuse was then placed between the two and was lit while the expectant boys and men, standing at a good distance, watched with eager anticipation as a small cloud of smoke indicated the progress of the fire through the fuse. Then suddenly there was a roar, a flash of orange flame, and a cloud of white, odoriferous smoke. Ringing merrily the upper anvil was thrown several feet into the air before falling to the ground. The whole exciting business was then repeated as long as the supply of black power lasted.*

It does sound exciting, does it not?

I had a chance to observe the "blowing" of an anvil when I visited Floyd Daniel, the dean of anvil "blowing" or "shooting," in the late 1980s. I asked him to write an article on the subject for this book and he was kind enough to do so.

Shooting The Anvil

> *Shooting the anvil, as we know it , developed from a ceremony honoring St. Clement, one of the patron saints of the blacksmiths. The blacksmiths had three patron saints, St. Eloi, St. Dunstan and St. Clement, as different groups of blacksmiths recognized different saints. All of these were connected to blacksmithing by various feats they were supposed to have performed in connection with the craft. Only one, however, was connected to the shooting of the anvil and that was St. Clement. St. Clement had an anvil as an emblem as it is said that he was martyred by being tied to an anvil and thrown into the sea. We do know that he is commemorated among the martyrs in the Canon of the Mass, but my research does not identify the method of his death. His feast day occurs in the Roman Calendar on November 23. The day of St. Clement, November 23, was observed by blacksmiths as a holiday and was celebrated with explosions of gunpowder on anvils and a procession with an effigy of St. Clement.*
>
> *It is not known when this tradition first began or why the explosion of gunpowder was adopted as St. Clement lived well before the invention of gun powder. My research did not give the date of his birth and death but reveals that he was the third successor to St. Peter and served as Pope from AD 92 to AD 101.*

It is not known when the custom of exploding powder on the anvil developed into using two anvils as we now do. But we know that it was a custom in the early days of our country to 'Shoot The Anvil' on holidays or special occasions. This added quite a bit to the festivities as it is an impressive sight to see the flash of fire, the cloud of smoke and hear the roar as the top anvil is hurled into the air by the explosion.

Almost all of the old anvils were made with a cavity in the bottom. Most of these cavities are handling holes which are approximately one inch square and approximately two inches deep. These handling holes were put into the anvil during its manufacture to allow the workers to move the anvil around by inserting an iron rod into the holes. There were usually two more handling holes in the body of the anvil, one under the horn and one under the heel. By the use of iron rods in these holes the anvils were easily manipulated by the workmen.

To shoot the anvil, one is placed upside down on the ground and the cavity is filled with black powder. The second anvil is placed right side up so that the two cavities coincide. A fuse is then placed between the anvils and into the powder cavity. Some of the anvils had rough bases so that you could not get a close seal between them. In cases such as this, clay was used as a sealant to prevent loss of power during the explosion due to the escape of the gases thru the openings caused by the rough places. Today caulking compound is used for the same purpose. Some anvils, for some unknown reason, were made with a large concave cavity in the bottom which would allow the use of more powder consequently giving a greater explosion. In the early days the emphasis was on the explosion with its resultant flash of flame and cloud of smoke. Today people are more impressed by the height to which the anvil is blown.

My shooting anvils each have a large concave cavity in the bottom which allows for the use of considerable amount of powder. The base anvil weighs 90 pounds and top anvil (or flyer) weighs 80 pounds. I have shot this anvil to a measured height of 104 feet, 6 inches.

This is not a how to do it article and the author and the Madison-Morgan Cultural Center disclaim all responsibility for any injury sustained or caused by anyone attempting to shoot the anvil.

In the early 1990s Gill Fahrenwald sent me a photo of a postcard from around the turn of the century. It shows a group of men standing at a distance while a line of six pairs of anvils are being blown, one after another. The first one has already exploded and the second is just exploding; the top anvil rising about six inches from the bottom. The other four pair have not been fired yet. An interesting observation occurred to me when I studied the photo. The top anvil is face down on the bottom of the base anvil. In both Bealer's and Daniel's description the anvils are placed base to base.

I suggest that the photo reveals the actual way the anvils were placed, not base to base. The reason is just common sense. The first anvil is placed face down, and the handling hole is filled with powder. If the second anvil's handling hole is matched up with the one on the bottom none of the powder is going to be in the hole above. If you want black powder to make the loudest boom possible you want it to be compressed as much as possible. By putting the face of the top anvil over the bottom hole, as the photo shows, this is accomplished.

As Mr. Daniel says, the idea was to make as much noise as possible, not to see how high the anvil would fly.

"The Anvil Shoot." *(Mr. Gill Fahrenwald photo)*

Here is a little piece my wife found in *The Southern Christmas Book* entitled "Blowing Anvils."[10]

> *In these postwar [Civil War] Christmas days " anvil shooting" came into its own. At the country stores on school grounds and at various Celebrations, intent men packed black powder into the holes of one anvil, put another anvil firmly over it and lit the powder. The double boom could be heard for many a mile. What if the anvils were badly damaged? Hadn't they all had a right good time?*

Harnett Kane apparently never saw it happen because blowing the anvil the way it was expedited in the past would not hurt a wrought anvil.

The South was not the only area in which anvil blowing occurred. It was a Northern practice too, but probably on a more limited scale. It was also practiced in part of England, but in a different way. The curator of the museum in our small town was sifting through old ordinances one day.[11] When he saw me later he said he thought he found something in the town ordinances about anvil blowing, and that I should check it out. You bet I did. Here is what I found in the 1877 *By-Laws and Ordinances of the Village of Berrien Springs.*

> *Sec. 5. Every person firing a cannon, anvil, gun or pistol within the corporate limits of the original plat of said village, or within fifteen rods of the outer line of said original plat, without a written permission of the Board of Trustees, signed by the Recorder and countersigned by the President, may be fined not less than one , nor more than five dollars for each and every offense...*

Apparently anvils were being fired at some time in or near the village or the mention of shooting the anvil would not have been in the ordinance. Berrien Springs, Michigan, is a very Northern village. In 1898 there were three backsmith shops in the village, supported by the village and surrounding farms. The 1839 Courthouse Museum in Berrien Springs has an anvil from one of these blacksmith shops, a *Peter Wright*, weighing 125 pounds. Could it have been used in shooting?

Here is part of a letter from England that I received from Mr. Michael Field. He says this about the story he relates in the letter: "... and I can assure you that the following material is absolutely true."

Firing the Anvil

In England it was quite common to find old traditional customs being practiced in their original form in the country well into the [19]30s, their origins being traced back over several hundred years to pagan days. Many of these are well known, the last load of corn leaving the harvest field for instance, the Harvest Festival sheaf of corn in church, the name 'The Green Man' given to many country pubs and many others; but there are some which are not as well known, one being the East Anglian custom of 'Firing the Anvil.'

When people speak of East Anglia they usually mean the counties of Norfolk and Suffolk there in many of the country villages it was traditional in the old days to 'Fire the Anvil' when one of the local maidens got married. In many country villages it is common to find the church built at one end of the street, and the pub at the other, and with the blacksmith's shop in the middle. Following the wedding ceremony the couple would leave the church and walk to the pub for their 'wedding breakfast' passing on the way the blacksmith's shop, and as they did so the blacksmith would 'fire the anvil.'

A traditional anvil has at one end a Beak or Bick (horn) on which the blacksmith shapes his hot metal. Below the beak there is usually a small square hole in the face (of the waist) which was formed during the making of the anvil. Shortly before the wedding ceremony was due to end the blacksmith would fill this hole with black powder heat up a metal rod in the fire and then, as the newly-weds passed the shop,he would ram the red hot tip of the rod into the hole and so 'fire the anvil.'

There is a variation to the method . Quite often if an anvil is turned upside down one will find a depression in the base and some blacksmiths would use this for the powder. Having filled it with the powder he would then plug it with a wooden bung through which a hole had been bored and when it was time to 'Fire the Anvil' he would ignite the powder by applying a lighted taper to the hole -but there was an element of danger involved when using this method as a Norfolk friend of mine remembers.

'When I was a small boy I used to like to watch the blacksmith 'fire the anvil'. On one occasion as I watched he filled the bottom of the anvil with gun-powder and then hammered the wooden plug in. He then discovered that he had forgotten to drill a touch hole in the plug, so as the sound of the wedding march sounded from the church he began to drill a hole in the wood, and I watched as the anvil blew up in his face and killed him outright.

The Big Bang

The following story was told to Mr. John Lemoine by Mr. Chester P. Hood who was an antique dealer in Bellingham, Massachusetts, for over forty years having come from Woodstock, Vermont, in 1943. Chester was born in Woodstock in 1914 and died in Bellingham in 1990. John says that he had but one arm, having lost the other in a farm accident, but could do more with his one arm than most men can with two.

I will relate this story as close as John wrote it with but slight editing and paraphrasing. John says that Chester told him this story in 1958.

There was a certain blacksmith in the Woodstock area in the 1920s who shot an anvil every Fourth of July. This particular Fourth, about 1922, he decided that he would make the biggest bang ever. He over-loaded the explosives (presumably black powder) between the anvils and loaded himself with drinks.[Maybe he loaded himself with drinks first and then loaded the anvils.] Everyone in the area was there as always for the big event and the smith told them that this would be the best anvil shoot yet. "KA BOOM!!" - The anvil flew up and side ways going over the trees heading for town. Everyone was cheering as the anvil went out of sight while they ran down the dirt road, excited and laughing, following the anvil's flight path. Then the merriment stopped. About 1500 feet from where the anvil took off (the measurement was made later) the people saw the hole in the roof of a house and everyone got quiet, that is of course except for the people who lived in the house. They were really excited. The anvil went through the roof of the kitchen, through the ceiling and the kitchen floor into the cellar.

The explosion broke the windows in the front of the blacksmith shop and some on the side. A small buggy close to the explosion was flipped on its side and its roof was torn to shreds. The women in the crowd later complained that they got all dirty from the dust thrown up by the explosion and everyone got dirt in their eyes. [It is a wonder that someone didn't loose their hearing.] Some guy named Burt got his bottle of whisky smashed out of his hand.

No one stayed upset for long, but pitched in to fix the damage to the property, saying that it was the best shoot yet, boy-oh-boy !

The local people talked about the event the rest of year and a few to follow, but there was never another shoot.

John asked Chester what became of the anvil. Chester said it was an old warn out junkie, maybe 100 pounds, but no more and had been used for shooting for many years. After the final shoot the anvil set out in front of the blacksmith shop. Passing kids would always point it out to strangers and tell the story of the grand shoot. For years dogs used it as they would a fire hydrant.

The blacksmith shop closed in the mid-1930s and when World War II broke out someone gave the anvil to the scrap drive. The blacksmith shop remained undisturbed until the 1950s when the tools were auctioned off and the shop burned down shortly afterward.

All in all the men of the area seemed to talk more about Burt losing his bottle from the explosion than anything else connected with the event. Chester told John that Burt never went walking anywhere without his bottle except for that small moment in time. For years men would say, "hey Burt, hang on, hang on."

John says he regrets now not finding out more about the shoot and the name of the town, but he was young and such details do not matter much then.

Mr. Daniel did not tell you how to set up for "blowing," "shooting" or "firing" the anvil, but I know that some of you who have not already tried it are going to do so.

I take no responsibility for what I am about to tell you, in case you blow yourself up, for as Mr. Field says, it can be a dangerous business. The way the Americans did it in olden times seems to me to be safe enough, though. You can get only about two ounces of powder in the bottom handling hole, and all the wrought anvils made had but a small hole in the bottom in most cases. For noise, the old method is probably as good as any.

In the English case the smith who fired the anvil using the hole in the waist probably did not want to unstrap the anvil from the block just to make a little noise for a wedding. Firing loose black power gives more of a "woof" than a "bang." The plugged method used by the other Englishman would certainly have been louder.

Many of the wrought anvils made after 1900, including *Trenton, Arm and Hammer* and *Hay-Budden,* have quite a depression under the base as well as the handling hole in the center. I know you can pack over a pound of powder in some of them. Some of the Swedish cast-steel anvils also have an oval depression under the base. If you are foolish enough to use a later anvil for shooting for height be sure it is a wrought anvil. Stay away from any cast-iron anvil. The metal in a wrought anvil will give some without breaking under an explosion, but a cast iron one will not and if a cast-iron anvil should break when fired it will most likely fly all to pieces with shrapnel going in all directions. A half a pound of black powder will lift an 80- to 100-pound anvil a good distance in the air if set up properly. The best anvil to use for both top and bottom are ones with what I call the "*Trenton* oval." The oval on a 100-pound anvil will hold about one-half pound of powder piled up. The top anvil's oval will compress it making a fairly tight chamber. Ovals are dish-shaped so that once the anvil is fired the explosive force is expended horizontally rather than into the body of the anvil as a straight hole would do. The top anvil should be lighter than the bottom one for best effect when you are shooting for height. The bottom one should be on a very hard surface. If it is on soft ground a half-pound of powder will push a 100-pound anvil several inches into the ground, and the top one will not go very high.

Most shooters shoot the anvil straight up, but I would recommend putting one or two degrees of trajectory on the anvil. At least you will know in what direction the top anvil will go. One going straight up could come down on the bottom anvil, possibly breaking one or both of them and throwing metal around. When a hundred-or-so-pound anvil falls from 50 to 100 feet it is going to put a dent into something, hopefully not you.

Before placing the top anvil on, put in a long fuse, and run a bead of caulking compound around the edge of the bottom anvil. The caulk will also help hold the fuse in place. Press the top anvil down firmly on the bottom one, sealing in the powder.

Light the fuse and make your getaway. When Mr. Daniel shot the one I watched, coward that I am, I hid behind a tree and peeked around it. The noise the blast made was a lot louder than I thought it would be. The bank in the town a mile away called Mr. Daniel to ask if he was shooting anvils again or had something blown up in town. He assured them that all was well.

Therefore, if you have to shoot an anvil, make sure you are in the wide open spaces and have some place to hide when the blast goes off.

Again, I say I take no responsibility for anyone trying to "shoot" the anvil, for the persons involved or the anvils. If done properly it is safe enough, but any explosion or working with explosives involves a certain amount of risk. Enjoy!

Appendix H

Treatise on Anvils from 1842 [12]

This is the complete text of the part mentioned in Chapter 3, Peter Wright and Chapter 4, Fisher & Norris.

Anvils

> *Anvils are made of scrap iron, of common iron, and even cast iron; and their relative value and prices are in the order of enumeration. The latter sort are chiefly used in the forges where iron is made, and for those other purposes in which the metal to be hammered upon them is either not very hard, or not required to be beat out very thin, so as materially to affect the surface or test the solidity of the anvil. In the formation of various articles of steel, and of table knife blades especially, none but anvils of the first rate manufacture are sufficiently firm to endure the hard and incessant beating to which they are subjected. For in the forging of the blades just named, there is a final hammer process called smithing, upon which much of the ultimate goodness of the article depends; and as this process takes place when the blade is comparatively cool and hard, the rebound of the heavy hammers with every stroke is so considerable, that the anvil must be perfect indeed to sustain it. Anvils even of the second description, when used for such purposes as that last mentioned are not infrequently seen to crack across their entire substance; though for general purposes and for exportation, they are in very large demand.*

[Author's note: Between 1830 and 1850 the largest variety of English makes were being imported into the U. S. These second-class anvils probably made up a large percentage of them.]

The blocks out of which iron anvils are formed are prepared at the power forges, and consist of what are known to the trade by the appellation of' 'butts.' These butts are of various sizes, being composed of three or four or more blooms, solidly welded together into a cubical mass: in this state they are transferred to the anvilmaker, who brings them into the shape required by immense labour, welding on large projecting pieces, and cutting off superfluous metal with chisels, according to the design he has in view. Some anvil - smiths, instead of uniting the beak and the quarter as well as the feet, by welding pieces to the block, forge the upper part of the body, including the beak and quarter, out of one piece of iron, and the lower portion, with its projections, out of another piece, as represented below.

These parts, represented in fig.19, are then welded together by the horizontal plane, in consequence of which mode, by avoiding all vertical junctures, the anvils are said to be less liable to fracture, as well as being by this course much simplified in the manufacture.

As it would be impossible either to manage in the fire a red hot anvil, weighing four or five cwt. (and some weigh much more), or to transfer it to the seat upon which it has to be formed merely by the aid of tongs, the workmen manage it by means of a long and stout iron rod, fixed in a hole made for that purpose in the anvil lump. This rod is suspended towards its farther end by a chain to a crane, the arm of which traversing between the hearth and the anvil-seat, enables the men to move the heated mass with comparative facility. This chain, which is suspended from the axle, of a collar, fig.20, moving on the superior surface of the horizontal arm, contains toward the lower end a click iron, upon which a ring is hung, and which, by being fastened higher or lower, elevates or depresses the loop on the chain to the height required. For some purposes, and where great nicety is an object, besides this click and ring there is also a screw regulator, which being turned by means of a spanner, admits of the adjustment of the length of chain with the greatest degree of required accuracy (fig.20). The seat upon which it is hammered is a sort of cast-iron bed, raised about a foot from the floor of the workshop, and having at intervals rising in the edges to prevent the anvil from being driven off during the application of lateral strokes. The fire is excited by the double bellows, made to rise and fall by means of a cross-staff or lever, the motion of which is communicated by a man who turns a fly-wheel, or, when power can be had, by an application to their under boards of that ingenious but simple contrivance the eccentric wheel.

When the iron body of the anvil is perfected in form, a plate of good steel, about an inch in thickness, is welded upon the face, and assimilated at the edges, till it becomes perfectly united with the inferior mass.

In hardening anvils (as well as large hammers) the mass is not suddenly immersed in water, as is the method with articles of steel generally; such a course with so large a body would be unsuccessful, as the ebullition thereby caused would be so great, that the steel would be prevented from coming into that immediate contact with the water which is necessary to produce hardness. Instead, therefore, of instantly plunging the anvil, it is suspended by chains, and lowered by means of a windless into a tank, and a stream of water from above is then suffered to run upon the steel face, until the latter be sufficiently cooled; and, as it becomes tempered by the heat remaining in the body, it is dropped into the water below, and quenched till it becomes quite cold.

The anvils upon which steel articles are to be forged are generally fashioned at the bottom merely in the form of cubes for insertion, by means of wedges, either in stone or wooden stocks: a blacksmith's anvil, on the contrary, as well as others, is formed with a projecting base, and buttresses at the corners, in order that it may stand firmly upon the surface of a stock which is always of wood. The two other descriptions of 'stocks', not in common use, were mentioned in the newspapers a few years ago; the first a female, who suffered a blacksmith's anvil to be placed upon her breast, while a horseshoe was actually turned upon it by the hammer! the other was formed of the butt end of an oak tree, and was fixed in a smith's shop, at a village called Gargrave; from this butt healthy stems and leaves continued to shoot for some time, though both the principal roots and the bark had been removed.

What do these last two items have to do with making anvils, and in a treatise such as this?

Two items stand out to me in this article. First, England was exporting inferior anvils to the United States and other countries and second, anvils were being made in two pieces at least ten years before Peter Wright's patent for a two-piece anvil.

Fig. 19.

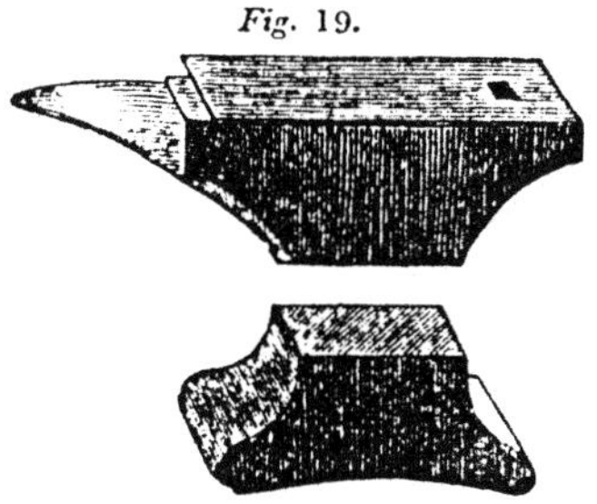

Fig. 20.

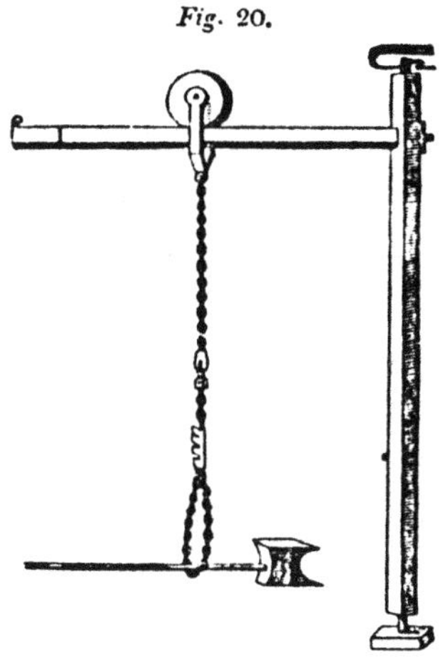

APPENDIX I

Ancient Bronze and Iron Anvils

The material presented here helps establish the fact that most of the features found on relatively modern anvils were on anvils of ancient origin; they are not necessarily features that came about in the past two centuries.

The material comes from two sources; *The Antiquaries Journal* [13] and *Notes on Prehistoric and Early Iron in the Old World.*[14]

The 'Summary' of the article, "The Anvils of Bronze Age Europe" from *The Antiquaries Journal* follows.

About three dozen bronze anvils are known from Bronze Age Europe. They are small in size, probably indicating that they were used for delicate work. Simple, Beaked and Complex forms are distinguished, based on the presence of special features such as work surfaces, beaks, swages and punching holes, which were used in various processes of manufacture and decoration of small bronze and gold artifacts. The anvils are widely distributed throughout north-west Europe, and dated examples belong to the later Middle and Late Bronze Age.(C.1200-700 B.C.) Thirty six anvils are catalogued and briefly described.

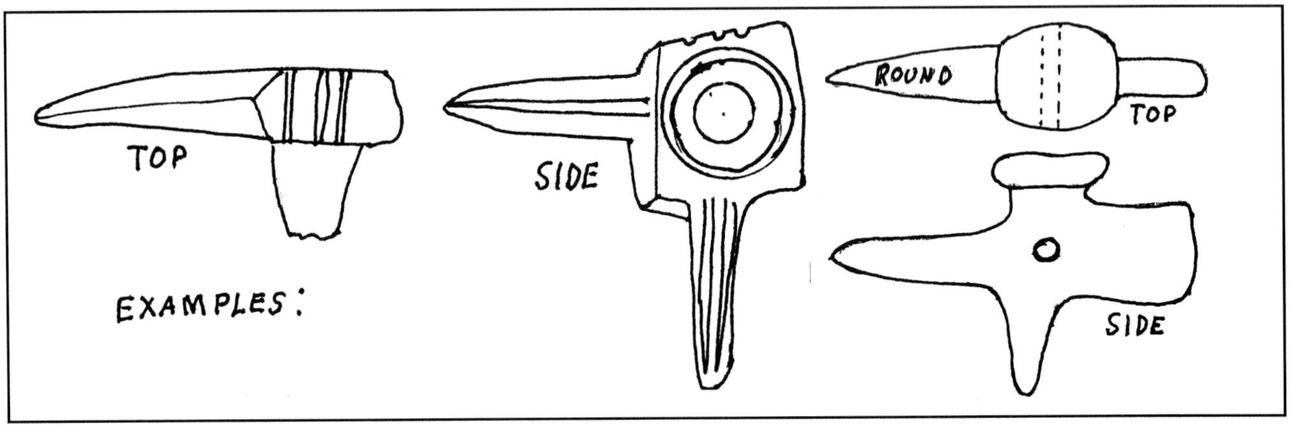

From an article entitled "Anvils" in *Notes on Prehistoric and Early Iron in the Old World* come a couple of interesting items.

Concerning the anvil, we have seen that the stone anvil has been used to the present day, and it must certainly date back to the very beginning form of metal working....

...For nail making, and large iron anvils, we must again turn to the Roman period. The Romans, and smiths working under Roman instruction, forged very substantial objects at times. Such work can be carried out on a heavy stone anvil, but a heavy iron one is better. These big anvils are often merely cubes of iron hollowed out underneath so as to give four feet which rest on the ground. [That would be a considerable anvil indeed.] In general such big tools are hornless, but have an extension of the working table to one side, and a nail heading, or other device, at the opposite side. Such anvils may be up to 30 cm. [almost 12 inches] in height. They are not common, for Ohlhaver cites only five examples. ...

I had guessed that stone anvils were used at various times in my chapter on Materials and Methods of Anvil Manufacturing, but rocks are a dangerous material to use as an anvil because most are porous to some extent and absorb moisture. A white-hot piece of metal laid on a stone could cause the stone to explode. Whatever the case, by the time the New World was discovered an iron anvil could be procured for the smith's needs.

In the next part of the passage cited, there is a drawing of a large Roman anvil with sort of a horn coming off of one end of the face and a pritchel hole going through the anvil's face on the opposite end. The author of this article refers to the hole as being used for heading nails. However, the author, or his or her source of information, is mistaken. Although the hole could be used as such for large spikes in a pinch, in the drawing the hole appears to be a standard 3/8- or 1/2-inch pritchel. The anvil face tapers at this part, allowing the hole to go all the way through. The anvil represented is 8 5/16 inches high and 7 1/8 inches across the face, which is slightly convex. The face continues to taper to the base, to about 3 1/8 inches square. It is my guess that this Roman anvil would weigh close to 100 pounds; not a nail maker's anvil, but a smith's.

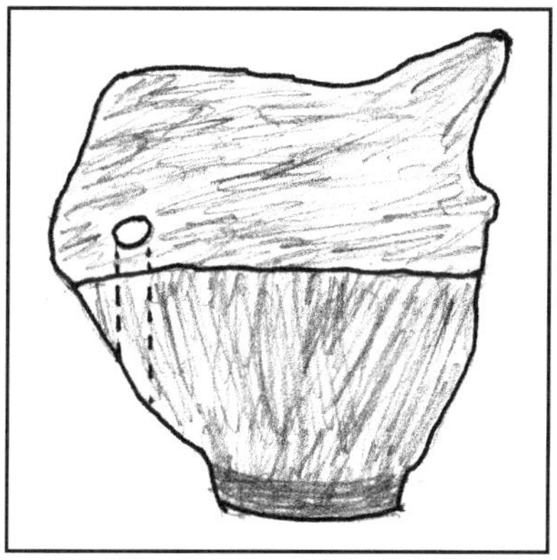

It is also of value to note that the pritchel hole went through a thick part of the anvil, giving the hole support, and not through the thinner part where the slightly flat horn comes off the face, where the heading hole would logically be. It would appear that the pritchel hole preceded the hardie hole, although the pritchel did not reappear until about 1830.

It is obvious from the drawing that this anvil must have fit into some kind of holding device, such as a heavy wooden block for support, as from its design it would not stand alone and be useable.

The article also shows two iron stake anvils with horns but does not give their size.

Appendix J

The Use of the Sawmaker's Anvil

After the circular saw was invented in the 19th century and put into use, it was found that a wavy cut resulted with much binding of the saw blade in the cut. The reason was that when the flat blade hit the wood at high speed, it tended to bend back on itself at the point of contact, thereby creating a waving motion in the blade. To remedy this, the saw blade was tensioned or stressed by hammering in the right places on a cold saw blade using a *sawmaker's* anvil. The larger the blade, the larger the anvil had to be up to a point. With this modification, now when the blade comes up to speed, its centrifugal force puts more tension on the blade, as it tries to force itself back to flatness. This holds the blade stiff as it cuts through the wood, and there is no wavy motion in the blade. Just to look at a circular saw blade, you cannot see this concave or stressed surface which is put in by a very skilled craftsman. Too much tension would create problems worse than none.

I found it hard to believe, but this is the only way, I was told, that the circular saw blades are tensioned today (1988) for blades down to as small as six inches in diameter. I visited a circular saw manufacturer a few years ago and they had four of these anvils in use. Two were *Fishers* and the others *Trentons*. They weighed about 300 pounds each. It is interesting to note that there was a canvas curtain around this area so that the *sawsmith* could not be observed while he worked.

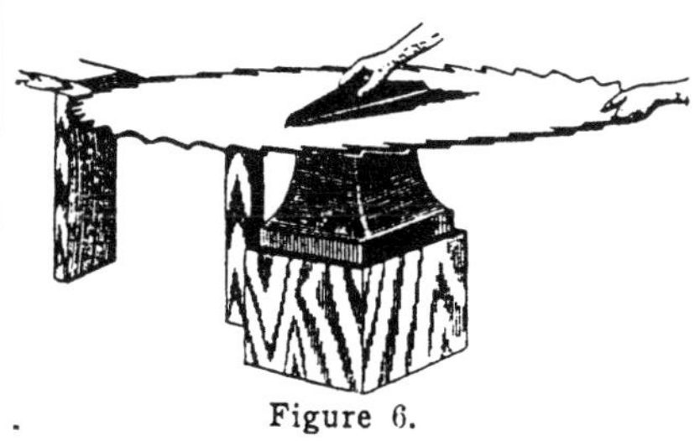

Figure 6.

A circular saw on the sawmaker's anvil. The American Blacksmith, May, 1921. [15]

APPENDIX K

Meteorite Anvils

The two best-known examples of meteorites being used as anvils were found in the southwestern United States. They are called the Tucson Meteorites. They were found being used by the Spanish garrison stationed in Tucson, Arizona, when the U.S. Army arrived and took possession of the area. The first is called the "Ring" meteorite and weighs 1370 pounds. It looks like a rough signet ring. The second is long and somewhat flat, and is known as the "Elephant Ear" or the "Carleton Meteorite." It weighs 633 pounds. General James H. Carleton was the commander of the California Column of Union troops who in 1862 recaptured Tucson from the Confederate forces and stole the "Elephant Ear."[16] A 45-page booklet, *The Tucson Meteorites*, by Richard R. Willey,[17] tells the history of these two meteorites. The "Ring" meteorite has been in the possession of the Smithsonian Institution since 1863. The other belonged to a blacksmith in Tucson for a while after the U.S. Army moved in, but an officer offered the smith a *real* anvil in exchange for the meteorite, and sent the meteorite to San Francisco. Of course, the smith never did get the *real* anvil; (sound familiar?). The Smithsonian Institution finally acquired this meteorite in 1939. The meteorites are both from the same metal mass. For their composition, I quote from the booklet.[18] "The Metallic matrix comprises about 92% of the total volume and consists of approximately 88.3% iron, 9.6% nickel, .9% silicon, .4% cobalt, and minor amounts of chromium, phosphorus, and copper. The other 8% is made up of various stony minerals." Excellent anvil material just fell right out of the sky. I examined and photographed both of them. Unfortunately, I had to lie on my back to see the working surface of the "Ring"meteorite (see photo).

Located on the working surface of the "Elephant Ear" meteorite, on one end, are a 7/8-inch square hole and a 3/8-inch round hole. Because these "anvils" are located in the Gem Section of the Museum, the personnel showing them had no idea what these holes were for so I explained. It must have taken a long time to drill and square up the hardie hole, as well as drilling the pritchel hole. You can even see where someone had tried to saw a piece off the "Ring" meteorite, but without success.

For my money neither of these meteorites would be as useful as a piece of heavy gauge railroad rail for an anvil. The booklet suggested that an anvil in that area in the nineteenth century was worth its weight in gold, as scarce as they were.[19] The book *Southwestern Colonial Ironwork* states that no manufactured Spanish anvils have been found in that area in recent years.[20] The only ones to be found are the smith-made wrought-iron , unplated block anvils. However, there is one steel-faced Spanish (possibly German made) anvil in Sonoma, California, with a date stamped on it of 1766, and of course, the meteorites. Twyla Thomas, of the meteorites section of the Smithsonian, sent me a list of 99 metallic meteorites and of these, eight were known to have been used as anvils.[21]

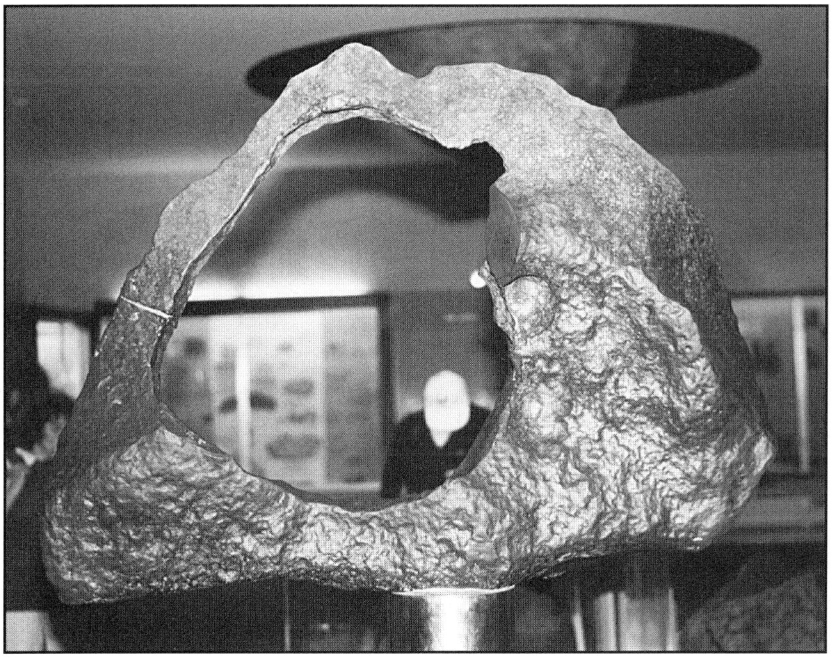

The "Ring" meteorite. At the Smithsonian Institution. (R.A.P. photo)

The working face of the "Ring" meteorite. At the Smithsonian Institution. (R.A.P. photo)

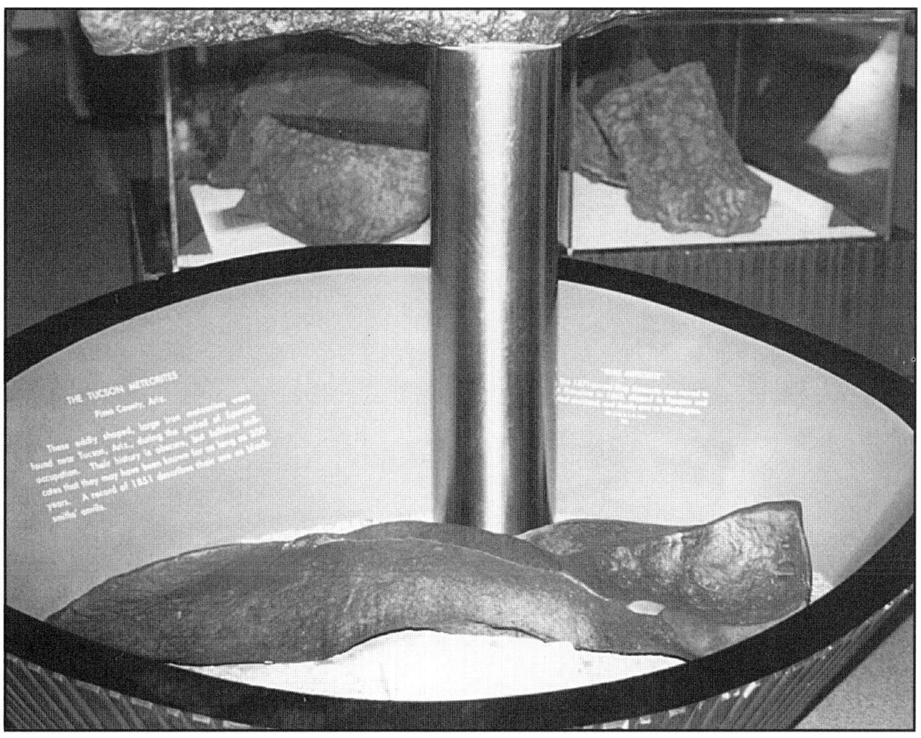

The Tucson Meteorite, "Elephant Ear," beneath the "Ring" meteorite. At the Smithsonian Institution. (R.A.P.)

APPENDIX L

Miniature and Craft Anvils

Miniature and craft anvils are those anvils generally below twenty-five pounds in weight. However, the miniatures generally weigh a pound or less. Many of both sizes look like and have every feature of their big brothers.

The craft or bench anvils generally weigh from five to twenty-five pounds, a few a little less and very few over twenty-five pounds; most are in the ten-pound range. These are mostly made of cast iron, a few forged, and a few of cast steel. I have an eight-pound piece of light gauge railroad track that was made into a beautiful craft anvil by Mr. James Penna's father and given to me shortly before James passed away.

Craft anvils were and are used for such things as riveting, very light forging, bending small pieces by hammering, etc. They seem to come in all shapes, as the photos will show. Most of those made today come from China and that area of the world. School shops have made them and one American major anvil maker. The Columbian Vise and Manufacturing Company, still makes one from eight to ten pounds, of cast iron. Before 1924 the same company, then known as the Columbian Hardware Company, made a ten-pound one that was of forged steel.

The following photo will illustrate various different shapes and sizes.

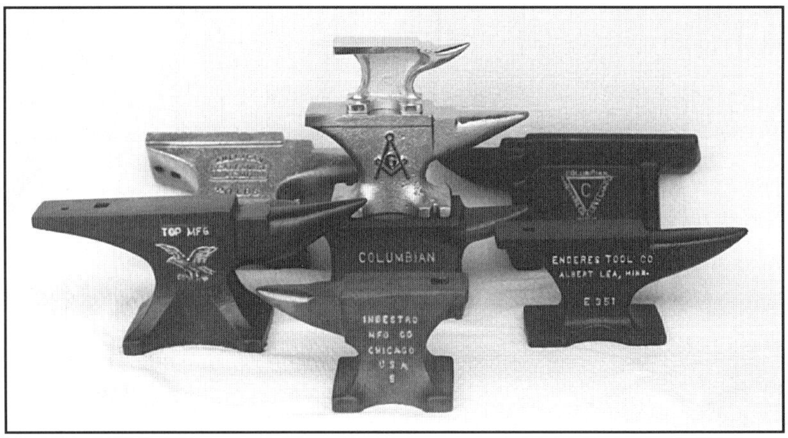

A group of craft anvils. (R.A.P. photo)

Miniatures are mostly meant as paper weights, having some company's advertisement on them; however, many are just plain with no markings at all. Jeweler's anvils on the other hand are miniature-like, but are real anvils and do not fall into this classification. Miniatures were and are used to commemorate some anniversaries or as souvenirs as well as for advertising. They can weigh from a few ounces to a couple of pounds, depending on what material they are made of.

A craft anvil make by the American Scale & Vise Co. Kansas City, Mo. (R.A.P. photo)

There are sub-miniatures of gold and silver on necktie chains and such, but I will leave these alone. Some of these weigh just a few grams or so.

Miniatures can be found made of glass and other ceramics, but the vast majority are of metals, including brass, copper, bronze, pewter, lead, cast iron, forged steel, cast steel and even gold and silver. Some are made of wood, or even ivory, but these are usually one-of-a-kind, carved by some craftsman. I have made a few out of black walnut and red cedar.

Miniatures are probably the most collectable item in the anvil field, not only because they are small and easy to handle and display, but because they come in such a vast variety of materials, shapes, and inscriptions. All the major anvil makers made them, or had them made, for advertising purposes. Although they have been referred to as salesmen's samples, they are not. Some of the anvil makers such as the makers of the *Trenton* anvil put the company's name and forging logo on their miniatures rather than their large anvil's trademark, meaning that the miniature was meant to advertise the forge company and not necessarily the anvil they made. Others such as Hay-Budden used the anvil's trademark.

In the case of Hay-Budden there is a reproduction in circulation since 1992. It is cast in bronze, and is actually nicer than the originals, which were either of plated lead or pewter and brass.

The following photos will give you some idea of the variety of miniatures produced.

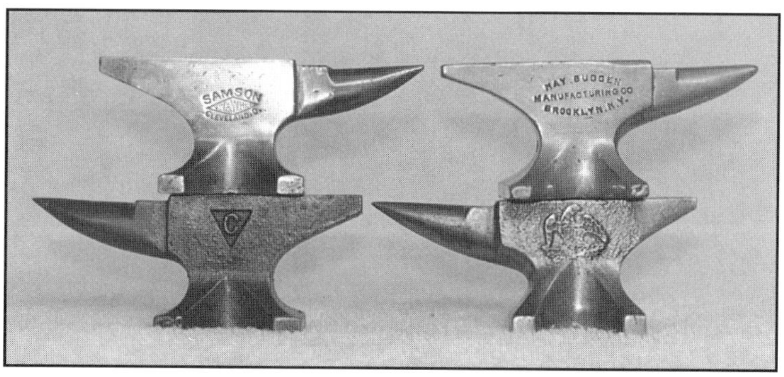

A group of miniature advertising anvils made for anvil manufacturers. (Mr. Gill Fahrenwald photo)

Miniature advertising anvil.
(Mr. Gill Fahrenwald photo)

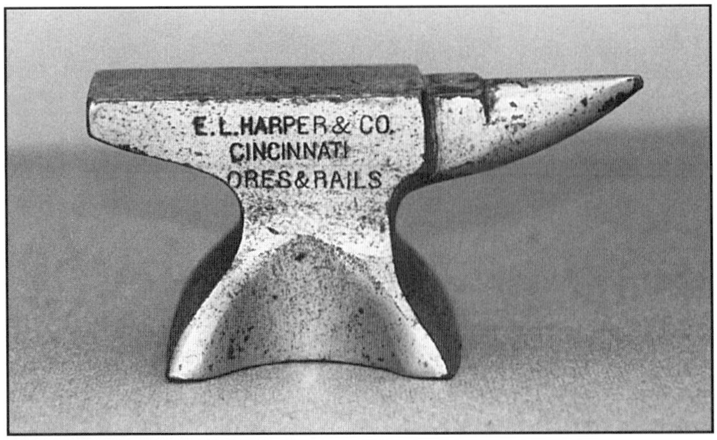

Miniature advertising anvil. (Mr. Gill Fahrenwald photo)

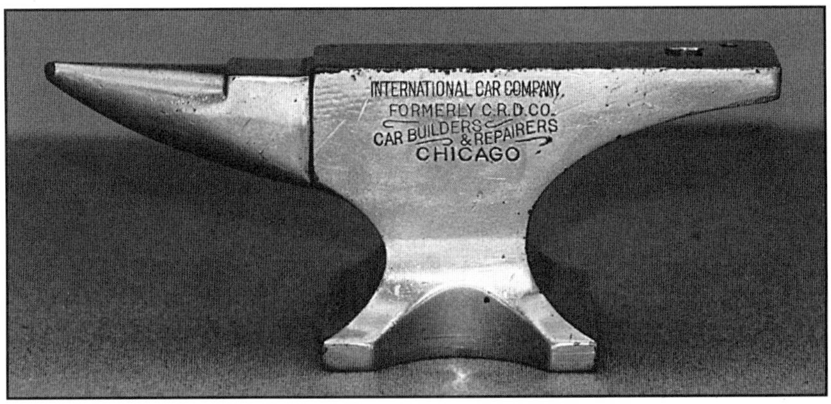

Miniature advertising anvil. (Mr. Gill Fahrenwald photo)

Miniature advertising anvil.
(Mr. Gill Fahrenwald photo)

A toothpick holder miniature glass anvil. These come in several colors. (Mr. Gill Fahrenwald photo)

APPENDIX M

The 1898 Hay-Budden Article [22]

The following article was found and sent to me by Mr. Gill Fahrenwald. It clinched my theory that Walter Ring of the American Wrought Anvil Company had been connected in some way with the Hay-Budden Manufacturing Company. The article is from the January, 1898, issue of *The Blacksmith and Wheelwright* journal.

> *To Whom it May Concern*
> *Brooklyn, N. Y. December 22, 1897*
>
> *The Hay-Budden Manufacturing Co. wish to caution all blacksmiths when purchasing a Hay-Budden anvil to carefully examine the anvil and see that their name is plainly stamped on same.*
>
> *Some time ago they sold a lot of seconds or damaged anvils from which they had entirely removed their name and all marks. These anvils are being offered as first quality Hay-Budden anvils, and this means is taken to post consumers.*
>
> *Hay-Budden Manufacturing Co.*
> *Walter F. Ring, Secretary and Treasurer.*

This article not only tells of the connection between Hay-Budden and Walter Ring, but also tells us that Hay-Budden dumped their seconds on the market. With or without markings someone who is familiar with *Hay- Budden* anvils would be able to spot one of the seconds and think the markings had just worn off, because Hay-Budden never did stamp their trademark in very deep. I rarely find one with a complete trademark and I have looked at hundreds.

APPENDIX N

ANVILS

> *Anvil manufacture in the United States, called a "dying industry" by some companies that have withdrawn from it, gives promise of being revived by the tariff. A flood of foreign anvils preceded the tariff revision in 1930, making it difficult for American manufacturers to compete. More than one producer gave up in despair, turning his attention to other products. Some that persisted found the market better in 1931, and are giving employment today to men making anvils who would be idle if it were not for this protection.*

Imports Reduced Sharply

Imports of anvils, 5 pounds or more in weight, from Jan. 1 to June 17, 1930, when the tariff was 1 5/8 cents per pound, amounted to 328,919 pounds. In the remainder of the year, when the tariff was 3 cents, the imports dropped to 71,168 pounds. For the full calendar year the imports were 400,087 pounds. In 1931, with the duty at 3 cents, the imports fell to 104,621 pounds.

American manufacturers have suffered not only by reason of importations, but also through the increasing use of welding apparatus. Nevertheless, anvils still find considerable use on farms, for repairing equipment; in blacksmith shops; in railroad shops, garages, and structural fabricating plants. One mail order house purchased 160 to 170 tons of anvils in 1930. The amount of steel consumed in this country in the past several years in the manufacture of anvils has averaged about 500 tons annually, according to the best available estimates from manufacturers.

A company in Sweden which claims to have been making anvils for more than 700 years says in its trade publication: "In the old days the anvil body was forged out of soft iron. The face was welded-on blister steel. After the introduction of the open-hearth process several makers cast the body, but the face still had to be welded on." This was part of its description of its anvil cast in one piece.

American manufacturers for some time have been successfully casting anvils in one piece, using alloy steels from electric furnaces, and heat treating. One such company is making all its anvils of vanadium steel.

Some makers forge the top part of the anvil, and weld it at the waist to the base, which may be a steel or iron casting. A face plate of special steel sometimes is welded to the top of the anvil. Some of smaller and lower priced anvils are cast entirely from iron, with the face against a chill, to temper it.

The better class of anvils, sold in large numbers, pass scientific tests, but from time immemorable [sic] the experienced user has applied his own test by hammering the anvil and noting the ring and rebound. A clear, bell-like note indicates to him that it is sound, and an elastic rebound, that it has just the right characteristics. This rebound test, used by blacksmiths when they made knight's armour, has been refined to the point where it is used as scientific measurement of hardness and elasticity.

Nomenclature Persists

Some old and curious names still are used to differentiate various portions of the anvil. The average person can readily see that there is the horn, the face, and the tail, but in between the face and the horn on many anvils there is a step-off, and this is the "blight", useful in certain forming operations. Nearly every anvil has two holes in its face, one square, the other round. The one that is square, to prevent square shanked tools from turning, has been known for ages as the "hardy"; the round one, over which the work is laid for punching, is the "pritchell".

Probably no other implement of industry has found such universal use in the past as the anvil. It has been developed into different shapes in various countries, the national craftsmen preferring their own type. In some countries of Europe two-pointed anvils are used; in some instances, the tail instead of being extended in the same width as the face, is narrowed to half or less the width of the face.

Weights of anvils in this country range from a few pounds to 800 pounds, but the most popular sizes are in the 70 to 450-pound classification, and those used on farms generally are 70 to 100 pounds. So-called anvil blocks, used in power machines, and not included in the estimate of steel consumption for anvils, have been cast as large as 6 feet wide and 7 feet high, 14 feet long, weighing 115 tons.

The article is from *Steel* magazine, November 14, 1932, p.34.[24] It was sent to me by Mr. Bill Hill.

APPENDIX O

Forging Anvils In Germany

The following article appeared in the Spring, 1989 issue of the *Anvils Ring*, the official publication of the Artist-Blacksmith's Association of North America (ABANA) and in the California Blacksmith, November, 1989 issue, an ABANA chapter. The article is used by permission. The article is reproduced in its entirety from the *Anvil's Ring*.

When I first saw this article I could hardly believe that anvils were still being produced by the built-up method in the 20th century. I am told that the photographs were taken after World War II. Figure 10 shows a workman forging using an air hammer which the text says did not come into use until 1931. To my knowledge the built-up method was dropped by all major anvil makers in England well before the 20th century and was never used in the United States. The modern operation this company did perform was to put the face plate on in one piece.

I wonder what became of the hole under the face plate? They must have filled it in or it would most certainly have become a weak spot in the face plate.

- When the text refers to (Fig. 3), where it is printed alone, it should read (Fig. 7) not (Fig. 3).
- The captions for figures 8 and 9 should be reversed.
- Figure 10 caption should read: "Face plate being drawn out on 130 kg. Nazal air hammer."
- These must have been printer errors.

Although this article has very little to do with anvils produced or shipped to this country, it does show in graphic form how anvils were made by the built-up method, therefore its inclusion.

FORGING ANVILS

BY JIM AUSTIN

This article is a report for the California Blacksmith Association in exchange for grant funds extended to Jim for his studies in Germany. The Anvil's Ring wishes to thank the CBA for forwarding this material to ABANA for its first publication.

In Germany, from the early days of its iron technology until the 1960's, anvils were forged by hand in small shops. I visited the very last shop to do this type of work, which is located near Cologne in the Federal Republic of Germany. There the last trained anvil smith still works, hardening modern cast-steel anvils in the old fires of the shop. Since most of the forges and old tools still exist, it is possible to get a rough idea of how this work was done. Moreover, the owner of the shop, Mr. Refflinghaus, was very helpful in explaining the work, which he watched as a boy, and invited me to watch the hardening of anvils. I also

A very old anvil form — similar to the old Berlin form. The two arches are referred to as "Church windows."

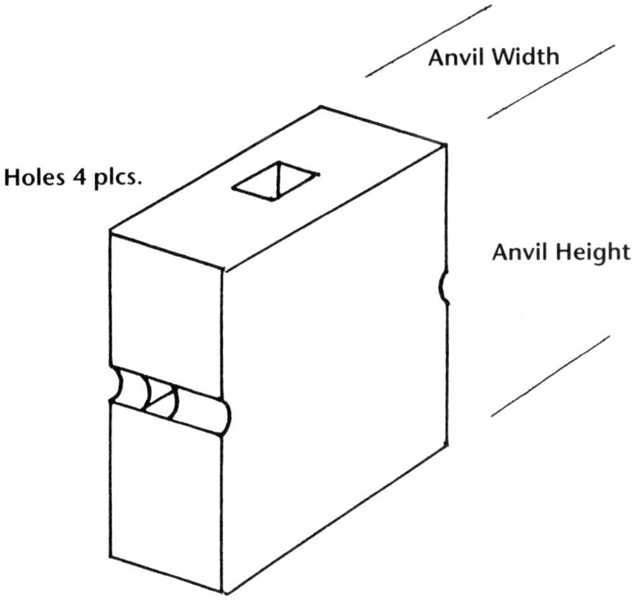

Fig. 1: Central block

Fig. 2: Luppen

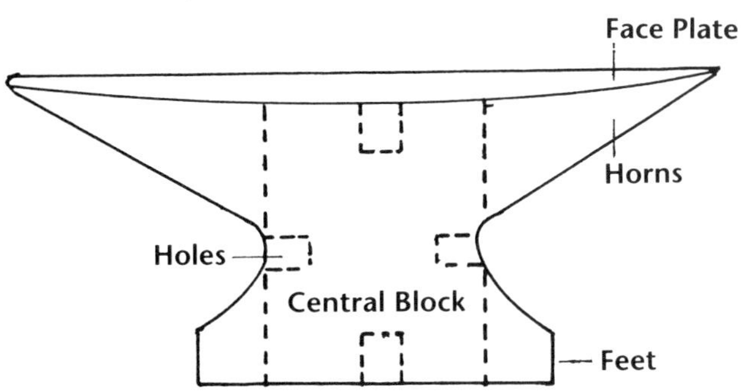

Fig. 3: Double-horned anvil showing welding configuration.

Fig. 4: Anvil parts held together.

brought two old anvils to his shop to have their faces and edges reforged and rehardened.

Anvils were traditionally welded together by a team of strikers out of many blocks of low-grade wrought iron. These blocks were delivered from the wrought iron works to the anvil smithies in various shapes and sizes, and were called Luppen (pl.) The largest block of the anvil was the central block. For a double-horned anvil this block amounted to approximately half of the finished weight. It defined the height and width of the anvil (Fig. 1 & 2). The central block had four square holes in its narrow faces which accepted the rods that were used to manipulate the growing anvil from start to finish. The other features of the anvil, such as feet, horns, clip horns, upsetting blocks, breasts, etc., also started as rectangular blocks and were each shaped and welded separately to the central block (Fig. 3 & 4).

The anvil forges (two for each anvil being made) were located on opposite sides of the shop, about 25 feet apart. Each consisted of a semicircular base, about 5′ across and 8″ high, onto which the coal was heaped (Fig. 5). Behind this was a free-standing brick wall braced with iron and surmounted by a hood to draw off the smoke. Through this wall, somewhat above the level of the base, came a cast iron tuyere with an air outlet in the form of a horizontal crescent about 1 3/4″ wide and 1/2″ tall. As the tuyere iron burned off it was pushed further through the wall. The air blast was created by an electric blower and controlled by means of a large stopcock.

On the right side of the wall stood a pivoted boom of heavy iron construction called a "Wensuhle", capable of taking the weight of a heavy anvil, say 400 kg. The boom was used to suspend the anvil in the fire and swing it to the forging block about four feet away. This block consisted of cast

Fig. 6: Tacken close-up.

Fig. 5:
Anvil forge overall.

iron about 7" thick, with raised edges on three sides, and was called the "Tacken" (Fig. 6). On it the anvil could be positioned in a wide variety of ways to complete the many welds.

To start an anvil, a long iron rod called an "Ambossbaum" (anvil tree) with a tapered square end was driven into one of the square holes of the central block and fixed with little wedges. The rod was slung on a hook and chain coming from the boom of the forge. By sliding the chain along the rod close to the central block, the heavy block could easily be balanced, rotated and swung using the lever arm of the rod. The block was brought to the forge just above the tuyere and positioned so as to heat the lower corner where, first of all, a foot would be welded on. The block and rod arrangement was held in a fixed position by hanging a weight on the end of the rod and putting a crossbar through a hole in the rod to keep it from rotating (Fig. 3). The block was partially buried with coal and heated.

Meanwhile, in the opposing fire, a small block for the foot was brought up to heat for welding to the central block. To judge the heat of the central block, a long thin iron rod was poked through a specially constructed channel leading from the core of the fire under the tuyere to one side of the base of the forge. At welding heat liquid iron dripped from the central block and collected in a small pool at the head of the channel, and shot out of the base of the forge when the channel was thus freed. At this point, the central block was heaved from the fire and laid in the correct position on the Tacken (cast iron block) with one smooth and lightening-fast motion. The foot block was simultaneously brought from the opposite fire and welded in place by heavy blows from a team of three to five strikers. Great care had to be exercised in laying on the foot block since it stuck fast to the central block in whatever

Fig. 7: Weight and rod on Ambossbaum.

Fig. 8: Rough horns notched for rod and with rod welded on.

Fig. 9: Central block with rough horn set on top.

position the two pieces first made contact. The welding of the foot being complete, the strikers continued to forge the foot to its final shape. Sometimes additional small pieces, called "sausages", were glut-welded into the inside corners of the joint to give the foot a smoother and more pleasing shape.

After both feet had been welded and finished the horns were prepared for welding to the central block. The blocks for the horns were first fullered and welded to rods for their manipulation during forging and welding (Fig. 8). Then the horns were roughly drawn out to their final shape under a large power hammer (water-driven until 1931, a 130 kg. Nazel hammer thereafter). Again the central block went into the fire to be brought to a welding heat simultaneously with the horn. To weld a horn the central block was brought to the Tacken and the horn set on top pointing straight up (Fig. 9). The first blows were applied to the tip of the horn and the rest so as to close up the weld seams. To buttress the welds of the horns, which occasionally and most distressingly broke off, vertical strips of iron called "Bandhitte" if flat, or "Kerben" if wedge shaped, were welded over the seams of the horns on both sides.

After the feet, horns, clip horns, upsetting blocks, etc., had been welded on in analogous fashion, the anvil was ready for the hardest step of all, the laying on of the Swedish steel face. Flat stock for the faces came in bars up to about 3/4" in thickness and about 6" in width. (An analysis of such material after the war gave the following results: C 0.63-0.68%; Si 0.15-0.25%; Mn 0.4-0.6%.) The face plates were cut and forged from this stock so that they completely covered the face of the anvil from tip to tip (Fig. 10 & 11). They added great strength to the horns by bonding them to the central block with a continuous layer of steel, which was subjected to ten-

Fig. 10: Face plate placed end-on in this chamber.

Fig. 11: Two face plates.

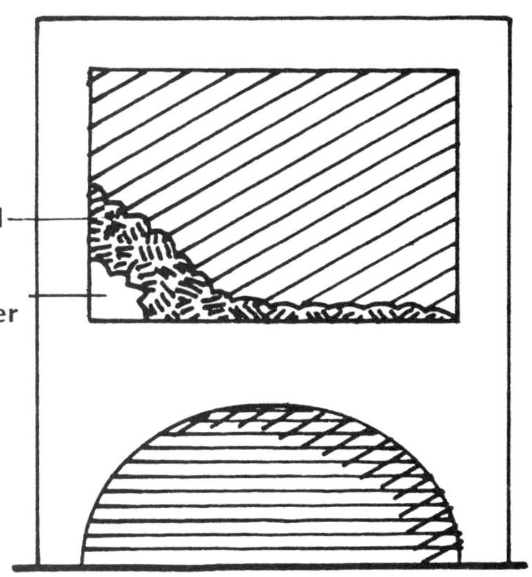

Fig. 12:
Front view of forge
for faceplate welding.

Fig. 13: Worker laying on anvil face.

sile stress in the anvil.

Over the body of the anvil the plate retained its full thickness and became progressively thinner toward the tips. A rod was welded to one tip of the plate for manipulation. The plate was placed in a specially constructed forge in the shape of a large box with an open front. Here the coal could be banked along one side so that it formed a long, low chamber when it burned out (Fig. 12). The face plate was heated in this chamber until it was dark red, then a mixture of anhydrous borax and high-carbon steel chips were heaped on it to give a layer about 1" thick. The plate was returned to the forge and brought to a welding heat in the middle section. Any section which threatened to burn was touched by a thick, long-handled iron spatula to shield it briefly from the flames and absorb some of the heat from it.

During this time the anvil body was heated in the corresponding section. When the two pieces were up to heat they were brought together on the Tacken. The anvil was placed upright and the steel face was carried to it with the chips and borax on top. It was carefully tipped on edge and the lower edge positioned over the corresponding edge of the anvil (Fig. 13). Then it was pivoted on this edge and dropped precisely onto the anvil body, whereupon strikers welded it in the middle with heavy, precise blows in a well-practiced rhythm and pattern (Fig. 14 & 15).

After the middle section was welded the plate was welded successively to each horn (Fig. 16). Here the anvil was placed in the fire so that the blast from the tuyere acted upon the bottom of the horn and did not burn the steel plate. During the course of laying on the face the top hole of the central block was covered. After the steel face was welded the anvil was ready to receive the square and round holes. These were

Fig. 14: Welding face in the middle.

punched and drifted through from the top, and cleaned up on the bottoms of the horns with curved drifts. The last step of forging the anvil was to level the face of the anvil with strikers' hammers and flatters. As the anvil cooled it was "binked", hammered and peened on the surface to even out the texture and close up porosity left by burned away metal.

When the anvil was cool it was ready to be ground on the face. To do this it was suspended on chains in front of a water-driven sandstone wheel about six feet in diameter and about one foot thick. Here it was ground flat and the edges trued by eye (Fig. 17). The process was finished with the two-man file, a heavy, coarse file about five feet long with handles on both ends. Later this work was performed on a planer.

The last major step in the production of an anvil was, and is, of course, to harden the face. To do this it was suspended upside down in the forge with the use of the boom and anvil tree (Fig. 18). The fire was banked and the draught controlled so that the anvil face was heated evenly from tip to tip to just over the critical temperature. This took about one hour of constant attention and fire tending for a 150 kg. anvil, and was quite a feat considering that such an anvil was over three feet long. The heat generally penetrated about 3" into the body of the anvil.

When the anvil was ready to be quenched things moved at a fast pace and a certain tension was present among the workers. After all, if the difficult welding of the steel face had not been performed perfectly, it might separate partially or completely during the quench, thereby ruining the anvil. While the anvil was still in the fire the water chute from the company millpond was opened and a solid 12" square column of water with an eight foot head gushed into the shop, falling with great force straight down into a sunken trough

Fig. 15

Fig. 16:
Welding face on horn (shoulder).

Fig. 18

Fig. 17: Anvil grinding.

against the old stone wall at the end of the shop, and there churning out through a hole into the millpond of the next company downstream. The anvil was heaved out of the fire and set upright near the quenching trough with a lightening-quick motion. A dish of water was dashed over the face to lift scale, which was instantly cleaned away with a broom of twigs and a heavy file. The anvil was then "walked", by means of a rod inserted into the square hole, to the edge of the trough, where a ramp lead to a spot underneath the water column.

A steel cover in the shape of a large flat-headed nail was placed into the round hole of the anvil. This kept the water from pouring through the hole and quenching the core of the horn. A long crook was then placed in the square hole of the anvil and it was slid down the ramp to a position exactly under the column of water from the chute. The quenching of the face happened very rapidly and was allowed to continue for a few minutes. The anvil was then pulled out of the trough with a winch and left to dry from its residual heat. Finally, the anvil was polished on the face and its weight, manufacturer and year of manufacture were stamped on the side.

The anvil smiths labored under extreme conditions at a job requiring both endurance and great skill. Their leather aprons were burned stiff from the heat of white-hot anvils. The only protection for their faces and eyes were the brims of their hats. Thus it is perhaps not hard to understand that the trade died out, not due to economic reasons, but because young people could not be found to learn this dangerous and difficult work. Today it can almost only be appreciated in the anvils of older workshops in Germany where these pieces are still prized by traditional blacksmiths and considered the tool of all tools.

APPENDIX P

Updated Information As of 1997

The manuscript for this book has been in the process of being written since 1990. With the information that I had at the time it appeared that the manuscript would produce a book of about two hundred pages. I finally closed the manuscript at the beginning of 1997 with about five hundred pages. This project would never be completed if the information that keeps coming in is added. However, since the beginning of 1997, important information has come to my attention which will add to and correct material already in the manuscript and I felt that it must be included. At this stage of the project new material other than a few word changes cannot be added as it would disrupt the whole page numbering etc. Therefore this information is added as an appendix.

English Anvils -Additions to chapter 3

MOUSEHOLE

It has been brought to my attention by Mr. John Hatfield (owner of the Mousehole Forge site) that the forge produced anvils as heavy as 9 cwt. (1008 pounds). I had assumed that with the equipment available at the forge a five hundred pound anvil was about the largest that could be produced.

Mr. Hatfield also sent me a copy of an anvil order memo to the Mousehole Forge that backs up the 9 cwt. unverified statement . In 1848 a company ordered six large anvils from the Mousehole Forge. One of the Armitages wrote up the order as follows:

Wanted immediately for Marshes of Shepherd, Sheffield. Order 197.

Next followed four crude sketches of anvils from the waist up with dimensions. The overall length of these anvils was 44 1/2 inches. The face was to be 22 1/2 x 6 1/2 inches. The sketches give a top and side view of two slightly different anvils, "A" and "B". Below the sketches it reads as follows:

All with 1 1/2 [inches] Square Holes with 5/8 in. Round holes. One square Hole to be countersunk in each.

5 Anvils to Sketch B *4 to 5 ?.* *Tops to be level from end to end.*
1 Anvil to Sketch A

Messers Armitage
Mouse Hole *Guaranteed for 12 Mos.*
? June-1848

The sketches showed two hardie holes in each anvil and the location of the hardies and pritchels plus other information on horn sizes and how the anvils were to be steeled.

If these anvils have a standard base for *Mousehole* anvils made at that time they will weigh six to seven hundred pounds or more.

In M.T. Richardson's book *Practical Blacksmithing* he mentions a 960-pound English anvil being displayed at the 1876 Centennial Exhibition in Philadelphia, Pennsylvania, but does not mention the maker. I assumed that the 960-pound anvil must have been in the Peter Wright display; although a photo of the Peter Wright display does not show an anvil of that size. It is very probable that the 960-pound anvil was in the Mousehole display.

PETER WRIGHT

Here is a selection from an English book entitled *Walks in the Black Country and its Border-Lands,* by Elihu Burrit, 1869, pp.144-145.

> *....Perhaps the distinctive speciality of Dudley is Wrights Anvil and Vice factory.* [Ed. note: Vice is the correct English spelling not vise]. *The anvil business has been carried on by the Wright family for 200 years. They probably have sent more anvils to the United States within this period than all other English makers put together, and there are few blacksmith's shops in America in which their name is not well known. During the last year, they turned out nearly 11,000, and also 9000 vices. The present head of the house, Peter Wright, introduced, some years ago, a great improvement, for which he obtained a patent. It simply consists in making the anvil of one solid piece of iron; whereas, by the old system, the different parts were made separately and then welded together. This was a difficult and unsatisfactory process, for frequently the weld would not be perfect in some place, and the hammer and sledge would ere long find out the defect, for the anvil would ring like a cracked bell under their strokes, and after awhile the horn or beck would go sheer by the board. The improvement in making them out of a solid block of iron is a very valuable one indeed, remedying all these defects of the old system. To accomplish this, the grains or threads of the iron, as Dr. Johnson would say, are 'reticulated with remarkable complications.' To use a simile which may help some to get an approximate idea of the process, a ball of iron wire as large as a bushel basket is welded in a solid mass; then that is again drawn into thin strips, which are again folded up and welded again,and hammered until a block is formed of the utmost tenacity of which the metal is capable. When the anvil is worked out to its perfect shape, as the French Marshall said of the old Imperial Guard, 'Elle meurt mais ne se rend pas;' it may wear out but never breaks. This is not however the exact process; I have used the ball of wire merely as a simile. The raw material is old scrap iron, like old horse-nails, hoops, and the like, that have been passed under the friction of wear and thus been purified and solidified for their new field of usefulness..."*

This portion of the book was sent to me by Mr. Vance Burns, who received it from England.

Mr. Burrit's simile certainly was a poor one. Here again is the mention of the Wrights being in the anvil business for 200 years. This would put their starting date at least back to 1669. As mentioned in the section on Peter Wright the oldest anvil of that make that I am sure of goes back only to 1830, at most. Maybe they did not put their name on their anvils before 1830.

In other material sent by Mr. Burns there is a listing of anvil makers in old directories as follows:

In the 1830 trade directory of Pigot and company, Dudley, there is no listing of a Wright anvil maker, although Wilkinson is listed. In the Robinson Guide of *ca.* 1839 Peter Wright is listed as a vice and anvil maker located on Prospect Row, Dudley. In the Post Office directory of *ca.* 1845 Peter Wright is listed as a vice and anvil maker located on Constitutional Hill. Constitutional Hill, Dudley, is where the company remained until going out of business as far as I know.

Slater's 1850 directory lists Peter Wright as makers of anvils "& vices with patent solid box." Peter Wright is listed in the Birmingham Post Office directories for 1854 and 1860. In the 1864 Post Office directory Peter Wright is listed as follows:

> *Wright, Peter & Sons, vice & anvil manufacturers & sole patentees of the solid box vice, solid horn anvil & parallel vice, manufacturers of railway carriage wheels & axles, Constitutional Hill Works & Broadwell Forge Oldbury.*

Where was Wright located before 1839? If they were anvil makers in about 1669 or 1700 they could not have moved very much. They probably leased various forges before moving to the Constitutional Hill works.

In an article written in the September 1903 issue of *The American Blacksmith* it gives almost word for word the same material appearing in an article in the July, 1904 issue of *The Blacksmith and Wheelwrigh*t (see chapter 3, Peter Wright). *The Blacksmith and Wheelwright* gives the source for the article, but *The American Blacksmith* does not. The significance of *The American Blacksmith* article is the last paragraph.

> *The last advices are that a change has just taken place in the ownership of the above firm,* [Peter Wright & Sons], *Mr. Isaac Nash of Belbroughton, the well-known manufacturer of Walden Scythes, having purchased the business outright. The manufacture of the anvil, however, will be carried on at Dudley, precisely as before by the new proprietors, who have had long experience in the same trade, and whose anvils have attained a high reputation in England and on the Continent. The two businesses of this latter firm will, of course, be conducted independently.*

It was thought by some as mentioned in chapter 3, and rightly so, that Peter Wright was bought out by Wilkinson. Could Wilkinson have bought Peter Wright & Sons from Isaac Nash?

WILKINSON

In chapter 3, in the section on Wilkinson anvils it gives the impression that *Wilkinson* anvils were produced by one manufacturer, J. Wilkinson or Joseph Wilkinson or J. Wilkinson & Sons. The truth is there were three and sometimes four Wilkinsons manufacturing anvils in Dudley, England, and at least two of them were exporting anvils to the United States.

From the material sent to me by Mr. Vance Burns, there were in the 1830 directory three different Wilkinson companies manufacturing anvils. They were: James Wilkinson, Junior, Tower

Street, Dudley; Joshua Wilkinson, Queens Cross, Dudley; and Joshua Wilkinson & Sons, Free Bodies, Dudley. In Robson's Guide of *ca.* 1839 they are listed as: J & A Wilkinson, Vice Handle Maker, Castle Street, Dudley; J. Wilkinson & Son, Vice & Anvil Maker, Dixon Green, Dudley; and J. Wilkinson, Vice & Anvil Maker & Gas Tool manufacturer, Queens Cross Dudley.

In the Post Office directory of *ca.* 1845 four Wilkinsons are listed as vice and anvil makers. They are: James Wilkinson, Castle Street; Joseph Wilkinson, Spring Gardens; Joshua Wilkinson Junior, Freebodies; and Joshua Wilkinson Junior, Vice & Anvil Maker & Gas Tube manufacturer, Golden Cross Inn & Queens Cross, Dudley.

All of these Wilkinson companies were not necessarily in Dudley as some of the directories were from Birmingham which is about ten miles east of Dudley. However I assume that they were all in Dudley as the streets or sections of the town do not indicate otherwise.

By 1864 there were three Wilkinsons producing anvils. They were: Joshua Wilkinson & Sons, Queens Cross, Dudley; Joseph Wilkinson, Spring Gardens; and Joshua Wilkinson, St. John's Road, Kates Hill.

The two anvil makers who, I am sure, shipped anvils to America are Joshua Wilkinson & Sons of Queens Cross, Dudley; and Joseph Wilkinson of Spring Gardens, possibly Dudley, as I have recorded numbers of both anvils (see chapter 3).

What was assumed by some to have been "swords" or "cannons" as part of the trademark of "J. Wilkinson & Sons, Queens, Dudley," I suggest are neither, but a type of cross representing "Queens *Cross*, Dudley." I have never found the word "Cross" as part of this Wilkinson trademark and crossed ovals represent that word. Also, it seems that none of the Wilkinson anvil makers made cannons, or swords as I mentioned in earlier parts of this book. The cannon and sword makers were other Wilkinsons.

The anvils I have recorded with "Joseph Wilkinson, warranted" as the trademark appear to be much older in style than J. Wilkinson & Sons. Probably by the 1890s or earlier only J. or Joshua Wilkinson & Sons were producing anvils in England because their latest trademark says only "Wilkinson" (see chapter 3). Also the few times Wilkinson is mentioned in any writing or advertisements in the late-19th and early-20th centuries only one Wilkinson is indicated and that is at Queens Cross, Dudley.

It would seem that further research is needed in reference to the Wilkinsons.

Additional *Old English* anvils to be added to the list.

- BOOTH & CO.
BOYD MILL
SHEFFIELD
0 • 3 • 16 (100lbs.) — This anvil does not have a pritchel hole and is sharp under the horn where it joins the body. It may have been made by the Mousehole Forge.

- SANDERSON
SHEFFIELD
0 3 23 (107lbs.) — This is a typical *Old English* style anvil and is also sharp under the horn where the horn joins the body. Shortly after locating this anvil I had a phone call from a person who had acquired an anvil with the same trademark.

HANDLING HOLES IN ANVILS - Addition to chapter 7.

The following information comes from Mr. John Lemoine.

All wrought anvils made before the 20th century have at least three handling holes. The early *Fisher, Eagle* cast anvils have two handling holes, one in the waist under the horn and one in the waist under the heel as does the *American Star* anvil. A 378 pound *Fisher* sawmaker's anvil dated 1921 has three handling holes. It seems apparent to me why wrought anvils would need handling holes to necessitate their manufacture, but why would handling holes be necessary in a cast anvil? Many of the operations performed on wrought anvils require manual manipulation whereas manual handling of a cast-iron anvil with a steel face would only require manual handling during heat treatment, if then.

The answer to some of these questions come by way of Mr. Lemoine. He knew a Mr. Charles Krull who was a blacksmith and who knew a Mr. George Bowles. Part of Mr. Bowles work, was as an anvil repairman for Douglas Axe Works in Douglas, Massachusetts. John says he bought his first anvil in 1948 from Mr. Krull. It was a 150-pound *Hay-Budden.* Mr. Bowles passed on much of his knowledge of anvil repair to Mr. Krull who in turn passed it on to Mr. Lemoine. Mr. Krull even had some of the tools used by Mr. Bowles and John saw these.

John says the Douglas Axe Works used many anvils of all makes, regular smithing anvils and hornless anvils. A hornless swordmaker's style anvil is probably what an axemaker's anvil looks like. The Douglas Axe Works repaired their own anvils using methods they had learned from the anvil manufacturers. They reworked, refaced and heat treated their anvils just as the manufacturers and those specializing in anvil repair did (see Smith & O'Leary).

The hole in the bottom of the anvil was the most important, John says. The anvil was held by a special pair of three handled tongs. The third leg was straight and extended into the hole in the bottom of the anvil. The other two clamped the anvil. These tongs were about seven feet long and were used to move the anvil "in and out of the fire on wooden rollers that caught on fire at times." They used wood because it was not slippery, John says. The anvils could also be rotated by this means to get an even heat on the face.

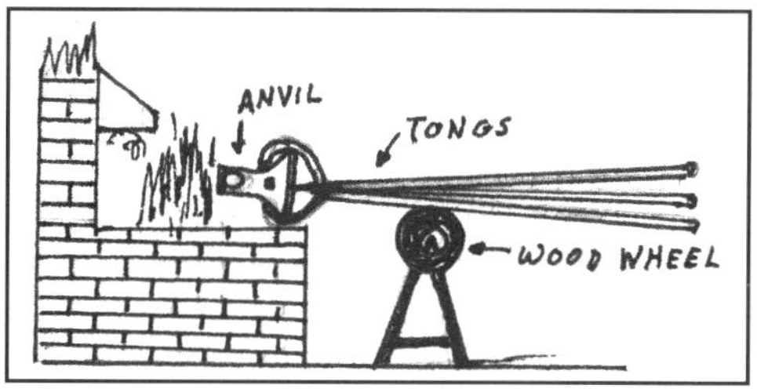

Three handled tongs in use.

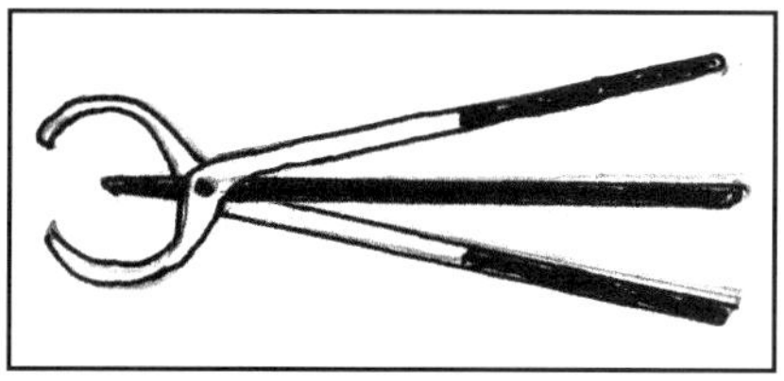
Three handled tongs

These three drawings are adaptations of John Lemoine's originals.

The hole in the front or back of the foot as in the case of large *Hay-Budden* anvils was used with a special tool to tip the anvil as it was being ground face down on the large grinding wheel (see *Peter Wright).* From about 1850 on *Peter Wright* anvils had all four handling holes

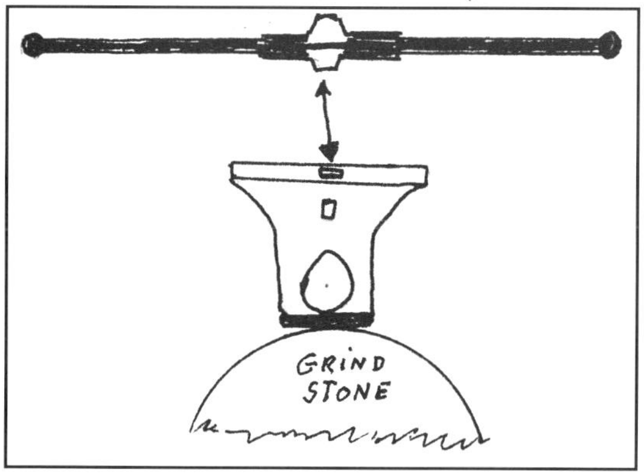

Special tool used with the anvil for grinding the face.

and the method used was similar to the manner described. All *Hay-Budden* anvils had at least one handling hole in the bottom. Both *Trenton* and *Arm and Hammer* anvils started out with four handling holes, but soon the companies found methods for manufacturing and manipulating their anvils without these holes.

Fisher probably had tools similar to these tools which plugged into the holes for final heat treating and possibly for grinding. Everyone who knew exactly how all of these handling holes were used by the different anvil makers has long since passed and John Lemoine is probably the only person living who has seen the tools depicted here.

Modern American Cast Anvils - addition to chapter 4.

I learned of the following anvil maker late in 1997, too late to be included.
The drawing does not show the hardie or pritchel holes, but I assume that they are provided.
Retail price as of October, 1997 is $551.25.

CALIFORNIA ELECTRIC STEEL

PATTERN MAKERS · FOUNDERS · MACHINISTS

PHONE: 209-736-0465 FAX: 209-736-9014 P.O. BOX 817

ANGELS CAMP, CALIFORNIA 95222

CALIFORNIA ELECTRIC STEEL IS THE OLDEST CONTINUOUSLY OPERATING FOUNDRY WEST OF THE MISSISSIPPI. IT WAS ESTABLISHED IN 1854 TO SERVICE PRIMARILY THE MINING INDUSTRY AND OTHER BUSINESSES IN THE MOTHER LODE.

BACK IN THE EARLY DAYS OF THE FOUNDRY, CALIFORNIA ELECTRIC STEEL ALSO OPERATED (AS PART OF THE FOUNDRY) A PATTERN SHOP, MACHINE SHOP, AND A BLACKSMITH SHOP. IN MEMORY OF THAT TRADITION WE ARE INTRODUCING OUR OWN LINE OF ANVILS, STARTING WITH OUR MODEL 200.

FOLLOWING IS A DESCRIPTION OF THE MODEL 200.

FINISHED WEIGHT: 188.0# MATERIAL: 4135 STEEL

HEAT TREATED TO: 395-425 BHN (ROCKWELL 42-45c)

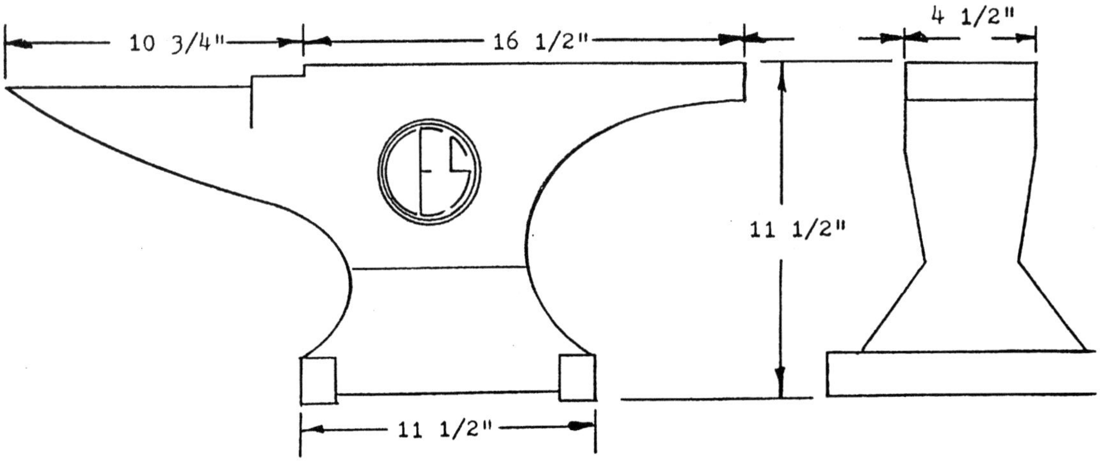

MODEL 200

Chapter Notes

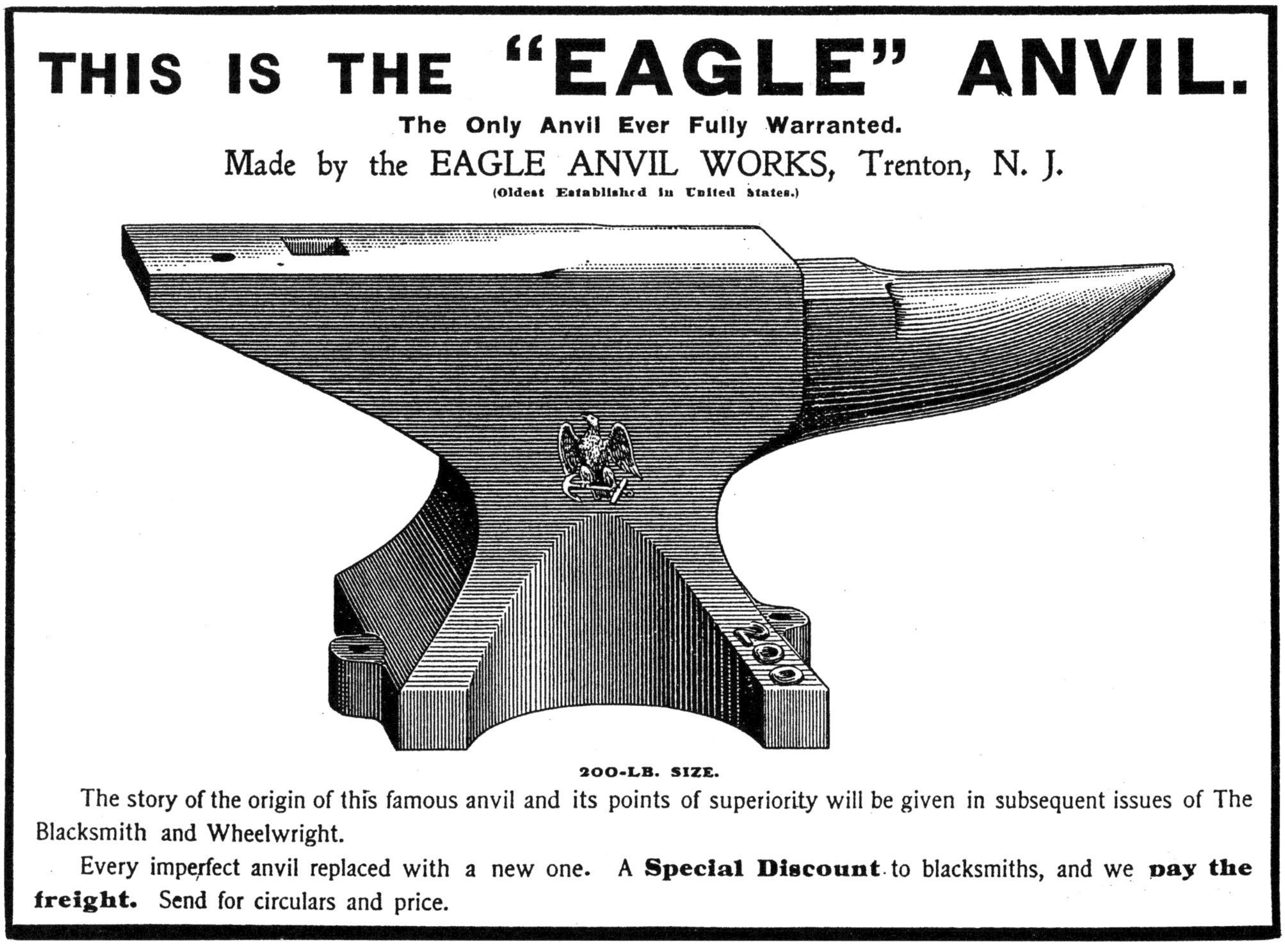

This manuscript has been written over a period of more than ten years with hardly any thought of putting in reference notes in the beginning. As the manuscript expanded to more than a few pages with information coming from many sources I felt the need for references and credits. I should point out that with very few exceptions none of the authors who have written about anvils have listed their sources and so when a mistake was made by the author of the original piece of information every author using that information has made the same mistake, although not one of them ever gives credit to the originator.

A few of the references listed came from material that I have read at one time or other, but had not noted. In most cases I have located these sources, however if you should find a blank space when checking out a reference number within the text it is one that I have not rediscovered yet.

Most of the references and advertisements taken from *The Blacksmith and Wheelwright* came from either Mr. Hunter Pilkinton or Mr. Larry Wood.

Introduction

1. *The Chronicle.* June, 1976. "Datable Characteristics of Anvils," by Robert G. Hill, pp. 21-24. The *Chronicle* is the journal of the Early American Industries Assn., Inc.

   The Bible verse mentioned is from Isaiah, 41:7. "The craftsman encourages the goldsmith, and he who polishes with the hammer encourages him who strikes on the anvil, saying of soldering 'That is good' and he fastens in with nails so that it may not be moved." *Modern Language Bible.*
2. *American Blacksmith, Toolsmith's and Steelworker's Manual.* John G. Holmstrom and Holford, Special Exclusive Edition, Printed by Frederick J. Drake & Co., especially for Sears, Roebuck & Co., Chicago, Illinois, 1915.

   The book in which I found the paragraph on the anvil was first published in 1904, called *Modern Blacksmithing, Rational Horseshoeing and Wagon Making,* by John G. Holmstrom, Frederick Drake & Co., Chicago, pp.33-34.

   The second book *Twentieth Century Toolsmith and Steelworker.* Holmstrom and Holford, was combined with the first in 1911, in which the material on *Hay-Budden* was found, pp.34-37.
3. *Practical Blacksmithing.* Compiled and edited by M. T. Richardson. Four volumes, originally published in 1889-91, combined in one. Crown Publishers, 1978, pp. 99-106.
4. *To Draw, Upset, & Weld. The Work of the Pennsylvania Rural Blacksmith 1742-1935.* Jeannette Lasansky, Oral Traditions Project of the Union County Historical Society, Lewisburg, Pa., 1980, p.7.
5. *Practical Blacksmithing.* pp. 99-106.

Chapter 1

1. *The Chronicle.* March, 1983. "Pieces of Eight, Anvils, and Bells," by Robert G. Hill, p.16.
2. Ibid.
3. *The Chronicle.* January, 1936, p.8.
4. *Early American Wrought Iron.* Albert H. Sonn, Charles Scribner's Sons, New York, 1928, p. 37.

5. *Southwestern Colonial Ironwork.* Marc Simmons and Frank Turly, Museum of New Mexico Press, Santa Fe, 1980, pp.184-185.
6. *Union.* A Rochester, New York newspaper. December 20, 1935. "Gift to Rochester Society Thought to Be Nation's Oldest Work Implement," by Joan Lynn Schild.
7. Ibid.
8. *The Chronicle.* September, 1934, p.6.
9. *Sheffield Directory, 1787.* "Manufactures of Anvils and Hammers; Alsop, Samuel, Little Sheffield; also Vices and Screws."
10. *Mousehole Forge.* Julia Hatfield. Unpublished Master's dissertation, University of Sheffield, 1991, p.51.
11. *The Chronicle.* June, 1974, pp. 35-36; September, 1974, p.56; December, 1974, pp.71-74, March, 1975, pp.16-17 and June, 1976, pp. 21-24.
12. *The Chronicle.* June,1976. "Datable Characteristics of Anvils," by Robert G. Hill, p.23.
13. *The Anvil's Ring.* Spring, 1989. "Forging Anvils," by Jim Austin, p.19.
14. *The American Blacksmith.* September, 1914. "History and Development of the Anvil 1," by James Cran, p.303.
15. Ibid.
16. I contribute this statement to the fact that there were many anvil manufacturers in England before about 1850, but afterward most of them disappeared.
17. *The Blacksmith and Wheelwright.* November, 1894. "Farrier's Perfect Anvil," p.174.
18. *The Holy Bible.* Dauay version of 1609. "Ecclesiasticus," chapter 38:29-31.
19. *The Antiquaries Journal.* 1981 Vol. 12 LXE, Part I, Oxford University. "The Anvils of Bronze Age Europe," by Margaret Ehrenbery. Sent to me by Mr. John Catchings.
20. Ibid.
21. *The American Blacksmith.* September, 1914, p.303.
22. *The Blacksmith and Wheelwright.* April, 1891.
23. *The American Blacksmith.* September, 1914, p.303.
24. Ibid.
25. Hanchett Manufacturing. Big Rapids, Michigan. In their advertising sheet of 1985.
26. From a photograph taken of the anvil with a plaque attached indicating the original owners.
27. *The American Blacksmith.* September, 1914, p.303.
28. *Mousehole Forge.* Julia Hatfield. Unpublished dissertation.
29. Iron Age Antiques. Mr. Bill Gichner.
30. *The National Geographic Magazine.* July, 1952, p.28.
31. *The Blacksmith and Wheelwright.* November, 1894, p.174.
32 I do not remember who made this statement, but from dating American farrier's anvils before and after 1910 by their serial numbers I came to the same conclusion.
33. From Mr. Kcn Mankcl of Mankel's Blacksmith Shop.
34. Snyder and Oaks. Patent No. 667,593, February 5, 1901. This was a cast-steel anvil. I have a number of slides taken of one.
35. *The Iron Age.* May, 1899. This article was reproduced in a booklet-form by the Hay-Budden Manufacturing Company.
36. *Pioneer Iron Works.* Mary Stetson Clarke, Chilton Book Company, Philadelphia, 1968, p.29.
37. *A Pictorial Encyclopedia of Trades and Industry.* Denis Diderot, 1751. *The Anvil's Ring.* September,1980, p.3.

38. *Practical Blacksmithing.* 1889-1891, p.102.
39. *The Tucson Meteorite.* Richard R. Willey, Smithsonian Institution, 1987.
40. Smithsonian Institution. Gem Section, Meteorites. Explained by Twyla Thomas of the Institution.
41. *The Antiquaries Journal.* 1981.
42. *Southwestern Colonial Ironwork.* 1980, p.43.
43. Ibid.
44. *Notes on Prehistoric and Early Iron in the Old World.* H.H. Cochlan, Pitt River Museum, University of Oxford, 1972.
45. *The Chronicle.* June, 1974. "The Pomeroy Anvil," by John S. Kebabian, p.36.
46. *Blacksmith's and Farrier's Tools at Shelburne Museum.* H.R. Bradley Smith, The Shelburne Museum, Inc., 1966, p.22.
    Wrought Iron. James Aston, A.M. Byers Company, 1939, p.9.
47. *Metals Handbook.* American Society for Metals, Cleveland, Ohio, 1938 edition. "The Bessemer Process of Steel Making," by Dr. G.B. Waterhouse.
48. *The Village Blacksmith.* Alden A. Watson, Thomas Y. Crowell, New York, 1968, p.21.
49. *The Art of Blacksmithing.* Alex W. Bealer, Funk & Wagnalls, New York, 1976, p.37.
50. *Machinery's Handbook.* 5th edition. The Industrial Press, New York, 1919, p.1170.
51. *Practical Blacksmithing.* p.100.
52. Ibid.
53. Ibid. p.102.
54.
55. *Wrought Iron.* James Ashton and Edward B. Story, A.M. Byers Company, Pittsburgh, Pennsylvania, 2nd edition, 1941, p.9.
56. *The Anvil's Ring.* Spring, 1989, pp.20-21.
57. *Metals Handbook.* 1939 edition.
58. *The Chronicle.* June, 1976. "More About Anvils," by John S. Kebabian, p.24.
59. *Blacksmith's and Farrier's Tools at Shelburne Museum.* p.65.
60. The last *Fisher* anvil was made in 1979. From a phone conversation with the owner of the Crossly Machine Shop in 1984. This was about 28 years after the last wrought anvil was made in the United States (*Trenton*) and about 10 years after the *Vulcan* steel-faced anvil was made.
61. Cast-steel anvils could not be produced economically until the "open hearth" method was perfected at about the close of the 19th century.
62. *Blacksmith's and Farrier's Tools at Shelburne Museum.* p.65. "They were made a great deal like the Swedish anvils, namely, from electric furnace cast steel." Electric furnace steel did not come into common use until well after the turn of the 20th century, so earlier anvils had to be made of open hearth steel.
63. Ibid.
64. Mr. J. Edward Finneran Jr. explained this to me during an interview. He was the last owner of the Columbus Anvil and Forging Company.
65. From the records of the Columbus Forge and Iron Company. From Mr. David Tilton the last owner of the company.
66. *The Chronicle.* September, 1974. "Comments on Beck Irons," by Alex W. Bealer III, p.56.

67. Hay-Budden patent of 1909 and Walter F. Ring's patent of 1908 (see Hay-Budden and American).
68. From interviews with Mr. J. Edward Finneran Jr. and Mr. David Tilton. Also from information gleaned from anvil manufacturer's literature down through history.
69. From about 100 B.C. to 1951, as far as I can tell from the evidence now available.
70. From recording over a thousand anvils myself.
71. From studying over thirty anvils of this type myself.
72. From the Peter Wright and Sons patent (see Peter Wright).
73. *A Pictorial Encyclopedia of Trades and Industry.* Denis Diderot, 1751.
74. *The Blacksmith and Wheelwright.* July, 1904.
75. *Carey-Best Scrap-Dudley Wood.* The name of an English anvil made between about 1830 to about 1850.
76. *The American Blacksmith.* October, 1914, p.17.
    Machinery's Handbook. 5th edition, 1919, p.1192.
    The Anvil's Ring. Spring, 1989, p.20.
77. *The Anvil's Ring.* Spring, 1989, pp.20-21.
78. *Practical Blacksmithing.* 1889, pp.100-101.
79. *The Blacksmith and Wheelwright.* July, 1909. "The Secret of Success of the Peter Wright Anvil."
    The Chronicle. September, 1975, pp.50-51.
80. *The Anvil's Ring.* Spring, 1989, pp.24-25.
81. *Practical Blacksmithing.* 1889, p.101.
82. *The Anvil's Ring.* Spring, 1989, p.22.
83. *The Blacksmith and Wheelwright.* January,1898. "RepairingAnvils," by Christian Schmidt, Trenton, New Jersey, December, 10, 1897. From Mr. Gill Fahrenwald.
84. *Practical Blacksmithing.* p.100.
85. *Transactions of the Society for the Encouragement of Arts, Manufactures and Commerce.* London,1821.
    "Improved Construction of Anvils," pp. 78-79. Sent by Mr. John Catchings.
86. Peter Wright patent No.13, 902, 1851. "Manufacture of Anvils." From the Smithsonian Institution.
87. I have never found the hole in the front of the foot in the hundreds of built-up anvils that I have inspected.
88. That is the explanation Mr. David Tilton gave me.
89. *Practical Blacksmithing.* 1889, p.100.
90. *The Iron Age.* May,11,1899.
91. *The American Blacksmith.* January, 1910, p.55. A Fisher and Norris advertisement.
92. United States Patent Office. Patent No.5331, October 16, 1847. William Martin Jr. and Mark Fisher, of Newport, Maine. "Improvement in Welding Cast-iron to Malleable Iron or Steel."

Chapter 2

1. *The Chronicle.* September, 1934, p.6.
2. *The Blacksmith in Eighteenth-Century Williamsburg.* Williamsburg Craft Series, Published by Colonial Williamsburg, 1978, p.3.

3. *Pioneer Iron Works.* Mary Stetson Clarke, Chilton Book Company, New York, 1968, p.45. *Ironworks on the Saugus.* F.M. Hartley, University of Oklahoma Press, 1957.
4. Ibid.
5. Ibid.
6. *Pioneer Iron Works.* 1968, p.8. *Ironworks on the Saugus.* 1957, p.165.
7. *Pioneer Iron Works.* 1968, p.4.
8. Ibid.
9. Owned by Mr. David Hutchison, Cordova, Maryland.
10. *Every Day Life in the Massachusetts Bay Colony.* George Francis Dow, Dover Edition, 1988, p.121.
11. Ibid. p.132.
12. South Bend Lathe. South Bend, Indiana, sells Japanese machine tools as well as machine tools that they produce.
13. *In Louisa County.* June, 26, 1959.
14. *To Draw, Upset & Weld.* Jeanette Lasansky, A Keystone Book, 1980, p.4.
15. *American Axes.* Henry J. Kauffman, Stephen Greene Press, 1972, p.18. *Encyclopedia of American History.* "British Regulation of Colonial Iron Industry." pp.5-31. *The Browns of Providence Plantation.* James B. Hodges, Harvard University Press, 1952, p.125.
16. *Southwestern Colonial Ironwork.* 1980, pp.43-46.
17. *Early American Wrought Iron.* Albert H. Sonn, Bonanza Books, New York, 1979, p.36.
18. From a letter from the curator of the Mission Museum. The anvil was donated to the Mission Museum in 1953. No history came with it. *European Tools, From the 17th to the 19th Century.* Richard J. Wattenmaker, Firch and Alain Joyaux, Flint Institute of Arts, April 26,-June 7, 1981. On page twenty of this publication an anvil is shown which is very similar to the Sonoma anvil and it is also dated "1766." This catalog says the anvil pictured is of French origin.
19. The history was passed on to me by Mr. Monroe Berman, a tavern-keeper in Venedy, Illinois.
20. *Practical Blacksmithing.* 1889, p.100.
21. Photos from Mr. Robert Hansen of Detroit, Michigan, 1995.
22. *The Art of Blacksmithing.* 1976, p.69.
23. *The Blacksmith and Wheelwright.* November, 1894, p.174.
24. I have a photograph of this anvil.
25. In 1988 Canada minted a silver dollar commemorating the 250th anniversary of the Saint-Maurice Ironworks in the province of Quebec.
26. *Kenneth Lynch.* Catalog of 1984.
27. *The Art of Blacksmithing.* 1976, p.69.
28. *Steel.* November 14, 1932, p.34.
29. From a phone call to the New York Secretary of State.
30. *Centaur Forge Ltd.* Catalog. Burlington, Wisconsin, 1985.
31. *Sheffield's Industrial Museum.* Moorland Publishing Co. Ltd., England, 1982. Sent to me by Mr.John Catchings.

32. *Horace T. Potts & Co.,Stock Book.* Philadelphia, Pennsylvania, 11th edition, 1915.
33. Ibid., pp. 28-31.

Chapter 3

1. *Sheffield Directories, 1787 and 1797.* "Manufacturers of Anvils and Hammers." Alsop, Samuel, Little Sheffield, also "Vices and Screws." "Anvil Makers." Alsop, William and John, "Sheffield- moor." Alsop, Samuel.
2. Ibid."Bullock, John, Smithfield. Darwin, Thomas, Shudchill also Rollers and Dyes. Hill, Thomas, Wicker, 1787. Darwen, Thomas, 28 Forge-Street. Hill, Thomas, Wicker, 1797."
3. *Mousehole Forge.* Julia Hatfield. Unpublished dissertation.
4.
5. Given to me by Mr. Howard Busse, St. Joseph, Michigan.
6. *Tom Wittte's Antiques.* Catalog of 1988, p.7. "T. Tillotson and Company, Columbia Place, Sheffield."
7. Mr. Floyd Daniel of Madison, Georgia, owns this anvil.
8. From Mr. Gill Fahrenwald of Olympia, Washington.
9. Mr. Bill Gichner of Iron Age Antiques owned this anvil.
10. Mr. Robert S. Cerny of Bainbridge, New York, owned this anvil.
11. Mine.
12. Mr. Steve Roth owned this anvil.
13. *Tom Witte's Antiques.* Catalog of 1988, p.7.
14. Mr. Floyd Daniel owned this anvil.
15. Ibid.
16. Mr. Ken Wilson of Montaque, Michigan, owns this anvil.
17. Mr. Frank Prillwitz of Eau Claire, Michigan, owned this anvil.
18. Mr. Benny Crenitt of Meridian, Mississippi, owned this anvil.
19. Mr. Hunter Pilkinton of Waverly, Tennessee, owns this anvil.
20. Mr. Ken Wilson owns this anvil.
21. *Mousehole Forge.* Julia Hatfield, unpublished dissertation, 1991.
22. *The American Blacksmith.* September, 1914.
23. *Mousehole Forge.* Julia Hatfield, unpublished dissertation, 1991.
24. *The American Blacksmith.* September,1914.
25. *Transactions of the Hunter Archaeological Society.* "The History of the Mousehole Forge," by D.T. Smith, 1975. Sent to me by Judith Anstee of Leeds, England.
26. *Water Power on the Sheffield Rivers.* David Crossley (ed.), 1989. Sent by Judith Anstee.
27. *Mousehole Forge.* Julia Hatfield, unpublished dissertation, 1991.
28. *Steel.* November14, 1932. "Large Uses of Steel in Small Ways," p.34.
29. *Transactions of the Hunter Archaeological Society.* 1975.
30. *The Grove of Eagles.* Winston Graham, Doubleday & Co.,Garden City, New York, 1963.
31. From a letter received from Mr. and Mrs. Hatfield of Sheffield, England. Owners of the Mousehole Forge site.
32. Ibid.

33. *The American Blacksmith.* September, 1914.
34. *The Cutting Edge, An Exhibition of Sheffield Tools.* The Ruskin Gallery, Sheffield, England, 4 July - 3 October , 1992.
35. *Water Power on the Sheffield Rivers.* 1989.
36. Ibid.
37. From my recording of *Mousehole* anvils as compared to other English makes.
38. From my observation of recorded *Mousehole* anvils it appears that the Forge used the built-up process until about 1875.
39. *The American Blacksmith.* September, 1914.
40. In 1910 the English government required all ceramics meant for export to be marked, "England." I assume that this also applied to anvils because all later model English anvils made after the 20th century have "England" as part of their trademark.
41. *The Blacksmith and Wheelwright.* July, 1904, p.320.
42. From a phone call to the Secretary of State of New York.
43. *The American Blacksmith.* October, 1914, p.17.
44. *Transactions of the Society, Instituted at London for the Encouragement of Arts, Manufactures and Commerce.* 1821. Actually the material was written in 1820 and published in 1821. pp.78-79. From Mr. John Catchings.
45. *The Cabinet of Useful Arts.* Rev. Dionysius, et. al. "A Treatise on the Progressive Improvement and Present State of the, Manufactures In Metal, Vol. I, Iron and Steel, London, 1842." Received from Mr. John Catchings.
46. *The American Blacksmith.* October, 1914, p.17.
47. Ibid. p.18.
48. *Practical Blacksmithing.* 1889, p.100.
49. *The American Blacksmith.* October, 1914, p.17.
50. *Charles A. Strelinger & Company.* Catalog, 1895, p.75.
51. From Mr. David Shayt of the Smithsonian Institution, Washington, D.C.
52. *The Blacksmith and Wheelwright.* July, 1904, p.320.
53. *The American Blacksmith.* October, 1914, p.17.
54. *The Blacksmith and Wheelwright.* February, 1891. From Mr. Hunter Pilkinton.
55. *Gardner's Photographic Sketch Book of the Civil War.* Alexander Gardner, Dover Publications Inc., New York, 1959. First published in 1866. Plate No.16.
56. Ibid.
57. Ibid. Plate No.77.
58. *McMaster Carr Supply Company.* Catalog #44, Chicago, Illinois, 1937, p.905.
59. *Practical Blacksmithing.* 1889, p.102.
60. *The Blacksmith and Wheelwright.* November,1894, p.174.
61. Ibid.
62. From Mr. Roy Price of Aliquippa, Pennsylvania.
63. From Mr. John Lemoine of Whitinsville, Massachusetts.
64. From Mr. Roy Price.
65. From Mr. Ken Mankel of Cannonsburg, Michigan.
66. From Mr. John Fairbanks of Staffordshire, England. The letter from "The Black Country Museum" was from Anne Lineen, museum services manager, by way of Mr. Fairbanks.
67. *The Blacksmith and Wheelwright.* February, 1891.

68. A letter from The Black Country Museum, October 13, 1995.

Chapter 4

1. *Wrought Iron.* 2nd edition, 1941, p.9.
2. *Technical Metals.* Harold V. Johnson, Charles Bennett Company, Peoria, Illinois, 1968.
3. Alamo Iron Works. 101 Montana, Box 231, San Antonio, 78291, founded 1878.
4. *The Blacksmith and Wheelwright.* April, 1896. Advertisement. Modern Equipment. Buffalo Forge Company, New York, 1916, p.28.
5. *Buffalo Forge Company.* Catalog of 1916, p.28.
6. *Simmons Hardware Company.* St. Louis, Missouri, catalog of 1921-22.
7. Phoenix Hardware Manufacturing Company, Buffalo, New York. In the *Manufactures Directory* under "Blacksmith's Anvils," 1941.
8. A phone call to the Secretary of State of Wisconsin, in 1994.
9. Brochure of the Illinois Iron & Bolt Company.
10. A phone call to the Secretary of State of Wisconsin, 1994.
11. Phone call to Fisher and Norris in 1984.
12. Ibid.
13. *The American Blacksmith.* April,1913, p.40.
14. Mr. David Tilton of the Columbus Forge and Iron Company.
15. Sent to me by Mr. John Catchings.
16. *Dictionary of American Hand Tools.* Compiled and published by Alvin Sellens, 1990, p.8.
17. *Early Tools of New Jersey and The Men Who Made Them.* Alexander Farnham, Kingwood Studio Publications, Stockton, New Jersey, 1984, pp.107-111.
18. *Blacksmith's and Farrier's Tools at the Shelburne Museum.* 1966, pp.64-65.
19. *Practical Blacksmithing.* 1889, pp.102-105.
20. Crossley Machine Shop. A phone call of 1984.
21. *Blacksmith's and Farrier's Tools at the Shelburne Museum.* 1966, p.65.
22. *Early Tools of New Jersey and the Men Who Made Them.* 1984, p.110.
23. *Blacksmith's and Farrier's Tools at the Shelburne Museum.* 1966, p.64.
24. *The American Blacksmith.* December, 1910, p.11.
25. From a phone conversation with the Crossley Machine Shop in 1984.
26. *Blacksmith's and Farrier's Tools at the Shelburne Museum.* 1966, p.64.
27. From a photograph of the 1876 hardware display of Fisher and Norris at the Centennial Exposition in Philadelphia, Pennsylvania. Photos provided by the Smithsonian Institution through Mr. David Shayt.
28. Every reference since *Practical Blacksmithing* was written refer to the Fisher and Norris anvil at the Centennial Exposition as weighing 1600 pounds. The wonder to me is why Clark Fisher or one of the family did not correct the error. They certainly must have known about it when *Practical Blacksmithing* was published or even before if it was published incorrectly in a newspaper.
29. *The Blacksmith and Wheelright.* May, 1883. Ibid. May 1893. Both from Mr. Hunter Pilkinton.
30. From Mr. John Lemoine.
31. *Gardiner's Photographic Sketch Book of the Civil War.* 1866, 1959, Plate #16.

32. *Practical Blacksmithing.* 1889, pp.102-105.
33. Ibid.
34. *The Blacksmith and Wheelwright.* July, 1888, p.106.
35. *Machinery's Handbook.* The Industrial Press, New York, 1919, p.299. *Welding: Principles & Practice.* Raymond Sacks, Charles A. Bennett Company, Peoria, Illinois, 1976.
36. *Practical Blacksmithing.* 1889, pp.99-106.
37. *The Blacksmith and Wheelwright.* July, 1888, p.106.
38. *The Cabinet of Useful Arts.* 1842. (see Appendix H.)
39. *The Blacksmith and Wheelwright.* July, 1888, p.106.
40. Ibid. January, 1898. From Mr. Gill Fahrenwald.
41. Ibid. March, 1896. From Mr. Hunter Pilkinton.
42. From Mr. Gill Fahrenwald.
43. *Early Tools of New Jersey and The Men Who Made Them.* 1984, p.119.
44. *Commissioner of Patents: Report of 1852-3.* "The American Star Anvil," p.380.
45. *Charles A. Strelinger.* Catalog of 1895, p.75.
46. The 1875 date comes from the patent date on the very early anvils. The 1969 date is assumed from *The Art of Blacksmithing,* 1st and 2nd editions, p.64. It is very possible that they did not go out of business until 1971, according to the Secretary of State of Wisconsin.
47. *The Blacksmith and Wheelwright.* May,1884. Under "Blacksmith's Tools."
48. Ibid.
49. From the advertisement flier from the "American Skein & Foundry Company a division of the Illinois Iron & Bolt Co." With a price list dated January 1, 1966.
50. *The Columbian Hardware Company.* Catalog No. 1901. "The Columbian Hardware Co., World's Largest Makers of Vises and Anvils, Cleveland, Ohio, U.S.A.," 1907.
51. Phone call to the Secretary of State of Ohio.
52. *Blacksmith's and Farrier's Tools at Shelburne Museum.* 1966. p.65.
53. From a phone call to "The Warren Group" in 1984. Their catalog.
54. I have recorded a number of *Söderfors* anvils with the Columbian trademark stamped on them dated 1923, 1924 and 1926.
55. *Blacksmith's and Farrier's Tools at Shelburne Museum.* 1966, p.65. Columbian's catalog of 1907 lists thirty- four sizes, 10-600 pounds.
56. *Blacksmith's and Farrier's Tools at Shelburne Museum.* 1966, p.65.
57. Ibid. p.31.
58. Ibid.
59. *Steel.* November 14, 1932, p. 32. From Mr. Bill Hill.

Chapter 5

1. *The American Blacksmith.* October, 1914, p.18.
2. *The Blacksmith and Wheelwright.* April, 1896, p.68.
3. Ibid.
4. *The Chronicle.* September, 1975, pp.50-51.
5. These are suggested dates on my part from what little evidence that has been discovered.
6. From the files of the Smithsonian Institution.

7. From material given to me by Mr. J. Edward Finneran Jr.
8. From an interview with Mr. J. Edward Finneran Jr. and confirmed by Mr. David Tilton.
9. From the auction sale papers of November 8, 1955.
10. From an interview with Mr. J. Edward Finneran Jr.
11. From a phone call to the Ohio Secretary of State.
12. Ibid.
13. *Steel.* November 14, 1932, p.32.
14. From the auction sale papers of November 8, 1955.
15. *The American Blacksmith.* October, 1914, p.18. Confirmed by Mrs. Warren B. Sage, granddaughter of Frederick C. Budden.
16. *American Blacksmithing, Toolsmiths' and Steelworker's Manual.* Holmstrom and Halford, Frederick J. Drake & Co., 1911&1913. "Twentieth Century Toolsmith and Steelworker," pp.34-35.
17. From a phone call to the New York Secretary of State.
18. *The American Blacksmith.* October, 1914, p.18.
19. *Charles A. Sterlinger & Co.* Catalog,1895, p.75.
20. *Blacksmith's and Farrier's Tools at Shelburne Museum.* 1966, pp.67-68.
21. *The Blacksmith and Wheelwright*. November, 1894. From the back cover of the magazine. Sent by sent by Mr. Hunter Pilkinton.
22. From Mrs. Warren B. Sage.
23. Ibid.
24. *The American Blacksmith and Motor Shop.* December, 1923, p.34. Formerly *The American Blacksmith* magazine.
25. *The Blacksmith and Wheelwright.* April, 1898.
26. From a phone call to the New York Secretary of State.
27. *The American Blacksmith.* October, 1909, p.38.
28. From the files of the Smithsonian Institution.
29. From Mr. John Lemoine.
30. *Blacksmith's and Farrier's Tools at the Shelbourne Museum.* 1966, p.69.
31. Ibid. Also from a number of advertisements in blacksmithing journals. For example, *The Blacksmith and Wheelwright*, July,1901, (back cover).
32. *Mousehole Forge.* Julia Hatfield, unpublished dissertation, 1991.
33. From communications with Mrs. Warren B. Sage.
34. *The American Blacksmith.* October, 1914, p.18.
35. *Blacksmith's and Farrier's Tools at Shelbourne Museum.* 1966, p.67.
36. I believe Mr. John Lemoine or a member of the Budden family mentioned this.
37. From the Columbus Forge and Iron Company serial number books.
38. 1925 South Bend Supply Company, South Bend, Indiana, catalog.
39. A copy of this brochure was sent to me by Mr. John Catchings in 1996. A reprint from *The Iron Age.* May 11,1899.
40. From Mr. David Tilton.
41. From a phone call to John Boker (great-great grandson of the founder of H. Boker and Company, New York) in 1991.
42. From a phone call to the Ohio Secretary of State.
43. *The Blacksmith and Wheelwright.* November, 1900, p.344.

44. From the Smithsonian Institution.
45. *Charles A Strelinger and Company.* Catalog of 1895, p.75.
 The Blacksmith and Wheelwright. May, 1901, p.609.
 Sears, Roebuck & Co. 1902 catalog, No.111, p.612.
46. *Sears, Roebuck & Co.* 1902 catalog.
47. From an inteview with Mr. David Tilton.
48. From an interview with Mr. Joseph King and from company records.
49. From an interview with Mr. Karl Wright (the last anvil maker).
50. Ibid.
51. From an interview with Mr. Joseph King.
52. From company records. Furnished by Mr. David Tilton.
53. Ibid.
54. Ibid.
55. From an interview with Mr. David Tilton.
56 Mr. John Lemoine.
57. Records sent to me by Mr. David Tilton.
58. From an interview with Mr. Joseph King.

Chapter 6

1. *Foxfire #5.* Eliot Wiggenton (Ed.)., Anchor Books, Garden City, New York, 1979, p.174.
2. *Schwabacher Hardware Company,* catalog. Seattle, Washington, 1920, p.139. Anvils sell for thirty cents per pound.
3. *The American Blacksmith.* June, 1907, p.13. Also in many other Fisher and Norris advertisements.
4. *Blacksmith's and Farrier's Tools at Shelburne Museum.* 1966, p.69.
5. *The Cutting Edge, an Exhibition of Sheffield Tools.* July 4 - October 3, 1992, p.71 and p.78.
6. The Harmonist Society, a Pietist religious group from Germany, existed in the United States from 1803 to about 1900. They were very industrious and established several communities.

Chapter 7

1.
2. As told to me by Mr. Ben Miller an Amish blacksmith.
3. *American Blacksmithing, Toolsmiths' and Steelworker's Manual.* 1915, p.36.
4. *Metalwork, Technology, and Practice.* 7th edition. Repp, McCarthy, and Ludwig, McKnight Publishing Company, 1982, p.303.
5. "Trivia." L.M. Boyd, Crown Syndicate Inc., 1995.
6. Information sent to me by Judith Anstee.
7. *The Village Blacksmith.* Ronald Webber, South Brunswick and New York, Great Albion Books, 1971, pp.83-92. From Mr. John Catchings.
 Gretna Green a Romantic History. Olgo Sinclair, Unwin Hyman Limited, London, 1988, pp.1-29. From Mr. John Catchings.
 The Gretna Blacksmith Story. A. Vinceny Jones, Photochrom Co. Ltd., England, 31 pages.

From Mr. Gill Fahrenwald.
8. *The Anvil's Ring.* Winter, 1992-93. "Anvil Town" by Harley Neff, Kailua, Hawaii.
9. *The Furrow.* Date unknown. John Deere and Company, pp.27-30. Sent to me by Mr. Jim Campbell of Uniontown, Pennsylvania.
10. Information sent to me by Judith Anstee.
11. *Best of the Hammer, Volume I.* Brian D. Flax, Raymond's Quiet Press, Albuquerque, New Mexico, 1983, p.36.
12. The photo copy showing the post card was from Mr. Gill Fahrenwald.
13. Sent to me by Barbara Johnson (my daughter).
14. Cartoon by Britt, The Morning News Tribune, Tacoma, Washington, 7/30/92. Sent by Mr. Gill Fahrenwald.
15. Cartoon by McCoy. Sent by Mr. Gill Fahenwald.
16. Cartoon by Stahler, The Cincinnati Post, 1992. Sent by Mr. Gill Fahrenwald.
17. "The Far Side," by Gary Larson, Chronicle Features, 1992.
18. "Off the Mark," by Mark Parise, Atlantic Feature, 1992. From Mr. Gill Fahrenwald.
19. Sent by Mr. John Catchings.
20. *Anvil* magazine. "Voice of the Farrier and Blacksmith," Georgetown, California.
21. *The Anvil's Ring.* Official journal of The Artist-Blacksmith's Association of North America.
22. From Mr. Gill Fahrenwald.
23. *The Blacksmith and Wheelwright.* July, 1891. Sent by Mr. Gill Fahrenwald.
24. Next to "The Village Blacksmith" by Longfellow, this is probably the best known poem about blacksmithing.

Appendix

1. National Agricultural Center, 630 Hall of Fame Drive, Bonner Springs, Kansas,66012. Approx. 300 anvils.
2. *The Anvil's Ring.* Spring, 1981. "The Anvil," by Karl Schroen, p.19.
3. *Twentieth Century Toolsmith and Steelworker.* 1911, p.35.
4. *Dictionary of American Hand Tools.* 1990, pp.5-6.
5. From a Columbus, Ohio, newspaper, written about 1950.
6. The article was found by Mike McGee and a copy was sent to me.
7. *The Anvil's Ring.* March, 1979. "Beware the Anvil's Ring," by Stephen A. Mackezie, p.40.
8. All of these patents are from the archives of the Smithsonian Institution.
9. *The Art of Blacksmithing.* 1969, pp.72-73.
10. *The Southern Christmas Book.* Harnett T. Kane, David McKay Co., New York, 1958, p.220.
11. The curator is Mr. Robert Myers of Berrien Springs, Michigan.
12. *The Cabinet of Useful Arts.* London, 1842. From Mr. John Catchings.
13. *The Antiquaries Journal.* Society of Antiquaries of London, 1981. Vol. LXI, Part I, Oxford University Press. "The Anvils of Bronze Age Europe," by Margaret R. Ehrenberg, p.27. From Mr. John Catchings.
14. *Notes on Prehistoric and Early Iron in the Old World.* pp.117-122. From Judith Anstee

and found by her father.
15. *The American Blacksmith.* May,1921. "Hammering Circular Saws." (author not given), pp.206-208.
16. *The Tucson Meteorites.* Richard R. Willey, Smithsonian Institution Press, Washington, D.C., 1987.
17. Ibid.
18. Ibid. p.40.
19. Ibid.
20. *Southwestern Colonial Ironwork.* 1980, pp.43-46.
21. *Handbook of Iron Meteorites.* Vol. 3. Vagn F. Buchwald. University of California Press, Berkely, 1975, pp.1235-1244 and 162-163. From Twyla B. Thomas, Museum Specialist, Division of Meteorites, Smithsonian Institution.
22. *The Blacksmith and Wheelwright.* January, 1988. From Mr. Gill Fahrenwald.
23. *The Blacksmith and Wheelwright.* May, 1901, p. 609. From Mr. James Moon of Stonyford, California.
24. *Steel.* November 14, 1932, P. 34. From Mr. Bill Hill.

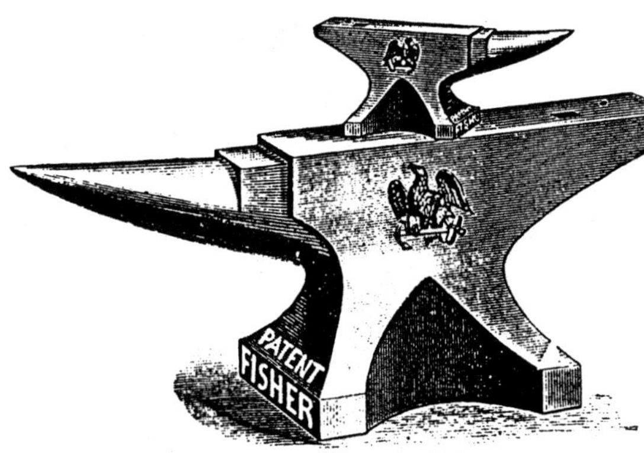

Eagle.

English

Peter Wright's Anvils.

Fig. 555.

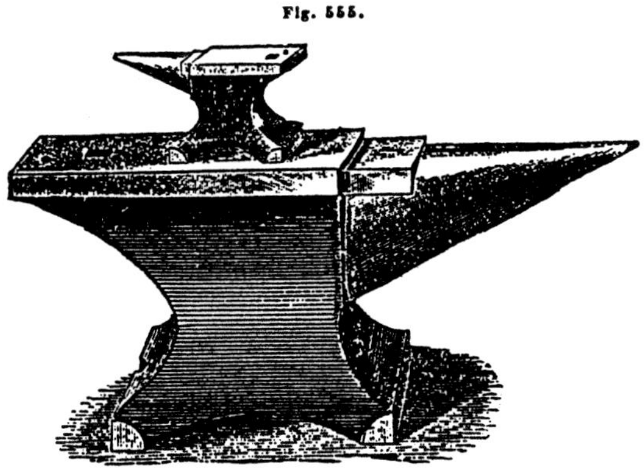

Eagle Anvils.

Fig. 556.

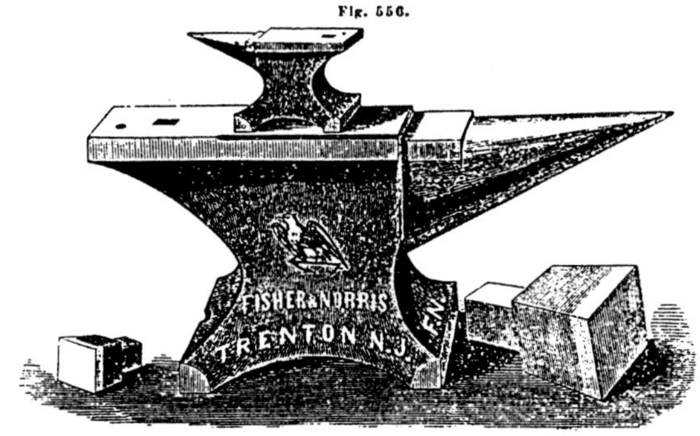

Glossary

The following is a selected list of terms related to anvils and their manufacture. It is intended primarily for those readers who are not familiar with metal working. It is arranged alphabetically. The definitions are my own, for the purpose of this publication.

Air hammer (hand-held) - Similar to an electric hand drill in appearance, except that the action is reciprocating (back and forth) rather than rotating. It is used to hammer in protrusions on anvils, such as the weld line at the waist, and to drive in wedging pins to lock up dies in large power hammers. Before the air hammer was invented and put to use an iron battering ram swung from a chain was used to drive in these wedges.

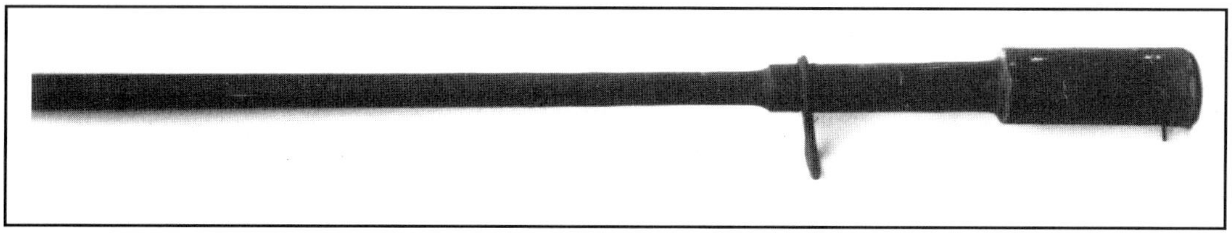

Hand forged ram used to lock up dies with wedge pins. (R.A.P.)

Air powered hammer - A fairly large upright single-frame power hammer driven by compressed air. The air is usually produced within the hammer body itself. Some large steam hammers have been converted to air in modern times, using a large external air compressor.

Alloy - A mixture of different metals. Example: copper + tin = bronze.

Anneal - To remove as much of the hardness from a metal as possible. The metal is heated to a specific temperature, then cooled very slowly.

Anvil - A stationary metal object that is acted upon by a moving object. In this work it is generally used in context as a tool that smiths use for forging purposes. The term *anvil* has an uncertain history and the spelling as we know it today came into common use sometime toward the end of the eighteenth century. The most likely meaning of the term *anvil* as far as I can figure out came from the English *Oxford Dictionary* and meant "to hand weld."

Blister steel - An iron product that will harden when heat treated. It is produced by packing wrought iron bars in a sealed container with a carbon source such as powdered charcoal. The container is then heated in a furnace to a high temperature for a specific period of time. When the iron bars are removed from the container after cooling they give the appearance of having blisters on them, therefore the name. They have also absorbed a certain amount of carbon. This was known as the "cementation process."

Blister steel was considered the lowest grade of hardenable steel, but was widely used for the *face plates* of anvils in earlier times.

Board hammer - An upright power forging machine that looks like a guillotine. The head or striking part is raised by one or more wooden boards (generally hard maple) depending on the weight of the head. The boards are raised by two counter-rotating mechanically driven rollers, pressing on the opposite sides of the boards. When the boards and head reach the top of their travel the rollers open and then the boards, head and top die free fall between guided ways onto the bottom die. The action is fairly quick, but one blow or many can be struck.

These board hammers are rated by the weight of the head. They range in weight from 300 to 1500 pounds.

After a time the boards start slipping and have to be replaced. The board hammer is a cheaper machine to operate than the steam hammer, but is limited to the size forgings it can produce.

Braze - Used when two metals are to be joined at a temperature above 800 degrees F., but below the melting point of the metals being joined . A metal such as brass with a melting point below that of the parent metals is used to join them. Silver soldering is really a brazing operation. Some anvils are refaced using brazing compounds to attach the new *face plate* to the anvil's body.

Broach - To enlarge, shape, or size a hole by either pushing or pulling a broaching tool through an existing hole in the metal. Example: Changing a round drilled hole in the *heel* of an anvil into a square *hardie hole.*

Bronze - Originally a mixture or alloy of copper and tin. Anvils were probably made of this metal before the "iron age", about 1500 B.C.

Cast - To melt a material and pour it into a mold or cavity to produce a desired shape.

Cast iron - An iron product containing above 1.5% carbon. It comes in a number of types depending on its carbon content and heat treatment. Best known are "pig iron" and "gray iron." It has a low melting point (about 1800 degrees F.), therefore it is used extensively for items not under stress, but where mass is an important factor.

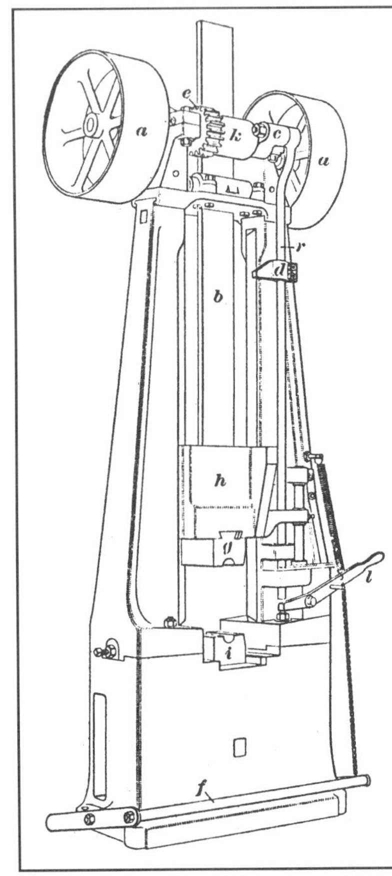

A board hammer. "b" is the board, "h" is the head, "g" is the upper die and "i" is the lower die.

Cast steel - (1) Before Bessemer steel was produced it meant "crucible steel." A superior hardening steel developed by the watchmaker Huntsman in Sheffield, England, about 1740. Blister steel bars were chopped up, placed in a sealed crucible, then placed in a furnace where the blister steel melts, thereby distributing the carbon evenly throughout the iron. The molten steel was then poured into a mold to cool, therefore the name. Tools marked "crucible cast steel" meant the best tools money could buy. Tools marked "cast steel" after Bessemer's system came into common use simply means the tool is made of cast steel rather than wrought iron or of steels made by the "cementation" process. All steels since the Bessemer system are cast before they are worked. (2) Steel that is melted and poured into a mold of some type, after which it can be worked by forging.

Chill-plate - A piece of metal, usually copper, that is placed in a mold to quickly cool the part of the metal being poured, thereby hardening the surface that comes in contact with the chill-plate.

Clips - Relatively thin metal used to hold metal rods together, as in a wrought-iron fence or gate. Also the raised metal on the front of a horse shoe to protect the horse's hoof and to help to keep the shoe in place. A *clip horn* on the farrier's anvil helps to facilitate the raising of this clip.

Closed dies - An impression in two blocks of metal, half of the finished shape being in each block. When the die blocks close on the material to be shaped and are then opened, the material has taken on the exact shape of the die impressions. Generally there is an upper and lower die, the upper being guided so that it matches perfectly with the lower fixed die when closed. If one of the die blocks slips the piece to be forged is ruined.

They are usually used in double frame forging machines. Closed dies are generally progressive; that is, the first set of impressions only approximate the shape and the last impressions complete it. On a complex shape there may be four sets of impressions across the die block to achieve an exact finished shape.

Cold-blast furnace - Cold air is blown into the furnace directly as it is received from the atmosphere without being preheated.

Cope - The upper half of the enclosure that contains the mold in casting.

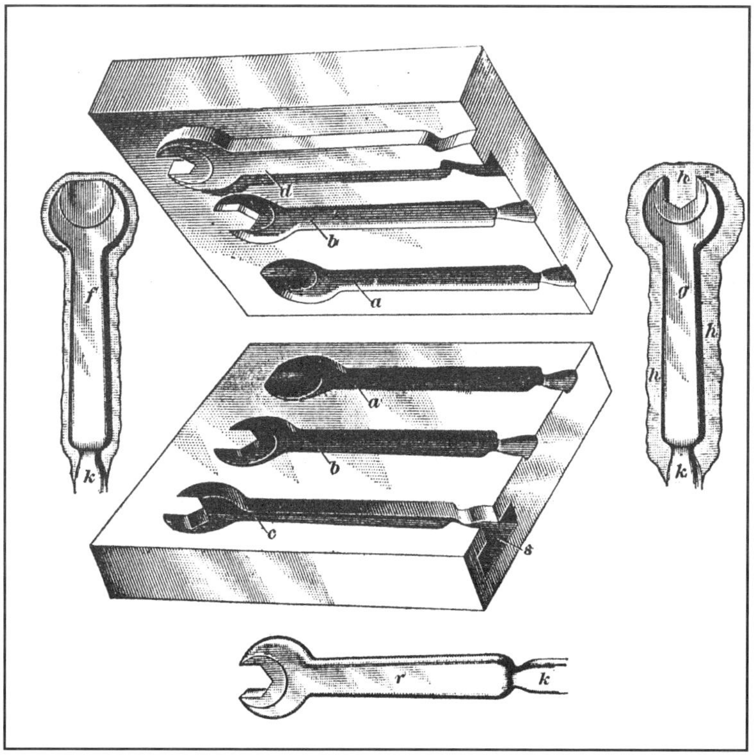

A set of closed dies.

Crucible - A ceramic container into which metal is placed, to be melted.

Ceramic crucible.

Dies - A die produces a shape either convex or concave. For example: a threading die produces screw threads either by removing metal by cutting threads, or by raising metal as in roll threading.

Drag - The lower half of the enclosure that contains the guide pins for the cope.

Draw - Is to stretch metal by forging. It can also mean to remove some of the hardness of a metal by heating.

Drift - Similar to a punch, but is either straight without any taper or is larger in the center of the drift than on either end. The straight drift is used to drive out pins, bolts, etc., and is generally softer than the material or object being driven. The tapered drift was used by the blacksmith to enlarge and size a punched hole in red hot iron.

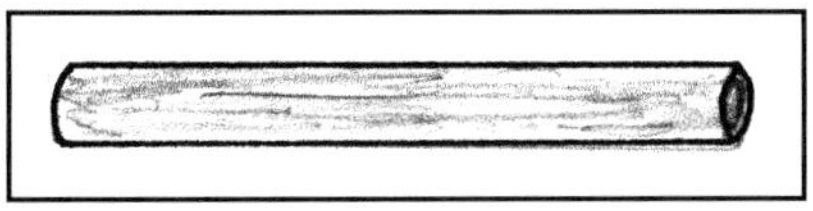

A straight drift.

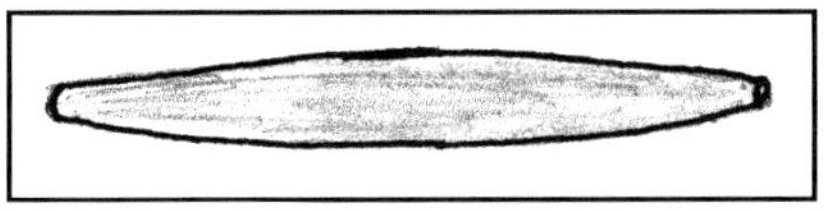

A hole sizing drift.

Drop forge - The term usually applies to a forging operation where the top die in the forging machine falls on the object to be made by gravity. A "board hammer" is an example. Early steam hammers worked the same way, the steam being used simply to raise the head of the ram and top die. However, the term has commonly come to mean any object that is forged in dies, and is not a casting or machined piece.

Ductile iron - Also known as nodular iron. A relatively new cast iron product that is replacing low carbon steel in some applications, as ductile iron is less expensive. It has about the same tensile strength as low carbon steel, and can also be hardened. It can be cast, but not forged. When cast into the shape of an anvil it has a fair ring when struck, not dead like cast iron. Therefore it may be mistaken for cast steel. ***BEWARE!*** I am told that some of the newer cast anvils made in this country are made of this metal. It tends to work harden and crack if not properly heat treated. It is not the best anvil material for heavy use, but much better than cast iron.

Ferrous metals - Any metal containing primarily iron and can contain or be alloyed with any number of other elements. All steel including non-magnetic stainless steels are ferrous metals.

Flask - When the cope and drag are put together they become the flask which contains the complete casting mold.

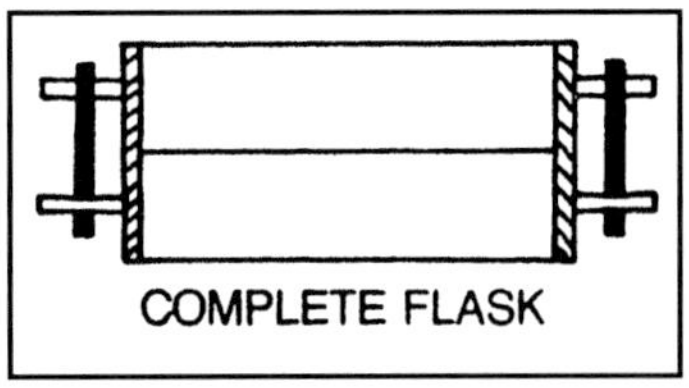

The molding. Flask, consisting of the cope (upper) and the drag (lower).

Flat dies or "smith's dies" - Also known as open dies, although open dies can have an impression in them, whereas flat dies do not. The shape produced on flat dies depends entirely on the skill of the operator. A smith working an object on the *face* of the anvil using only his hammer to form the shape would be using flat dies. However, a power hammer is generally being referred to when flat dies are to be used.

Flatter - A smithing tool used to smooth out rough flat surfaces on hot metal left by hammer blows. Used in former times to smooth out the rough flat surfaces on wrought anvils as they were being manufactured. It takes two people to use the flatter, one to hold and move the flatter about, and the other to strike the flatter with a sledge hammer.

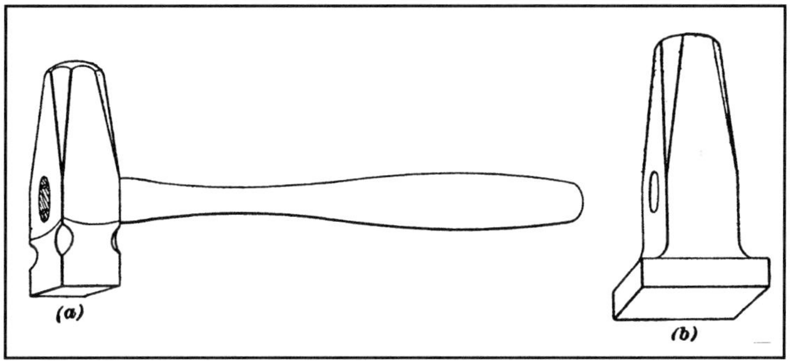

"a" is a set hammer. "b" is a flatter without its handle. They are both used for the same purpose.

Forge - As a noun, the term generally refers to the facility where forging of metal takes place, or to the piece of equipment used for heating the metal if an open fire is used. If the equipment for heating metal is a closed, box-like affair, it is usually referred to as a furnace. As a verb, forge means to shape a piece of metal by hammering, rolling or pressing.

Forging - Forming a metal shape by applying some type of pressure. The operation can be performed cold, warm or hot .Generally, forging is performed when the metal is very hot, in a soft plastic state. Cold and warm forging is performed on certain metals when not much movement of the metal is required. Forging produces the strongest item as compared to casting or machining the same item from the same material. In many cases, such as for "**I**" beams or railroad rack, hot roll forging is also the cheapest way to produce these items.

A forged wrench.

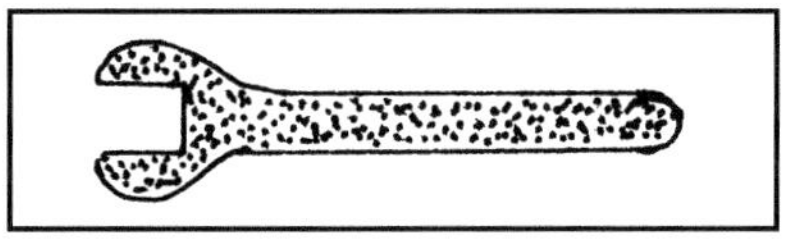

A cast wrench.

A machined wrench.

Notice the grain structure in each wrench.

Forge press - A large forging machine using continuous pressure to form a metal object in closed or open dies. The pressure is applied in one continuous motion, not blows, as applied by the board or steam hammers. However, the press cannot perform the detailed work on a piece of metal as well as the hammers can. The forge press is the largest of the forging machines and are rated by the pressure they exert, not by the weight of the head.

Press forging is to be preferred on large thick pieces, as the slower build up of pressure on the metal allows the forging force to go to the center of the item. Hammer forging is so rapid that the force applied does not always reach the center portion, leaving the center in its original unforged state. This creates stresses in the metal.

Fuller and fullering - Fullers have a convex face and are used to draw metal faster than with a flat surface such as a hammer. They are also used to smooth concave surfaces such as the *waist* of an anvil. They can be in the form of machine dies or they can be hand-held tools. Generally when used as an anvil tool there is the upper and lower fuller, one stationary in the hardie hole and the other hand-held and struck with a large hammer or sledge.

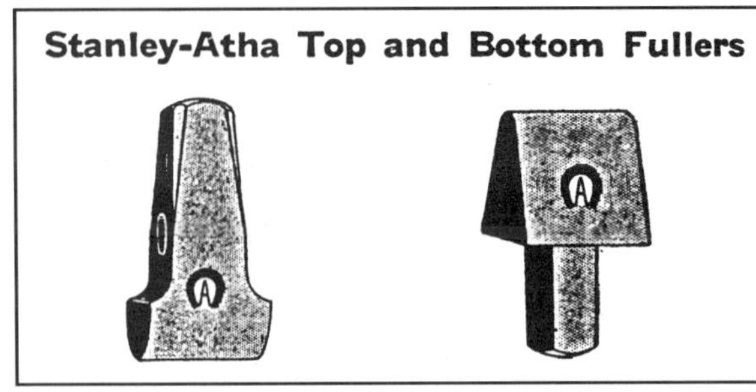

The top and bottom fullers. The top fuller needs a handle.

Heat sink - Any material that absorbs heat quickly, such as copper or aluminum or the face of an anvil.

Hot-blast furnace - The atmospheric air is preheated before it is blown into the furnace, thereby producing hotter temperatures in the furnace.

Indirect method - In metal working this term refers to the manufacture of wrought iron. In very early times wrought iron was made directly from the iron ore. The direct method wasted fuel and metal with only small amounts of wrought iron being produced at one time. With the indirect method the iron ore is first rendered cast iron by the blast furnace and then the pieces of cast iron were turned into wrought iron by a refining process. The indirect process produced wrought iron in larger quantities at lower cost.

Iron - Today when this term is used we tend to think of cast iron or at least I thought so. Before this present age the term meant “wrought iron.”

Normalize - After some metals are worked, stresses are built up within the piece. To relieve these stresses and bring the piece into a uniform state, the piece is heated to a specific temperature and then cooled slowly, but not annealed.

Open dies - These forging dies can be flat, convex, or concave. They can have the approximate finished shape of the piece, but the finished shape depends on the skill of the machine operator who manipulates the piece being forged. Pieces such as large shafts, large anvils, etc., were forged in open dies, allowing the piece to be moved back and forth and rotated.

Open dies are usually used in power hammers of the single frame type, although very large double-framed hammers can use these dies to advantage at times. Most wrought anvils were forged in open dies. Many smiths considered open die forging superior to closed die work; however, open die work is more costly for large quantities of the same work and also requires a much higher degree of skill. Open dies, like flat dies, are often referred to as “smith’s dies.”

Pattern - As used in a metal foundry, it consists of the shape of a finished item, in former times made of wood, used to make an impression of the shape in a sand mold. It is then removed from the mold so that hot liquid metal can be poured into the impression or cavity in the mold with the resulting piece being the same as the pattern.

Porter bars - Long metal rods used for handling hot anvils while they are being manufactured. They are inserted into indentations called *handling holes* or *mouse holes* punched into the anvil body, allowing the anvil to be handled and manipulated while work is performed on it.

Quench - To cool heated metal quickly, usually in water or oil. A water quench is quicker than an oil quench, as water removes the heat from the metal faster. The choice depends on the composition of the metal. A brine solution (salt and water) was usually used in the blacksmith shop. The salt not only kept the water from freezing in the winter, but also helped the water to cool the metal quicker. For quenching large objects such as anvils, a falling stream of water was necessary in order to break the steam barrier that built up around the metal object as it was plunged into the water. The steam barrier prevented the metal from cooling quickly. Therefore, you cannot harden the face of an anvil by submerging it in a large tank of water.

Shear steel and double shear steel - A hardenable iron produced by forge-welding blister steel bars together and then drawing them into a single bar again. When performed again "double shear steel" was produced. This process distributed the carbon more uniformly throughout the iron, making a superior steel. Both were used in making cutting tools, therefore the name. Some of the better English anvil makers used both for the *face plates* of their anvils.

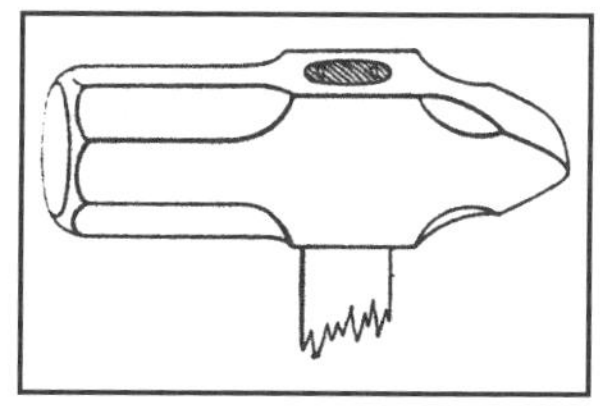

A blacksmith's sledge.

Sledge hammer - A heavy hammer held with both hands. They were made in weights from six to forty pounds. A forty-pound sledge might have been used to weld the *face plate* on to the anvil body. Various size sledge hammers were used in the manufacture of anvils.

Smith - The term refers to the craftsman who works metal. This craftsman or artisan may be a blacksmith, a white-smith (pewter), goldsmith, silversmith, etc. The term is usually applied to the blacksmith now. If your last name is Smith, someone in your ancestry worked metal.

Smithy or blacksmith shop - The facility or building where the blacksmith worked, not the person. Sometimes referred to as a forge or forge shop. "Under the spreading chestnut tree the village *smithy* stands," [emphasis added] from Henry Longfellow's poem, "The Village Blacksmith."

Steam hammer - A large power hammer of both single and double frame, the double-framed hammers being the larger. Steam power from a boiler is piped to the hammer and used to raise the head and top die which are attached to a piston rod and piston which is within a cylinder at the top of the frame. In the first hammers the steam merely raised the head, but later hammers became double-acting, thereby forcing the head and top die along with their weight against the work, which is held on the lower die. Steam hammers are rated by their head weight, starting at 300 pounds and going up to many tons. However, the driving force of the steam increases the head's striking power tremendously. Most steam hammers use guide ways on the frame to guide the head. In the hands of a skilled operator very delicate work can be performed even on the larger steam hammers. The photo inside the cover of the book shows an older single-frame steam hammer without guide ways in use. Another more modern single-frame hammer can be seen under Trenton, p.341 and a double-frame hammer under Trenton p. 346.

Steel - Before the Bessemer converter came into common use an iron that could be forged, hardened, and tempered was called "steel." The steels before 1860 were those made by the cementation and crucible processes.

Today almost all of the iron products except the cast irons are called steels, even most of those that cannot be hardened. The common steels containing only carbon as their alloy are called "plain" or "carbon" steel. The carbon content is the determining factor as to how the metal can be heat-treated. Plain steel today is classified into three categories: low, medium, and high-carbon steel. Low-carbon steel contains up to three-tenths of one percent carbon (.03) or 30 points of carbon. It is called "mild steel," "cold rolled" or "hot rolled" steel. It cannot be hardened,although it will work-harden to some extent and can be case-hardened, which is essentially the cementation process. If the carbon content falls below one-tenth of one percent (.01) or 10 points of carbon, it becomes "soft iron" and is no longer called steel.

Medium-carbon steel contains from three-tenths of one percent (.03) or 30 points of the carbon to one-half of one percent (.05) or 50 points of carbon and can be hardened and tempered to some degree.

High-carbon steel contains from one-half of one percent (.05) or 50 points of carbon to one and one-half percent (1.5) or 150 points of carbon, and is also known as "tool steel." It can be made very hard by heat treatment. When iron has more than 150 points of carbon the carbon in the steel begins coming out of a very fine distribution and the carbon begins sticking together, producing a brittle iron we call "cast iron."

There are many other types of steel, but these are the ones anvil makers dealt with mostly shortly before and after 1900.

Swage - A tool with a concave surface used to reduce any size metal. It is the reverse of the fuller on its working end. They are made for hand and power hammer use. The hand-held swage has two parts. The fixed part fits in the *hardie hole* of the anvil and the matching part is held by a handle in the hand and is struck by a sledge. Swages are actually types of "open dies."

Swedish steel - This was made from a very good grade of iron ore found in Sweden. It was considered superior to many other steels. It was used by a number of anvil makers to face their anvils.

Tongs - They are similar in action to pliers, but with long handles or "rains." The gripping end is shaped and sized to the material to be held while hot. They can be from about a foot long to as much as seven or eight feet and quite heavy depending upon the material to be held. Anvil tongs were made especially to hold various parts of the anvil while it was being forged under the power hammer.

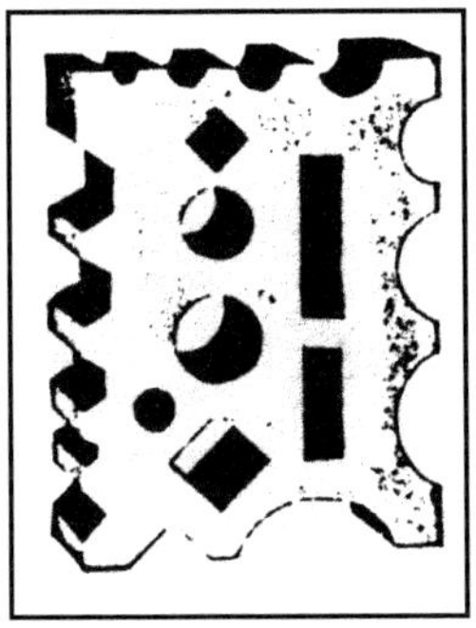

A swage block.

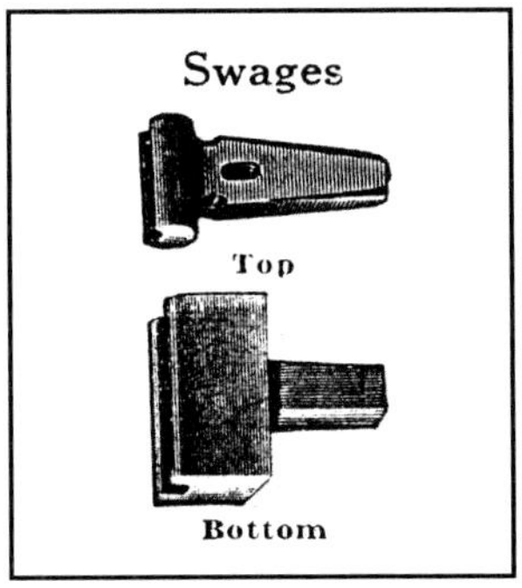

A top and bottom swage. The top swage needs a handle. These come in many sizes.

Temper - As a noun, a certain hardness of a metal.

Tempering - To remove some of the degree of hardness of a metal after it is heated and quenched. This can be accomplished in a number of ways, but basically the metal is reheated to about 700 degrees F or less and then quenched again. The tempering temperature and quench depends on the type of metal and the purpose for which it is to be used. This is also known as "drawing the temper."

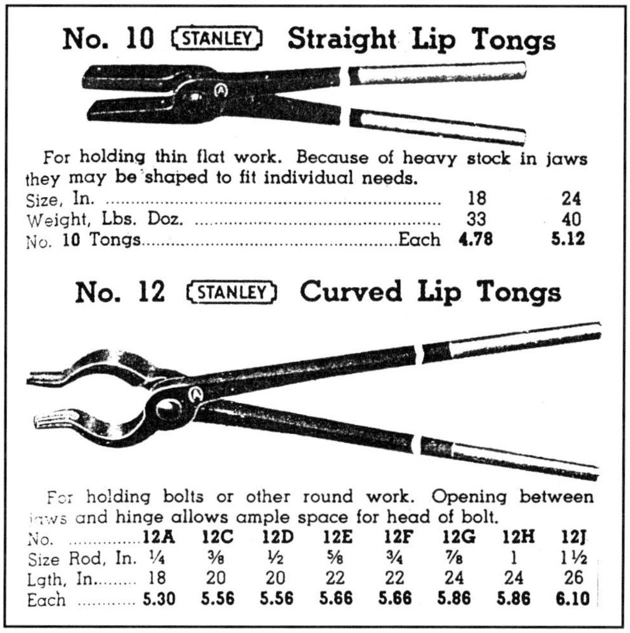

No. 10 (STANLEY) Straight Lip Tongs

For holding thin flat work. Because of heavy stock in jaws they may be shaped to fit individual needs.

Size, In.	18	24
Weight, Lbs. Doz.	33	40
No. 10 Tongs ... Each	**4.78**	**5.12**

No. 12 (STANLEY) Curved Lip Tongs

For holding bolts or other round work. Opening between jaws and hinge allows ample space for head of bolt.

No.	**12A**	**12C**	**12D**	**12E**	**12F**	**12G**	**12H**	**12J**
Size Rod, In.	¼	⅜	½	⅝	¾	⅞	1	1½
Lgth, In.	18	20	20	22	22	24	24	26
Each	**5.30**	**5.56**	**5.56**	**5.66**	**5.66**	**5.86**	**5.86**	**6.10**

Barrett Christie Co. catalog, 1954, Chicago, Il.

Trip hammer - Also known as a helve hammer, tilt hammer, or a beam hammer. This power hammer was the forerunner of the steam and board hammers. They were generally operated by a water wheel which turned a shaft. At the end of the round shaft there were cogs, cams, or teeth spaced evenly around the shaft. These cogs raised an almost horizontal wooden beam, then let the beam drop before the next cog was engaged. The beam could be hinged on the end or in the middle, as there were any number of variations. On one end of the beam an iron head (generally cast-iron) was attached similar to a carpenter's hammer on its handle. The head rested on a square or rectangular anvil when not in motion. The action was usually very fast, but the teeth or cogs did not lift the beam very far. Therefore, the head weighing as much as several hundred pounds did not strike with great force. To remedy this, a fixed wooden beam was attached above the hammer beam. Before the head reached its highest point it engaged the fixed beam which acted like a spring, thus forcing the hammer beam and head down with greater force.

The expression "quick as a trip hammer" came from this machine. A number of these beam hammers were used well into the 20th century. Some even had changeable upper and lower dies.

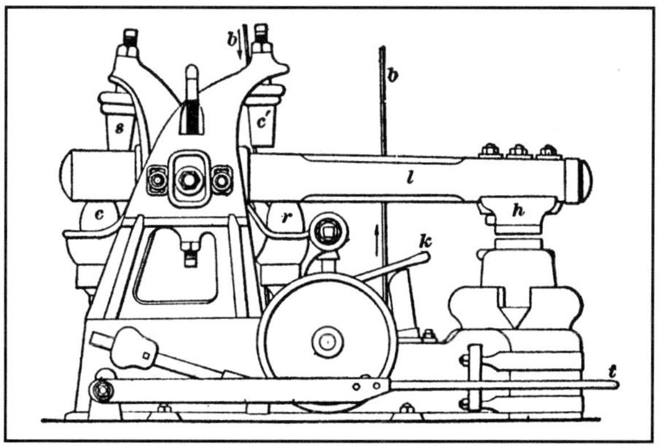

A Bradley helve hammer which is similar to a trip hammer in action. "h" is the hammer head attached to the wooden helve or arm "l".

Upset - When a piece of metal is enlarged at some point. The end of an auto lug wrench is first upset on the end and then the nut size impression is punched in.

Weld - When two or more pieces of material are brought to their melting point and then joined together to create one piece. Actually when two ice cubes are melted together by using a little heat from a lit match, you have a weld.

It generally refers to welding metals together at some point.

Weld, arc or electric weld - This usually refers to what is called "stick welding," where a consumable electrode or rod is to be used to conduct electric current to the spot to be welded. The rod arcs, creating an electric flame which melts the pieces to be joined while the rod furnishes the filler material. Before "stick welding," it meant using one or two carbon electrodes to do the melting and a filler metal was fed into the flame to help join the two points. There are a number of different types of arc welding today, but they all use electricity.

Weld, butt - When two pieces of metal are joined end to end thus:

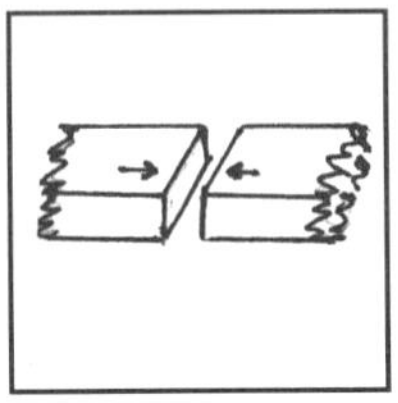

The butt weld.

Weld, forge - The wrought iron or steel items to be joined are heated in a fire to just below the melting point, quickly placed together and pressure is applied. If this is performed correctly the two pieces will become one. This was the only practical way iron and steel could be welded until well into the 20th century. This method of welding, using an open fire, is rarely used in industry today.

Weld, horizontal - A weld in the horizontal plane. This is the best type when the pressure is applied vertically on the item, such as on an anvil.

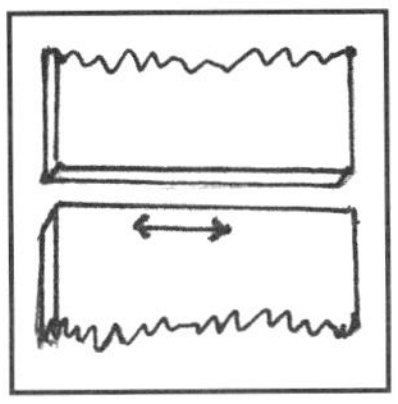

The horizontal weld.

Weld, jump - A butt weld, and usually the weakest of the forge welds.

Weld, lap - When one piece of metal is placed over another and then welded either with pressure as in forge welding.

Weld, vertical - A weld in the vertical plane. In a forged weld this is the weakest, when pressure is applied vertically.

Work-harden - Most commercial metals, especially steel, tend to surface-harden when worked cold, either by rolling, rubbing or hammering.

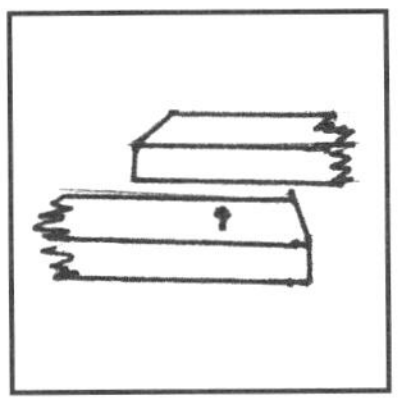

The lap weld.

Wrought - To work metal by a mechanical action either by hammering, pressing, rolling or bending.

Wrought iron - An iron product containing practically no carbon, but containing differing amounts of slag, ferrous silicate. This is a glass-like material. The iron containing the slag has to be worked (wrought) considerably after being removed from the refiner's furnace where it is made. After the iron is worked sufficiently the remaining slag becomes thread-like throughout the iron. This thread-like slag imparts a tough fibrous property to the iron. The more wrought iron is worked, to a point, the better it becomes. That is one reason anvil makers forged scrap iron in earlier times rather than use new wrought iron, often called "muck bars," because of its low quality. Good

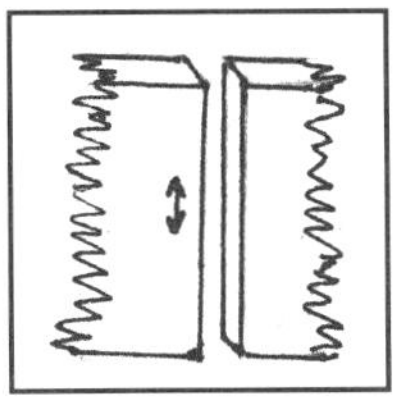

The vertical weld.

wrought iron can have 250,000 or more of these slag fibers per square inch of iron. Although wrought iron does not have the tensile strength of mild steel, it resists rust better and will stand much more bending stress than steel.

Before Bessemer steel came into use wrought iron was the only iron that could be forged other than the tool steel made from wrought iron.

The production of commercially made wrought iron ceased in this country in the late 1950s. Wrought iron has been replaced by low carbon steel which is cheaper to produce, but in many ways is not as good as high quality wrought iron. Wrought iron is the metal in most forged anvils except for the steel *face plate*. If a piece of wrought iron is partly sawn through and bent, you can actually see the fibrous structure. The fact that the ferrous silicate is inert and does not rust allows the fibrous structure to be seen on a rusty piece of wrought iron.

A broken piece of wrought iron. Notice the fibrous structure. (R.A.P. photo)

A piece of wrought iron bent cold. This photo was used in the Columbus Anvil and Forging Co. brochure to show the quality of wrought iron they used in their anvils.
(photo courtesy of J.E. Finneran Jr.)

BIBLIOGRAPHY

American Blacksmith, The. Walter O. Bernhardt (Ed.) *A Practical Journal of Blacksmithing and Wagonmaking*. Buffalo, N.Y. Feb.1902 - Dec.1923.

American Blacksmithing, Toolsmith and Steelworker's Manual. John Holmstrom and Henry Holford Frederick Drake & Co., 1911 and 1913.

Antiquaries Journal, The. The Journal of the Society of Antiquaries of London, Vol. LXI, Part 1, Oxford University Press, 1981.

Anvil's Ring, The. Journal of the Artist-Blacksmith's Association of North America.

Art of Blacksmithing, The. Alex Bealer, Funk & Wagnalls Publishing Co., 1969.

Best of the Hammer, Brian D. Flax (Ed.), Raymond's Quiet Press, Volumes: 1,2,3,4, &5.,1983.

Blacksmith's and Farrier's Tools at Shelburne Museum, H.R. Bradley Smith, Shelburne Museum, 1966.

Blacksmith and His Art, The. J.E. Hawley, Pioneer Arizona Foundation, 1976.

Blacksmith in Eighteenth-Century Williamsburg, The Published by Colonial Williamsburg, MCMLXXVIII [1978].

Blacksmith & Wheelwright, The. "Devoted to the Interests of Blacksmiths," M.T. Richardson (Ed.), M.T. Richardson Co., N.Y. Dec. 1882 -Jan. 1925.

Book of Trades, The. Jost Amman and Hans Sachs, Dover Publications, N.Y. 1973.

Centaur Forge Ltd., Burlington, Wis., Catalogs 1982 - 1985.

Canton Hardware Co.,The. Jobbers of Hardware, Canton, Ohio, 1896.

Chains and Chainmaking, Charles Fogg, Shire Publications Lld., 1981.

Charles A. Strelinger & Co., Catalog of 1895, Detroit, Michigan.

Channon Company, H. Catalogue Number 33, H. Channon Co., Chicago, Ill., 1908.

Chronicle, The. Early American Industries Association, Inc., June 1974, September 1974, December 1974, June 1976, March 1983.

Cutting Edge on Exhibition of Sheffield Tools, The. Ruskin Gallery, Sheffield, England, 4 July, - 3 October 1992.

Cyclopedia of Modern Shop Practice, Howard M. Raymond, American Technical Society, 1907.

Decorative and Sculptural Ironwork Tools, Techniques Inspiration for Modern Blacksmithing, Dona Z. Meilach, Crown Publishers, N.Y. 1977.

Dictionary of American Hand Tools, Alvin Sellens, Published by Alvin Sellens 1990.

Diderot Pictorial Encyclopedia of Trades and Industry, A. Dover Publications, 1959.

Early American Wrought Iron, Albert H. Sonn, Bonanza Books, N.Y., 1979.

Early Tools of New Jersey and the Men Who Made Them, Alexander Farnham, Kingwood Studio Publications, Stockton, New Jersey, 1984.

Eisen Schmiedehandwerk Kunsthandwerk Kunit, Max Fievogel & Valentin Zandonella, Verag Peter Meiti Schaffhausen, [no date].

Every Day Life in the Massachusetts Bay Colony, George Francis Dow, Dover Publications, N.Y., 1988.

Forging and Forming Metals, S.E. Rusinoff, American Technical Society, 1952.

Forging Practice, Carl G. Johnson, American Technical Society, Chicago, Ill., 1940.

Foxfire 5, Eliot Wigginton, (Ed.), Anchor Books, 1979.

Gardner's Photo graphic Sketch Book of the Civil War, Alexander Gardner, Dover Publications, N.Y., 1959.
Grove of Eagles, The. Winston Graham, Doubleday & Co.,1963.
Guns and Gunmaking Tools of Southern Appalachia, John Rice Irwin, The Museum of Appalachia Press, Norris, Tenn., 1980.
Historical Guide to Wagon Hardware and Blacksmith Supplies, A.. Towana Spivey (Ed.), Museum of the Great Plains, Lawton, Ok., 1979.
Horace T. Potts & Co., *Stock Book,* Philadelphia, Pa., 1915.
Ironworking, W.K.V. Gale, Shire Publications Lld., 1981.
Ironworks on the Sauges, E. N. Hartly, University of Oklahoma, Norman, Ok., 1957.
McMaster-Carr Supply Co., Catalog 44, Chicago, Ill., 1937.
Metallurgy, Johnson-Weeks 5th Edition, American Technical Society, 1977.
Millwrights & Mechanics Guide, Audel & Co., N.Y., 1940.
Modern Machine Shop Practice, Brooks and Hand, Frederick J. Drake & Co., 1911.
Montgomery Ward & Co., Catalogue No. 57, Chicago, Ill.,1895.
Montgomery Ward & Co., Catalogue No. 103, Chicago, Fall and Winter, 1925-26.
Museum of Early American Tools, A. Eric Sloan, Funk & Wagnalls, N.Y., 1964.
National Geographic Magazine, The. Today on the Delaware, Penn's Glorious River, Albert W. Atwood & Robert F. Sisson, Washington, D.C., July, 1952.
Notes on Prehistoric and Early Iron in the Old World, H.H. Cohlan, T.K. Penniman and Blackwood (Ed.) University of Oxford Press, [no date].
Pioneer Iron Works, Mary Stetson Clark, Chilson Book Co., Philadelphia, Pa., 1968.
Practical Blacksmithing, M.T. Richardson (Ed.), M.T. Richardson Publishers, N.Y., 1889.
Rogers Advanced Machinist, William Rogers, Theo. Audel & Co., N.Y. 1902.
Rogers Progressive Machinist, William Rogers, Theo. Audel & Co., N.Y., 1901-1903.
Russell Erwin Manufacturing Company, American Hardware, 1865, facsimile editors, Association for Preservation Technology, 1980.
Search for Early New Jersey Toolmakers, Alexander Farnham, Kingwood Studio Publications, Stockton, N.J., 1992.
Sears Roebuck and Co., Catalogue No. 104, Chicago, Ill., 1897.
Sears Roebuck and Co., Catalogue No. 111, Chicago, Ill., 1902.
Sears Roebuck and Co., Catalogue No. 117, Chicago, Ill., 1908.
Sears Roebuck and Co., Catalogue No. ? Chicago, Ill.,1922.
Shop and Foundry Practice, International Textbook Company, N.Y., 1903.
Southwestern Colonial Ironwork, Marc Simmons and Frank Turly, Museum of New Mexico Press, 1980.
Steel, journal, November 14, 1932.
Technical Metals, Harold V. Johnson, Charles A. Bennett Co., Peoria, Ill., 1968.
To Draw, Upset, & Weld, The Work of the Pennsylvania Rural Blacksmith 1742-1935, Jeannette Lasansky, Keystone Books, Oral Traditions of Union County Historical Society, Court House, Lewisburg, Pa. 1980.
Tools, Machinery, Blacksmith's Supplies, Sears Roebuck and Co., Chicago, Ill., (Reprint by Mid-West Tool Collectors Assn. and ABANA, February 1985; Catalogue, Ca. 1915.)

Tool Steel Simplified, Frank R. Palmer, The Carpenter Steel Co., Reading, Pa., 1937.
Tucson Meteorite, The. Richard R. Willey, Smithsonian Institution, 1987.
Village Blacksmith,The. Ronald Webber, Great Albion Books, 1974.
Welding: Principles & Practices, Raymond Sacks, Chas. A. Bennett Co., Peoria, Ill., 1976.
With Hammer in Hand, Charles F. Hummel, University Press of Virginia, Charlottesville, Va. 1968.
Wrought Iron, Its Manufacture, Characteristics, and Applications, James Aston and Edward B. Story, A.B. Byers Co., Pittsburgh, Pa., 1936.

INDEX OF ANVILS

AMERICAN

MODERN AMERICAN

ENGLISH

OTHER

General Index

A

B

F

G

H

I

J

Q

R

S

T

U

V

W

Y

Z